Companion to Euclid

BERKELEY MATHEMATICS
LECTURE NOTES

Volume 9

Companion to Euclid

A course of geometry, based on
Euclid's *Elements*
and its modern descendants

Robin Hartshorne

American Mathematical Society

Berkeley Center for
Pure and Applied Mathematics

In Memory of
Jonathan Churchill Hartshorne
1972–1992

1991 *Mathematics Subject Classification.* Primary 51-01;
Secondary 51Mxx, 51N20, 12F10.

∞ The paper used in this book is acid-free and falls within the guidelines established to ensure permanence and durability.
Visit the AMS home page at URL: http://www.ams.org/.

10 9 8 7 6 5 4 3 2   02 01 00 99 98

Contents

Introduction 1

Chapter 1. Euclid's Geometry 6

1. A first look at Euclid's *Elements* 6
2. Ruler and compass constructions 16
3. Euclid's axiomatic method 25
4. Construction of the regular pentagon 41
5. Some newer results 46

Chapter 2. Hilbert's Axioms 56

6. Axioms of incidence 56
7. Axioms of betweenness 62
8. Axioms of congruence for line segments 70
9. Axioms of congruence for angles 76
10. Hilbert planes 81
11. Intersection of lines and circles 87
12. Euclidean planes 93

Chapter 3. Geometry over Fields 97

13. The real Cartesian plane 97
14. Abstract fields and incidence 104
15. Ordered fields and betweenness 111
16. Congruence of segments and angles 116
17. Rigid motions and SAS 122
18. Non-Archimedean geometry 131

Chapter 4. Segment Arithmetic 135

19. Addition and multiplication of line segments 135
20. Similar triangles 142
21. Introduction of coordinates 149

Chapter 5. Area 154

22. Area in Euclid's geometry 154
23. Measure of area functions 163
24. Dissection 169
25. Quadratura Circuli 176
26. Euclid's theory of volume 179
27. Hilbert's third problem 185

Chapter 6. Construction Problems and Field Extensions 193

28. Three famous problems 193
29. The regular 17-sided polygon 199
30. Constructions with compass and marked ruler 207
31. Cubic and quartic equations 214
32. Appendix: Finite field extensions 222

Chapter 7. Non-Euclidean Geometry 233

33. History of the parallel postulate 234
34. Neutral geometry 241
35. Behavior of parallels 252
36. Non-Euclidean area 256
37. Circular inversion 263
38. Digression: Circles determined by three conditions 273
39. The Poincaré model 283
40. Hyperbolic geometry 296
41. Hilbert's arithmetic of ends 312
42. Hyperbolic trigonometry 325
43. Characterization of hyperbolic planes 335

Chapter 8. Polyhedra [not yet written]

44. The five regular solids
45. The thirteen Archimedean solids
46. Semiregular solids

Notes 341

References 347

Index of Axioms and Acronyms 353

Index of Euclid's Propositions 354

Index 356

Preface

For the last five years, I have been teaching a Junior–Senior level course on the classical geometries. These lecture notes have grown out of that teaching experience. I assume only high school geometry and some abstract algebra. The course begins in Chapter 1 with a critical examination of Euclid's *Elements*. Students are expected to read concurrently Books I–IV of Euclid's text, which must be obtained separately, for example in the inexpensive Dover edition. Then in Chapter 2 we study Hilbert's axiom system to bring the subject up to a modern standard of rigor.

For a one-semester course such as I teach, that forms the core material, which takes six to eight weeks. Then, depending on the taste of the instructor, one can follow a more geometric path by going directly to non-Euclidean geometry in Chapter 7, or a more algebraic one, exploring the relation between geometric constructions and field extensions, by doing Chapters 3, 4, and 6. For me, one of the most interesting topics is the introduction of coordinates into an abstractly given geometry, which is done for a Euclidean plane in §21, and for a hyperbolic plane in §41.

Throughout the notes, I have attempted to choose topics that are accessible to undergraduates and that are interesting in their own right. The exercises are meant to be challenging, to stimulate a sense of curiosity and discovery in the student. I purposely do not indicate their difficulty, which varies widely.

I hope this material will become familiar to every student of mathematics, and in particular to those who will be future teachers.

As this is a work in progress, I welcome feedback, comments, and criticisms.

Robin Hartshorne
Mathematics Department
University of California
Berkeley, CA 94720
e-mail: robin@math.berkeley.edu

Introduction

In ancient Greece, a little after the time of Plato, but before Archimedes, a man named Euclid wrote the *Elements of Geometry*, gathering and improving the work of his predecessors Pythagoras, Theaetetus, and Eudoxus into one magnificent edifice. This book soon became the standard for geometry in the classical world. With the decline of the great civilizations of Athens and Rome, it moved eastward to the center of Arabic learning in the court of the caliphs at Baghdad. In the late Middle Ages it was translated from Arabic into Latin, and since the Renaissance it has not only been the most widely used textbook in the world, but has had an influence as a model of scientific thought which extends way beyond the confines of geometry. As Billingsley said in his preface to the first English translation (1570), "Without the diligent studie of Euclides Elementes, it is impossible to attaine unto the perfecte knowledge of Geometrie, and consequently of any of the other Mathematical Sciences". Even today, though few schools use the original text of Euclid, the content of a typical high school geometry course is the same as what Euclid taught more than two thousand three hundred years ago.

In this course we will take Euclid's *Elements* as the starting point for a study of geometry from a modern mathematical perspective.

To begin, we will become familiar with the content of Euclid's work, at least those parts which deal with geometry (Books I–IV, VI, and XI–XIII). Here we find theorems that should be familiar to anyone who has had a course of high school geometry, such as the fact (I.4) that two triangles are congruent if they have two sides and the included angle equal, or the fact (III.21) that a given arc of a circle subtends the same angle at any point of the circle from which it is seen. (Throughout these notes, references such as (I.4) or (III.21) refer to the corresponding Book and Proposition number in Euclid's *Elements*.)

Many of Euclid's propositions pose construction problems, such as (I.1): to construct an equilateral triangle, or (IV.11): to construct a regular pentagon inscribed in a circle. Euclid means to construct the required figure using only the ruler, which can draw a straight line through two points, and the compass, which can draw a circle with given center and given radius. These **ruler and compass constructions** are often taught in high-school geometry. Note that Euclid casts these problems in the form of constructions, whereas a modern mathematician would be more likely to speak of proving the existence of the required figure.

At a second level, we will study the logical structure of Euclid's presentation. Euclid's *Elements* has been regarded for more than two thousand years as the prime example of the **axiomatic method**. Starting from a small number of self-evident truths, called postulates, or common notions, he deduces all the succeeding results by purely logical reasoning. Euclid thus begins with the simplest assumptions, such as Postulate 1: to draw a line through any two given points, or Postulate 3: to draw a circle with given center and radius. He then proceeds step by step to the culmination of the work in Book XIII, where he gives the construction of the five regular solids: the tetrahedron, the cube, the octahedron, the icosahedron, and the dodecahedron.

Upon closer reading, we find that Euclid does not adhere to the strict axiomatic method as closely as one might hope. Certain steps in certain proofs depend on assumptions that, however reasonable or intuitively clear they may seem, cannot be justified on the basis of the stated postulates and common notions. So, for example, the fact that that the two circles in the proof of (I.1) will actually meet at some point seems obvious, but is not proved. The **method of superposition** used in the proof of (I.4), which allows one to move the triangle ABC so that it lies on top of the triangle DEF, cannot be justified from the axioms. Also various assumptions about the relative position of figures in the plane, such as which point lies between the others, or which ray lies in the interior of a given angle, are used without any previous clarification of what such notions should mean.

These lapses in Euclid's logic lead us to the task of disengaging those implicit assumptions that are used in his arguments, and providing a new set of axioms from which we can develop Euclid's geometry according to modern standards of rigor. The logical foundations of geometry were widely studied in the late 19th century, and led to a set of axioms proposed by Hilbert in his lectures on the Foundations of Geometry in 1899. We will examine Hilbert's axioms, and we will see how these axioms can be used to build a solid base from which to develop Euclid's geometry pretty much according to the logical plan which he first laid out. We will also cultivate an awareness of what additional axioms may be required for certain portions of the theory.

At our third level of reading Euclid's *Elements*, involving rather broader investigations than the first two levels mentioned above, we will consider various mathematical questions and subsequent developments that arise naturally from Euclid's presentation.

For example, the modern reader quickly becomes aware that Euclid does not use numbers in his geometry. He speaks of **equality** of line segments, and a notion of one segment being added to another to form a third segment, but he does not mention the **length** of a line segment. When it comes to **area** (I.35 ff), though Euclid does not say explicitly what he means by equality of area, we can infer from his proofs he means a notion generated by cutting figures in pieces and adding or subtracting congruent figures. He does not use any number to measure the area of a triangle. So we may note with surprise that the famous Pythagorean theorem (I.47) does not state that the square of the length of the hypotenuse (a number) is equal to the sum of the squares of the lengths of the two sides of a right triangle; rather it says that the area of a square built on the hypotenuse is equal to the area formed by the union of the two squares built on the sides.

The absence of numbers may seem curious to a student educated in an era when the real numbers are all-powerful, when an interval is measured by its length (which is a real number), and an area by a certain integral (another real number). In fact some modern educators have gone so far as to build the real numbers into the axioms for geometry with the "ruler postulate", which says that to each interval is assigned a real number, its **length**, and two intervals are congruent if they have the same length. However, this use of the real numbers at the foundational level of geometry is far from the spirit of Euclid.

So we may ask, what role do numbers play in the development of geometry? As one approach to this question we can take the modern algebraic structure of a field (which could be the real numbers, for example), and show that the **Cartesian plane** formed of ordered pairs of elements of the field forms a geometry satisfying our axioms. But a deeper investigation shows that the notion of number appears intrinsically in our geometry, since we can define purely geometrically an **arithmetic of line segments**. We will show that (up to congruence) one can add two segments to get another segment, and one can multiply two segments (once a unit segment has been chosen) to get another segment. These operations satisfy the usual associative, commutative, and distributive laws so that we obtain an **ordered field**, whose positive elements are the congruence equivalence classes of line segments.

Thus we establish a connection between the abstract geometry based on axioms and the methods of modern algebra.

I would like to emphasize throughout this course how methods of modern algebra help to understand classical geometry and its associated problems.

For example, in the theory of area, one can formalize Euclid's notion of equality based on adding and subtracting congruent figures. However, we do not know any purely geometric proof that this theory of area is non-trivial, so that, for example, one figure properly contained in another will have a smaller area. Euclid just cites Common Notion 5, "the whole is greater than the part." But unless we are willing to accept this as an axiom, we should give a proof. Such a proof can be provided using algebraic arguments in the field of segment arithmetic.

In the realm of ruler and compass constructions, algebraic methods have led to notable results. For example, Gauss achieved an extraordinary result in 1796, when he used roots of unity to show that it was possible to construct a regular 17-sided polygon—the first new polygon construction since Euclid's constructions of the pentagon, the hexagon, and the decagon. On the other hand, field theory, in particular the Galois theory of finite field extensions of $\mathbb{Q}$, has provided proofs of the impossibility of certain ruler and compass constructions such as the regular 7-sided polygon, the trisection of the angle, or the doubling of the cube. For in the algebraic interpretation, one can construct with ruler and compass points whose coordinates lie in successive quadratic extensions of $\mathbb{Q}$, while the three problems just mentioned require the solution of cubic equations. We will see however that these three problems can be solved if one allows the use of a **marked ruler**. In fact, constructions using the marked ruler, in addition to ordinary straight-edge and compass, correspond exactly to the solution of equations of the third and fourth degrees.

Euclid bases his treatment of similar triangles (Book VI) on a complicated theory of proportion (developed in Book V) where ratios of given quantities are compared by seeing if arbitrary rational multiples of the one exceed or fall short of the other. This method foreshadows Dedekind's 19th-century definition of a real number as a division ("Dedekind cut") of the rational numbers into two subsets, namely those greater than and those less than the given real number. The theory of proportion depends on **Archimedes' axiom**, which states that given any two segments there is an integer multiple of the first which will exceed the second. Using the field of segment arithmetic mentioned above we can give (following Hilbert) an alternative development of the theory of similar triangles which is simpler and does not depend on Archimedes' axiom.

In developing the theory of volume of three-dimensional figures in Books XI and XII, remarkably Euclid abandons the finite dissection methods used for the area of plane figures, and brings in an infinite method, the "method of exhaustion" attributed to Eudoxus, which suggests the limiting process used to define the Riemann integral. Gauss (1844) expressed his regret that such an infinite method should be used for something so apparently elementary as the volume of a triangular pyramid (XII.5). Hilbert, in his famous list of problems stated in 1900, asked if this infinite limiting process was really necessary, and Dehn in the same year provided an answer by showing that a pyramid cannot be dissected into a finite number of pieces and reassembled into a cube. In Dehn's proof, abstract algebra again provides a solution to a geometric problem.

While discussing the foundational and theoretical questions mentioned above, we also have a practical side to this course. We make a point of carrying out many ruler and compass constructions, for example Euclid's elegant construction of the regular pentagon (IV.11), and carefully counting our steps to heighten awareness of the process. At the same time we will find explicit expressions for various lengths constructed using nested square roots to emphasize the connection with field extensions of $\mathbb{Q}$. When studying area, we will make explicit dissections of figures to show equality, such as for the Pythagorean theorem (I.47) or Dudeney's brilliant dissection in four pieces of an equilateral triangle into a square. When we come to Euclid's construction of the five regular solids, we will make models of these and we will also explore the thirteen Archimedean solids and the other "semi-regular" convex polyhedra whose faces are all regular polygons.

Also on the practical side, we will study results that belong to the domain of "Euclidean geometry" although they do not appear in Euclid's *Elements.* Some of these were discovered long ago, such as the fact that the three altitudes of a triangle meet in a point, which was known to Archimedes, while others were found more recently, such as the Euler line and the nine-point circle associated to a triangle. The technique of circular inversion, which became popular in the second quarter of the 19th century, provides an example of the modern transformational approach to geometry, and gives a convenient tool for the solution of classical problems such as the problem of Apollonius: to find a circle tangent to three given circles.

Finally, the investigation of the role of the parallel postulate has led to some of the most important developments arising out of Euclid's geometry. Already from the time of Euclid onward, commentators noted that this postulate was much less elementary than the others, and they questioned whether it might not be a consequence of the other postulates. Two millennia of efforts to prove the parallel postulate by showing that its negation led to absurd (but not contradictory) results were considered failures until, in the mid-19th century, a brilliant shift of perspective, with lasting consequences for the history of mathematics, admitted that these "absurd" conclusions were merely the first theorems in a new, strange, but otherwise consistent geometry. Thus were born the various non-Euclidean geometries which have been so valuable in the modern theory of topological manifolds, and in the development of Einstein's theory of relativity, to mention just two applications. In this course we will discuss the beginnings of **neutral geometry** (assuming no parallel axiom) and **non-Euclidean geometry** (assuming the negation of the Euclidean parallel axiom). We give an analytic model of non-Euclidean geometry over a field due to Poincaré. Then we give an axiomatic treatment of **hyperbolic geometry** based on the axiom of existence of limiting parallel lines. The two approaches are brought together by constructing an abstract field out of the geometry, and showing that any abstract hyperbolic plane is isomorphic to the Poincaré model over its associated field. Once again algebraic methods help us to understand geometry.

CHAPTER 1

Euclid's Geometry

In this chapter we create a common experience by reading portions of Euclid's *Elements*. We discuss the nature of proof in geometry. We introduce a particular way of recording ruler and compass constructions so that we can measure their complexity. We discuss what are presumably familiar notions from high school geometry as it is taught today. And then we present Euclid's construction of the regular pentagon and discuss its proof.

Throughout this chapter proofs are informal. We do not presuppose any particular knowledge, and yet we assume familiarity with everything in high school geometry. The purpose of this chapter is to create a common base and language with which to begin our more formal study of geometry in the following chapters.

In the last section of this chapter we present some newer results which do not appear in Euclid's *Elements* but nevertheless belong to the subject of "Euclidean geometry".

Note: Reading this chapter should be concurrent with reading Euclid Books I–IV, so as to understand all proofs and constructions. Exercises given here will reinforce this reading.

1 A First Look at Euclid's *Elements*

When we first open Euclid's *Elements* to see what is in this famous book, we find familiar facts about the geometry of lines, triangles, and circles in the plane. I say familiar, because almost every elementary or high school curriculum has some geometry in it, and what has been taught for thousands of years and still is commonly taught as "geometry" is material from Euclid's *Elements*.

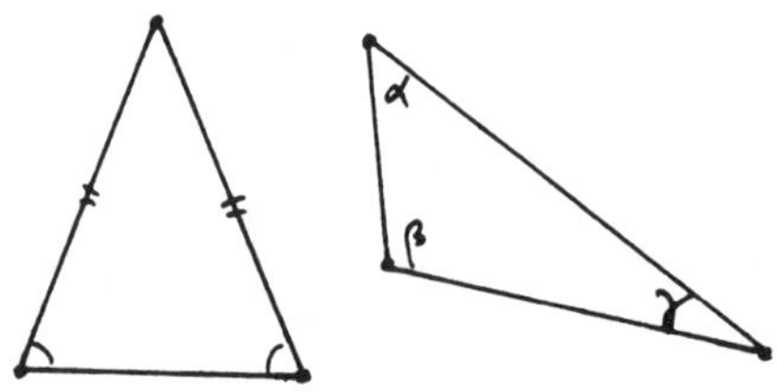

We find for example that a triangle is called *isosceles* if two of its sides are equal, and in that case it follows (I.5) that the two base angles of the triangle are also equal.

We find the theorem (I.32) which says that the sum of the three angles of a triangle is $180^\circ : \alpha + \beta + \gamma = 180^\circ$. However we note that Euclid does not use degree measure for angles. Instead he says the sum of the angles of a triangle is equal to two right angles.

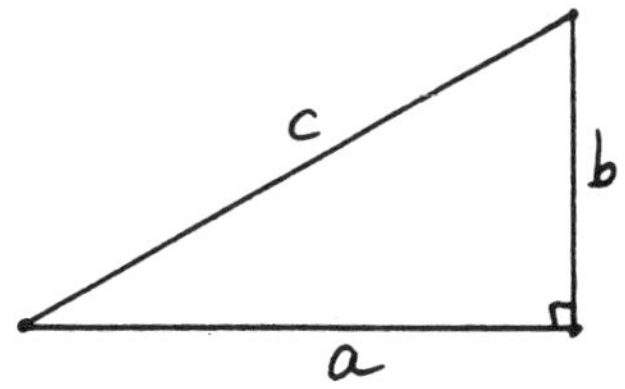

We find the famous "Pythagorean theorem" (I.47), which says that in a right triangle the sum of the squares of the legs is equal to the square of the hypotenuse:

$$a^2 + b^2 = c^2.$$

We note, however, that Euclid does not use algebraic notation to express this result. Instead he shows that the area of the square on the hypotenuse is equal to the combined area of the squares on the two sides.

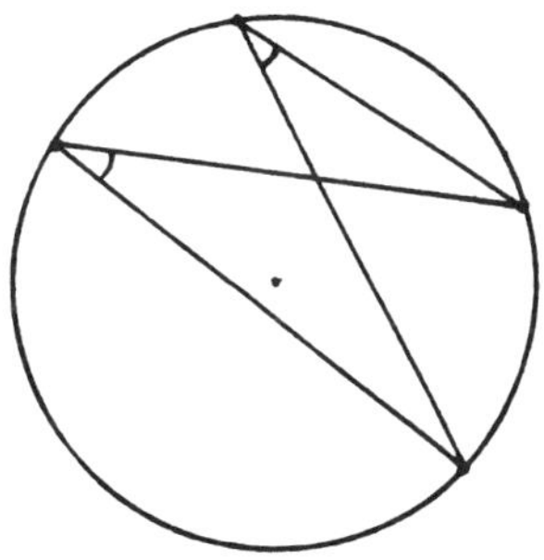

In Book III, which deals with circles, we find the result (III.21), which I hope will be familiar to most readers, that an arc of a circle subtends the same angle at different points of the circle from which it is seen.

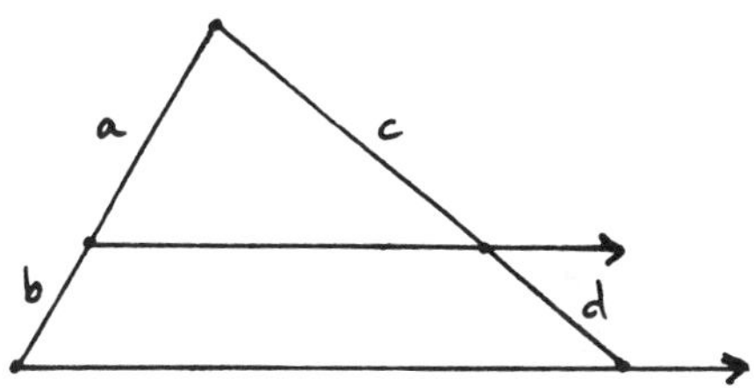

Then in Book VI, which deals with similar triangles and the theory of proportion, we find the familiar result (VI.2) that a line parallel to the base of a triangle will cut the sides proportionately, namely $a : b = c : d$.

Euclid's *Elements*, written circa 300 BC, is a systematic account of the geometry and number theory of his time. What is remarkable is that these same propositions still form the basis of teaching geometry today.

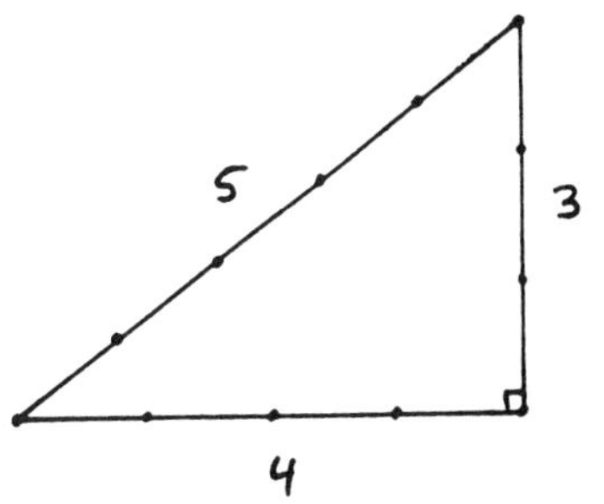

Historically speaking, most of these results were known long before Euclid. Within the realm of Greek mathematics, the theorem on isosceles triangles is attributed to Thales, and the theorem on the sum of the angles of a triangle and the theorem on the sides of a right triangle are attributed to Pythagoras. Both men lived three hundred years before Euclid.

Before the Greeks, Eves [History, p. 35] points out that the theorem of Pythagoras was known to the ancient Babylonians (1900–1600 BC). Also, there are reports (see Eves, [History, p. 49]) that the ancient Egyptians used a rope knotted in twelve equal segments, which could be stretched out to form a triangle with sides 3, 4, 5, to construct right angles for laying out fields.

The great contribution of Euclid, for which he is justly renowned, is that he organized the geometrical knowledge of his time into a coherent logical framework, whereby each result could be deduced from those preceding it, starting with only a small number of "postulates" regarded as self-evident.

To appreciate Euclid's achievement, let us try to put this in perspective.

The most naive approach to geometry is to regard it as a collection of facts, or truths about the real world. Ancient geometry began as a set of useful rules for measuring fields, laying out cities, building buildings or constructing altars.

By the time of Euclid, we can detect two important changes in the perception of geometry.

One concerns the nature of geometrical truth. There is a distinction between the real world with all its imperfections, and some kind of abstract or ideal existence which people in this world strive to attain. This point of view is evident in the writings of Plato, who was born about one hundred years before Euclid. Speaking of the geometers, he says (near the end of Book VI of *The Republic*)

> "although they make use of the visible forms and reason about them, they are not thinking of these, but of the ideals which they resemble; not the figures which they draw, but of the absolute square and the absolute diameter, and so on..."

Thus geometry is elevated from the status of a practical science to the study of relationships in this ideal existence, with a consequent shift of emphasis to the mathematically exact solution of a problem as opposed to an accurate approximate solution which would be sufficient for practical purposes. Because geometry engages the mind in contemplation of these ideal relationships, Plato also recognizes its value in education, for we find a little later (*Republic*, Book VII) the following exchange:

> "... the knowledge at which geometry aims is knowledge of the eternal, and not of ought perishing and transient.
>
> That, he replied, may be readily allowed, and is true.
>
> Then, my noble friend, geometry will draw the soul toward truth, and create the spirit of philosophy; and raise up that which is now unhappily allowed to fall down.
>
> Nothing will be more likely to have such an effect.
>
> Then nothing should be more sternly laid down than that the inhabitants of your fair city should by all means learn geometry."

Euclid's geometry is the geometry of this ideal world in the sense of Plato, with its emphasis on exact relationships. In this sense it can be regarded as abstract mathematics. From the point of view of the modern mathematician, however, Euclid's geometry is still tied to the real world because it concerns the unique ideal world of Plato's philosophy of which the real world is a reflection. For example, Euclid does not hesitate to use arguments from time to time (we will look at specific cases later) which seem perfectly acceptable in view of our experience of the real world, yet which are not logical consequences of his initial assumptions.

The modern mathematician goes one step further, by trying to make explicit *all* of his assumptions, and creating a consistent mathematical structure which no longer derives its validity from the real world. The "truth" of a particular result in the real world is then no longer relevant. The only question is whether that result is consistent with or can be logically deduced from the assumptions of this particular theory. The modern point of view allows for many different equally valid abstract mathematical theories, whereas for Euclid there was only one geometry.

The other most important change in perspective from that of naive geometry which we see in Euclid's *Elements* is the emphasis on proofs. It is no longer sufficient to say such-and-such is true, or even to give many instances where its truth is evident. The Greeks since Pythagoras had been concerned with justifying their geometrical results, and Euclid's *Elements* is the ultimate expression of this trend, where all the propositions are proved in one grand logical sequence.

So what exactly is a proof?

The answer to this question depends on the context. Suppose for example we are discussing one of those famous hard problems which circulate informally among amateurs, such as the following: Let ABC be a triangle. Let BD and CE be the angle bisectors at B and C. Suppose that BD is equal to CE. Then show that the triangle ABC is isosceles. This statement is eminently reasonable, but a proof using the usual methods of high school geometry is surprisingly elusive.

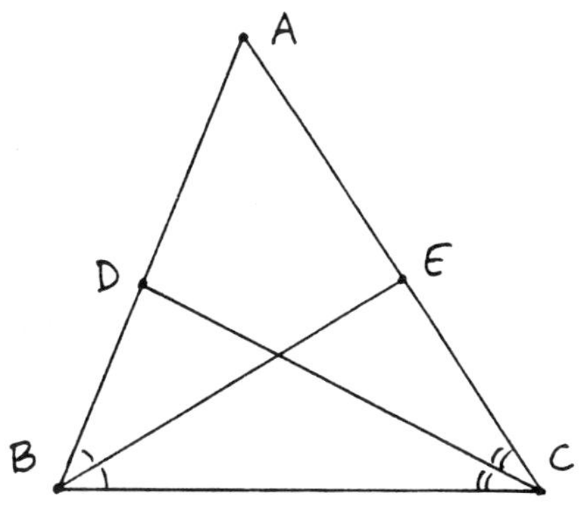

With a hard problem like this, most people would accept as proof any demonstration of its truth based on well-known results which can be found in books, whether the methods used were from geometry, trigonometry, analytic geometry, or even calculus. A purist might increase the difficulty of the problem by insisting on a purely geometric solution. Among experienced mathematicians, there would be little disagreement about what constituted a valid proof, once it was found.

In another context, a proof can be characterized simply as a convincing argument. Suppose you are explaining a result to another person who has a similar general background, but who has not seen this particular result. For example, I wish to inscribe a hexagon with six equal sides in a circle with center O.

I choose a point A on the circle, and with my compass centered at A, with radius AO, I mark off a point B on the circumference. Then with center B and radius BO I mark off another point C on the circumference. I repeat this process, always with radius equal to the radius of the original circle, to get further points D, E, and F. Then I draw AB, BC, CD, DE, EF, all of which have the same length, equal to OA, by construction. I claim FA also has the same length, so that $ABCDEF$ will be an equilateral hexagon inscribed in the circle.

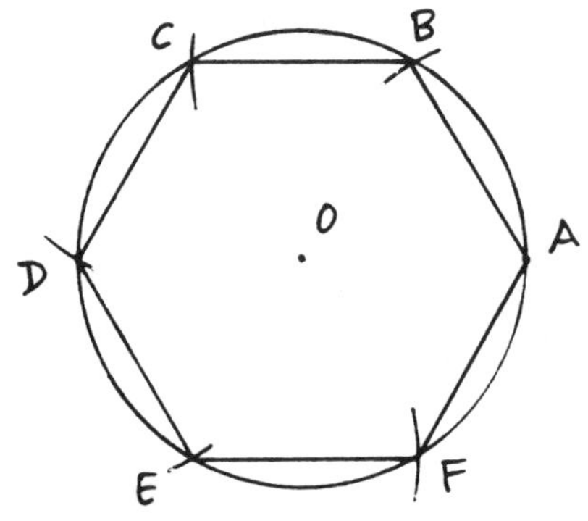

Why does this work? How would you explain this so as to convince another person? To get a real-life answer, I put this question to my 17-year-old son, now a high school senior. His first response was "I have done it myself, so I know it works". — "Yes", I said, "from a practical point of view it works. But how do you know this is an exact solution and not just a very good approximation?"

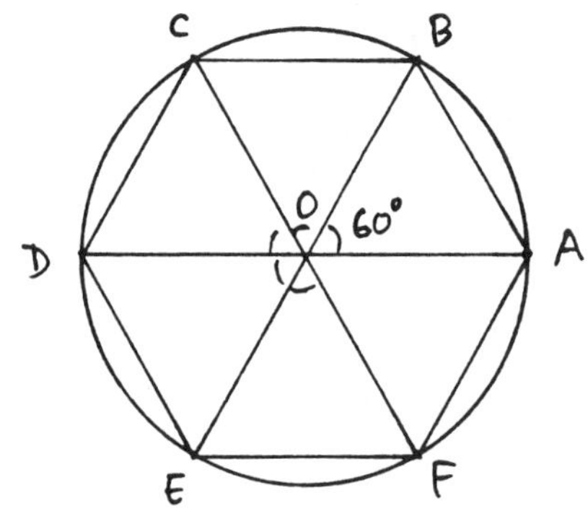

After a few minutes of thought he drew the lines from O to A, B, C, D, E, F, and then explained that OAB is an equilateral triangle by construction. Therefore the angle $\angle AOB$ at the center is $60°$. The same is true for the next four triangles $BOC, \dots, EOF$. Thus we have five $60°$ angles, so the remaining angle $\angle AOF$ must also be $60°$. Then the triangle AOF having two sides the same and the same central angles, must be the same as the triangle AOB, and so $FA = AB$.

"Fine", I said, "that is very convincing, assuming that your listener knows the angle of an equilateral triangle is $60°$, and the angle of one total revolution is $360°$. It seems your listener would have to know the theorem that the sum of the three angles of a triangle is $180°$. What if he asked you to explain why that is true?"

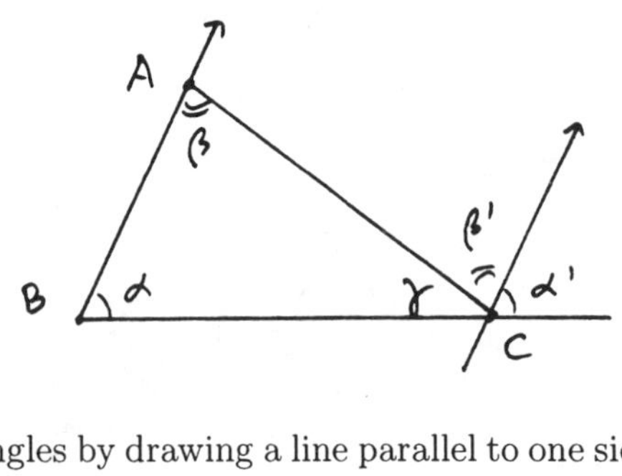

I mentioned a proof of the sum of the angles by drawing a line parallel to one side AB of a triangle through the third vertex C. Then $\alpha = \alpha'$ because of the parallel lines, and $\beta = \beta'$ because of the parallel lines, so $\alpha + \beta + \gamma = \alpha' + \beta' + \gamma = 180°$ because it is a straight angle. — "But then you have to know theorems about the angles formed when a line cuts two parallel lines." There ensued a discussion about proliferation of questions, like the endless "why"'s of a three-year-old, and the danger of getting into circular arguments.

So we see that the notion of proof as a convincing argument may work well, but depends on who your listener is, and is also subject to the danger of infinite regress if your listener is uncooperative. (At this point you might like to look at (IV.15) to see how Euclid solves this same problem.)

A third and much stricter notion of proof applies to the writer of a mathematical treatise such as Euclid's Elements. A proof must deduce the result in question by a series of logical steps based only on those results which have already been proved earlier in the book, and on those definitions, postulates, and common notions which have been set out as self-evident at the beginning, and which form the starting point for the logical chain of deductions. Even this notion of proof is not absolute, however, because what constitutes an acceptable proof for a given result will depend on where that result is situated in the logical sequence. So for example, Proclus says that Euclid devised an entirely new proof of the theorem of Pythagoras (I.47). We can infer that he had to do so, because he placed it at the end of Book I and therefore could not use the well-known proof by similar triangles (cf. (25.5)), (which was most likely the method used by Pythagoras), since similar triangles do not appear until Book VI.

It is for this logical structure, perhaps even more than for its mathematical content, that Euclid's *Elements* is famous. The *axiomatic method* of sequential logical deduction, starting from a small number of initial definitions and assumptions, has become the basic structure of all twentieth century mathematics. Euclid's *Elements* is the first great example of this method. The importance of the axiomatic method in modern times was emphasized at the turn of the century by David Hilbert, whose work we will study later in this course.

And now, dear reader, it is time for you to open your copy of Euclid and start reading. Abraham Lincoln, speaking of his scanty formal education, says "He studied and nearly mastered the six books of Euclid since he was a member of Congress".[1] You need not go so far as that, but I do urge you to read at least as much as is suggested in the exercises below.

A note on the exercises in this book

One of my students, in an essay discussing the suitability of Euclid's *Elements* as a text for teaching geometry today, suggested that it would be better to use no text at all, so that students could have the excitement of rediscovering geometry for themselves. If we lived in ancient Athens, when the study of geometry was synonymous with reading Euclid's *Elements*, then I would agree. But we do not live in ancient Athens, and mathematics, including geometry, has developed a great deal since the time of Euclid.

So I propose instead that we take Euclid's *Elements* as a starting point, a touchstone to provoke questions and further investigation, and that we set out to rediscover modern mathematics for ourselves.

My philosophy of mathematics is that you learn by doing. To study mathematics is to do mathematics, not just to learn what other people have done. Therefore

[1]Abraham Lincoln, Autobiographical Writings, Kingsport, Tenn. (1947), p. 9.

the exercises in this book are designed (to the best of my ability) to stimulate mathematical activity. There are very few routine exercises. Most require some puzzling, some experimentation. Many offer a challenge of exposition: once you understand what is happening, how do you explain it clearly in writing? Many allow room for creativity. There may be several ways to give a correct proof or a correct construction. In fact, one of the pleasures of teaching this course has been to see the multitude of imaginative methods with which students have solved the more open-ended problems. I encourage students to work together in groups, to share ideas, and to defend to each other the solutions they have found.

So perhaps the best way to use this book is treat it as no-text. Go directly to the exercises and start to work, collecting terminology and hints from the text only as needed!

Exercises

1.1 See what you can remember from high school geometry. Make a list of definitions and theorems. Do you remember the "side-angle-side" criterion for congruent triangles? Could you prove it? Can you prove that the three angle bisectors of a triangle meet in a point? Can you prove that the three altitudes of a triangle meet in a point? Do you remember the definition of similar triangles and facts about them?

1.2 Read Euclid's *Elements*, Book I, Propositions 1–34. Be prepared to explain the statements and present the proofs of (I.4), (I.5), (I.8), (I.15), (I.26), (I.27), (I.29), (I.30), and (I.32).

1.3 Discuss the structure of Euclid's proofs.

a) Proclus describes six parts of a theorem (see Proclus (1970), p. 159): the *enunciation*, which states what is given and what is sought, the *exposition*, which says again what is given, often in a more specific form; the *specification*, which makes clear what is sought; the *construction*, which adds what is needed; the *proof*, which infers deductively what is sought from what has been previously demonstrated; and the *conclusion* which confirms what has been proved. Identify these parts (some of which may be missing) in (I.1), (I.4), and (I.5).

b) Discuss Euclid's habit of presenting only one case of a proposition and leaving the others to the reader. For example, in (I.7) what other cases should we consider, and how would you complete the proof in those other cases?

c) Discuss the method of *reductio ad absurdum* (arguing to an absurdity) as a method of proof. How does this work in (I.6)? Can you think of a *direct* proof of this result (i.e. without assuming the contrary)?

For the following problems 1.4–1.9, present proofs in the style of Euclid, using any results you like from Book I, 1–34 (excluding the theory of area which starts

PROP. VII. TH. IV.

Si des extremitez de quelque ligne droite, on mene deux autres lignes droites, ſe rencontrans à un point; des mêmes extremitez, on n'en pourra pas mener deux autres égales à icelles, chacune à la ſienne, & de même part, ſe rençontrans à un autre point.

Soit la ligne AB, des extremitez de laquelle ſoient menées deux lignes droites AC & BC ſe rencontrant à quelconque

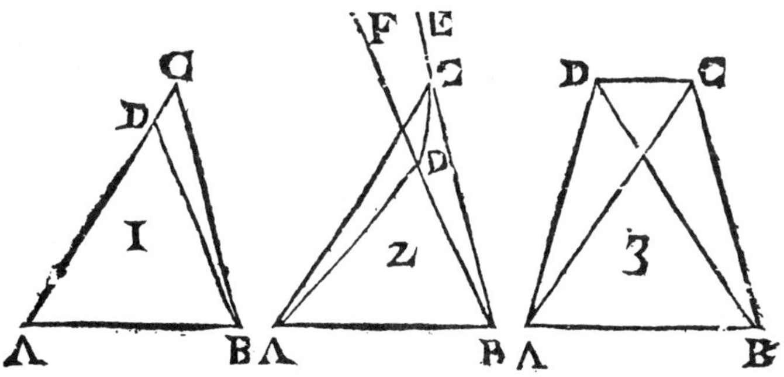

point C. Je dis que des mêmes extremitez A & B, & de la même part que C, on ne peut mener deux autres lignes droites égales à icelles AC & BC chacune à la ſienne, qui ſe rencontrent à un autre point que C; c'eſt à dire que ſi de l'extremité A on mene la ligne AD égale à AC, & de l'extremité B la ligne BD égale à BC, il ne peut être que le point de rencontre D, ſoit autre que le point de rencontre C.

Car ſi faire ſe peut, que le point de rencontre D, tombe ailleurs qu'au point C: où iceluy point D tombera ſur l'une ou l'autre des lignes AC, BC; ou dans le triangle ACB; ou hors iceluy.

Premierement, iceluy point de rencontre D, ne peut être ſur la ligne AC, comme en la premiere figure: car il faudroit que les deux lignes AD, & AC fuſſent égales entr'elles, ſçavoir eſt la partie au tout; ce qui eſt abſurde: partant la rencontre D, ne ſe fera point ſur AC, ny auſſi ſur BC, à cauſe de la même abſurdité.

Plate I. A page from Henrion's Euclid (1677) showing (I.7) and the three different cases of the proof.

with (I.35)). Be sure to refer to Euclid's definitions, postulates, common notions, and propositions by number whenever you use one.

1.4 A *rhombus* is a figure with four equal sides. Show that the diagonals of a rhombus meet at right angles, and that the four small triangles thus formed are congruent to each other.

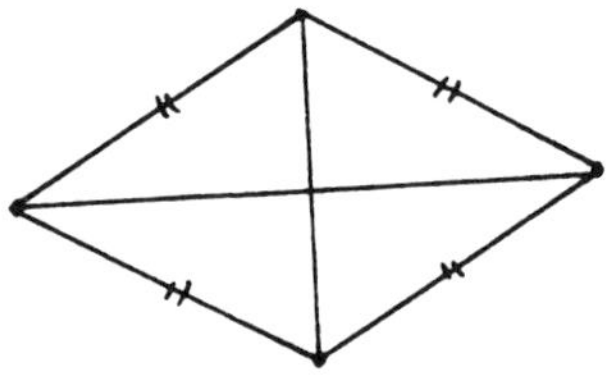

1.5 A *rectangle* is a four-sided figure with four right angles. Show that the two diagonals of a rectangle are congruent and bisect each other.

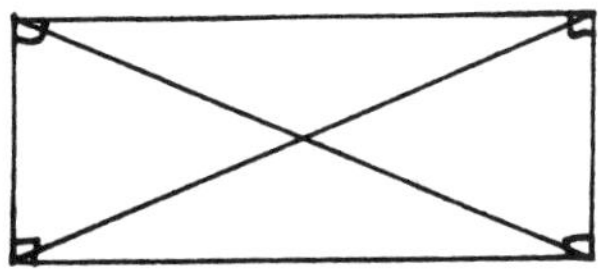

1.6 The exterior angles of a pentagon, with sides extended, add up to four right angles.

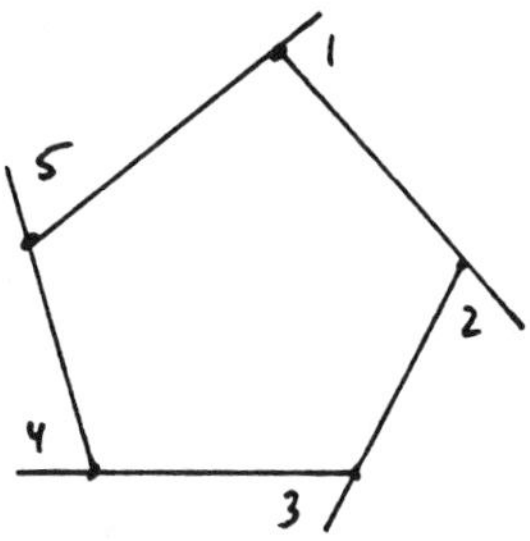

1.7 If two right triangles have one side and the hypotenuse respectively congruent, then the triangles are congruent. (We call this the right-angle-side-side theorem (RASS). Note in general that "ASS" is false: if two triangles have an angle and two sides equal, they need not be congruent.)

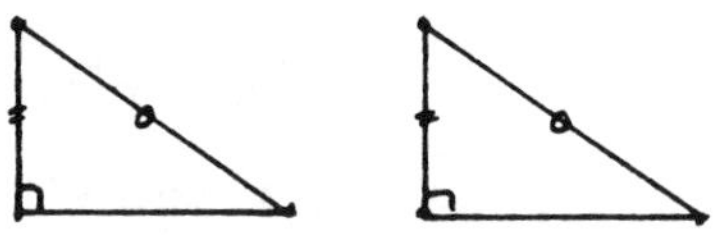

1.8 Show that the three angle bisectors of a triangle meet in a point. Be careful how you make your construction, and in what order you do the steps of your proof. (If you need a hint, look at (IV.4).)

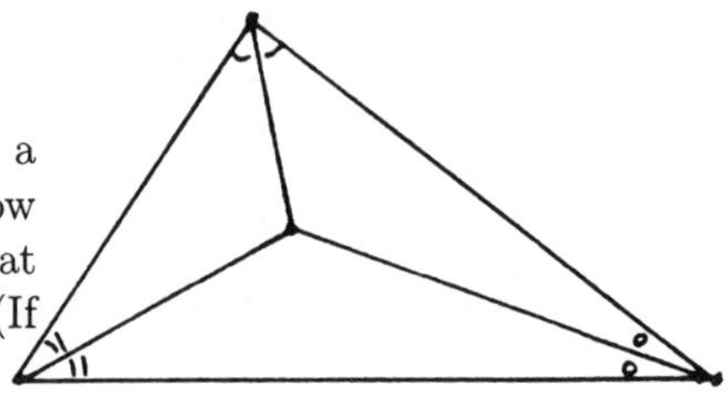

1.9 The three perpendicular bisectors of the sides of a triangle meet in a single point. Be sure to give a reason why they should meet at all. For a hint, look at (IV.5).

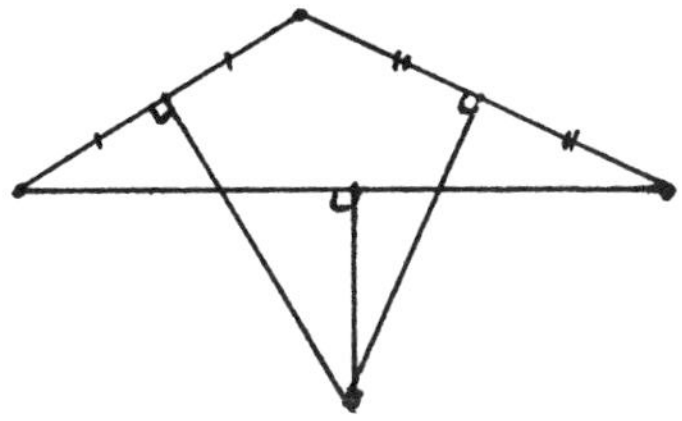

1.10 Read the *Elements*, Book III, Propositions 1–34. Be prepared to present statements and proofs of (III.16), (III.18), (III.20), (III.21), (III.22), (III.31), and (III.32).

For the following problems, present proofs in the style of Euclid, using any results you like from (I.1)–(I.34) and (III.1)–(III.34) (still excluding the theory of area).

1.11 Let AB and AC be two tangent lines from a point A outside a given circle. Show that $AB \cong AC$.

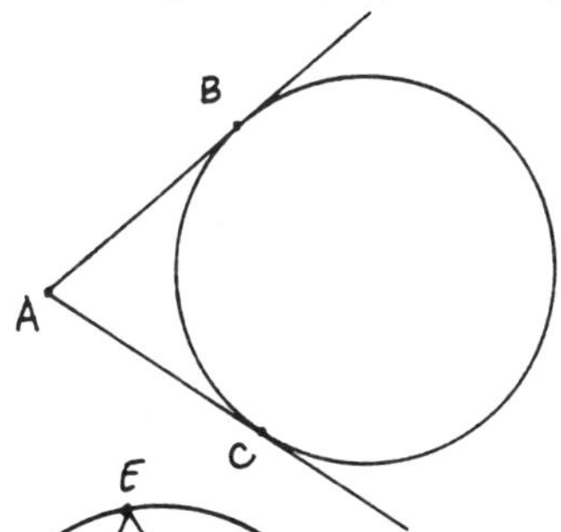

1.12 Let two circles be tangent at a point A. Draw two lines through A meeting the circles at further points B, C, D, E. Show that BC is parallel to DE.

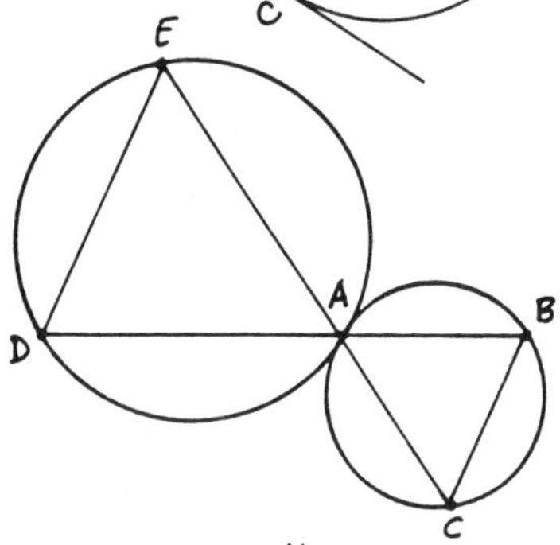

1.13 Suppose that three circles are mutually tangent at points A, B, C. Show that the tangent lines to the circles at the points A, B, C all meet in a single point.

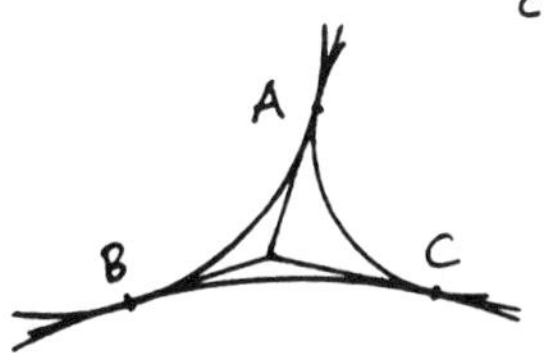

1.14 Given a pentagon $ABCDE$. Assume that all five sides are equal, and that the angles at A, B, C are equal. Prove that in fact all five angles are equal (so it is a *regular* pentagon).

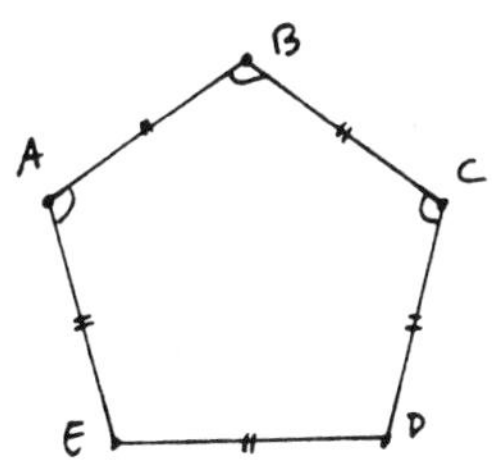

1.15 Let two circles γ and δ meet at a point P. Let the tangent to γ at P meet δ again at B, and let the tangent to δ at P meet γ again at A. Let θ be the circle through A, B, P. Let the tangent to θ at P meet γ and δ at C, D. Prove that $PC \cong PD$. *Hint*: Draw lines joining P and the centers of the three circles, and look for a parallelogram.

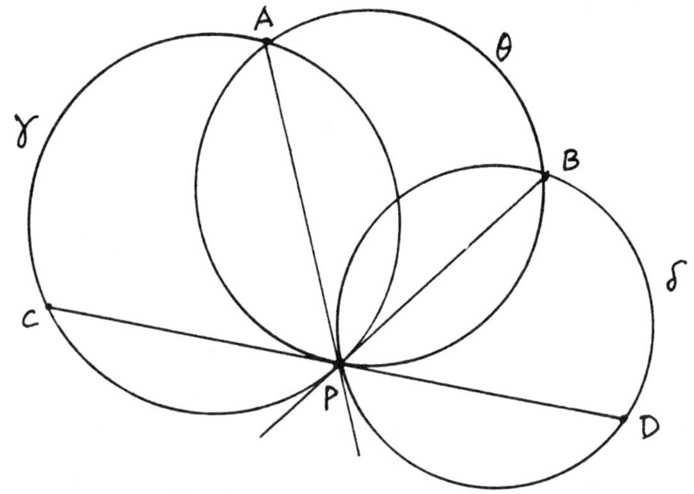

2 Ruler and Compass Constructions

One of the notable features of Euclid's *Elements* is his constructive approach to geometry. Many of his propositions are not theorems in the usual sense, that under certain hypotheses a certain result is true. Rather they are construction problems: given certain data, to construct a certain figure. For example, the first proposition of Book I is to construct an equilateral triangle. We could regard these constructions as existence proofs. But they are existence proofs of a very special kind: they are constructive, and the constructions are carried out with specified tools, the ruler (or straight-edge) and compass. Almost one third of the propositions in Book I, and all of the propositions in Book IV, are constructions. The constructive approach is even embedded in the initial assumptions of Euclid's geometry, because Postulate 1 says "to draw a straight line from any point to any point", and Postulate 3 says "to describe a circle with any center and distance". A modern mathematician would be more likely to say there exists a line through any two points, and replace Postulate 3 by a definition of a circle as the set of points equidistant from a given point.

This constructive approach pervades Euclid's *Elements*. There is no figure in the entire work which cannot be constructed with ruler and compass,[2] and this limits the world of subjects to be discussed to those which are constructible. So for example in Book IV, where Euclid discusses regular polygons inscribed in a circle, we find the triangle, the square, the pentagon, the hexagon, and the regular 15-sided polygon, all of which can be constructed. But there is no mention of a regular

[2]For the three-dimensional figures of Books XI–XIII we must allow also a tool which can draw a plane through three given points, and a tool which can rotate a semicircle about its diameter as axis to construct a sphere.

7-sided polygon, for example, and there are no theorems about regular n-gons such as one might find in a modern text. A modern mathematician would never doubt the existence of a regular 7-gon: just make angles of $2\pi/7$ at the center of the circle, and join corresponding points on the circumference. He would rather formulate the question, is it possible to construct the regular 7-gon with ruler and compass? But for Euclid, it seems he cannot discuss a figure until he has shown how to construct it. Look for example at (I.46), to construct a square on a given line segment. In terms of what is needed for the proof, this result could have been placed immediately after (I.34). Why is it here? Presumably because in the next proposition, the famous Pythagorean theorem (I.47) he needs to talk about the squares on the three sides of the right triangle, and he does not want to do this until he has shown that a square can be constructed on any given line-segment.

This brings us to the thorny question of what exactly does it mean to say a certain mathematical object exists? For some of the structures considered by modern-day mathematicians, this is indeed a difficult question. But for Euclid there was no doubt. I believe we will not be far from the truth if we say simply, that in Euclid's geometry only those geometrical figures exist which can be constructed with ruler and compass.

So now let us examine more closely how these ruler and compass constructions work. Look at (I.1) for example, to construct an equilateral triangle on a given line segment. We are given a line segment AB. We draw a circle with center A and radius AB, and another circle with center B and radius BA. These two circles met at a point C (and also at another point C' which we do not need). With the ruler we draw the lines AC and BC. Then ABC is the required equilateral triangle.

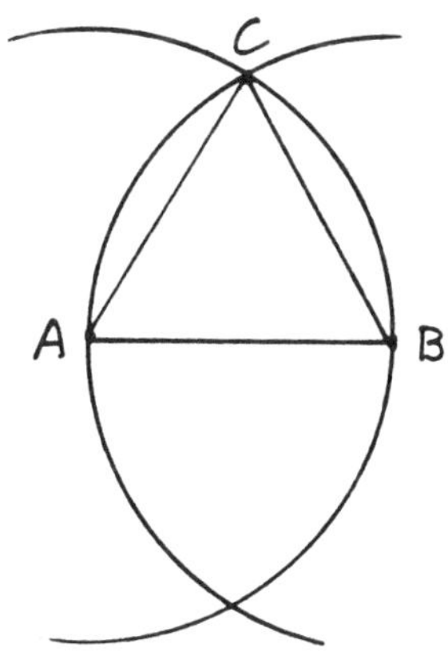

Thus the *construction* consists of a finite number of lines and circles drawn with the ruler and compass, starting from the initial data, obtaining new points as intersections along the way, and ending with the desired figure.

We will distinguish the construction, which is a series of applications of the ruler and compass to create a certain figure, from the *proof* that the figure constructed has the desired properties. The construction can be described, and makes sense, independently of any other constructions or proofs we may have made previously. But the proof that a certain construction works depends on its position in the logical sequence of propositions. In the case of (I.1), there are no previous propositions, so Euclid's proof depends only on the definitions, postulates, and common notions set out at the beginning of Book I. His proof says, in substance, that $AC = AB$ because they are both radii of the first circle, and $BA = BC$ because they are radii of the second circle, so $AB = AC = BC$ and hence the triangle is equilateral.

Next, let us look at (I.2). Given a point A and a line segment BC, we must construct a line segment AF originating at A, equal to BC. Euclid's method is as follows: Draw AB. Construct the equilateral triangle ABD using the construction of (I.1). Then with center B and radius BC draw a circle to meet DB extended at E. With center D and radius DE draw a circle to meet DA extended at F. Then AF is the required line segment.

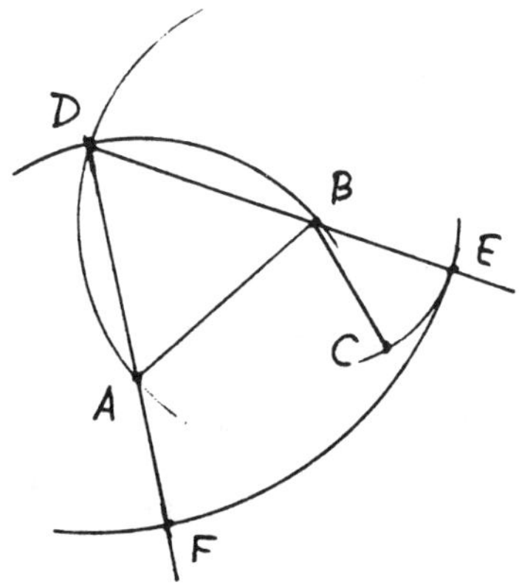

The proof is natural enough: $BC = BE$ by construction; $DE = DF$ by construction; $DB = DA$ by construction, so by subtraction $AF = BE = BC$ as required.

But the question which immediately arises is, why did Euclid go to all this trouble when he could have made a much simpler construction: set the compass points to the distance BC, then draw a circle with center A and radius BC, choose F any point on that circle, and join A to F? We must infer from the presence of this construction that Euclid allowed himself to use the compass only in its narrow sense to draw a circle with a given center and passing through a given point. It could not be lifted off the paper and used to transport a given distance to another location. So some people call Euclid's compass a *collapsible compass*: when you lift it off the paper the points fall together and do not preserve the radius they were set at. However, the function of this construction (I.2) is to show that with the collapsible compass one can still accomplish the same result, *as if the compass had not been collapsible*, namely to transport a distance to another point in the plane. So from now on, we will allow ourselves to use the compass in this stronger sense, to draw a circle with given center and radius equal to any given line segment.

To increase our awareness of the process of ruler and compass constructions, let us make precise exactly how the tools can be used, and let us set up a way of counting our steps as a measure of the complexity of the construction. The number of steps needed for a construction is not really important of itself, but by counting our steps we become more conscious of the process. This is one of the practical aspects of this course, to have some fun while we are pondering the deeper theoretical questions.

In any construction problem, there are usually some points, lines, or circles given at the outset. The ruler may be used to extend a given or previously constructed line in either direction. The ruler may be used to draw a new line through two distinct points either given or constructed earlier. The ruler may not be used to measure distances, and it may not have any markings on it (hence the frequently used term *straight-edge* to emphasize that it may only be used to draw straight lines).

The compass may be used to draw a circle with center a given or previously constructed point, and with radius equal to the distance between any two given or previously constructed points.

In addition, at any time one may choose a point at random, or subject to conditions such as that it should lie on a given line or circle, or be on the other side of a line from a given point, etc.

Each time a new line or circle is drawn, those points in which it intersects previously given or constructed lines and circles will be considered to be constructed also.

For counting, we consider each use of the ruler to construct a new line as one step, and each use of the compass to construct a new circle as one step. Extending lines previously given or constructed, choosing points at random, and obtaining new points as intersections, do not count as separate steps.

Thus for example, the construction of the equilateral triangle (I.1) above takes four steps:

The line segment AB is given

1. Draw circle with center A and radius AB.
2. Draw circle with center B and radius BA. Get C.
3. Draw AC.
4. Draw BC.

Then ABC is the required triangle.

When performing more complicated constructions, we will count all of the steps required to perform the entire construction, so that each construction is self-contained and independent of other constructions (though inevitably each construction will contain elements of other constructions). This imposes a different notion of economy of construction from Euclid's. For while Euclid in his sequential development of the propositions finds it most economical to utilize previous constructions, we will find that minimizing the total number of steps will often lead us to different constructions.

Look at (I.9), for example, to bisect a given angle. The angle is given by a point A and two rays l, m emanating from A. Euclid's method is this: Choose B on l at random. Find C on m so that $AB = AC$ (I.3). Draw BC. Construct the equilateral triangle BCD (I.1). Join AD. Then AD is the angle bisector.

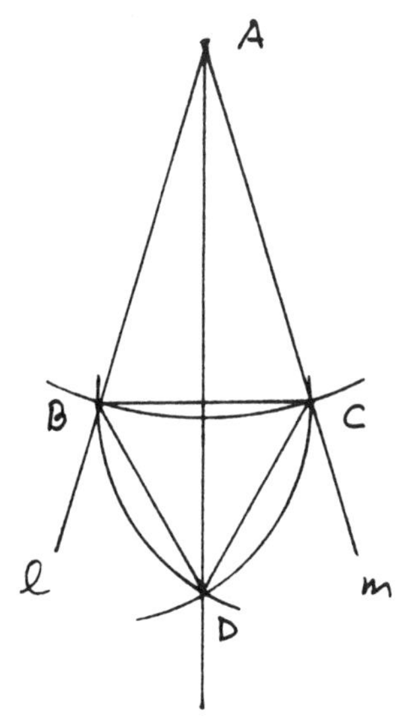

Euclid's method is economical for him because it makes use of previously described constructions (I.3) and (I.1). If we count the number of steps to carry out this construction, we find seven:

Choose B at random on l (no step)

1. Circle center A radius AB, get C.

2. Draw BC.
3. Circle center B radius BC.
4. Circle center C radius CB, get D.
5. Draw BD.
6. Draw CD.
7. Draw AD, which is the angle bisector.

If we are concerned only with making an independent construction for the angle bisector, there is no need to draw the lines BC, BD, CD. Thus the construction reduces to four steps. In order to prove that this construction works, we might want to draw the lines BC, BD, CD and argue as Euclid did. The lines then become part of the proof. But they are not part of the construction, so the construction still requires only four steps.

For another example, look at Euclid's construction (I.10) to bisect a given line segment. He first appeals to (I.1) to construct an equilateral triangle, and then to (I.9) to bisect the angle at its vertex. This is an elegant method, making use of what he has done before. But in terms of numbers of steps, it is not efficient. If we add the numbers of steps used in the two previous results, we get 11 steps. If we make use of points already constructed in (I.1) when we do the construction of (I.9), this reduces to 9. But it is possible to give a direct construction of the midpoint of a segment in only three steps (see Exercise 2).

A note about accuracy and exactness of constructions

When carrying out ruler and compass constructions, we attempt to make our drawings as *accurate* as possible. Using a sharp pencil we draw fine lines and make them pass through given points as closely as possible. Nevertheless, there is always a small error in each step, and those errors will compound throughout a long construction, so that the final figure does not always do just what you want. For example in constructing the circle circumscribed about a given triangle (Exercise 2.10), you may find that your circle passes nicely through two of the points but misses the third one slightly. This error is inevitable in any drawings we make.

But, to paraphrase the quote from Plato in §1, it is not the line and the circle drawn on the paper that we are thinking of, it is the absolute line and the absolute circle. And in this sense, our construction must be mathematically *exact*. In other words, it must be possible to prove using the reasoning of abstract geometry, that this construction in its ideal form gives the exact result we are seeking.

This distinction has caused considerable confusion among amateur angle trisectors and circle squares through the ages. For many of their constructions are remarkably accurate, while failing to be mathematically exact. (See for example Underwood Dudley's interesting book "A budget of trisections".)

Exercises

For each of the following problems, carry out a ruler and compass construction as accurately as you can. Number and label each of your steps as in the text. Feel free to use abbreviations such as "AB" for "draw a line AB"; "$\odot AB$" to draw a circle with center A and radius AB; or "$\odot cArBC$" to draw a circle with center A and radius BC. Label each new point as it is constructed and mention it (e.q. "get F") in the appropriate step. For the time being, we are not concerned with the proofs. Just do the construction. You should however be able to give an informal proof (= convincing argument) of why it works, if asked.

After you make your construction, locate the corresponding proposition in Euclid (Books I, III, or IV) and compare. How many steps does his method require? What do you think is the least number of steps possible? I will sometimes give a *par value* for a construction, which is the typical number of steps an experienced constructor would need. By trying harder, you can sometimes succeed with fewer steps.

2.1 Given an angle, construct the angle bisector. (Par = 4).

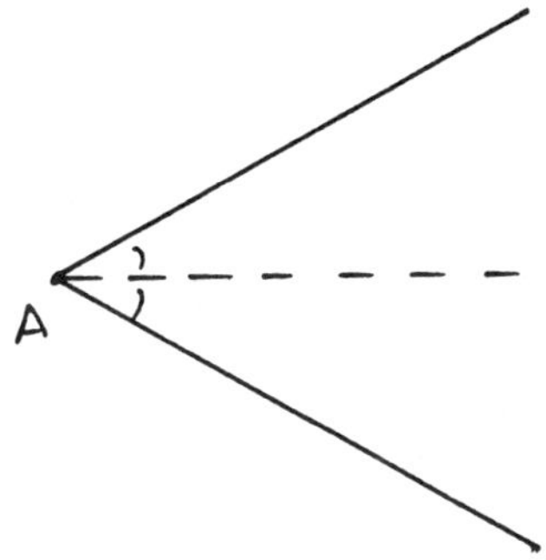

2.2 Given a line segment, find the midpoint of that segment. (Par = 3).

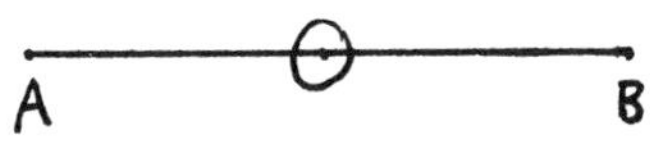

2.3 Given a line l and a point A on l, construct a line perpendicular to l through A. (Par = 4).

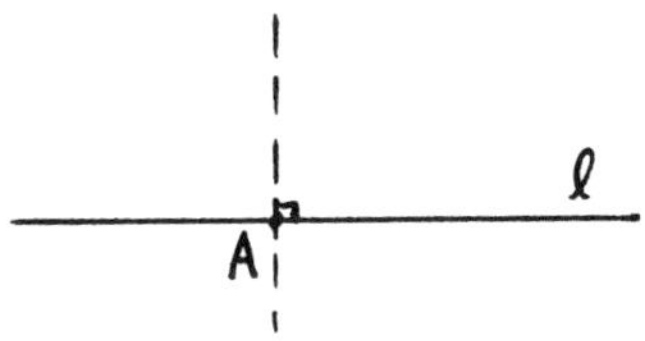

2.4 Given a line l and a point A not on l, construct a line perpendicular to l passing through A. (Par = 4).

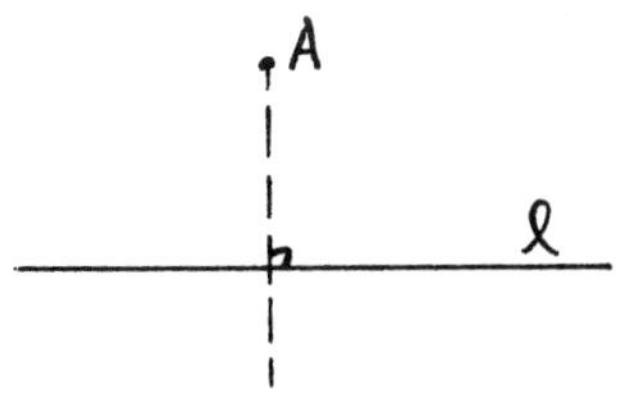

2.5 Given an angle at a point A, and given a ray emanating from a point B, construct an angle at B equal to the angle at A. (Par = 4).

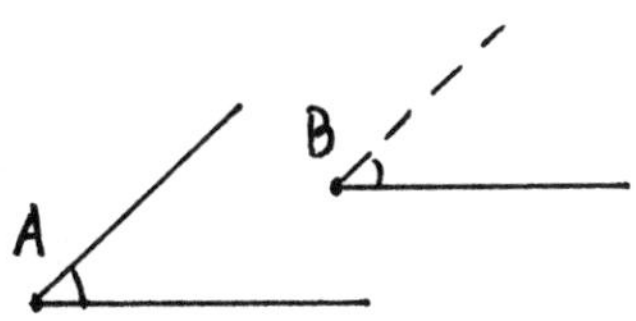

2.6 Given a line l and a point A not on l, construct a line parallel to l, passing through A. (Par = 4).

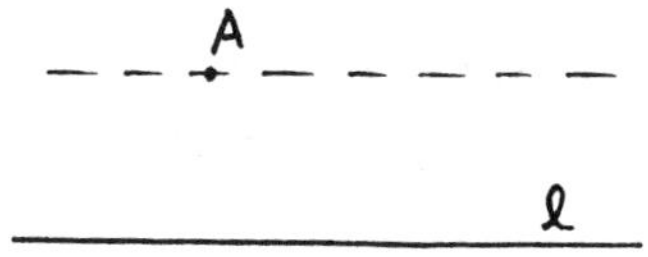

2.7 Given a circumference of a circle, find the center of the circle. (Par = 5).

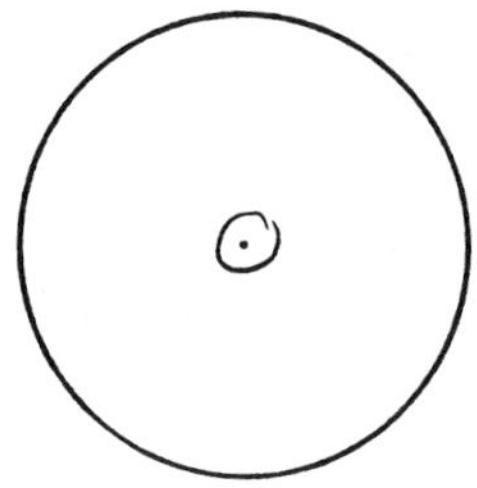

2.8 Given a circle with its center O, and given a point A outside the circle, construct a line through A tangent to the circle. (*Warning*: You may not slide the ruler until it seems to be tangent to the circle. You must construct another point on the desired tangent line before drawing the tangent.) (Par = 6).

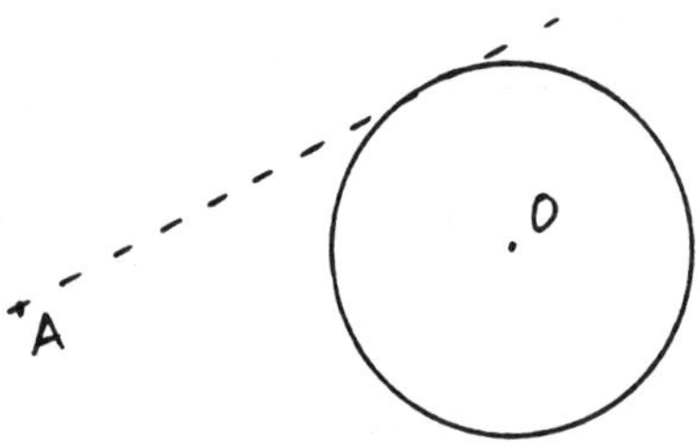

2.9 Construct a circle inscribed in a given triangle ABC. (Par = 13).

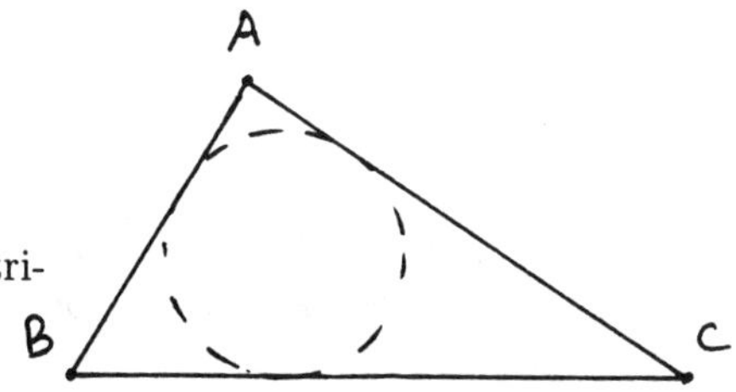

2.10 Construct a circle circumscribed about a given triangle ABC. (Par = 7).

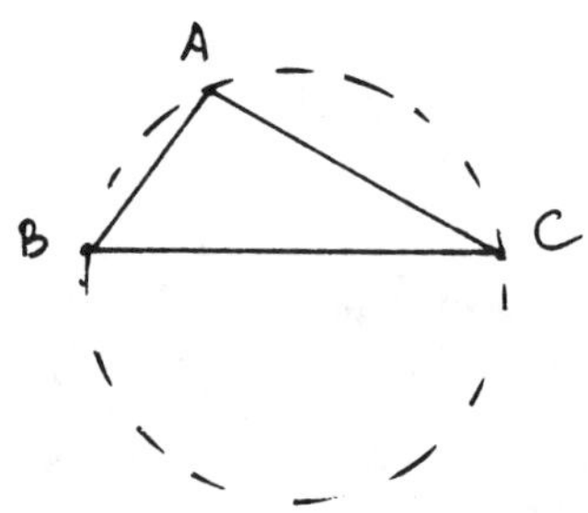

2.11 Given a line l, a line segment d, and a point O, construct a circle with center O which cuts off a segment congruent to d on the line l. (Par = 9).

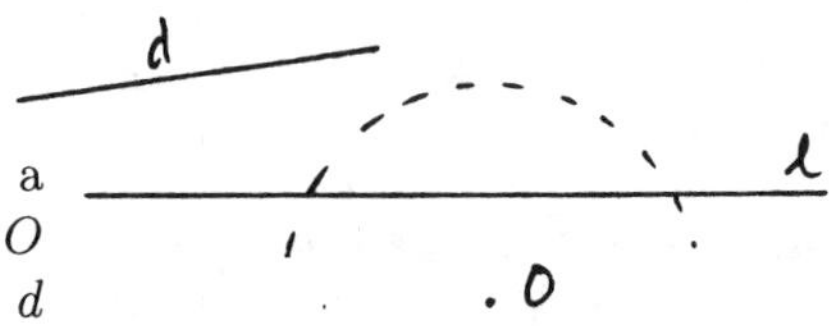

2.12 Given a point A, a line l, and a point B on l, construct a circle which passes through A and is tangent to the line l at B. (Par = 8).

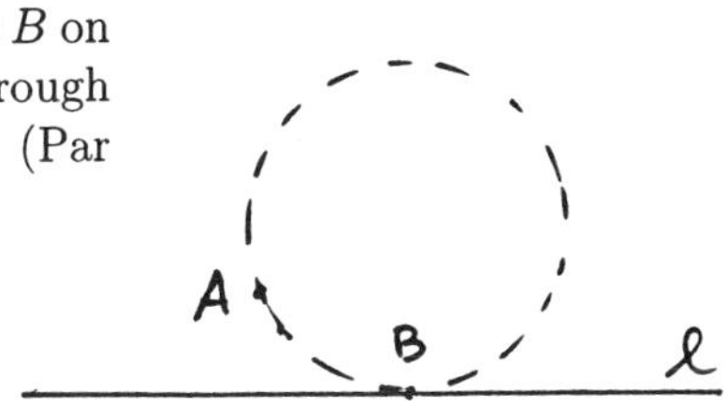

2.13 Construct three circles, each one meeting the other two at right angles. (Par = 10).

2.14 Given a line segment AB, divide it into three equal pieces. (Par = 9).

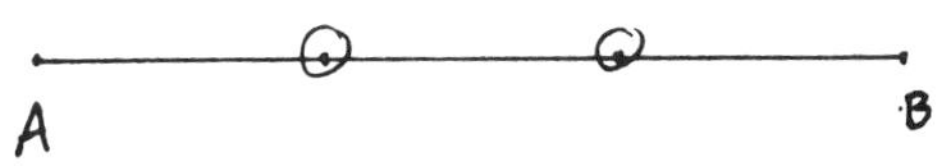

2.15 (The one-inch ruler.) Suzie's ruler broke into little pieces, so she can only draw lines one inch long. Fortunately her compass is still working. She has two points on her paper approximately 3 inches apart. Help her construct the straight line joining those two points.

2.16 (The rusty compass.) Joe's compass has rusted into a fixed position, so it can only draw circles whose radius is one inch. Fortunately his ruler is still working. Help him construct an equilateral triangle on a segment AB which is approximately 2-1/2 inches long. (Par = 6).

2.17 Using a ruler and rusty compass (cf. # 16), construct the perpendicular to a line l at a point A on l. (Par = 6).

2.18 Using a ruler and rusty compass, given a line l and a point A more than 2 inches away from l, construct the line through A and perpendicular to l.

2.19 Using a ruler and rusty compass, given a segment AB and given a ray AC, construct a point D on the ray AC such that $AB \cong AD$.

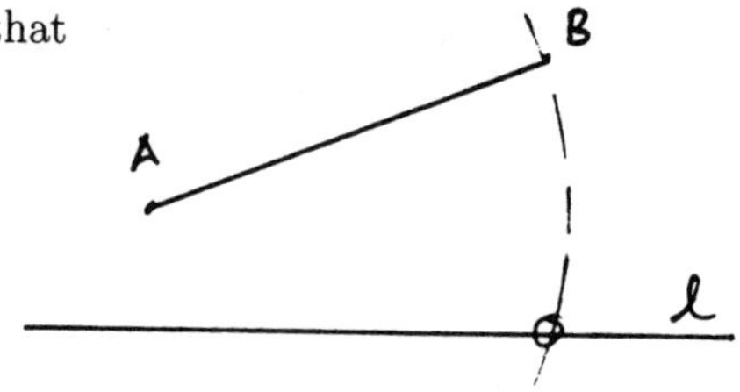

2.20 Using a ruler and rusty compass, given a line l and given a segment AB, more than one inch long, construct one of the points C in which the circle of center A and radius AB meets l.

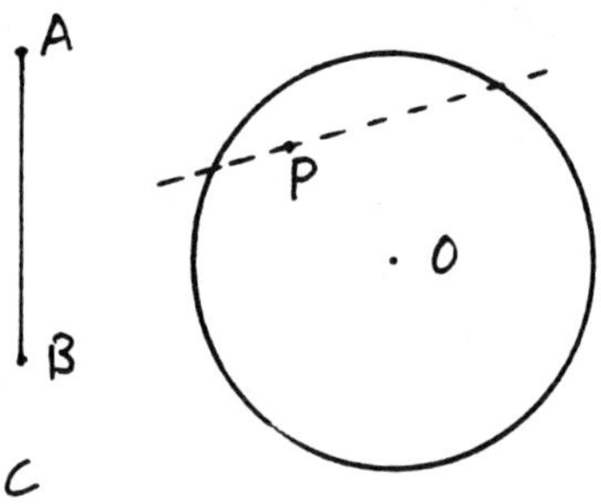

2.21 (Back to regular ruler and compass construction.) Given a segment AB, given a circle with center O, and given a point P inside O, construct (if possible) a line through P, on which the circle cuts of a segment congruent to AB. (Par = 5).

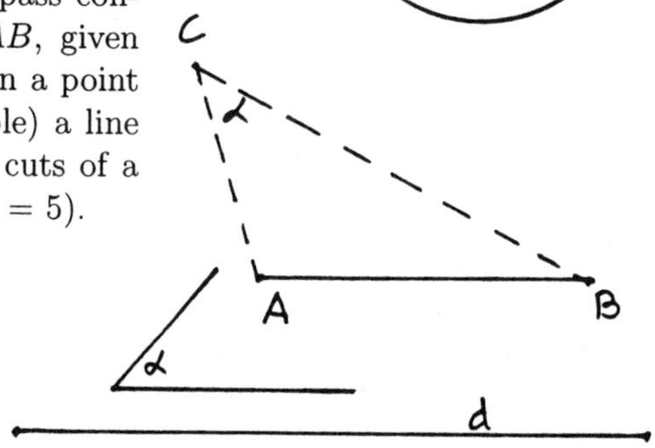

2.22 Given a segment AB, given an angle α, and given another segment d, construct a triangle ABC with base AB, angle α at C, and such that $AC + BC = d$. *Hint*: First think about how to construct all possible triangles with base AB and opposite angle α. (Par = 23).

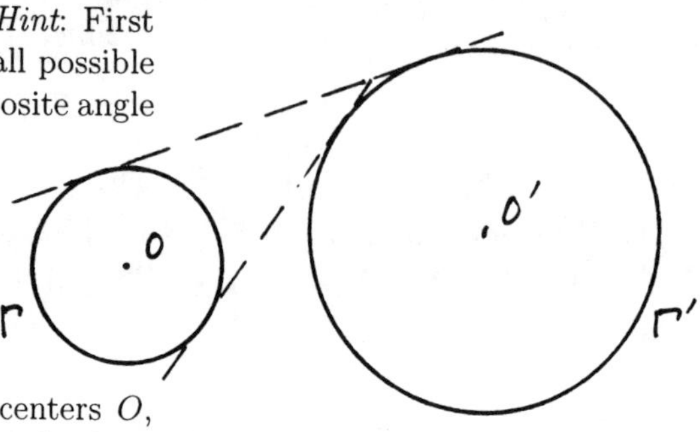

2.23 Given two circles Γ, Γ', with centers O, O', construct a line tangent to both of them.

3 Euclid's Axiomatic Method

One of the remarkable features of Euclid's *Elements* is its orderly logical structure. Euclid took the great mass of geometrical material which had grown in the previous two or three centuries, and organized it into one coherent logical sequence. This is what we now call the *axiomatic method*: starting from a small number of definitions and assumptions at the beginning, all the succeeding results are proved by logical deduction from what has gone before. Euclid's text has been a model of mathematical exposition, unchallenged for two thousand years, and only recently (in the last hundred years or so) replaced by newer mathematical systems which we consider more rigorous. As we read Euclid, let us observe how he organizes his material, let us be curious why he does things the way he does, and let us explore the questions which come to mind when we as modern mathematicians read this ancient text.

Definitions

Euclid begins with definitions. Some of these definitions are akin to the modern notion of definition in mathematics, in that they give a precise meaning to the term being defined. For example, the tenth definition tells us that if a line segment meets a line so that the angles on either side are equal, then these are called right angles. This tells us the meaning of the term right angle, assuming that we already know what is meant by a line, a line segment, an angle, and equality of angles. Similarly the fifteenth definition, rephrased, defines a circle to be a set of points C, such that the line segments OA from a fixed point O to any point A of the circle C, are all equal to each other, and the point O is called the center of the circle. This tells us what a circle is, assuming we already know what is a line segment, and what is meant by equality of line segments.

On the other hand, some of Euclid's other definitions, such as # 1, "a point is that which has no part", or # 2, "a line is breadthless length", or # 3, "a straight line is a line which lies evenly with the points on itself", give us no better understanding of these notions than we had before. It seems that Euclid, instead of giving a precise meaning to these terms, is appealing to our intuition, and alluding to some concept we may already have in our own minds of what a point or a line is. Rather than defining the term, he is appealing to our common understanding of the concept, without saying what that is. This may have been very well in a society where there was just one truth and one geometry and everyone agreed on that. But the modern consciousness sees this as a rather uncertain way to set up the foundations of a rigorous discipline. What if we say now, oh yes, we agree on what points and lines are, and then later it turns out we had something quite different in mind? So the modern approach is to say these notions are *undefined*, that is, they can be anything at all, *provided* that they satisfy whatever postulates or axioms may be imposed on them later. In the algebraic definition of an abstract group, for example, you never say what the elements of the group are, nor what the group operation is. Those are undefined. However, they must satisfy the group axioms that the operation is associative, there exists an identity, and that there exist inverses. The elements of the group can then be anything which satisfies these axioms. They could be integers, or they could be cosets of a subgroup of the integers, or they could be rotations of a geometrical object such as a cube, or anything else. So in our reading of Euclid, perhaps we should regard "point" and "line" as undefined terms.

It may be worth noting some differences of language between Euclid's text and modern usage. By a *line* he means something which may be curved, which we would call a curve. He says *straight line* for what we call line. And then he says a *finite straight line* (as in the statement of (I.1)) for what we would call a line segment. For Euclid, a *plane angle* is where two curves meet, and a *rectilineal plane angle* is when two line segments meet. Note that Euclid requires the two sides of an angle not to lie in a straight line. So for Euclid there is no zero angle, and there is no straight angle (180°). So we should think of Euclid's concept of angle as meaning an angle of α degrees, with $0 < \alpha < 180°$ (though Euclid makes no mention of the degree measure of an angle).

Euclid's notion of *equality* requires special attention. He never defines equality, so we must read between the lines to see what he means. In Euclid's geometry there are various different kinds of magnitudes, such as line segments, angles, and later areas. Magnitudes of the same kind can be compared: they can be equal, or they can be greater or lesser than one another. Also they can be added and subtracted (provided one is greater than the other) as is suggested by the common notions.

Euclid's notion of equality corresponds to what we commonly call *congruence* of geometrical figures. In high school geometry one has the length of a line segment, as a real number, so one can say that two segments are congruent if they have the same length. However, there are no lengths in Euclid's geometry, so we must regard his equality as an undefined notion. Because of common notion # 1, "things which are equal to the same thing are also equal to one another", we may regard equality (which we will call *congruence* to avoid overuse of the word equal) to be an equivalence relation on line segments. Similarly we will regard congruence of angles as an equivalence relation on angles.

Postulates and Common Notions

The postulates and common notions are those facts which will be taken for granted and used as the starting point for the logical deduction of theorems. If you think of Euclid's geometry in the classical way as being the one true geometry which describes the real world in its ideal form, then you may regard the postulates and common notions as being self-evident truths for which no proof is required. If you think of Euclid's geometry in the modern way as an abstract mathematical theory, then the postulates and common notions are merely those statements which are arbitrarily selected as the starting point of the theory, and from which other results will be deduced. There is no question of their "truth", because one can begin a mathematical theory from any hypotheses one likes. Later on, however, there may arise a question of relevance, or importance of the mathematical theory constructed. The importance of a mathematical theory is judged by its usefulness in proving theorems which relate to other branches of mathematics or to applications. If you begin a mathematical theory with weird hypotheses as your starting point, you may get a valid logical structure which is of no use. From that point of view the choice of postulates is not so arbitrary. In any case, we can regard Euclid's postulates and common notions collectively as the set of *axioms* on which his geometry is based.

Some commentators say that the postulates (as in Heath's edition) are those statements which have a geometrical content, while the common notions are those statements of a more universal nature, which apply to all the sciences. Other commentators divide them differently, calling "postulates" those statements which allow you to construct something (Post. 1–3 in Heath), and calling "axioms" those statements which assert that something is always true (Post. 4, 5 and the Common Notions in Heath). One should also note that some editors give extra axioms not listed in Heath's edition, such as that "halves of equals are equal", which is used by Euclid, e.g. in the proof of (I.37), or "two straight lines cannot contain a space".

The constructive nature of Euclid's approach to geometry is evident in Postulates 1–3: he states them as rules of construction, whereas a modern mathematician would more likely say "through any two distinct points there exists a line", and # 3 would simply become the definition of a circle. By the way, Euclid makes no explicit statement about the uniqueness of the line mentioned in Postulate 1, though he apparently meant it to be unique, because in the proof of (I.4) he says "otherwise two straight lines will enclose a space: which is impossible".

In the list of Postulates and Common Notions, Postulate 5 stands out as being much more sophisticated than the others. It sounds more like a theorem than an axiom. We will have more to say about this later. For the moment let us just observe that two thousand years of unsuccessful efforts to prove this statement as a consequence of the other axioms have vindicated Euclid's genius in realizing that it was necessary to include Postulate 5 as an axiom.

Intersections of Circles and Lines

As we read Euclid's *Elements* let us note how well he succeeds in his goal of proving all his propositions by pure logical reasoning from first principles. We will find at times that he relies on "intuition", or something that is obvious from looking at a diagram, but which is not explicitly stated in the axioms. For example, in the construction of the equilateral triangle on a given line segment AB (I.1) how does he know that the two circles actually meet at some point C? While the 5th Postulate guarantees that two lines will meet, under certain conditions, there is nothing in the definitions, postulates, or common notions that says that two circles will meet. Nor does Euclid offer any reason in his proof that the two circles will meet.

If you carry out the construction with ruler and compass on a piece of paper, you will find that they do meet. Or if you look at the diagram, it seems obvious that they will meet. However, that is not a proof, and we must acknowledge that Euclid is using something here which is not explicitly guaranteed by his axioms and yet which is essential to the success of his construction.

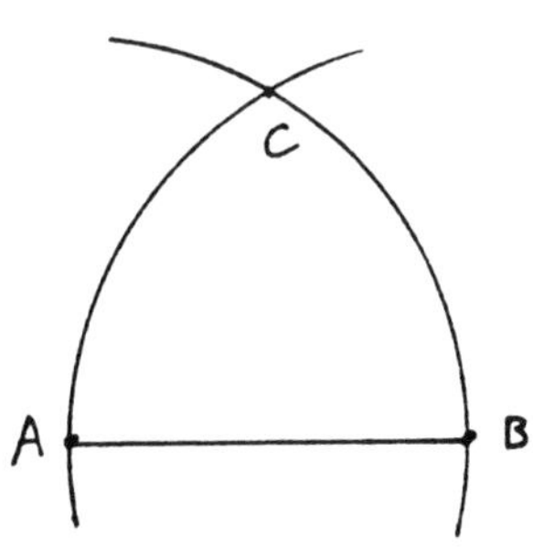

There are two separate issues here. One is the relative position of the two circles. Two circles need not always meet. If they are far apart from each other, or if one is entirely contained in the other, they will not meet. In the present case, part of one circle is inside the other circle, and part outside, so it appears from the diagram that they must cross each other.

The second issue is, assuming they are in a position so that they appear to meet, does the intersection point actually exist? A modern mathematician will immediately think of continuity and the intermediate value theorem: if $y = f(x)$ is a real-valued continuous function defined on the unit interval $[0, 1]$ of the real numbers, and if $f(0) < 0$ and $f(1) > 0$, then there is some point $a \in [0, 1]$ with $f(a) = 0$. In other words, the graph of the function must intersect the x-axis at some point in the interval.

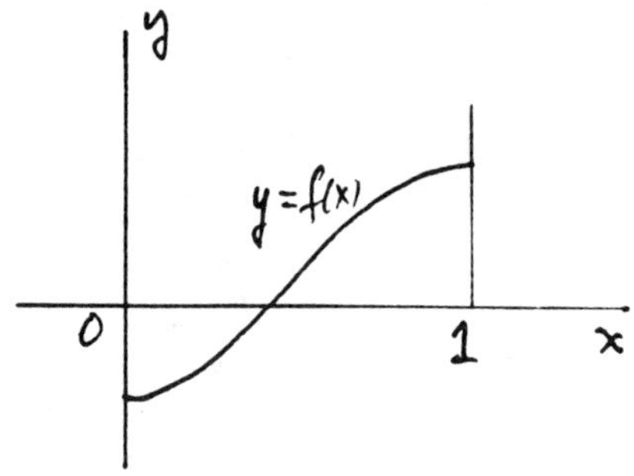

However, we must bear in mind that the concepts of real numbers and continuous functions were not developed until the late 19th century, and that this kind of mathematical thinking is foreign to the spirit of Euclid's *Elements*.

To make the same point in a different way, suppose we consider the *Cartesian plane* over the field of rational numbers $\mathbb{Q}$, where points are ordered pairs of rational numbers, and let AB be the unit interval on the x-axis. Then the vertex C of the equilateral triangle, which would have to be the point $\left(\frac{1}{2}, \frac{1}{2}\sqrt{3}\right)$, actually does not exist in this geometry.

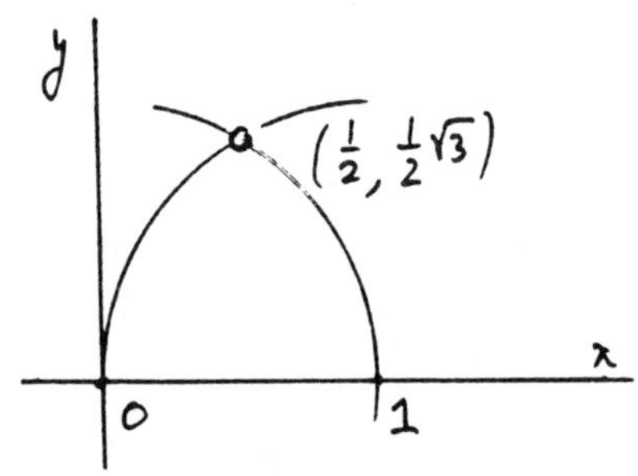

So later on, when we set up a new system of axioms for Euclidean geometry, we will have to include some axiom which guarantees the existence of the intersection points of circles with other circles, or with lines, at least those which arise in the ruler and compass constructions of Euclid's *Elements*. Some modern axiom systems (such as Birkhoff or the SMSG geometry) build the real numbers into the axioms with a postulate of line measure, or include Dedekind's axiom which essentially guarantees that we are working over the real numbers. In this book, however, we will reject such axioms as not being in the spirit of classical geometry, and we will introduce only those purely geometric axioms which are needed to lay a rigorous foundation for Euclid's *Elements*.

The issue of intersecting circles arises again in (I.22), where Euclid wishes to construct a triangle whose sides should be equal to three given line segments a, b, c. This requires that a circle with radius a at one endpoint of the segment b should meet a circle of radius c at the other end of the segment b. Euclid correctly puts the necessary and sufficient condition that this intersection should exist in the statement of the proposition, namely that any two of the line segments should be greater than the third. However, he never alludes to this hypothesis in his proof, so that we do not see in what way this hypothesis implies the existence of the intersection point. While some commentators have criticized Euclid for this, Simson ridicules them, saying "For who is so dull, though only beginning to learn the Elements, as not to perceive... that these circles must meet one another because FD and GH are together greater than FG." Still, Simson has only discussed the position of the circles and has a not addressed the second issue of why the intersection point exists.

The Method of Superposition

Let us look at the proof of (I.4), the side-angle-side criterion for congruence of two triangles (SAS for short). Suppose that $AB = DE$, and $AC = DF$, and the included angle $\angle BAC = \angle EDF$. We wish to conclude that the triangles are congruent, that is to say the remaining side and two angles are congruent to each other, respectively. Euclid's method is to "apply the triangle" ABC to the triangle DEF. That is, he imagines moving the triangle ABC onto the triangle DEF, so that the point A lands on the point D, and the side AB lands on the side DE. Then he goes on to argue that the ray AC must land on the ray DF, because the angles are equal, and hence C must land on F because the sides are equal. From here he concludes that the triangles coincide entirely, hence are congruent.

This is another situation where Euclid is using a method which is not explicitly allowed by his axioms. Nothing in the Postulates or Common Notions says that we may pick up a figure and move it to another position. We call this the *method of superposition.*

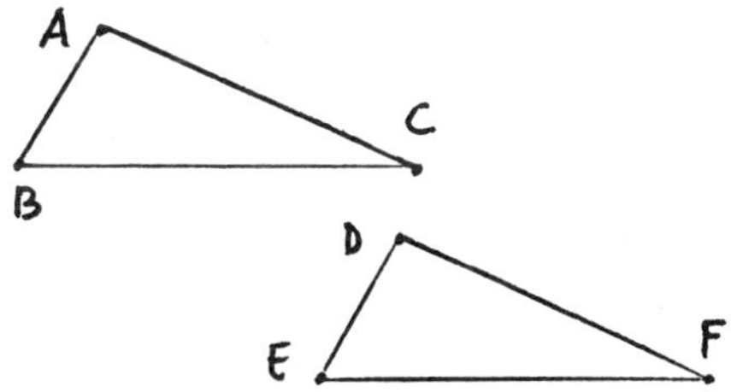

Euclid uses this method again in the proof of (I.8), but it appears that he was reluctant to use it more widely, because it does not appear elsewhere. If it were a generally accepted method, for example, then Postulate 4, that all right angles are equal to each other, would be unnecessary, because that would follow easily from superposition.

If we think about the implications of this method, it has far-reaching consequences. It implies that one can move figures from one part of the plane to another without changing their sides or angles. Thus it implies a certain homogeneity of the geometry: the local behavior of figures in one part of the plane is the same as in another part of the plane. If you think of modern theories of cosmology, where the curvature of space changes depending on the presence of large gravitational masses, this is a non-trivial assumption about our geometry.

To state more precisely what assumptions the method of superposition is based on, let us define a *rigid motion* of the plane to be a one-to-one transformation of the points of the plane to itself, which preserves straight lines, and such that segments and angles are carried into congruent segments and angles. To carry out the method of superposition, we need to assume that there exist sufficiently many rigid motions of our plane so that

a) we can take any point to any other point,

eſſe. Nam quum de re aliqua ſermonem inſtituimus: ea nobis tacitè per definitionem ſubit in animum: Non enim duos angulos æquales eſſe cogitabo, niſi quid ſit æquales eſſe angulos concipiam. Quod reſpiciens Euclides, angulorum æqualitatem proponere, atque eadem opera definire voluit: vt hoc Theorema pro Definitione haberemus. Nemo enim ſignificantius explicabit angulorum æqualitatem, quàm ſi dixerit duos angulos æquales fieri, quum duo latera vnum angulum continentia, duobus alterum angulum continentibus fiunt æqualia, & baſes quæ latera connectunt, æquales. Conſtat enim angulum tantum eſſe, quanta eſt duarum linearum ipſum continentium apertio, ſeu diductio, hanc verò tantam eſſe, quanta eſt baſis, hoc eſt, linea ipſas connectens. Atque vt clarè dicam, tantus eſt angulus B A C, quanta eſt remotio lineæ A C ab ipſa A B: tanta verò efficitur remotio, quantam exhibet linea B C. Hoc autem in Iſoſcelibus eſt euidentius. Sint enim duo Iſoſcelia A B C & D E F: quorum vnius duo latera A B & A C duobus D E & D F alterius ſint æqualia: angulusq́; A angulo D. Ac poſitis centris in A & D punctis, ducantur duo Circuli: prior ſecundum A B, alter ſecundum D E ſpatium. Horum prior manifeſtò tranſibit per B & C: alter verò per E & F puncta: quum A B & A C, itemq́; D E & E F ſint æqualia, & à centro vtrinq; exeuntia. Atque, ex definitione æqualium angulorum, erunt arcus B C & E F æquales. Angulorum enim magnitudo deſignatur ex arcubus Circulorum qui per extremas lineas quæ angulos continēt, tranſeunt. Ac conuerſo modo, æquales anguli atque æqualibus lineis comprehenſi, æquales ſubtendunt peripherias. Quum enim æqualia ſint ſpatia B C & E F, ea æqualibus rectis lineis claudi oportet: propterea quòd recta linea, eſt à puncto ad punctum via breuiſſima. Atque haud diſſimili iudicio, ex laterum ratione & baſium, quanta ſit angulorum magnitudo æſtimabimus. Quur ergo Euclides hoc inter Theoremata repoſuit, non inter Principia præmiſit? Nimirùm, quum ſpeciem quodammodo mixtam Principij & Theorematis præ ſe ferret: Principij, quòd in communi animi iudicio conſiſteret: Theorematis, quòd ſpeciatim Triangula Triangulis comparanda proponeret: maluit Euclides inter Theoremata referre: præſertim quum multa haberet capita, Principium verò ſimplex ac velut nudum eſſe debeat. Ex hoc præterea Axiomate tanquam ex locupletiſſimo Demonſtrationum themate, multæ Propoſitiones conſequi debebant, eiuſdem propè facilitatis & iudicij: quas, quia erant notiſſimæ, inter Principia annumerari non conueniebat. Paucis enim Principijs Geometriam contentam eſſe oportebat: immò multa Principia conſultò ſupprimuntur, ne ſit oneroſa multitudo: vt etiam quæ exprimuntur, tantùm ad exemplum exprimi videantur. Hùc accedit, quòd primum Theorema facile, perſpicuum, ac ſenſui obuium eſſe debebat, pro Geometriæ lege, quæ ex paruis humilibusq́; initijs, in progreſſus mirabiles ſeſe extollit.

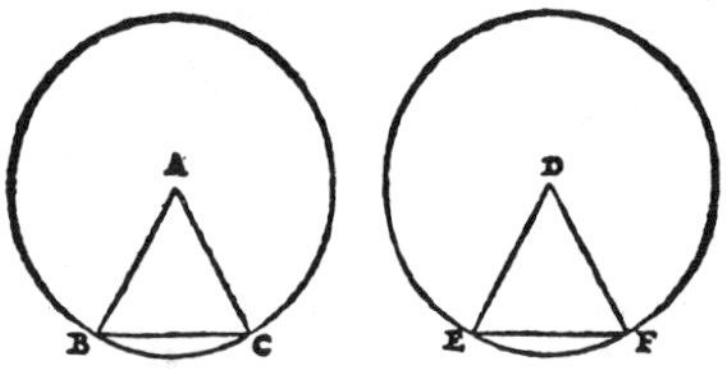

→ Huius itaque Propoſitionis veritatem non aliunde quàm à communi iudicio petemus: cogitabimusq́; Figuras Figuris ſuperponere, Mechanicum quippiam eſſe: intelligere verò, id demùm eſſe Mathematicum. Iam verò quum fuerit confeſſum duo Triangula inuicem eſſe æquilatera, ipſa quoque inter ſe æqualia fateri erit neceſſarium. Etenim nulla euidentiori ſpecie æqualitas Figurarum dignoſcitur, quàm ex laterum æqualitate: quanquam Circulorum æqualitas ex diametris definitur: ſed non aliam ob cauſam, quàm quòd linea obliqua ſui copiam adeò apertè non facit vt recta: Cuius menſuram facilè capimus, ac per eam, obliquarum inter ſe comparationem facimus.

Plate II. Commentary on (I.4) from Peletier's Euclid (1557). He says this proposition belongs among the common notions, because to superimpose one figure on another is Mechanics, not Mathematics.

b) we can rotate around any given point, so that one ray at that point is taken to any other ray at that point, and

c) we can reflect in any line so as to interchange points on opposite sides of the line.

If we were working in the real Cartesian plane $\mathbb{R}^2$ with coordinates x, y, we could easily show the existence of sufficient rigid motions by using *translations*, *rotations* and *reflections* defined by suitable formulas in the coordinates.

For example, a translation taking the point $(0,0)$ to (a,b) is given by

$$\begin{cases} x' = x + a \\ y' = y + b, \end{cases}$$

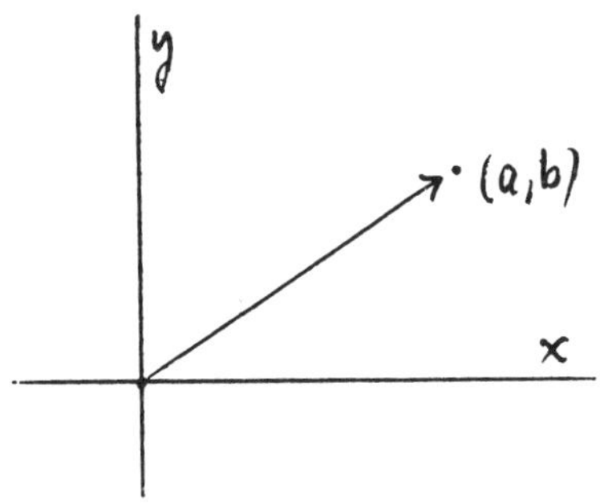

and a rotation of angle α around the origin is given by

$$\begin{cases} x' = x\cos\alpha - y\sin\alpha \\ y' = x\sin\alpha + y\cos\alpha. \end{cases}$$

Thus we can easily justify the use of the method of superposition in the real Cartesian plane. However, since there are no coordinates and no real numbers in Euclid's geometry, we must regard his use of the method of superposition as an additional unstated postulate or axiom.

To formalize this, we could postulate the existence of a group of rigid motions acting on the plane and satisfying the conditions $a), b), c)$ mentioned above. Indeed, there is an extensive modern school of thought, exemplified by Felix Klein's *Erlanger Programm* in the late 19th century, which bases the study of geometry on the groups of transformations which are allowed to act on the geometry. This point of view has had wide-ranging applications in differential geometry and in the theory of relativity, for example.

We will discuss the rigid motions in Euclidean geometry in greater detail later. For the moment let us just note that the proof of the (SAS) criterion for congruence in (I.4) requires something more than what is in Euclid's axiom system. Hilbert's axioms for geometry actually take (SAS) as an axiom in itself. This seems more in keeping with the elementary nature of Euclid's geometry than postulating the existence of a large group of rigid motions.

Finally let us note that Euclid's use of the method of superposition in the proof of (I.4) gives us some more insight into his concepts of "equality" for line segments and angles. In Common Notion 4 he says that things which coincide with one another are equal (congruent) to one another. In the proof of (I.4) he also uses the converse, namely, if things (line segments or angles) are equal to one another (congruent), then they will coincide when one is moved so as to be superimposed on the other. So it appears that Euclid thought of line segments or angles being congruent if and only if they could be moved in position so as to coincide with each other.

Betweenness

Question of betweenness, when one point is between two others on a line, or when a line through a point lies inside an angle at that point play an important, if unarticulated role in Euclid's *Elements*. To explain the notion of points on a line lying between each other, one could simply postulate the existence of a linear ordering of the points. Similarly for angles at a point one could talk of a circular ordering.

But when a hypothesis of relative position of points and lines in one part of a diagram implies a relationship for other parts of the figure far away, it seems clear that something important is happening, and it may be dangerous to rely on intuition.

For example, how do you know that the angle bisector at a vertex A of a triangle ABC meets the opposite side BC between the points BC and not outside? Of course it is obvious from the picture, but what if you had to explain why without drawing a picture?

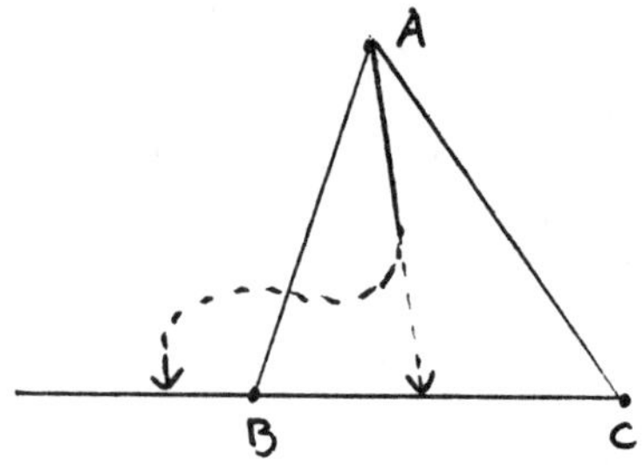

We have already seen that the relative position of two circles may affect whether they meet or not. Let us look at some other instances where betweenness plays an important role in a proof.

Consider (I.7), which is used in the proof of the side-side-side (SSS) criterion for congruence of triangles (I.8). In (I.7) Euclid shows that it is not possible to have two distinct triangles ABC and ABD, on the same side of a segment AB, and having equal sides $AC = AD$ and $BC = BD$.

The proof goes like this. Since $AC = AD$, the triangle ACD is isosceles, and so the base angles are equal (I.5). In the diagram $\angle 1 = \angle 4$. On the other hand, since $BC = BD$, the triangle BCD is isosceles, so its base angles are equal (I.5)—in our diagram $\angle 2 = \angle 3$. But now $\angle 2$ is less than $\angle 1$, which is equal to $\angle 4$, which is less than $\angle 3$. So $\angle 2$ is much less than $\angle 3$. But they are also equal, and this is impossible.

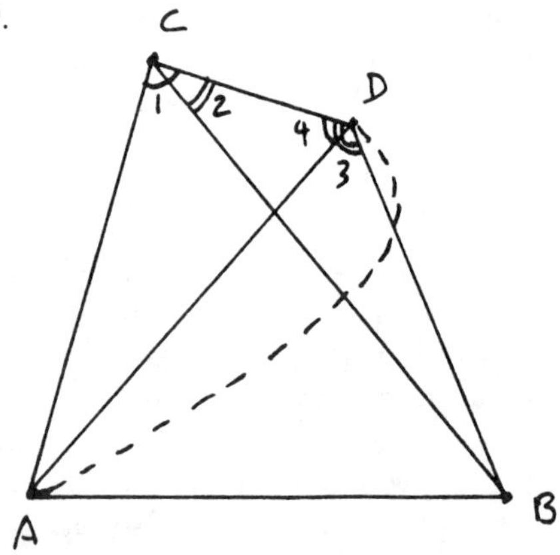

Note that this proof depends in an essential way on the relative position of the lines meeting at C and D, which determines the inequalities between the angles. If the line AD should reach the point D outside of the triangle BCD, as in our second (impossible) picture, then $\angle 2 < \angle 1$ and $\angle 3 < \angle 4$, and there is no contradiction. Thus the original proof depends on a certain configuration of lines being inside certain angles, which in turn depends on some global properties of the entire two-dimensional figure, and these relationships would be hard to explain convincingly without using a diagram. So as soon as we realize that we are depending on a diagram for part of our proof, a mental red flag should pop up to alert us to the question, what exactly is going on here, and what unstated assumptions are we using?

For another example where similar questions arise, look at the proof of (I.16), to show that an exterior angle of a triangle is greater than the opposite interior angle.

Let ABC be the given triangle. Bisect AC at E, draw BE, and extend that line to F so that $BE = EF$. Draw CF. Then by SAS (I.4), Euclid shows that the triangle BEA is congruent to the triangle FEC, and so the angle at A is equal to the angle $\angle ACF$. He then says that the angle $\angle ACF$ is less than the exterior angle $\angle ACD$, which proves the result.

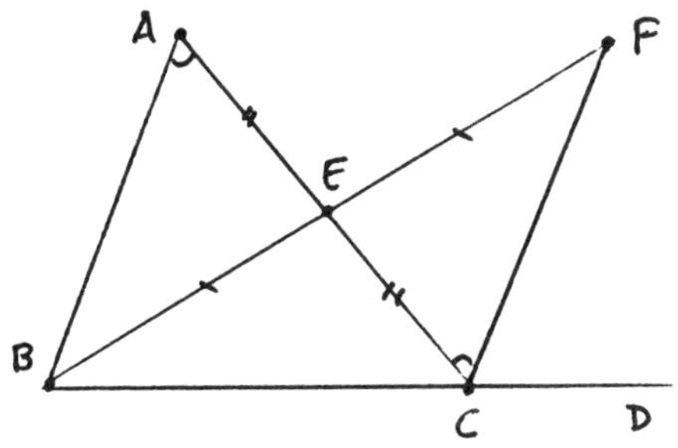

How do we know this relation among the angles? Because the line CF lies inside the angle ACD. But why is it inside? Since the line CF was constructed using the point F, which in turn was constructed using the point E, this is a global property of the whole figure, which is clear from the diagram, but would be hard to explain without a diagram.

To illustrate the danger of relying on diagrams in geometrical proofs, we will present a well-known fallacy due to W. W. Rouse Ball (1940). The following purports to be a proof that every triangle is isosceles. See if you can find the flaw in the argument.

Example 3.1 Let ABC be any triangle. Let D be the midpoint of BC. Let the perpendicular to BC at D meet the angle bisector at A at the point E. Drop perpendiculars EF and EG to the sides of the triangle, and draw BE, CE. The triangles AEF and AEG have the side common and two angles equal, so they are congruent by AAS (I.26). Hence $AF = AG$ and $EF = EG$. The triangles BDE and CDE have DE common, two other sides equal, and the included right angles equal. Hence they are congruent by SAS (I.4). In particular $BE = CE$.

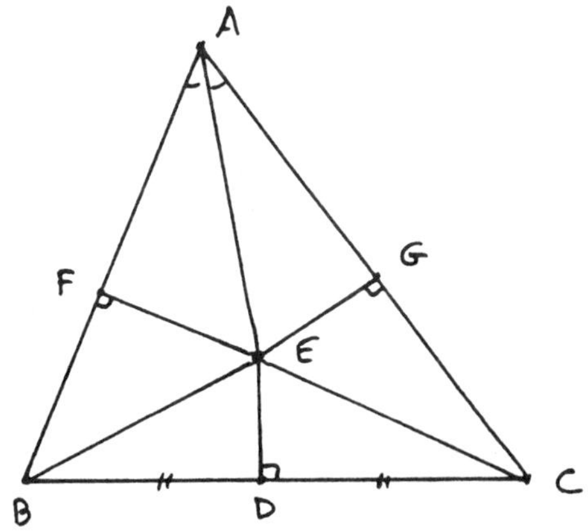

Now the triangles BEF and CEG are right triangles with two sides equal, so they are congruent (see lemma below), hence $BF = CG$. Adding equals to equals we find $AB = AF + FB$ is equal to $AC = AG + GC$. So the triangle ABC is isosceles.

There are several other cases to consider. If the point E lies outside the triangle, one can use this second figure, and exactly the same proof to conclude that AB and AC are the difference of equal segments $AF = AG$ and $BF = CG$, hence equal.

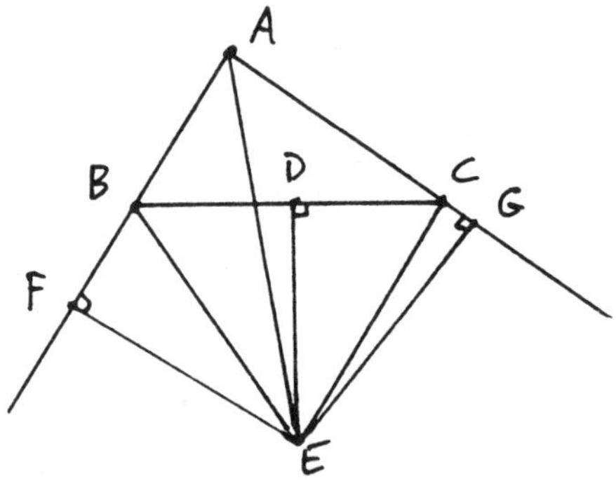

If E lands at the point D, or if the angle bisector at A is parallel to the perpendicular to AB at D, the proof becomes even easier, and we leave it to the reader.

We still need to prove the following

Lemma 3.2 (Right Angle Side Side) (RASS) *If two right triangles have two sides equal, not containing the right angle, they are still congruent.*

Proof. This result, though not stated by Euclid, is often useful. We give two proofs. The first method is to use (I.47) to conclude that the square on BC is equal to the square on EF. Then $BC = EF$ and we can apply (SSS) (I.8).

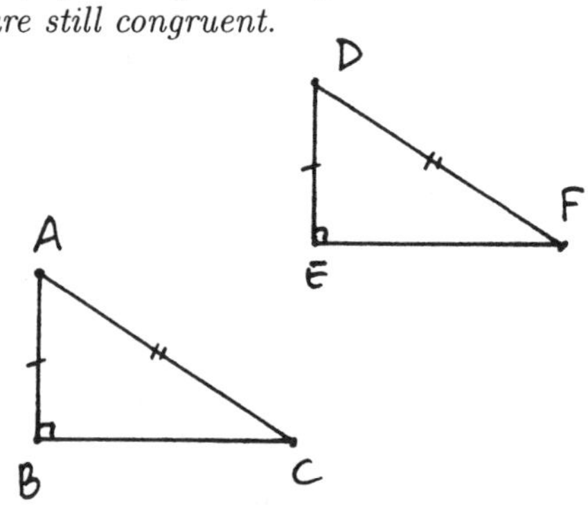

The second proof does not make use of (I.47) and the theory of area. Extend FE to G and make $EG = BC$. Then the triangles ABC and DEG are congruent by SAS (I.4). Therefore $AC = DG$. It follows that $DF = DG$, so the triangle DFG is isosceles. Therefore the angles at F and G are equal. Then the triangles DEG and DEF are congruent by AAS (I.26). But DEG is congruent to ABC, so the two original triangles are congruent.

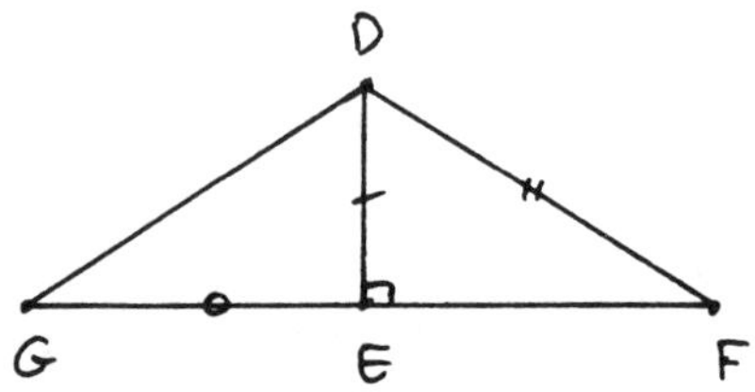

The Theory of Parallels

Book I of Euclid's *Elements* can be divided naturally into three parts. The first part ((I.1)–(I.26)) deals with triangles and congruence. The second part ((I.27)–(I.34)) deals with parallel lines and their applications, including the well-known (I.32) that the sum of the angles of a triangle is two right angles. The third part ((I.35)–(I.48)) deals with the theory of area.

Two lines are *parallel* if they never meet, even if extended indefinitely in both directions (Definition 23). The fifth postulate gives a criterion for two lines to meet under certain conditions, hence to be not parallel, so we often refer to the fifth postulate as the *parallel postulate*. Euclid postponed using this postulate as long as possible so that in fact the first part of Book I about triangles and congruence does not use the parallel postulate at all. It is first used in (I.29). Let us examine closely Euclid's theory of parallels and his use of the parallel postulate.

The first result about parallel lines, (I.27), says that if a line falling on two other lines makes the alternate interior angles equal, then the lines are parallel. This is proved using (I.16): if not, the lines would meet on one side or the other, and would form a triangle having an exterior angle equal to one of its opposite interior angles, which is impossible.

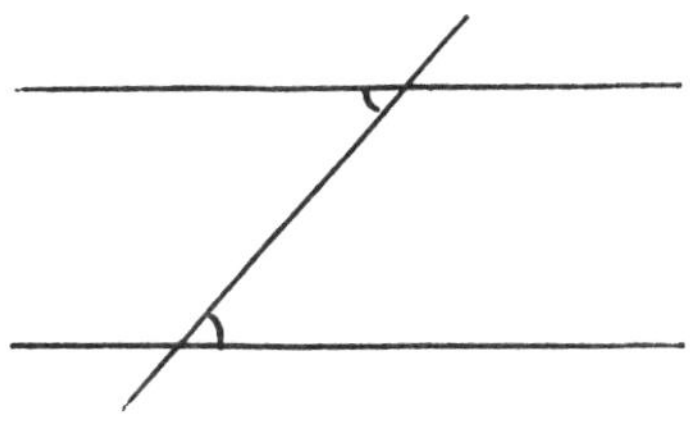

The next result (I.28) is similar, and follows directly from this one using vertical angles (I.15) or supplementary angles (I.13).

The fifth postulate is used to prove the converse of (I.27) which is (I.29): if the lines are parallel, then the alternate interior angles will be equal. For if not, then one would be greater than the other, and so the sum of the interior angles on one side of the transversal would be less than two right angles. In this situation, the fifth postulate applies and forces the lines to meet, which is a contradiction.

As for the existence of parallel lines, Euclid gives a construction in (I.31) for a line through a point P, parallel to a given line l. Draw any line through P, meeting l, and then reproduce the angle it makes with l at the point P (I.23). It follows from (I.27) that this line is parallel to l.

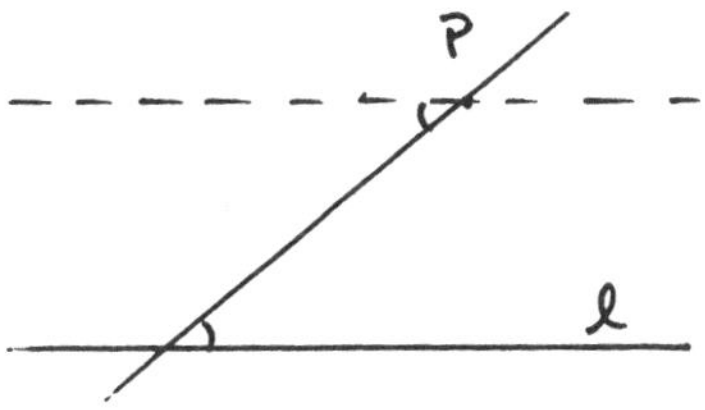

Why does Euclid place this construction after (I.29), even though it does not depend on (I.29) and does not make use of the parallel postulate? Presumably the answer, although Euclid does not say so, is that using (I.29) one can show that this parallel just constructed is *unique*. If there were any other line parallel to l through P, it would make the same angle with the transversal (by (I.29)) and hence would be equal to this one. Thus using the parallel postulate we can prove the statement

P. Given a point P not on a line l, there exists at most one line through P, parallel to l.

This statement (P) is often called "Playfair's axiom", after John Playfair (1748–1819), even though it already appears in the commentary of Proclus. Of course in Euclid's development of geometry, this is not an axiom, but a theorem which can be proved from the axioms. Some authors, however, like to take the statement (P) as an axiom instead of using Euclid's fifth postulate. So I would like to explain in what sense we can say that Euclid's fifth postulate is equivalent to Playfair's axiom (P).

Since the parallel postulate plays such a special role in Euclid's geometry, let us make a special point of being aware when we use this postulate, and which theorems are dependent on its use. Let us call *neutral geometry* the collection of all the postulates and common notions *except* the fifth postulate together with all theorems which can be proved without using the fifth postulate. Thus (I.1)–(I.28) and (I.31) all belong to neutral geometry, while for example (I.32) and (I.47) do not belong to neutral geometry.

If we take neutral geometry and add back the fifth postulate then we recover ordinary Euclidean geometry, and we can prove (P) as a theorem as we did above.

But now suppose we take neutral geometry and add (P) as an extra axiom. We will show that in this geometry we can prove Euclid's fifth postulate as a theorem.

Indeed, suppose given two lines l, m and a transversal n, so that the two interior angles 1, 2 on the same side are less than two right angles. Let P be the intersection of the lines m and n, and draw a line l' through P, making the alternate angle 3 equal to 1. This is possible by (I.23) which belongs to neutral geometry. Then by (I.27), which also belongs to neutral geometry, l' is parallel to l.

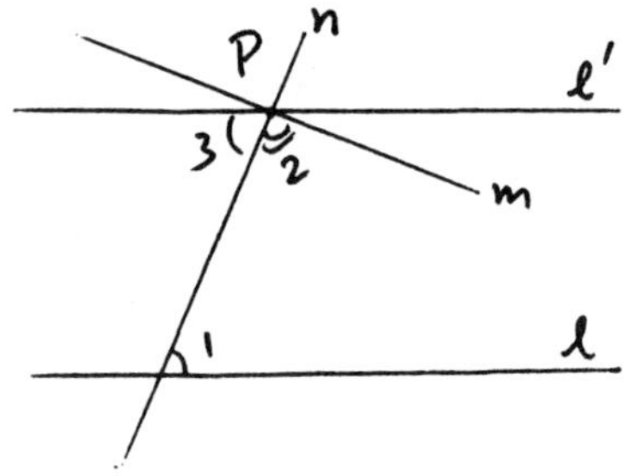

Now since 1 + 2 is less than two right angles, it follows that 2 + 3 is less than two right angles, and hence the line l' is different from m (I.13). Now we can apply (P). Since l' passes through P and is parallel to l, it must be the only line through P which is parallel to l. In particular, the line m which is different from l' cannot be parallel to l, and so by definition it must meet l. This proves the fifth postulate.

Thus in the presence of all the results of neutral geometry, we can use Euclid's fifth postulate to prove Playfair's axiom, or we can use Playfair's axiom to prove Euclid's fifth postulate. In this sense we can say that in neutral geometry, Euclid's fifth postulate is equivalent to Playfair's axiom. This means that adding either one of them as an additional assumption to neutral geometry will give the same body of theorems as consequences.

The Theory of Area

In (I.35), Euclid says two parallelograms on the same base and in the same parallels are equal to each other. In the figure, the parallelogram $ABCD$ is equal to the parallelogram $BCEF$. Clearly the parallelograms are not congruent.

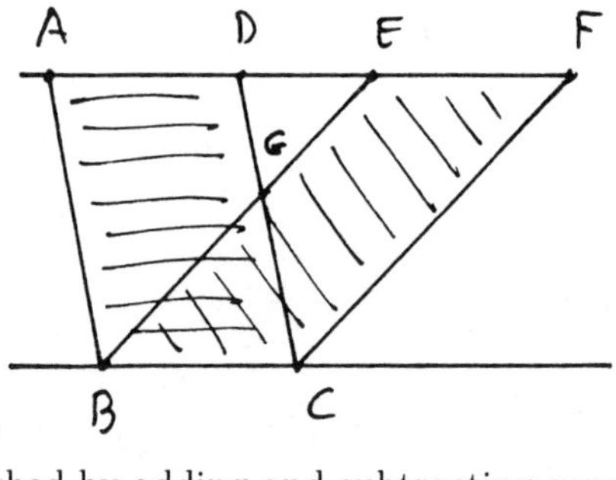

Looking at the proof, which is accomplished by adding and subtracting congruent figures, we conclude that Euclid must be referring to the area of the parallelograms when he says they are equal. But he has not said what is the area of a figure, so we must reflect a bit to see what he means.

Our intuitive understanding of area comes from high school geometry, where we learn that the area of a rectangle is the product of the lengths of the sides, the area of a triangle is one half the product of the lengths of the base and the altitude, etc. The "area" of high school geometry is a function which attaches to each plane figure a real number; the area of a non-overlapping union of figures is the sum of the areas, and so forth. Most likely no one ever told you the definition of area, nor did they prove that such an area function exists. Using calculus, you can define the area of a figure in the real Cartesian plane using definite integrals, and in that way it is possible to prove that a suitable area function exists. But in Euclid's geometry, there are no numbers and we certainly do not want to use calculus to define the concept of area in elementary geometry.

So what did Euclid have in mind? Since he does not define it, we will consider this new equality as an undefined notion, just as the notions of congruence for line segments and angles were undefined. We will call this new notion *equal content*, to avoid confusion with other notions of equality or congruence. We do not want to use the word area, because this notion is quite different from our common understanding of area as a function associating a number to each figure.

From the way Euclid treats this notion, it is clear that he regards it as an equivalence relation, satisfying the common notions. In particular

a) congruent figures have equal content.

b) If two figures each have equal content with a third, they have equal content.

c) If pairs of figures with equal content are *added* in the sense of being joined without overlap to make bigger figures, then these added figures have equal content.

d) Ditto for subtraction, noting that equality of content of the difference does not depend on *where* the equal pieces were removed.

e) Halves of figures of equal content have equal content (used in the proof of (I.37)).

f) The whole is greater than the part, which in this case means if one figure is properly contained in another, then the two figures cannot have equal content (used in the proof of (I.39)).

In terms of the axiomatic development of the subject, at this point Euclid is introducing a new undefined relation, and taking all the properties just listed as new axioms governing this new relation. In a later section of this book we will discuss Hilbert's reinterpretation of the theory of area where the relationship of having equal content is defined, and all its properties proved, so that it does not require the introduction of new axioms.

Now let us see what Euclid does with this purely geometric notion of equal content of plane figures. In (I.35), he proves that the two parallelograms have equal content by first showing that the triangle ABE is congruent to the triangle DCF, so they have equal content. Then by subtracting the triangle DGE from each (in different positions!) and adding the triangle BGC to each, he obtains the two parallelograms, which therefore have equal content.

In (I.37) he shows that two triangles ABC and DBC on the same base and in the same parallels have equal content. The method is to double ABC to get a parallelogram $EABC$, and to double DBC to get a parallelogram $DFBC$.

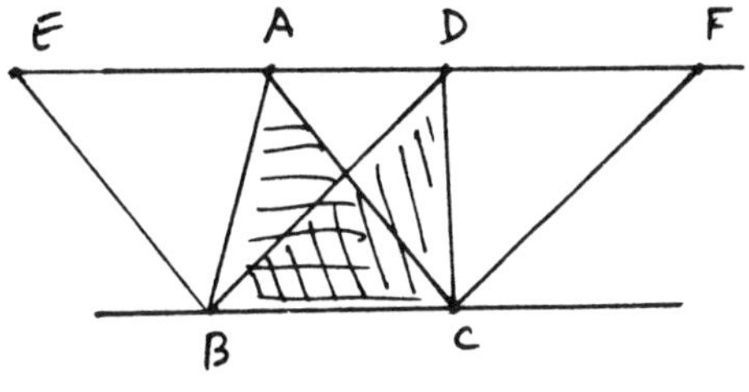

By (I.35) the two parallelograms have equal content, and then he applies the axiom that halves of equals are equal to conclude the triangles have equal content.

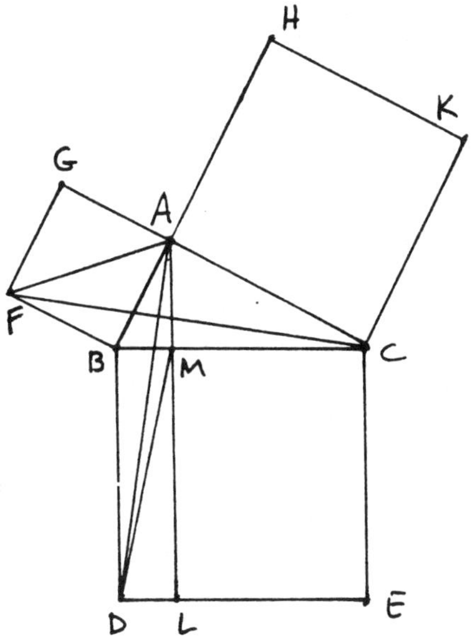

This is all that is needed to explain Euclid's beautiful proof of (I.47), the theorem of Pythagoras. The statement of the theorem is that if ABC is a right triangle, then the squares on the two legs together have equal content to the square on the hypotenuse. The proof goes like this. The triangle ABF is one half of the square $ABFG$. This triangle ABF has equal content with the triangle BFC by (I.37). The triangle BFC is congruent to the triangle BAD. And BAD has equal content to the triangle BMD by (I.37). This latter triangle is equal to one half of the rectangle $BDLM$. Hence the square $ABFG$ has equal content to the rectangle $BDLM$. Doing the same construction on the other side and adding, one has the result.

Euclid's statement of (I.47) in terms of equal content of the squares constructed on the sides of the triangle may come as a surprise to the modern student who remembers the formula $a^2 + b^2 = c^2$ (which I suppose in the minds of the general public is rivalled in fame only by Einstein's famous formula $E = mc^2$). We are used to thinking of a, b, c as the lengths of the sides of the triangle, in which case the theorem becomes an equation among numbers. How can we reconcile these two points of view?

The modern answer to this question, which we will discuss in more detail later, is that after introducing coordinates in our geometry we can prove the existence of an area function. The area of a square of side a will be a^2. Furthermore, we will show that having equal content in the sense of Euclid is equivalent to having equal area in the sense of the area function. Then the two formulations of the theorem of Pythagoras become equivalent.

This answer makes sense only when we are able to assign numerical lengths to arbitrary line segments, which the Greeks could not do. Yet there is ample evidence that the Greeks did know special cases of this formula when a, b, c are integers. The equation $3^2 + 4^2 = 5^2$ was known to the Egyptians, and Proclus in his note on (I.47) mentions two general formulas for generating such "Pythagorean triples" of integers, which he ascribes to Plato and to Pythagoras. So we can presume that the Greeks knew a number of right triangles with integer sides, in which case (I.47) can be represented by the equation among integers $a^2 + b^2 = c^2$. But the geometrical proof given by Euclid is then more general, because it applies to all triangles, and not just those for which one can find integers to fit the sides.

Euclid's theory of area plays an important role in the succeeding books of the *Elements*. It appears not only in results which correspond to our modern notion of area, but also in results, such as the construction of the regular pentagon (IV.11) which at first sight appear to have nothing to do with area. Roughly speaking,

Euclid uses arguments involving areas in places where we would expect to see a quadratic equation in analytic geometry. He can add two line segments to get another line segment, but there is no way to multiply line segments so as to get another line segment. Instead, one can regard the rectangle with sides equal to segments AB, CD as a product of these two segments. The results (I.42)–(I.45) on application of areas and all the results of Book II give a certain flexibility in manipulating and comparing different areas. This creates a sort of "algebra of areas", and one can regard results such as (II.14) as equivalent to the solution of certain quadratic equations. Note also the essential use of area in the proof of (VI.1) which is the cornerstone of Euclid's theory of similar triangles.

Exercises

3.1 Explain what is wrong with the "proof" in (3.0.1). (*Hint*: Draw an accurate figure.)

3.2 Read Euclid (I.35)–(I.48), Book II, and (III.35)–(III.37). Be prepared to present proofs of (I.35), (I.41), (I.43), (I.47), (II.6), (II.11), and (III.36).

3.3 Given a triangle ABC and given a segment DE, construct a rectangle with content equal to the triangle ABC, and with one side equal to DE.

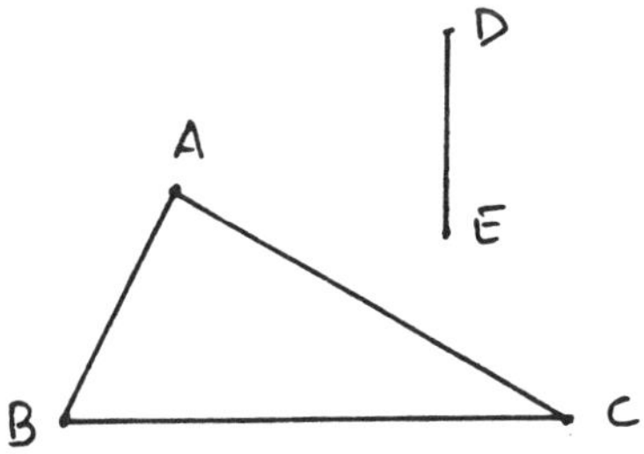

3.4 Given a rectangle, construct a square with the same content.

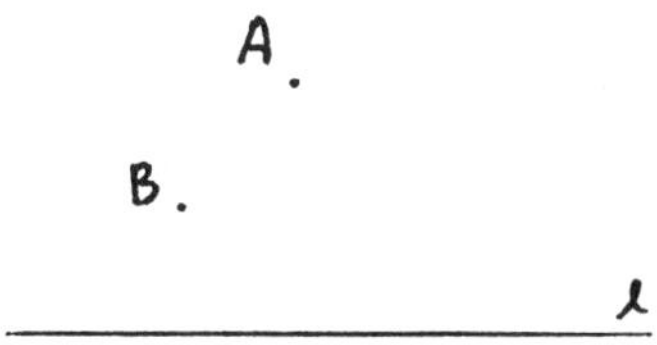

3.5 Given a line l and given two points A, B not on l, construct a circle passing through A, B and tangent to l. (*Hint*: Use (III.36).)

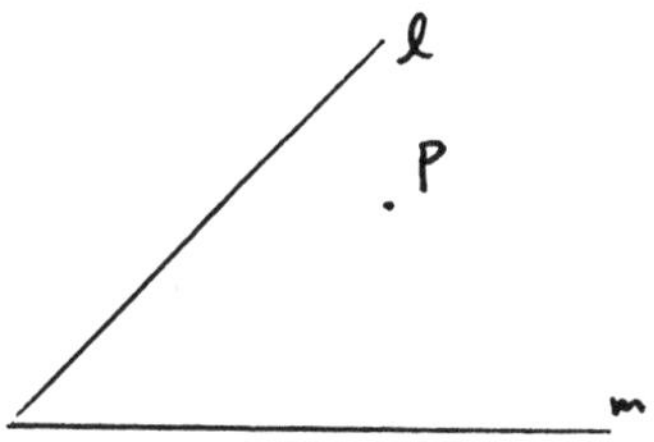

3.6 Given two lines l, m and a point P not on either line, construct a circle passing through P and tangent to both l and m.

3.7 Given a triangle ABC, let DE be a line parallel to the base BC, let F be the midpoint of DE, and let AF meet BC in G. Prove that G is the midpoint of BC, using only results from Euclid Books I–III. (*Hint*: Draw some extra lines to make parallelograms, and use (I.43).)

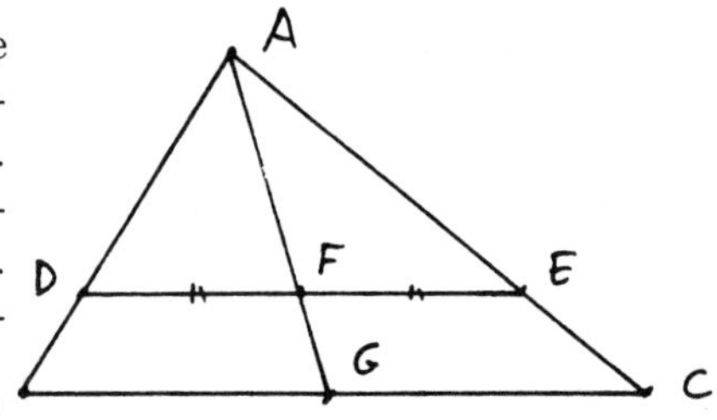

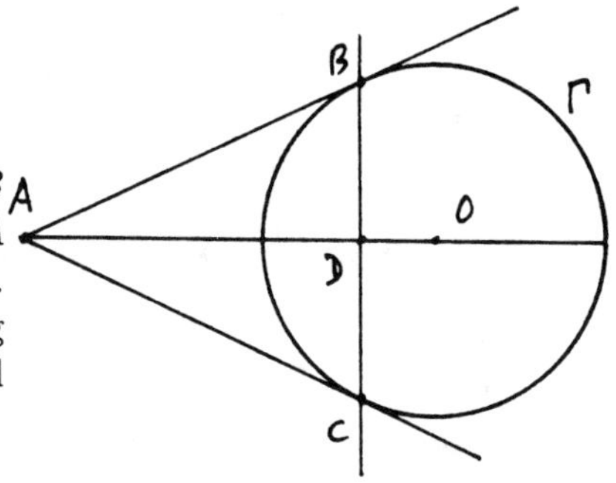

3.8 Let Γ be a circle with center O. Let AB and AC be tangents to Γ from a point A outside the circle. Let BC meet OA at D. Prove that $OA \times OD = OB^2$ (meaning the rectangle on OA and OD has equal content to the square on OB).

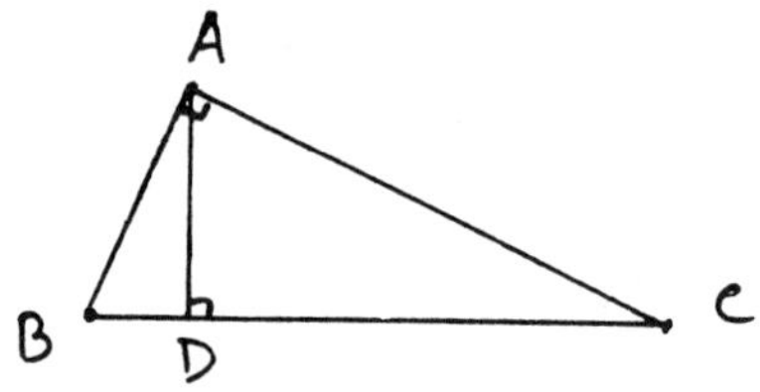

3.9 Let ABC be a right triangle, and let AD be the altitude from the right angle A to the hypotenuse BC. Prove that $AD^2 = BD \times DC$ (in the sense of content).

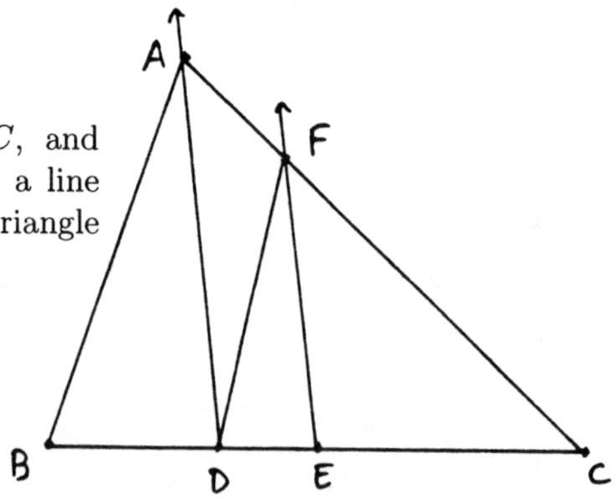

3.10 Problem. Given a triangle ABC, and given a point D on BC, to draw a line through D which will divide the triangle into two pieces of equal content.

Solution (Peletier). Let E be the midpoint of BC. Draw AD, draw EF parallel to AD. Then DF divides the triangle in half.

Prove that the content of the quadrilateral $ABDF$ is equal to the content of the triangle DFC.

3.11 (Campanus). Use the theory of content to show that the line DE joining the midpoint of two sides of a triangle is parallel to the third side. (*Hint*: Draw BE and DC. Show that the triangles BDC and BEC have the same content and then apply (I.39).)

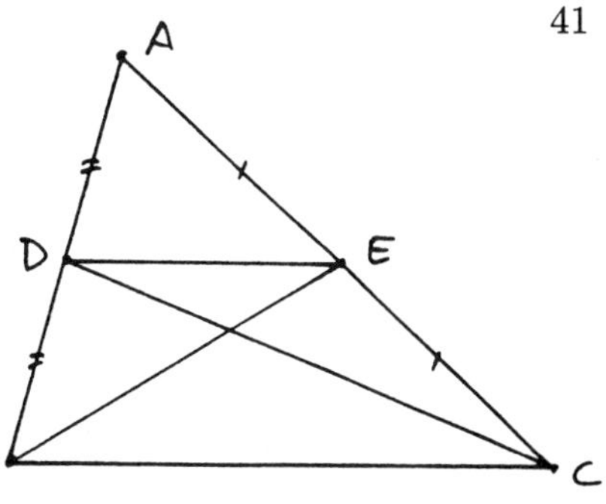

4 Construction of the Regular Pentagon

One of the most beautiful results in all of Euclid's *Elements* is the construction of a regular pentagon inscribed in a circle (IV.11). The proof of this construction makes use of all the geometry he has developed so far, so that one could say to understand fully this single result is tantamount to understanding all of Euclid's geometry. It also raises questions of exposition which are central to our modern examination of Euclid's methods. For example, why does Euclid use the theory of area in proving a result about a polygon?

In this section we will present Euclid's construction of the regular pentagon, and begin discussing the issues raised by its proof. In a later chapter we will give other proofs using similar triangles or the complex numbers. Euclid's original geometric proof must be regarded as a tour de force of classical geometry. It depends on the theory of area, which we will discuss in more detail later. So this section can be regarded as a taste of things to come: a first meeting with one of the deeper topics which is central to Euclid's geometry.

The key point of the construction of the pentagon is the following problem.

Problem 4.1 To construct an isosceles triangle whose base angles are equal to twice the vertex angle.

Construction ((II.11), (IV.10)). Let A, B be two points chosen at random.

1. Draw line AB.

Next, construct a perpendicular to AB at A, as follows

2. Circle AB, get C
3. Circle BC
4. Circle CB, get D
5. Line AD, get E.

Next, we bisect AE as follows

6. Circle EA, get F, G
7. Line FG, get H.

Now comes the unusual part of the construction:

8. Circle HB, get J
9. Circle AJ, get K

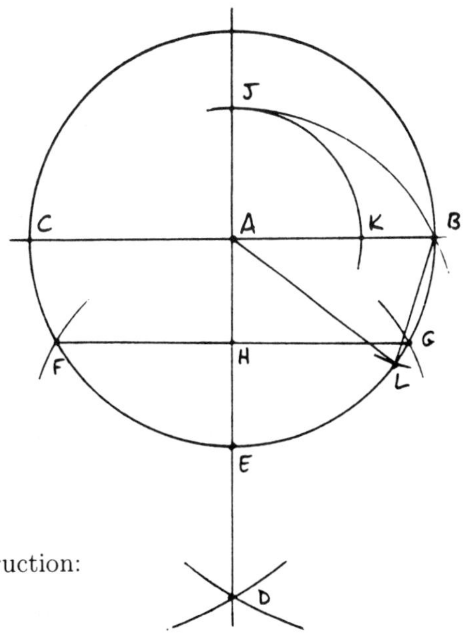

10. Circle center B, radius AK, get L
11. Line AL
12. Line BL.

Then ΔABL is the required triangle. The angles at B and at L will be equal to twice the angle at A.

Proof. From a modern point of view, it would seem that some theory of quadratic equations is essential for the proof. Euclid did not have any algebra available to him, but he was able to deal with quantities essentially equivalent to quadratic expressions via the theory of area. We can think of a rectangle as representing the product of its sides, or a square as the square of its side. These areas, without even assigning a numerical value to them, can be manipulated by cutting up and adding or subtracting congruent pieces. In this way Euclid establishes a "geometrical algebra" for manipulating these quantities (always by geometrical methods), which acts as a substitute for our modern algebraic methods.

Let us then trace the steps by which Euclid proves (IV.10), which is the key point in the construction of the regular pentagon. In Book I, especially (I.35)–(I.47) he discusses the areas of triangles and parallelograms, leading up to the famous Pythagorean theorem (I.47) which is stated in terms of area: the square built on the hypotenuse of a right triangle has area equal to the area of the squares on the two sides. The theorem is proved by cutting these areas into triangles, and proving equality of areas using the cutting and pasting methods just developed.

Book II contains a number of results of geometrical algebra, as described above, all stated and proved geometrically in terms of areas. In particular, (II.5), (II.6) and (II.11) are used in the proof of (IV.10). Note that (II.11), which is sometimes called the division of a segment in extreme and mean ratio, states that the interval AB is divided by a point K (in our notation (4.1) above) such that the rectangle formed by BK and AB has area equal to the square on AK. In this way the property of extreme and mean ratio is expressed using area.

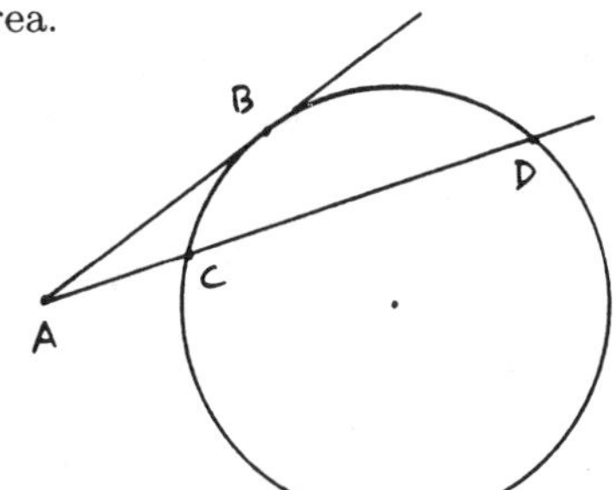

From Book III we need (III.36) and its converse (III.37). Proposition (III.36) says that if a point A lies outside a circle, and if AB is tangent to the circle at B, and if ACD cuts the circle at C and D, then the rectangle formed by AC and AD has area equal to the square on AB. This result is proved by several applications of (II.6) and (I.47).

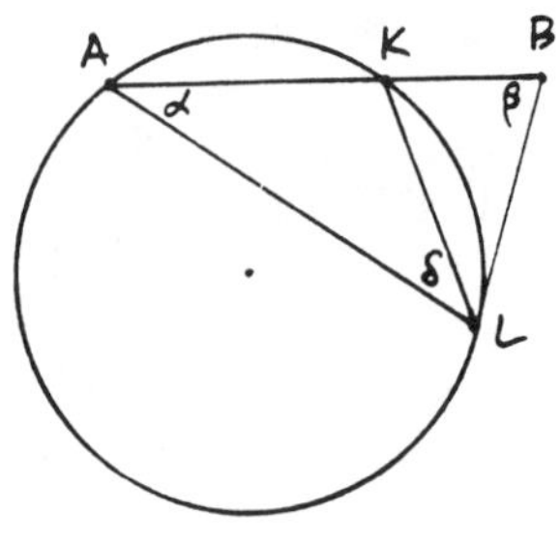

Now Euclid can prove (IV.10) by a brilliant application of (III.37). Let A, K, B, L be as in the construction (4.1) above. Then by (II.11), the rectangle with sides BK and BA has area equal to the square on AK. Since BL was constructed equal to AK, this is also equal to the square on BL.

Now consider the circle passing through the three points A, K, L. Since the rectangle on BK and BA is equal to the square on BL, it follows that BL is tangent to this circle (III.37)!

Hence the angle $\angle BLK$ formed by the tangent BL and the line LK is equal to the angle α at A, which subtends the same arc (III.32). Let $\angle KLA = \delta$. Then $\angle BKL$ is an exterior angle to the triangle ΔAKL, so $\angle BKL = \alpha + \delta$ (I.32). But $\angle BLK = \alpha$, so $\alpha + \delta = \angle BLA$, and this angle is β because ΔABL is isosceles. Hence $\angle BKL = \beta$. Now it follows that ΔBKL is isosceles, so $KL = BL = AK$. Hence ΔAKL is also isosceles, so $\delta = \alpha$. Now $\beta = \angle BLA = 2\alpha$ as required. □

Once we have the isosceles triangle constructed in (4.1), the construction of the pentagon follows naturally. The idea is to inscribe in the circle a triangle equiangular with the given triangle, and then to bisect its two base angles.

Problem 4.2 Given an isosceles triangle whose base angles are equal to twice its vertex angle, and given a circle with its center, to construct a regular pentagon inscribed in the circle.

Construction ((IV.2) and (IV.11)). Let ΔABC be the given triangle and let O be the center of the given circle. The first part of the construction is to obtain a tangent line to the circle. Let D be any point on the circle.

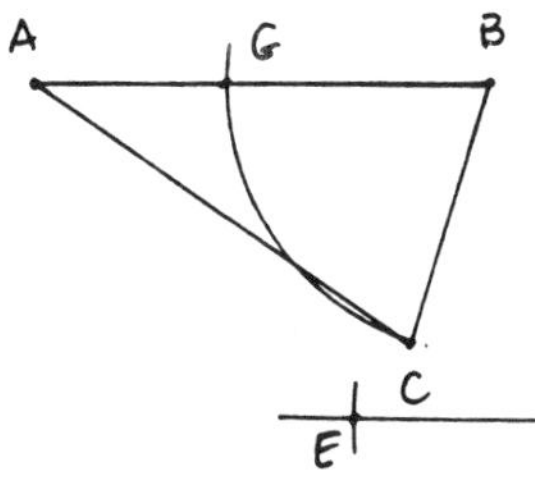

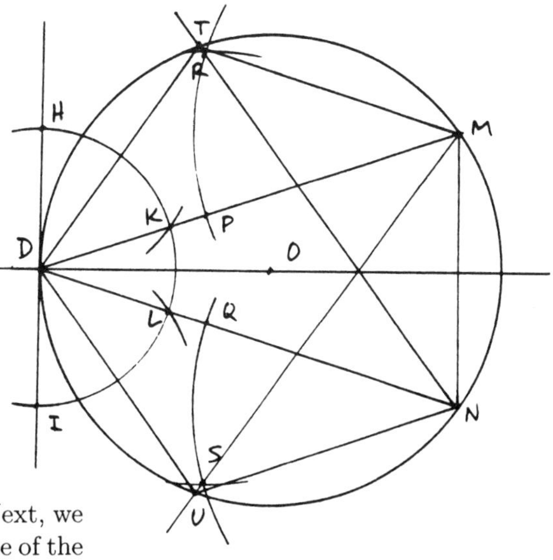

1. Line OD.
2. Circle DO, get E.
3. Circle EO.
4. Circle OE, get F.
5. Line DF.

Then DF will be a tangent line. Next, we reproduce the angle β from the base of the isosceles triangle at D, on both sides.

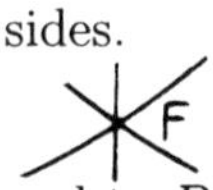

6. Circle BC, get G.
7. Circle at D with radius equal to BC, get H, I.
8. Circle center H radius CG, get K.
9. Circle center I, radius CG, get L.
10. Line DK, get M.

11. Line DL, get N.
12. Line MN.

Then ΔDMN is a triangle inscribed in the circle, equiangular with ΔABC. Next we bisect the angles at M, N.

13. Circle MN, get P.
14. Circle NM, get Q.
15. Circle QN, get R.
16. Circle PM, get S.
17. Line NR, get T.
18. Line MS, get U.

Then D, M, N, T, U will be the vertices of the pentagon.

19. Line DT.
20. Line TM.
21. Line DU.
22. Line UN.

Then $DTMNU$ is the required pentagon.

Proof. We follow the geometric proof given by Euclid. First of all, the line DF is constructed perpendicular to a diameter of the circle, so it is a tangent line to the circle (III.16). Next, the triangles ΔDHK and ΔDLI are constructed so that their three sides are equal to the three sides of ΔBCG. Hence by (SSS) = (I.8), it follows that $\angle KDH$ and $\angle LDI$ are both equal to the angle β of the triangle ΔABC at B. From there it follows that the angles of ΔDMN at M and N are both equal to β, because they subtend the same arcs cut off by the tangent line and the angles β just constructed (III.32). Since the sum of the three angles of a triangle is constant $= 180°$ (I.32), it follows that the triangle ΔDMN is equiangular with the triangle ΔABC. In particular, if α is the angle at D, then $\beta = 2\alpha$.

The points T, U are constructed by taking the angle bisectors of ΔDMN at M and N. Since the angles at M and N are β, their halves are equal to α. Thus the arcs DT, TM subtend angles α at N; the arc MN subtends an angle α at D; and the arcs DU, UN subtend angles α at M. Hence these five arcs are all equal (III.26), and the line segments on them are also equal. So we have constructed an equilateral pentagon inscribed in the circle. The angle subtended by each side at the center of the circle will be $2\alpha = \beta$. It follows that the angles of the pentagon are also equal, so the pentagon is *regular* in the sense that its sides are all equal and its angles are all equal. □

This completes the presentation of Euclid's construction of the pentagon. As usual, his method is adapted to economy of proof, not economy of steps used. The whole construction, as we have presented it here, takes $12 + 22 = 34$ steps. By collapsing separate parts of the construction, in particular, by constructing the triangle of (4.1) on a radius of the given circle, one can make a construction with less than half as many steps (cf. (4.3)). Note also that Euclid's construction of the points T, U by bisecting the angles at M, N makes possible his elegant proof that the five sides of the pentagon are equal. However in retrospect we see that MN

is actually one side of the pentagon, so T and U could have been constructed in a single step by a circle with center D and radius MN.

If there is such a thing as beauty in a mathematical proof, I believe that this proof of Euclid's for the construction of the regular pentagon sets the standard for a beautiful proof. In the words of Edna St. Vincent Millay, "Euclid alone has looked on beauty bare".

Now let us use the ideas of Euclid's method to construct a pentagon in as few steps as possible.

Problem 4.3 Given a circle with center O, construct a regular pentagon inscribed in the circle in as few steps as possible.

1. Draw any line through O. Get A, B.
2. Circle A, any radius.
3. Circle B, same radius, get C.
4. OC get D.
5. Circle DO. Get E, F.
6. EF, get G.
7. Circle GA, get H.
8. Circle center A, radius OH, get I, J.
9. Circle center B, radius IJ, get K, L.

10–14. Draw BK, KJ, JI, IL, LB.

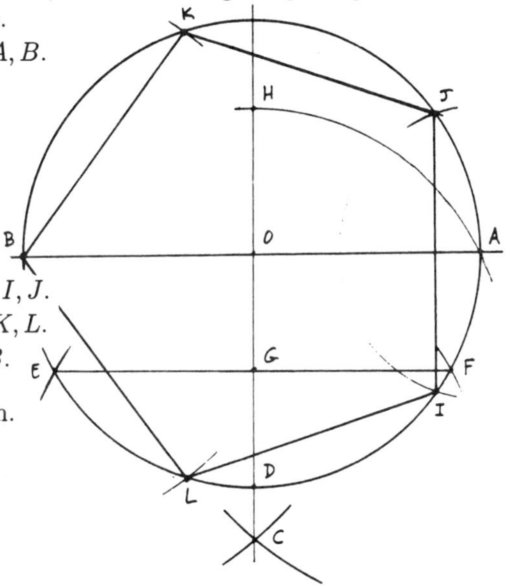

Then $BKJIL$ is the required pentagon.

Exercises

4.1 Read Euclid, Book IV.

4.2 Explain why the construction of (4.3) gives a regular pentagon.

4.3 Given a circle, but not given its center, construct an inscribed equilateral triangle in as few steps as possible. (Par = 10).

4.4 Construct a square in as few steps as possible. (Par = 9).

4.5 Given a line segment AB, construct a regular pentagon having AB as a side.

4.6 In the construction of (4.3) in the text, show that AH is equal to the side of the pentagon.

4.7 Let OA and OB be two perpendicular radii of a circle. Let C be the midpoint of OB. Let CD be the angle bisector of $\angle ACO$. Let DE be perpendicular to OA. Then prove that AE is a side of the regular pentagon inscribed in the circle.

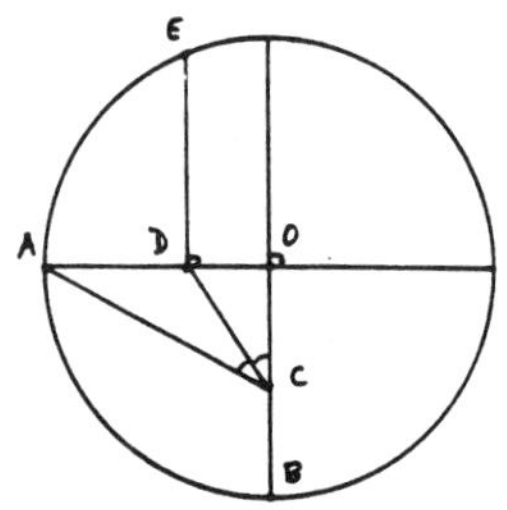

How many steps would it take to construct the pentagon by this method (given the circle and its center)?

5 Some Newer Results

In this section we mention some results of plane geometry which do not appear in Euclid's *Elements*, but which can be proved using the methods developed in Books I–IV. Some of these, such as the three altitudes of a triangle meeting in a point, were known to the Greeks. Others, such as the Euler line and the nine-point circle, were discovered only in the 18th and 19th centuries.

In some textbooks, these results are proved using similar triangles. In Euclid's *Elements*, similar triangles do not appear until Book VI, using the theory of proportion developed in Book V. In modern texts, similar triangles are defined by comparing the lengths of the sides. Since we have not yet discussed either of these methods, we will use only the pure geometric methods of Euclid Books I–IV in this section.

Two theorems taught in modern high school geometry are that the angle bisectors of a triangle meet in a point (the *incenter* of the triangle), and the perpendicular bisectors of the sides of a triangle meet in a point (the *circumcenter* of the triangle). Although not explicitly stated by Euclid, these two results are implicitly contained in (IV.4) and (IV.5).

On the other hand, the theorems about the three medians and the three altitudes of a triangle do not appear in Euclid, though they were known to Archimedes, so we will start with them.

Proposition 5.1 *Let ABC be a triangle, and let D, E be the midpoints of AB and AC, respectively. Then the line DE is parallel to the base BC, and equal to one half of it. In other words, if F is the midpoint of BC, then $DE \cong BF$.*

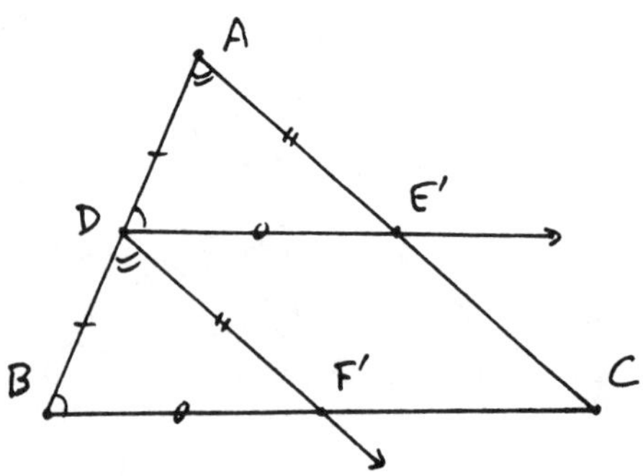

Proof. We begin with a slightly different construction. Let D be the midpoint of AB, and draw lines through D parallel to AC and BC. Let them meet the opposite sides in points E', F'. Since DE' is parallel to BC, the angles at B and D are congruent ((I.29): here we use the parallel postulate). Similarly, since AC is parallel to DF', the angles at A and D are equal.

Now $AD \cong DB$, and the angles of the triangle ADE' and DBF' at A and D are equal to those at D and B, respectively, so by (ASA = (I.26)), the triangles ADE' and DBF' are congruent. We conclude that $AE' \cong DF'$ and $DE' \cong BF'$.

Now look at the parallelogram $DE'F'C$. By (I.34) the opposite sides are equal. So $DF' \cong E'C$ and $DE' \cong F'C$. Thus we see that E' and F' are the midpoints of the sides AC and BC. So $E' = E$, the line DE' is equal to the line DE, and therefore DE is parallel to BC as claimed. Furthermore, we have seen that $DE' \cong BF'$, and F' is the midpoint of BC, so DE is equal to one-half of BC.

Corollary 5.2 *Let ABC be a triangle, and let D, E, F be the midpoints of the three sides. Then the sides of the triangle DEF are parallel to the sides of ABC, and the four small triangles formed are all congruent to each other.*

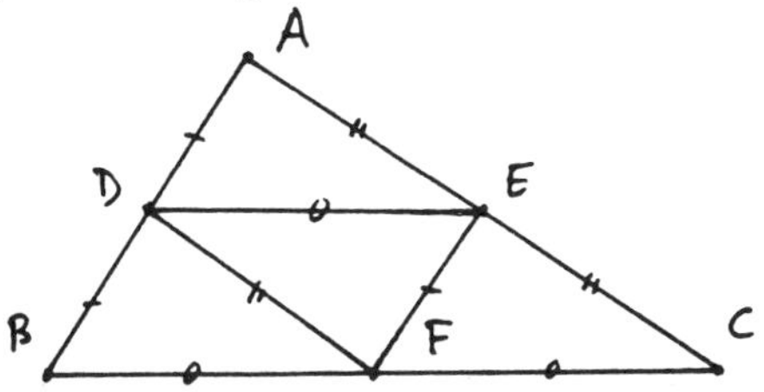

Proof. From the proposition it follows that each side of the triangle DEF is parallel to and equal to one half of a side of the triangle ABC. Then by (SSS = (I.8)) all four small triangles are congruent.

Definition. We say a triangle ABC is congruent to the *double* of a triangle FED, in symbols $ABC \cong 2FED$, if as in the diagram above, the three sides of ABC are double the sides of FED, and the three angles of ABC are equal to the three angles of FED.

Proposition 5.3 (2ASA) *Let ABC and $A'B'C'$ be two triangles, and assume that the angles at B and C are equal to the angles at B and C are equal to the angles at B' and C', and that $BC \cong 2B'C'$. Then the triangle ABC is congruent to the double of $A'B'C'$.*

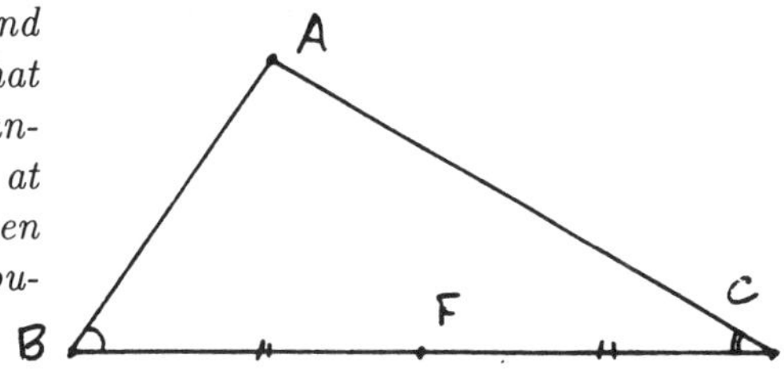

Proof. Let D, E, F be the midpoints of the sides of ABC, and draw the triangle DEF. Then from (5.2) we see that $DE \cong \frac{1}{2}BC \cong B'C'$. Furthermore, because DE is parallel to BC, the angles of the triangle ADE at D and E are equal to the angles at B' and C'. Now by (ASA), the triangle ADE is congruent to $A'B'C'$. But ABC is a double of ADE, so $ABC \cong 2A'B'C'$.

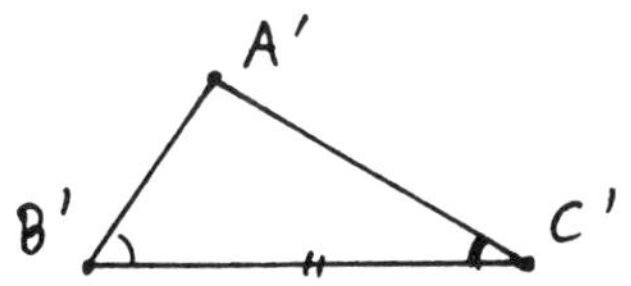

Remark. One can easily prove other double congruence theorems corresponding to (SAS) and (SSS): see exercises. Of course these are special cases of more general theorems on similar triangles which we will mention later.

Proposition 5.4 *The* medians *(= lines from a vertex to the midpoint of the opposite side) of a triangle meet in a single point (called the* centroid *of the triangle).*

Proof. Let ABC be the triangle, let D, E be the midpoints of AB and AC, and draw DE. Let the two medians BE and CD meet at a point G. Since DE is parallel to BC (5.1), we find that $\angle DEG = \angle CBG$ and $\angle EDG = \angle BCG$. On the other hand, $BC = 2DE$ (5.1). Therefore we can apply the previous result (5.3) and find that $\Delta BGC \cong 2\Delta EGD$.

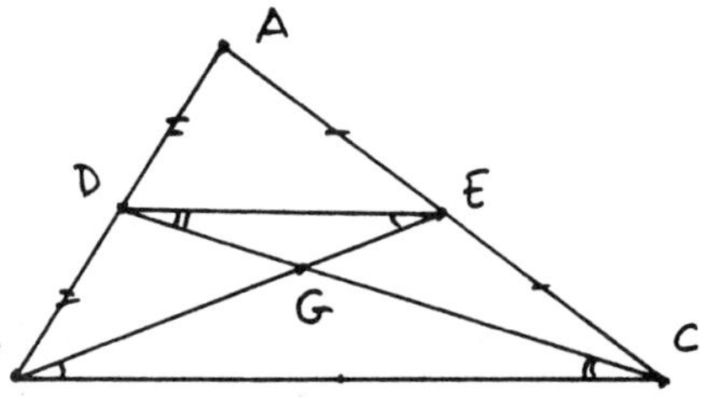

In particular, $BG = 2GE$. Thus G can be described as the point on the median BE which is $2/3$ of the way from B to E. Reversing the roles of A and C would therefore show that the third median AF also passes through G. Thus all three medians meet in the point G.

Corollary 5.5 *The centroid G lies on each median $2/3$ of the way from the vertex to the midpoint of the opposite side.*

Proof. Follows from the proof of (5.4).

Proposition 5.6 *The three* altitudes *(= lines through a vertex, perpendicular to the opposite side) of a triangle meet in a single point (the* orthocenter *of the triangle).*

Proof. Let ABC be the given triangle. Draw lines through the vertices A, B, C, parallel to the opposite sides, to form a new triangle $A'B'C'$. By (I.34) applied to the parallelograms $BCAC'$ and $BCB'A$ we see that $C'A = BC = AB'$. Thus A is the midpoint of $B'C'$, and similarly for the other two sides of $A'B'C'$.

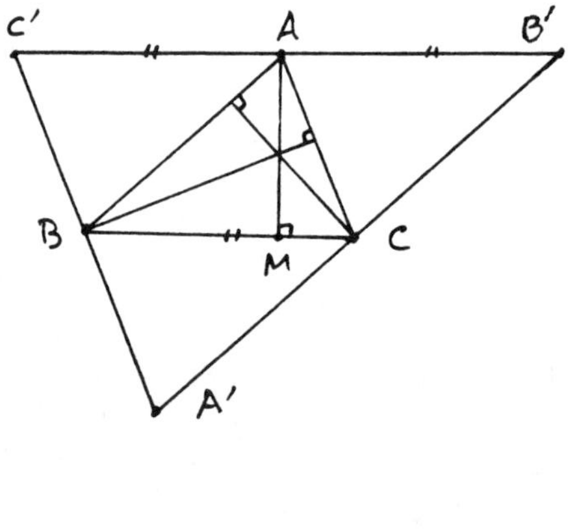

On the other hand, the altitude AM of the triangle ABC is perpendicular to BC, and hence also perpendicular to $B'C'$. Thus we see that the altitudes of the triangle ABC are equal to the perpendicular bisectors of the sides of the triangle $A'B'C'$. Hence they meet in a single point ((IV.5), cf. Exercise 1.9).

Proposition 5.7 (The Euler line) *In a triangle ABC, let O be the* circumcenter, *let G be the* centroid, *and let H be the* orthocenter. *Then O, G, H lie on a line (called the* Euler line *of the triangle) and $GH \cong 2OG$.*

Proof. For the proof, let F be the midpoint of BC, draw the median AF, and let the line OG meet the altitude AM in a point H'. Note that OF is perpendicular to BC, since O is the circumcenter. Hence OF is parallel to AM. Therefore $\angle GAH' \cong \angle GFO$. Also $\angle AGH' \cong \angle FGO$ since they are vertical angles (I.15). By our previous result on the medians (5.5), $AG \cong 2GF$. Thus we can apply (2ASA = 5.3) to conclude that $\Delta AGH' \cong 2\Delta FGO$. It follows that $GH' = 2OG$.

Thus the point H' is characterized as that point on the ray OG such that $GH' = 2OG$. Now permuting the roles of A, B, C, it follows that H' also lies on the other altitudes of ABC, so $H' = H$ is the orthocenter and our conclusions O, G, H collinear and $GH \cong 2OG$ follow.

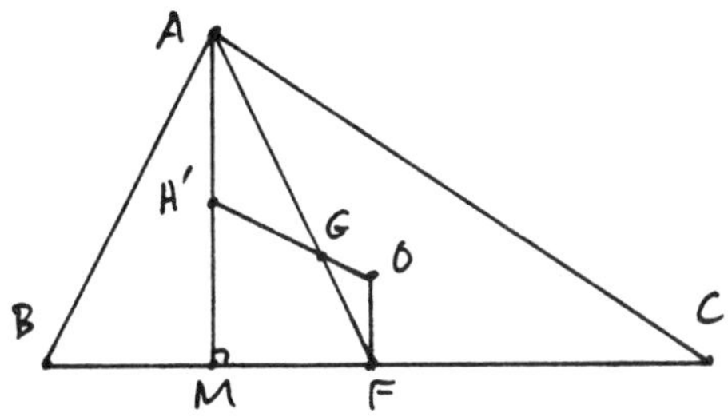

By the way, this argument provides another independent proof of the fact that the three altitudes meet in a point (5.6).

For our next results, we will introduce the very useful method of cyclic quadrilaterals.

Definition. A *cyclic quadrilateral* is a set of four points A, B, C, D lying in that order on a circle, together with the lines AB, BC, CD, DA joining them. The lines AC and BD are the *diagonals* of the cyclic quadrilateral.

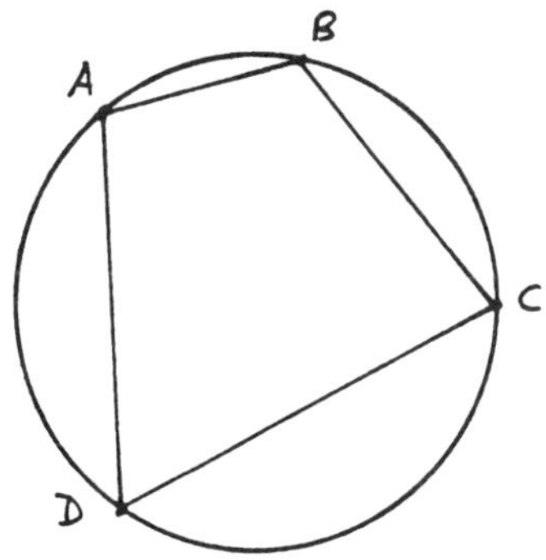

The importance of cyclic quadrilaterals comes from the relationships between the various angles of the figure, which characterize the property of the four points A, B, C, D lying on a circle.

Proposition 5.8 *Let A, B, C, D be four points in the plane, with A, B both on the same side of the line CD. Then A, B, C, D lie on a circle if and only if the angles $\angle DAC$ and $\angle DBC$ are equal.*

Proof. If A, B, C, D lie on a circle, then Euclid's (III.21) tells us that the angles at A and B are the same, since they both subtend the same arc DC.

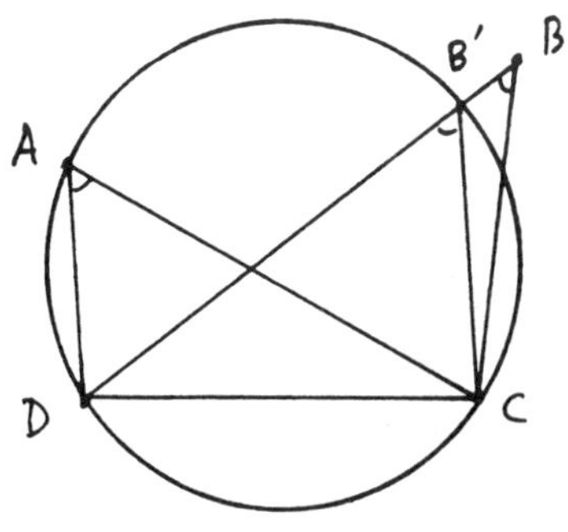

Conversely, suppose the angles at A and B are equal. Draw the circle through A, D, C (IV.5) and let it meet the line BD at B'. (In our figure, B lies outside the circle, but the argument will be similar if B lies inside the circle.)

Then by (III.21), the angle at B' is also equal to the angles at A and at B. If $B \neq B'$, this contradicts (I.16), because the angle $\angle DB'C$ at B' is an exterior angle to the triangle BCB', and so must be greater than the opposite interior angle at B. Hence $B = B'$ and all four points lie on the circle.

Theorem 5.9 (The nine-point circle, due to Brianchon and Poncelet (1820)) *In any triangle, the midpoints of the three sides, the feet of the three altitudes, and the midpoints of the segments joining the three vertices to the orthocenter, all lie on a circle.*

Proof. Let ABC be the given triangle. Let D, E, F be the midpoints of the sides, let K, L, M be the feet of the altitudes, let H be the orthocenter, and let P, Q, R be the midpoints of the segments joining the three vertices of H. We must show that $D, E, F, K, L, M, P, Q, R$ all lie on a circle.

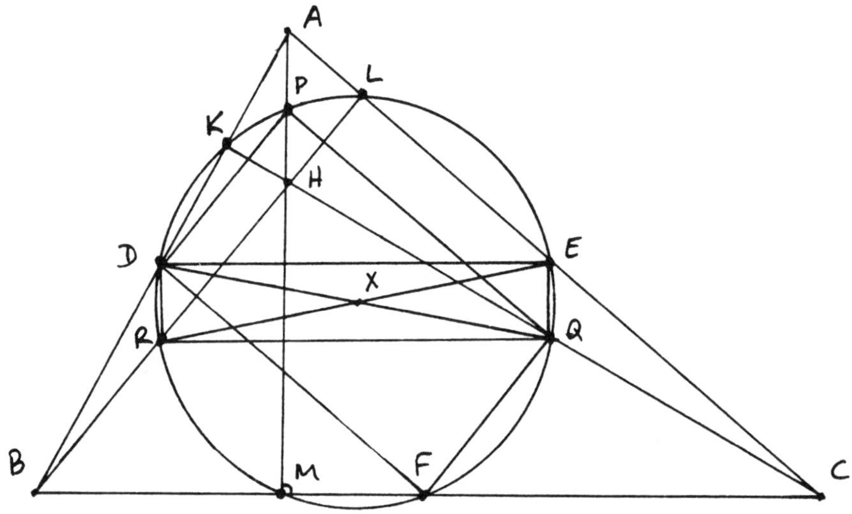

We make several applications of (5.1). Applied to the triangle ABC, we find that DE is parallel to the base BC. Applied to the triangle BCH, we find that RQ is parallel to the base BC. Hence DE is parallel to RQ. Now apply (5.1) to the triangle ACH. We find that EQ is parallel to the base AH. Similarly, using the triangle ABH, DR is parallel to AH. Hence EQ and DR are parallel. Furthermore, EQ and DR are perpendicular to DE and RQ since AH is perpendicular to BC. Thus $DEQR$ is a rectangle. If X is the center of this rectangle, then X is equidistant from the four corners. Thus D, E, Q, R lie on a circle Γ with center X. We will show that this circle contains the other required points.

By (5.8), $DLER$ is a cyclic quadrilateral, because the angles at D and L subtending ER are both right angles. Since a circle is determined by three points (III.10), this circle is the same as the circle Γ, in other words, L also lies on Γ.

A similar argument shows that $DKEQ$ is a cyclic quadrilateral, and so K lies on Γ.

Now, shifting perspective so that AC is regarded as the base of the triangle, the same argument shows that $DPQF$ is a rectangle, with the same center X, since X is the midpoint of DQ. Therefore P and F also lie on Γ.

Finally, $MDPF$ is a cyclic quadrilateral for the same reasons as above, so M is also on Γ. □

Proposition 5.10 (The orthic triangle) *Let ABC be any acute triangle, and let K, L, M be the feet of the altitudes of ABC. Then the altitudes of ABC are the angle bisectors of the* orthic triangle *KLM.*

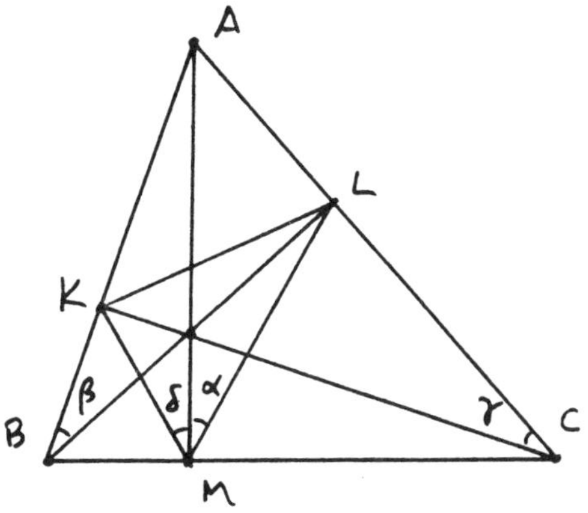

Proof. We use cyclic quadrilaterals. First $LMAB$ is a cyclic quadrilateral, because the angles at L and M are right. Hence $\alpha = \angle AML$ is equal to $\beta = \angle ABL$.

Next, $KLCB$ is a cyclic quadrilateral because the angles at K and L are right. Hence β is equal to $\gamma = \angle KCL$.

Finally, $MKAC$ is a cyclic quadrilateral because the angles at M and K are right, so γ is equal to $\delta = \angle AMK$.

Thus $\alpha = \delta$, so that the altitude AM of ABC is the angle bisector of the angle $\angle KML$ in the orthic triangle. The same argument of course applies to the other two altitudes.

We end this section with an ingenious construction given by Pappus in his commentary on the lost book of Apollonius "On tangencies".

Problem 5.11 Given a circle Γ and two points A, B, find a point C on the circle, such that if the lines CA, CB meet Γ in further points D, E, then DE is parallel to AB.

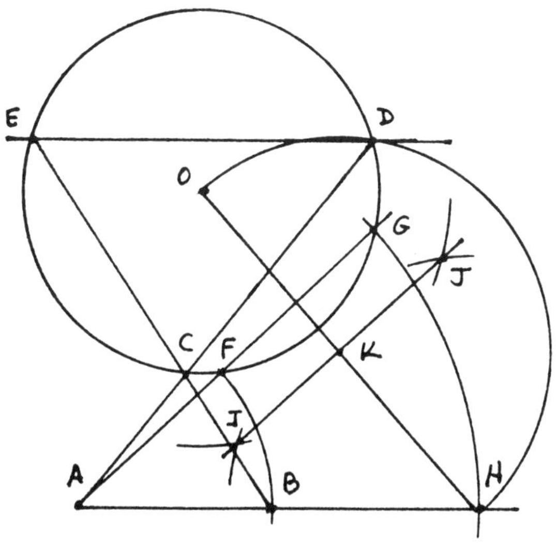

Construction. Let Γ be the given circle, with center O, and let A, B be the given points.

1. Line AB.
2. Circle AB, get F.
3. Line AF, get G.
4. Circle AG, get H.
5. Line OH.
6. Circle O, any radius.
7. Circle H, same radius, get I, J.
8. Line IJ, get K.
9. Circle KO, get D.
10. Line AD, get C.
11. Line CB, get E.
12. Line DE is parallel to AB, as required.

The proof of this construction is in the exercises.

Exercises

5.1 (2SAS) Suppose given two triangles ABC and $A'B'C'$. Assume that $AB \cong 2A'B'$ and $AC \cong 2A'C'$, and the angles at A and A' are equal. Prove that $\Delta ABC \cong 2\Delta A'B'C'$.

5.2 (2SSS) Suppose given two triangles ABC and $A'B'C'$ and assume that $AB \cong 2A'B'$, $AC \cong 2A'C'$, and $BC \cong 2B'C'$. Prove that $\Delta ABC \cong 2\Delta A'B'C'$.

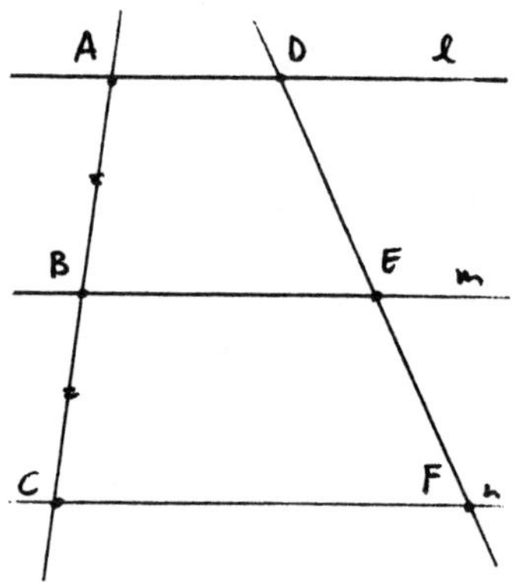

5.3 Let l, m, n be three parallel lines. Suppose they cut off equal segments $AB \cong BC$ on a transversal line. Show that the segments DE, EF cut off by any other transversal line are equal.

5.4 Given three line segments, make a ruler and compass construction of a triangle whose medians are congruent to the three given segments. What condition on the segments is necessary for this to be possible?

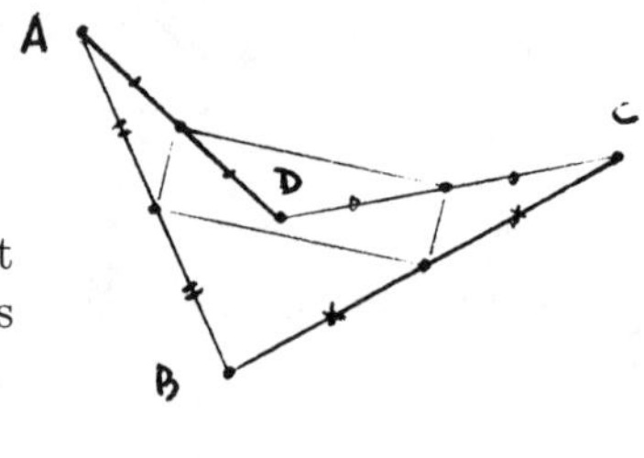

5.5 Let $ABCD$ be a quadrilateral. Show that the figure formed by joining the midpoints of the four sides is a parallelogram.

5.6 In any triangle, show that the center X of the nine-point circle is the midpoint of the segment OH joining the circumcenter O to the orthocenter H.

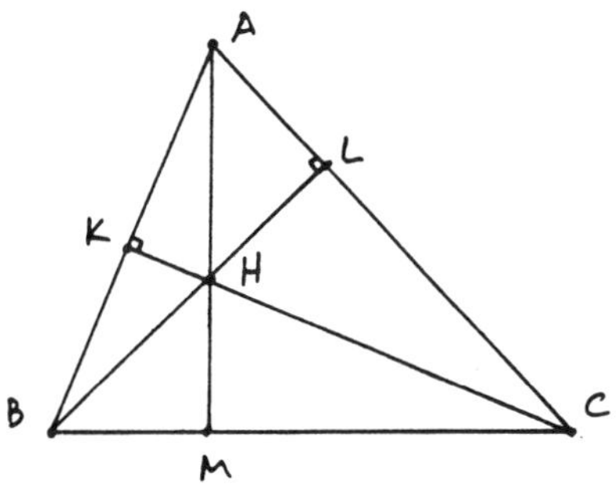

5.7 Use cyclic quadrilaterals to give another proof of (5.6), as follows. Let ABC be the given triangle. Let the altitudes BL and CK meet at H. Let AH meet the opposite side at M. Show that the two angles at M are equal, hence right angles, so that in fact AM is the third altitude. (This proof is probably the one known to Archimedes.)

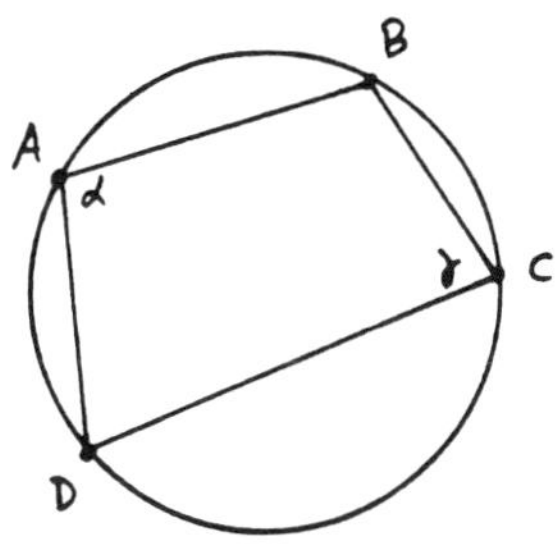

5.8 Show that the opposite angles α, γ of a quadrilateral $ABCD$ add to two right angles if and only if A, B, C, D lie on a circle.

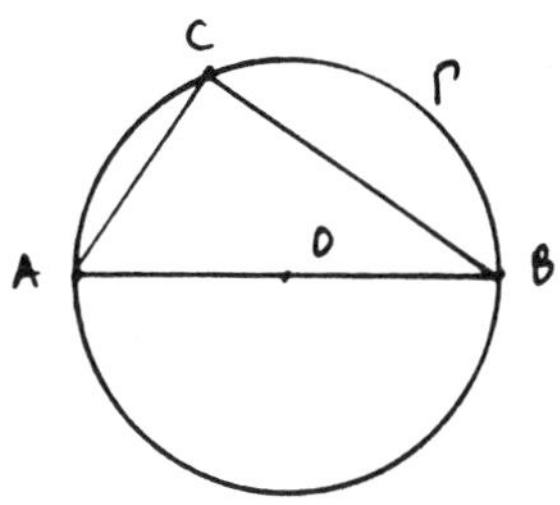

5.9 Let AB be the diameter of a circle Γ. Show that a triangle ABC has a right angle at C if and only if C lies on the circle Γ.

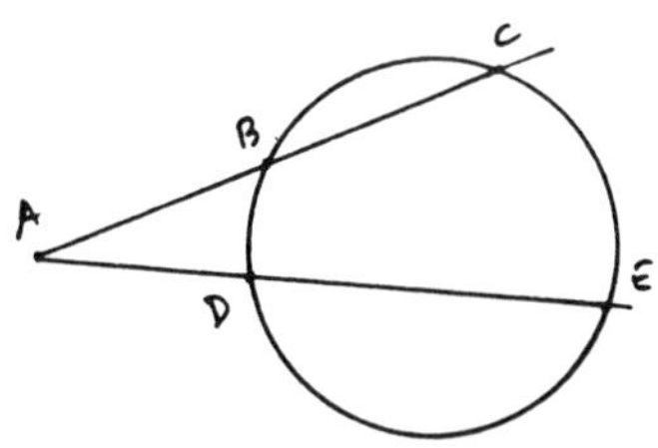

5.10 Let B, C and D, E lie on two rays emanating from a point A. Show that B, C, D, E lie on a circle if and only if $AB \times AC = AD \times AE$ (in the sense of content).

5.11 In the construction to problem (5.11), prove that DE is parallel to AB. *Hint*: First show that $BCDH$ is a cyclic quadrilateral. Then draw DH, and compare angles using (III.22) and (III.32).

5.12 In the construction to problem (5.11), show that the circle through A, B, C is tangent to Γ. Thus this construction solves the problem, "given a circle Γ and given two points A, B, to find a circle passing through A, B, and tangent to Γ".

5.13 (The Simson line). Let ABC be any triangle. Let P be a point on the circumscribed circle of ABC. Let D, E, F be the feet of the perpendiculars from P to the sides of the triangle (extended as necessary). Then D, E, F lie on a line. [First proved by W. Wallace, 1799.]

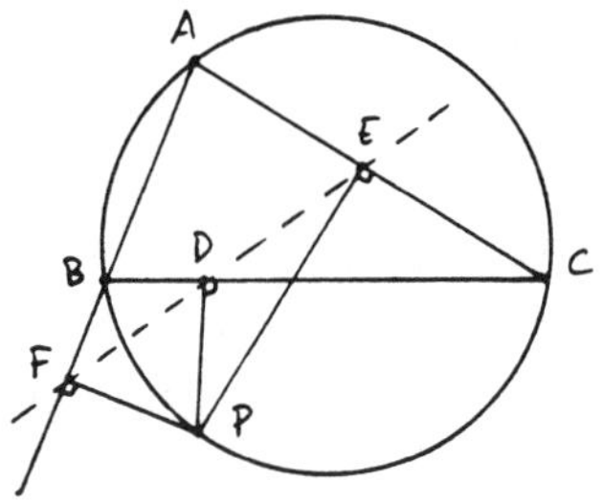

5.14 (The Miquel point). Let ABC be a triangle. Let D, E, F be points on the sides of the triangle. Show that the circles through ADE, BDF, and CEF all meet in a common point G. *Hint*: Let G be the intersection of the first two circles, then show that $CEGF$ is a cyclic quadrilateral [due to A. Miquel, 1838].

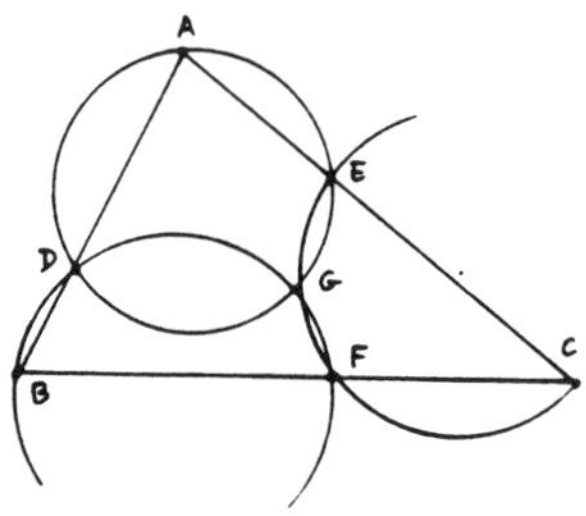

5.15 (Pappus' theorem). Let A, B, C be points on a line l, and let A', B', C' be points on a line m. Assume that $AC' \| A'C$ and $B'C \| BC'$. Show that $AB' \| A'B$. *Hint*: Draw a circle through A, B', C' meeting l in D. Then use cyclic quadrilaterals [cf. Hilbert, Foundations, §14].

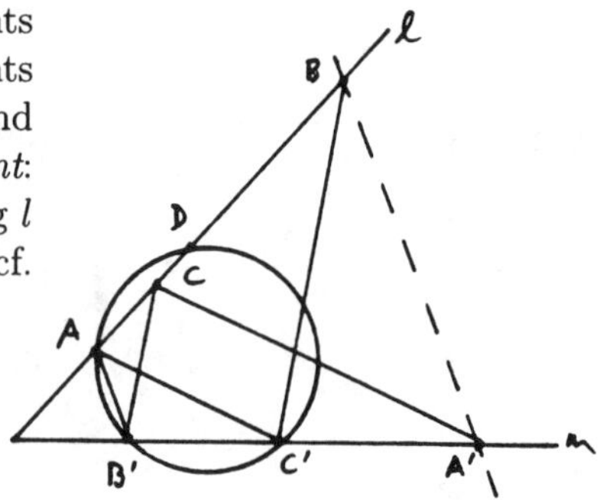

5.16 Construct three circles of different radii, each one tangent to the other two, with noncollinear centers, in as few steps as possible. (Par = 7).

5.17 Let A, B, C, D be four points on a circle Γ. Let four more circles pass through AB, BC, CD, DA respectively, meeting in further points A', B', C', D'. Show that $A'B'C'D'$ is a cyclic quadrilateral.

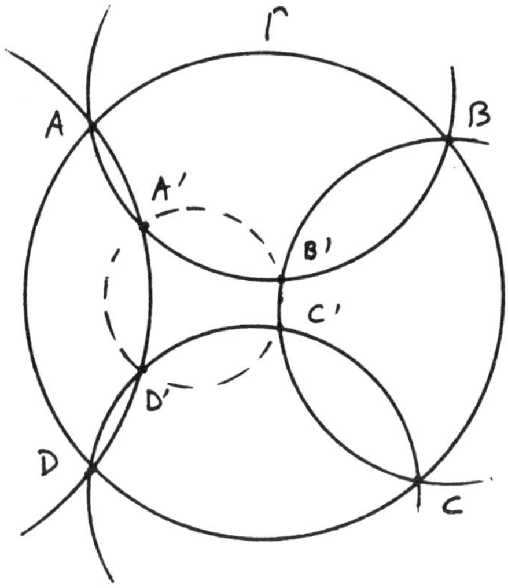

It was on that night that he dreamed his dream of titanic basalt towers—dripping with slime and ocean ooze and fringed with great sea mats—their weirdly proportioned bases buried in gray-green muck and their non-Euclidean-angled parapets fading into the watery distances of that unquiet submarine realm.

– from *Rising with Surtsey*
by Brian Lumley
in Tales of the Cthulhu Mythos
by H. P. Lovecraft
& divers hands
Arkham, Sauk City (1990)
p. 315

CHAPTER 2

Hilbert's Axioms

Our purpose in this chapter is to present (with minor modifications) a set of axioms for geometry proposed by Hilbert in 1899. These axioms are sufficient by modern standards of rigor to supply the foundation for Euclid's geometry. This will mean also axiomatizing those arguments where he used intuition, or said nothing. In particular, the axioms for betweenness, based on the work of Pasch in the 1880's, are the most striking innovations in this set of axioms.

Another choice has been to take the SAS theorem as an axiom, and thus bypass the method of superposition. It is possible to go the other route, and use motions of figures as a basic building block of geometry. This is what Hadamard does in his Leçons de Géométrie Elémentaire, but the result is a step backward in logical clarity, because he never makes precise exactly what kind of motions he is allowing.

The first benefit of establishing the new system of axioms is of course to vindicate Euclid's *Elements*, and thus establish "Euclidean" geometry as a vigorous mathematical discipline. A second benefit is to pose carefully those problems which have bothered geometers for centuries, such as the question of the independence of the parallel postulate. Unless one has an exact understanding of precisely what is allowed, one risks going around in circles in discussing these questions. So in the development of our geometry with the new axioms, we will keep the parallel postulate separate and note carefully what depends on it and what does not.

Besides presenting the axioms, this chapter will also contain the first consequences of the axioms, including different proofs of some of Euclid's early propositions, until we have established enough so that Euclid's later results can be deduced without difficulty from the new foundations we have established.

6 Axioms of Incidence

The axioms of incidence deal with points and lines and their intersections. The points and lines are undefined objects. We simply postulate a set, whose elements are called *points*, together with certain subsets, which we call *lines*. We do not say what the points are, nor which subsets form lines, but we do require that these undefined notions obey certain axioms:

I1. For any two distinct points A, B, there exists a unique line l containing A, B.

I2. Every line contains at least two points.

I3. There exist three noncollinear points (that is, three points not all contained in a single line).

Definition. A set whose elements are called points, together with a set of subsets called lines, satisfying the axioms (I1), (I2), (I3), will be called an *incidence plane.*

From this modest beginning we cannot expect to get very interesting results, but just to illustrate the process, let us see how one can prove theorems based on these axioms.

Proposition 6.1 *Two distinct lines can have at most one point in common.*

Proof. Let l, m be two lines, and suppose they both contain the points A, B, with $A \neq B$. According to axiom (I1), there is a unique line containing both A and B, so l must be equal to m.

Note that this fact, which was used by Euclid in the proof of (I.4) with the rather weak excuse that "two lines cannot enclose a space", follows here from the uniqueness part of axiom (I1). This should indicate the importance of stating explicitly the uniqueness of an object, which was rarely done in Euclid's *Elements.*

Having set up an axiom system, consisting of the undefined sets of points and lines, satisfying the axioms (I1)–(I3), a *model* of that axiom system is a realization of the undefined terms in some particular context, such that the axioms are satisfied. You could also think of the model as an *example* of the *incidence geometry* defined above.

Example 6.1.1 The real Cartesian plane $\mathbb{R}^2$. Here the set of points is the set $\mathbb{R}^2$ of ordered pairs of real numbers. The lines are those subsets which satisfy a linear equation $ax + by + c = 0$ in the variables x, y. To verify that the axioms hold, for (I1) think of the "two-point formula" from analytic geometry: given two points $A = (a_1, a_2)$ and $B = (b_1, b_2)$. They lie on the line

$$y - a_2 = \frac{b_2 - a_2}{b_1 - a_1}(x - a_1)$$

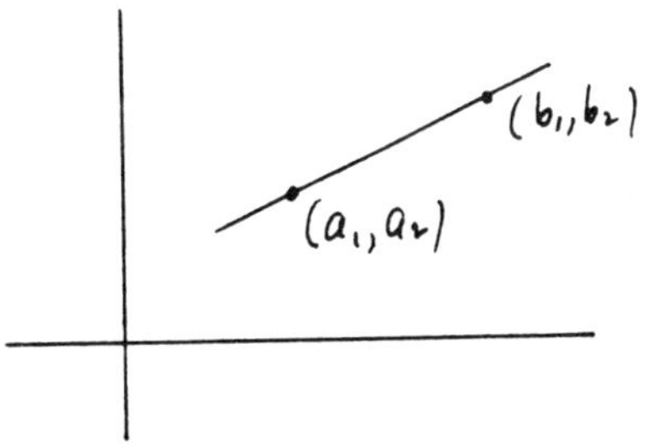

if $a_1 \neq b_1$; if $a_1 = b_1$ they lie on the line $x = a_1$. To verify (I2), take any linear equation involving y. Substitute two different values of x, and solve for y. This gives two points on the line. If the equation did not involve y, say $x = c$, take the points $(c, 0)$ and $(c, 1)$. To verify (I3), consider the points $(0, 0)$, $(0, 1)$, $(1, 0)$. One sees easily there is no linear equation with all three points as solutions.

Example 6.1.2 One can also make models out of finite sets. For example, let the set of points be a set of three elements $\{A, B, C\}$, and take for lines the subsets $\{A, B\}$, $\{A, C\}$, and $\{B, C\}$. We represent this symbolically by the diagram, where the dots represent the elements of the set, and the lines drawn on the page show which subsets are to be taken as lines.

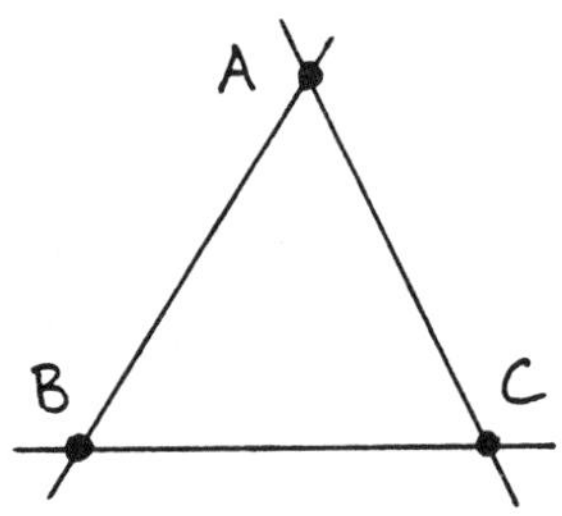

This diagram should be understood as purely symbolic, however, and has nothing to do with a triangle in the ordinary Cartesian plane. The verification of the axioms in this case is trivial.

Definition. Two lines are *parallel* if they have no points in common.

The parallel postulate, in its equivalent form given by Playfair, can be stated as a further axiom about incidence of lines. However, we do not include this axiom in the definition of incidence geometry. Thus we may speak of an incidence geometry which does or does not satisfy Playfair's axiom.

P. (Playfair's axiom). For a line l, and a point A not on l, there is at most one line containing A which is parallel to l.

Note that the real Cartesian plane (6.1.1) satisfies (P), as you know, and the three-point geometry (6.1.2) also satisfies (P) vacuously, because there are no parallel lines at all. Next we give an example of an incidence geometry which does not satisfy (P).

Example 6.1.3 Let our set consist of five points A, B, C, D, E, and let the lines be all subsets of two points. It is easy to see that this geometry satisfies (I1)–(I3). However, it does not satisfy (P), because, for example, AB and AC are two distinct lines through the point A, and parallel to the line DE.

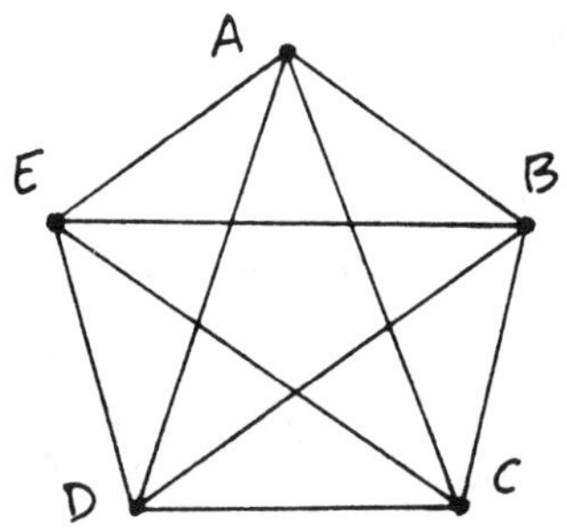

Remember that the word *parallel* simply means the two lines have no points in common. It does not say anything about being in the same direction, or being equidistant from each other, or anything else.

We say that two models of an axiom system are *isomorphic* if there exists a 1–1 correspondence between their sets of points in such a way that a subset of the first set is a line if and only if the corresponding subset of the second set is a line. For short, we say "the correspondence takes lines into lines". So for example, we see that (6.1.1), (6.1.2), and (6.1.3) are non-isomorphic models of incidence geometry, for the simple reason that their sets of points have different cardinality: there are no 1–1 correspondences between any of these sets.

On the other hand, we can show that any model of incidence geometry having just three points is isomorphic to the model given in (6.1.2). Indeed, let $\{1,2,3\}$ be a geometry of three points. By (I3), there must be three non-collinear points. Since there are only three points here, we conclude that there is no line containing all three. But by (I1), each subset of two must be contained in a line. Thus $\{1,2\}$, $\{2,3\}$, and $\{1,3\}$ are lines. Now by (I2), every line contains at least two points, so these are all the possible lines. In other words, the lines are just all subsets of two elements. Since (6.1.2) also has this property, any 1–1 correspondence between the sets $\{A,B,C\}$ and $\{1,2,3\}$ will give an isomorphism.

By the way, this proof shows that the isomorphism just found is not unique. There are six choices. This leads to the notion of automorphism.

Definition. An *automorphism* of a geometry is an isomorphism with itself, that is, it is a 1–1 mapping of the set of points onto itself, preserving lines.

Note that the composition of two automorphisms is an automorphism, and so is the inverse of an automorphism, so that the set of automorphisms forms a group. In the example above, any 1–1 mapping of the set of three elements onto itself gives an automorphism of the geometry, so we see that the group of automorphisms of this geometry is the symmetric group on three letters, S_3.

An important question about a set of axioms is whether the axioms are *independent* of each other. That is to say, that no one of them can be proved as a consequence of the others. For if one was a consequence of the others, then we would not need that one as an axiom. To try to prove directly that axiom A is not a consequence of axioms $B, C, D, \ldots$ is usually futile. So instead, we search for a model in which axioms $B, C, D, \ldots$ hold but axiom A does not hold. If such a model exists, then there can be no proof of A as a consequence of $B, C, D, \ldots$, so we conclude that A is independent of the others. This process must be repeated with each individual axiom, to show that they are all independent of each other. With a long list of axioms this can become tedious and difficult, so we will forego the process with our full list of axioms. But as an illustration of what is involved, let us show that the axioms (I1), (I2), (I3), and (P) are independent.

Proposition 6.2 *The axioms* (I1), (I2), (I3), (P) *are independent of each other.*

Proof. We have already seen that (6.1.3) is a model satisfying (I1), (I2), (I3), and not (P). Hence (P) is independent of the others.

For a model satisfying (I1), (I2), (P), and not (I3), take a set of two points, and the one line containing both of them. Note that (P) is satisfied vacuously, because there are no points not on the line l.

For a model satisfying (I1), (I3), (P), and not (I2), take a set of three points A, B, C, and for lines take the subsets $\{A, B\}$, $\{A, C\}$, $\{B, C\}$, and $\{A\}$. The existence of the one-point line $\{A\}$ contradicts (I2). Yet (P) is still fulfilled, because that one-point line is then the unique line through A parallel to $\{B, C\}$.

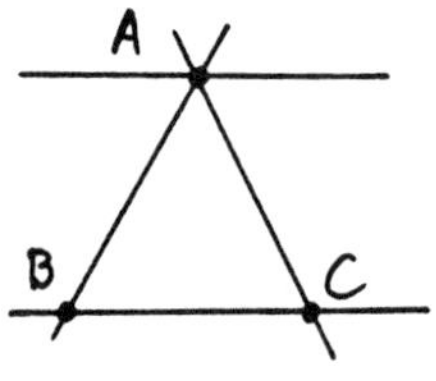

For a model satisfying (I2), (I3), (P) and not (I1), just take a set of three points and no lines at all.

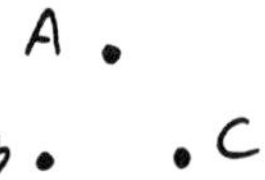

While we are discussing axiom systems, there are a few more concepts we should mention. An axiom system is *consistent* if it will never lead to a contradiction. That is to say, if it is not possible to prove from the axioms a statement A and also to prove its negation not-A. This is obviously a highly desirable property of a system of axioms. We do not want to waste our time proving theorems from a system of axioms which one day may lead to a contradiction. Unfortunately, however, the logician Kurt Gödel has proved that for any reasonably rich set of axioms, it will be impossible to prove the consistency of that system. (Any discussion of what this means precisely, or how he proved his result, is beyond the scope of this book!) So we will have to settle for something less, which is *relative consistency*. As soon as you can find a model for your axiom system within some other mathematical theory T, it follows that if T is consistent, then also your system of axioms is consistent. For any contradiction which might follow from your axioms would then also appear in the theory T, contradicting its consistency. So for example, if you believe in the consistency of the theory of real numbers, then you must accept the consistency of Hilbert's axiom system for geometry, because all of these axioms will hold in the real Cartesian plane. That is the best we can do about the question of consistency.

Another question about a system of axioms is whether it is *categorical*. This means does it describe a unique mathematical object? Or in other words is there a *unique* model (up to isomorphism) for the system of axioms? In fact, it will turn out that if we take the entire list of Hilbert's axioms, including the parallel axiom (P) and Dedekind's axiom (D), the system will be categorical, and the unique model will be the real Cartesian plane. (We may be able to prove this result later.) However, from the point of view of this book, it is more interesting to have an axiom system which is not categorical, and then to investigate the different possible geometries which can arise. Therefore we will almost never assume Dedekind's axiom (D), and we will only sometimes assume Archimedes axiom (A), or the parallel axiom (P).

Finally, one can ask whether the axiom system is *complete*, which means, can every statement which is true in every model of the axiom system be proved as a consequence of the axioms? Again, Gödel has shown that any axiomatic system of reasonable richness cannot be complete. We must refer the reader elsewhere for a fuller discussion of this question.

Exercises

6.1 Let F be a field, and let $V = F^3$ be a three-dimensional vector space over F. Let Π be the set of 1-dimensional subspaces of V. We will call the elements of Π "points". So a "point" is a 1-dimensional subspace $P \subseteq V$. If $W \subseteq V$ is a 2-dimensional subspace of V, then the set of all "points" contained in W will be called a "line". Show that the set Π of "points" and the subsets of "lines" form an incidence geometry satisfying (I1)–(I3).

6.2 A *projective plane* is a set of points and subsets called lines which satisfy the following four axioms:

P1. Any two distinct points lie on a unique line.

P2. Any two lines meet in at least one point.

P3. Every line contains at least three points.

P4. There exist three noncollinear points.

Note that these axioms imply (I1)–(I3), so that any projective plane is also an incidence geometry. Show the following:

a) The model described in Exercise 6.1 above is also a projective plane.

b) Every projective plane has at least seven points, and there exists a model of a projective plane having exactly seven points.

c) The axioms (P1), (P2), (P3), (P4) are independent.

6.3 An *affine plane* is a set of points and subsets called lines satisfying (I1), (I2), (I3), and the following stronger form of Playfair's axiom.

P′. For any line l, and any point A not on l, there exists a unique line m containing A and parallel to l.

a) Show that any two lines in an affine plane have the same number of points (i.e. there exists a 1–1 correspondence between the points of the two lines).

b) If an affine plane has a line with exactly n points, then the total number of points in the plane is n^2.

c) If F is any field, show that the *Cartesian plane* over F, that is the set F^2 of ordered pairs of elements of F, with lines being defined by linear equations, is a model of an affine plane.

d) Show that there exist affine planes with 4, 9, 16, or 25 points. (The nonexistence of an affine plane with 36 points is a difficult result.)

6.4 Let Π be an affine plane (Exercise 6.3). A *pencil* of parallel lines means all the lines parallel to a given line (including that line itself). We call each pencil of parallel lines an "ideal point", or a "point at infinity", and we say that an ideal point "lies on" each of the lines in the pencil. Now let Π' be the enlarged set consisting of Π together with all these new ideal points. A *line* of Π' will be the subset consisting of a line of Π plus its unique ideal point, or a new line, called the "line at infinity" consisting of all the ideal points.

a) Show that this new set Π' with subsets of lines as just defined forms a projective plane (Exercise 6.2).

b) If Π is the Cartesian plane over a field F (Exercise 6.3c), show that the associated projective plane Π' is isomorphic to the projective plane constructed in (Exercise 6.1).

6.5 Kirkman's schoolgirl problem (1850) is as follows: In a certain school there are 15 girls. It is desired to make a seven-day schedule such that each day the girls can walk in the garden in five groups of three, in such a way that each girl will be in the same group with each other girl just once in the week. How should the groups be formed each day?

To make this into a geometry problem, think of the girls as points, think of the groups of three as lines, and think of each day as describing a set of five lines, which we call a pencil. Now consider a *Kirkman geometry*: A set, whose elements we call *points*, together with certain subsets we call *lines*, and certain sets of lines we call *pencils*, satisfying the following axioms:

K1. Two distinct points lie on a unique line.

K2. All lines contain the same number of points.

K3. There exist three noncollinear points.

K4. Each line is contained in a unique pencil.

K5. Each pencil consists of a set of parallel lines whose union is the whole set of points.

a) Show that any affine plane gives a Kirkman geometry where we take the pencils to be the set of all lines parallel to a given line. (Hence by Exercise 6.3 there exist Kirkman geometries with 4, 9, 16, 25 points.)

b) Show that any Kirkman geometry with 15 points gives a solution of the original schoolgirl problem.

c) Find a solution for the original problem.

7 Axioms of Betweenness

In this section we present axioms to make precise the notions of betweenness (when one point is in between two others), on which is based the notion of sidedness (when a point is on one side of a line or the other), the concepts of inside and outside, and also the concepts of *order*, when one segment or angle is bigger than another. We have seen the importance of these concepts in reading Euclid's geometry, and we have also seen the dangers of using these concepts intuitively, without making their meaning precise. So these axioms form an important part of our new foundations for geometry. At the same time, these axioms and their consequences may seem

Plate III. Frontispiece to Archimedes (1792). "There is hope. I see traces of men".

difficult to understand for many readers, not because the mathematical concepts are technically difficult, but because the notions of order and separation are so deeply ingrained in our daily experience of life that it is difficult to let go of our intuitions and replace them with axioms. It is an exercise in forgetting what we already know from our inner nature, and then reconstituting it with an open mind as an external logical structure.

The geometrical notions of betweenness, separation, sidedness, and order will all be based on a single undefined relation, subject to four axioms. We postulate a *relation* between sets of three points A, B, C, called "B is between A and C". This relation is subject to the following axioms.

B1. If B is between A and C, (written $A*B*C$), then A, B, C are three distinct points on a line, and also $C*B*A$.

B2. For any two distinct points A, B, there exists a point C such that $A*B*C$.

B3. Given three distinct points on a line, one and only one of them is between the other two.

B4. (Pasch). Let A, B, C be three noncollinear points, and let l be a line not containing any of A, B, C. If l contains a point D lying between A and B, then it must also contain either a point lying between A and C or a point lying between B and C, but not both.

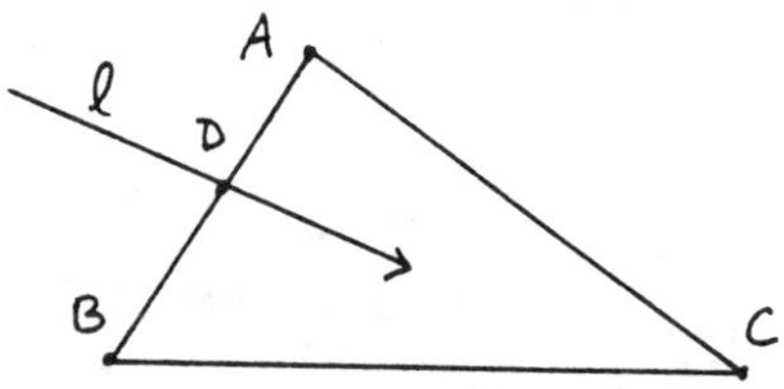

Definition. If A and B are distinct points, we define the *line segment* $\overline{AB}$ to be the set of points A, B, and all points lying between A and B. We define a *triangle* to consist of the three line segments $\overline{AB}$, $\overline{BC}$, and $\overline{AC}$ whenever A, B, C are three noncollinear points. The points A, B, C are the *vertices* of the triangle, and the segments $\overline{AB}$, $\overline{BC}$, $\overline{AC}$ are the *sides* of the triangle.

With this terminology, we can rephrase (B4) as follows: If a line l, which does not contain any of the vertices A, B, C of a triangle, meets one side $\overline{AB}$, then it must meet one of the other sides $\overline{AC}$ or $\overline{BC}$, but not both.

From these axioms together with the axioms of incidence (I1)–(I3) we will deduce results about the separation of the plane by a line, and the separation of a line by a point.

Proposition 7.1 (Plane separation) *Let l be any line. Then the set of points not lying on l can be divided into two nonempty subsets S_1, S_2 with the following properties:*

a) *Two points A, B not on l belong to the same set (S_1 or S_2) if and only if the segment $\overline{AB}$ does not intersect l.*

b) *Two points A, C not on l belong to the opposite sets (one in S_1, the other in S_2) if and only if the segment $\overline{AC}$ intersects l in a point.*

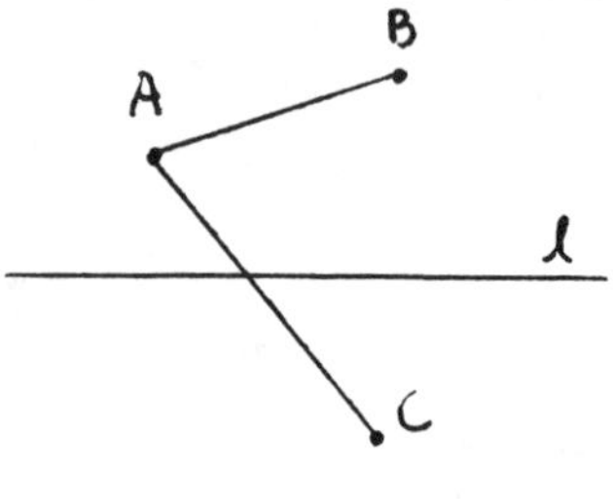

We will refer to the sets S_1, S_2 as the two sides *of l, and we will say "A and B are on the* same side *of l", or "A and C are on* opposite sides *of l".*

Proof. We start by defining a relation $\sim$ among points not on l. We will say $A \sim B$ if either $A = B$ or if the segment $\overline{AB}$ does not meet l. Our first step is to show that $\sim$ is an equivalence relation. Clearly $A \sim A$ by definition, and $A \sim B$ implies $B \sim A$, because the set $\overline{AB}$ does not depend on the order in which we write A and B (using axiom (B1)). The non-trivial step is to show the relation is transitive: if $A \sim B$ and $B \sim C$, we must show $A \sim C$.

Case 1. Suppose A, B, C are not collinear. Then we consider the triangle ABC. Since $A \sim B$, l does not meet $\overline{AB}$. Since $B \sim C$, l does not meet $\overline{BC}$. Now by Pasch's axiom (B4) it follows that l does not meet $\overline{AC}$. Hence $A \sim C$.

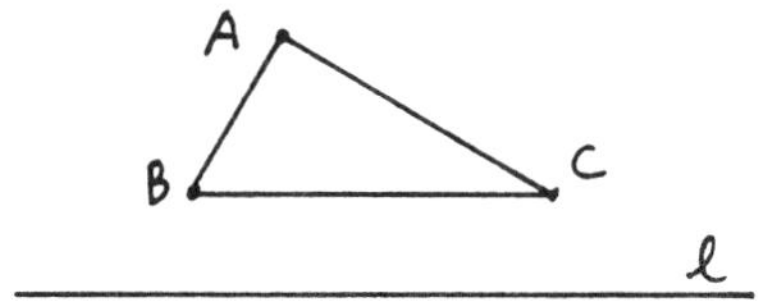

Case 2. Suppose A, B, C lie on a line m. Since A, B, C do not lie on l, the line m is different from l. Therefore l and m can meet in at most one point (6.1). But by (I2) every line has at least two points. Therefore there exists a point D on l, D not lying on m.

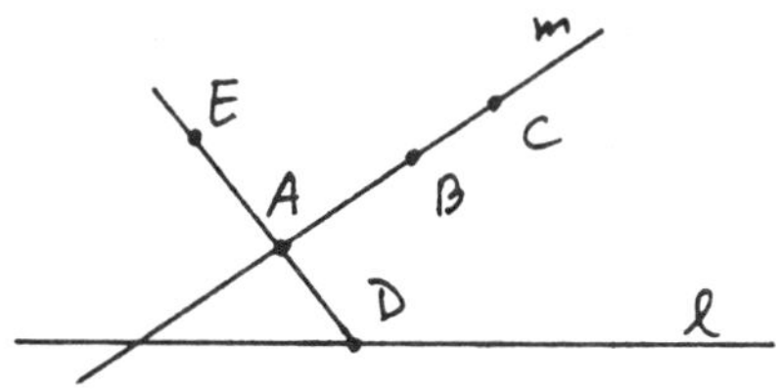

Now apply axiom (B2) to find a point E such that $D * A * E$. Then D, A, E are collinear (B1) hence E is not on l, since A is not on l, and the line DAE already meets l at the point D. Furthermore, the segment $\overline{AE}$ cannot meet l. For if it did, the intersection point would be the unique point in which the line AE meets l, namely D. In that case D would be between A and E. But we constructed E so that $D * A * E$, so by (B3), D cannot lie between A and E. Thus $\overline{AE} \cap l = \emptyset$, so $A \sim E$. Note also that E does not lie on the line m, because if E is on m, then the line AE is equal to m, so D would lie on m, contrary to our choice of D. Therefore A, B, E are three noncollinear points. Then by Case 1 proved above, from $A \sim E$ and $A \sim B$ we conclude $B \sim E$. Applying Case 1 a third time to the three noncollinear points A, C, E, from $A \sim E$ and $C \sim E$ we conclude $A \sim C$ as required.

Thus we have proved that $\sim$ is an equivalence relation. An equivalence relation on a set divides that set into a disjoint union of equivalence classes, and these equivalence classes will satisfy the property a) by definition. To complete the proof it will be sufficient to show that there are exactly two equivalence classes S_1, S_2 for the relation $\sim$. Then to say $\overline{AC}$ meets l, which is equivalent to $A \not\sim C$, will be the same as saying A, C belong to the opposite sets.

By (I3) there exists a point not on l, so there is at least one equivalence class S_1. Given $A \in S_1$, let D be any point on l, and choose by (B2) a point C such that $A * D * C$. Then A and C do not satisfy $\sim$, so there must be at least two equivalence classes S_1 and S_2.

The last step is to show that there are at most two equivalence classes. To do this, we will show that if $A \not\sim C$ and $B \not\sim C$, then $A \sim B$.

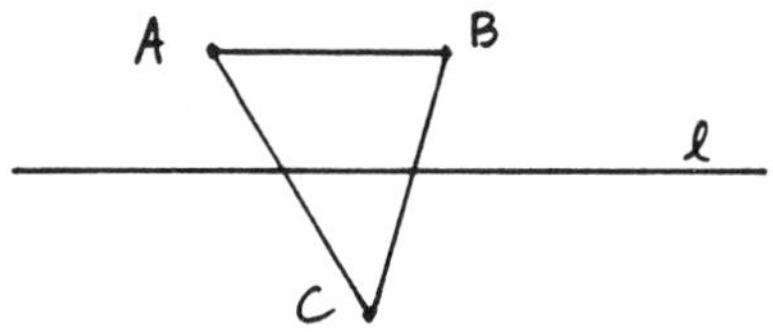

Case 1. If A, B, C are not collinear, we consider the triangle ABC. From $A \not\sim C$ we conclude that $\overline{AC}$ meets l. From $B \not\sim C$ we conclude that $\overline{BC}$ meets l. Now by Pasch's axiom (B4) it follows that $\overline{AB}$ does not meet l. So $A \sim B$ as required.

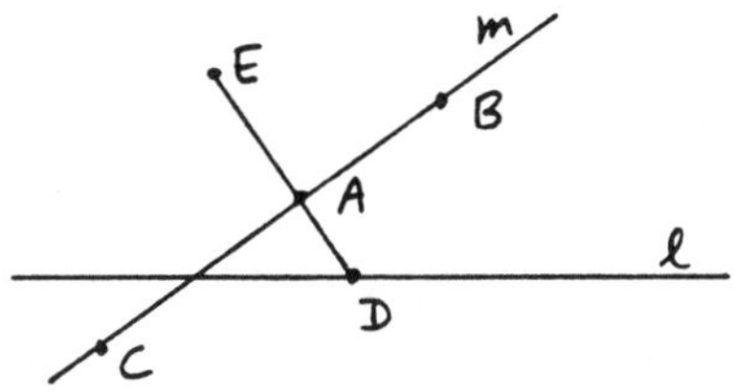

Case 2. Suppose A, B, C lie on a line m. As in Case 2 of the first part of the proof above, choose a point D on l, not on m, and use (B2) to get a point E with $D * A * E$. Then $A \sim E$ as we showed above.

Now $A \not\sim C$ by hypothesis, and $A \sim E$, so we conclude that $C \not\sim E$, since $\sim$ is an equivalence relation (if $C \sim E$, then $A \sim C$ by transitivity: contradiction). Looking at the three noncollinear points B, C, E, from $E \not\sim C$ and $B \not\sim C$ we conclude using Case 1 that $B \sim E$. But also $A \sim E$, so by transitivity, $A \sim B$ as required.

Proposition 7.2 (Line separation) *Let A be a point on a line l. Then the set of points of l not equal to A can be divided into two nonempty subsets S_1, S_2, the two* sides *of A on l, such that*

a) *B, C are on the same side of A if and only if A is not in the segment $\overline{BC}$, and*

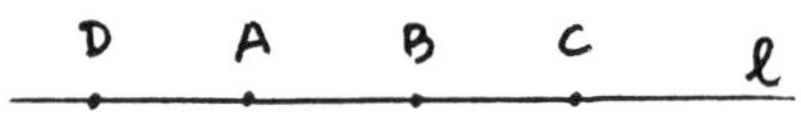

b) *B, D are on opposite sides of A if and only if A belongs to the segment $\overline{BD}$.*

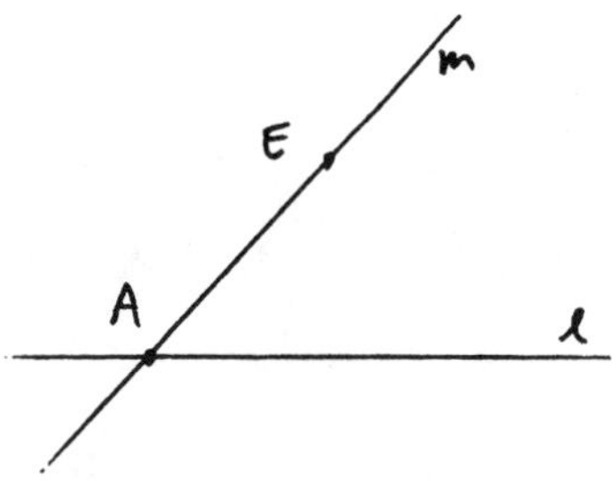

Proof. Given the line l and a point A on l, we know from (I3) that there exists a point E not on l. Let m be the line containing A and E. Apply (7.1) to the line m. If m has two sides S_1', S_2', we define S_1 and S_2 to be the intersections of S_1' and S_2' with l. Then properties a) and b) follow immediately from the previous proposition.

The only mildly non-trivial part is to show that S_1 and S_2 are nonempty. By (I2), there is a point B on l different from A. And by (B2) there exists a point D such that $B * A * D$. Then D will be on the opposite side of A from B, and will lie on l, so both sides are nonempty.

Now that we have some basic results on betweenness, we can define rays and angles.

Definition. Given two distinct points A, B, the *ray* $\overrightarrow{AB}$ is the set consisting of A, plus all points on the line AB which are on the same side of A as B. The point A is the *origin* or *vertex* of the ray. An *angle* is the union of two rays $\overrightarrow{AB}$ and $\overrightarrow{AC}$ originating at the same point, and not lying on the same line. (Thus there is no "zero angle", and there is no "straight angle" (180°).)

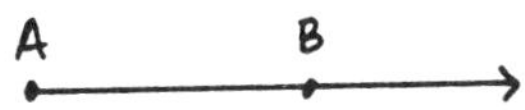

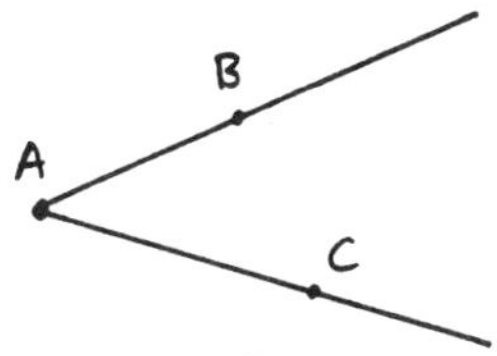

The *inside* (or *interior*) of an angle $\angle BAC$ consists of all points D such that D and C are on the same side of the line AB, and D and B are on the same side of the line AC. If ABC is a triangle, the *inside* (or *interior*) of the triangle ABC is the set of points which are simultaneously in the insides of the three angles $\angle BAC$, $\angle ABC$, $\angle ACB$.

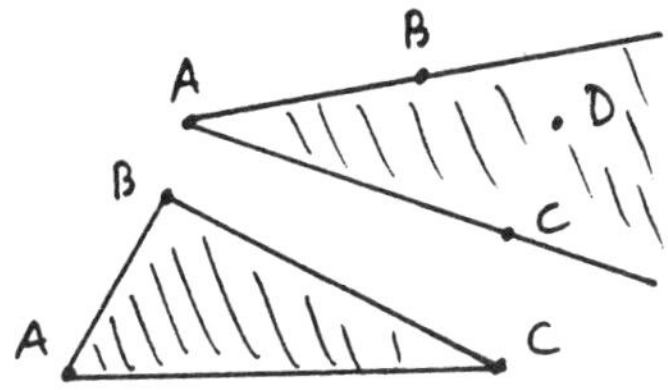

Proposition 7.3 (Crossbar theorem) *Let ABC be a triangle, and let D be a point inside the triangle. Then the ray $\overrightarrow{AD}$ must meet the opposite side $\overline{BC}$.*

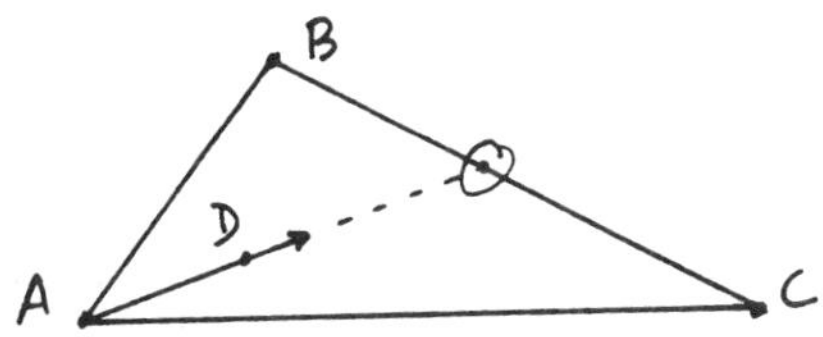

Proof. This is similar to the statement of Pasch's axiom (B4), except that we must consider a line AD which passes through one of the vertices of the triangle. We will prove it with several applications of (B4). To begin with, there exists a point E such that $A * E * B$ (Exercise 7.4). Then the line DE meets one side of the triangle. This line may pass through the point C. If not, then by (B4) it must meet one of the other sides. So we separate into three cases.

Case 1. The line DE passes through C. Then we can apply (B4) to the triangle BCE. Note that the line AD cannot contain B, C, or E, because D is inside the triangle, hence not on the lines AB or AC. Furthermore, since D and C are on the same side of the line AB, the point E cannot be between C and D. Similarly, B and D are on the same side of AC, and also B and E are on the same side of AC by choice of E, so D and E are on the same side of AC, and this implies that C cannot be between D and E.

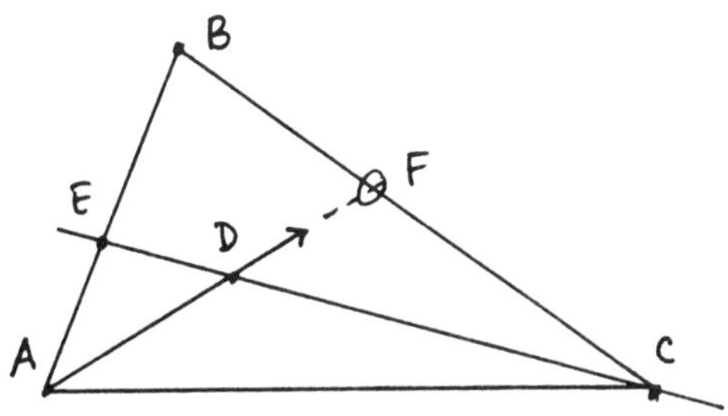

It follows from (B3) that D is between C and E. Thus the line AD meets one side of the triangle BCE. It cannot meet $\overline{BE}$, so by (B4) we conclude that it meets

$\overline{BC}$ in some point F. Since $F \in \overline{BC}$, it is also in the interior of the angle $\angle BAC$, so D and F are on the same side of A on the line AD, in other words, F is in the ray $\overrightarrow{AD}$, as required.

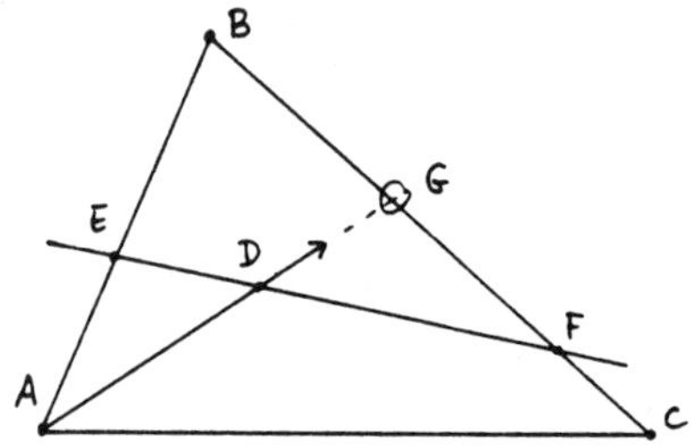

Case 2. The line DE meets $\overline{BC}$, in a point F. We apply (B4) to the triangle BEF. First verify (as above) that $E * D * F$. Then it will follow that AD meets $\overline{BF}$ in a point G, and G will lie in the segment $\overline{BC}$ (Exercise 7.1).

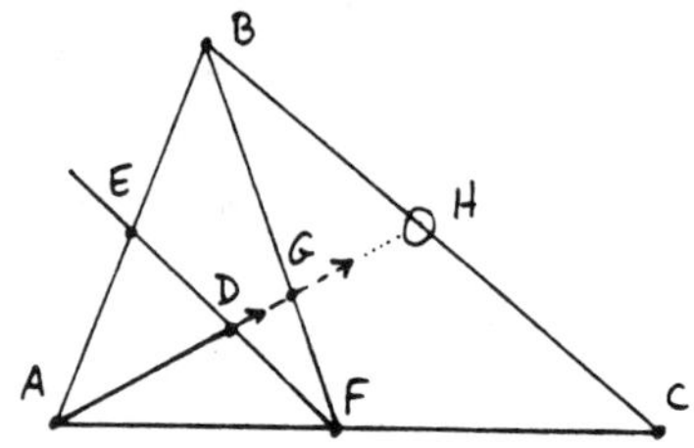

Case 3. The line DE meets $\overline{AC}$ in a point F. We first apply (B4) to the triangle BEF to conclude that AD meets $\overline{BF}$ in a point G. Then we apply (B4) to the triangle BCF to conclude that AD meets $\overline{BC}$ in a point H. The details are as in the earlier cases above.

Example 7.3.1 We will show that the real Cartesian plane (6.2), with the "usual" notion of betweenness, provides a model for the axioms (B1)–(B4).

First, we must make precise what we mean by the usual notion of betweenness. For three distinct real numbers $a, b, c \in \mathbb{R}$, let us define $a * b * c$ if either $a < b < c$ or $c < b < a$. Then it is easy to see that this defines a notion of betweenness on the real line $\mathbb{R}$ which satisfies (B1), (B2), and (B3).

If $A = (a_1, a_2)$, $B = (b_1, b_2)$ and $C = (c_1, c_2)$ are three points in $\mathbb{R}^2$, let us define $A * B * C$ to mean that A, B, C are three distinct points on a line, and that either $a_1 * b_1 * c_1$ or $a_2 * b_2 * c_2$, or both. In fact, if either the x- or the y-coordinates satisfy this betweenness condition, and if the line is neither horizontal nor vertical, then the other coordinates will also satisfy it, because the points lie on a line, and linear operations (addition, multiplication) of real numbers either preserve or reverse inequalities. Thus linear operations preserve betweenness. So we can verify easily that this notion of betweenness in $\mathbb{R}^2$ satisfies (B1), (B2), and (B3).

For (B4), let l be a line, and let A, B, C be three noncollinear points not on l. The line l is defined by some linear equation $ax + by + c = 0$. Let $\varphi : \mathbb{R}^2 \to \mathbb{R}$ be the linear function defined by $\varphi(x, y) = ax + by + c$. Since φ is a linear function, φ will preserve betweenness. For example, if l meets the segment $\overline{AB}$, then 0 will lie between $\varphi(A)$ and $\varphi(B)$. In other words, one of $\varphi(A), \varphi(B)$ will be positive and the other negative. Suppose $\varphi(A) > 0$ and $\varphi(B) < 0$. Consider $\varphi(C)$. If $\varphi(C) > 0$, then l will meet $\overline{BC}$ but not $\overline{AC}$. If $\varphi(C) < 0$, then l will meet $\overline{AC}$ but not $\overline{BC}$. This proves (B4).

Exercises

7.1 Using the axioms of incidence and betweenness and the line separation property, show that sets of four points A, B, C, D on a line behave as we expect them to with respect to betweenness. Namely, show that

a) $A * B * C$ and $B * C * D$ imply $A * B * D$ and $A * C * D$.

b) $A * B * D$ and $B * C * D$ imply $A * B * C$ and $A * C * D$.

7.2 Using (I1)–(I3) and (B1)–(B4) and their consequences, show that every line has infinitely many distinct points.

7.3 Show that the line separation property (7.2) is not a consequence of (B1), (B2), (B3), by constructing a model of betweenness for the set of points on a line, which satisfies (B1), (B2), (B3) but has only finitely many points. (Then by Exercise 7.2, line separation must fail in this model.)

7.4 Prove directly from the axioms (I1)–(I3) and (B1)–(B4) that for any two distinct points A, B, there exists a point C with $A * C * B$. (*Hint:* Use (B2) and (B4) to construct a line which will be forced to meet the segment $\overline{AB}$, but does not contain A or B.)

7.5 Be careful not to assume without proof statements which may appear obvious. For example, prove the following:

a) Let A, B, C be three points on a line with C in between A and B. Then show that $\overline{AC} \cup \overline{CB} = \overline{AB}$ and $\overline{AC} \cap \overline{CB} = \{C\}$.

b) Suppose given two distinct points A, B on a line l. Show that $\overrightarrow{AB} \cup \overrightarrow{BA} = l$ and $\overrightarrow{AB} \cap \overrightarrow{BA} = \overline{AB}$.

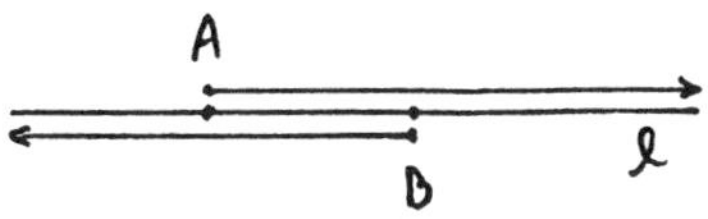

7.6 Show that the interior of a triangle is non-empty.

7.7 Suppose that a line l contains a point D which is in the inside of a triangle ABC. Then show that the line l must meet (at least) one of the sides of the triangle.

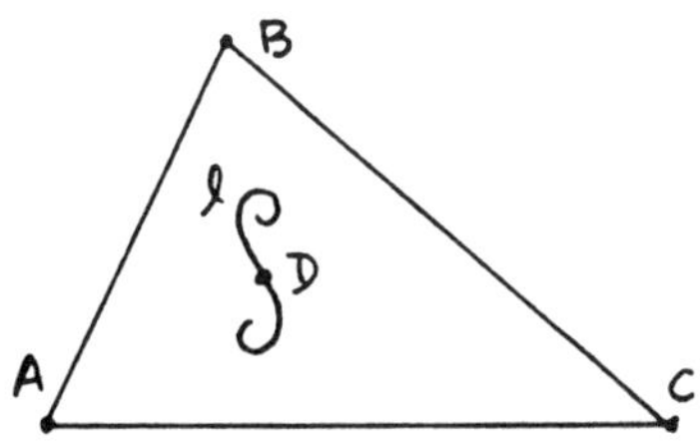

7.8 A set U of points in the plane is a *convex set* if whenever A, B are distinct points in U, then the segment $\overline{AB}$ is entirely contained in U. Show that the inside of a triangle is a convex set.

7.9 (*Linear ordering*) Given a finite set of distinct points on a line, it is possible to label them $A_1, A_2, \ldots, A_n$ in such a way that $A_i * A_j * A_k$ if and only if either $i < j < k$ or $k < j < i$.

7.10 Suppose that lines a, b, c through the vertices A, B, C of a triangle meet at three points inside the triangle. Label them

$$X = a \cdot c$$
$$Y = a \cdot b$$
$$Z = b \cdot c$$

Show that one of the two following arrangements must occur:

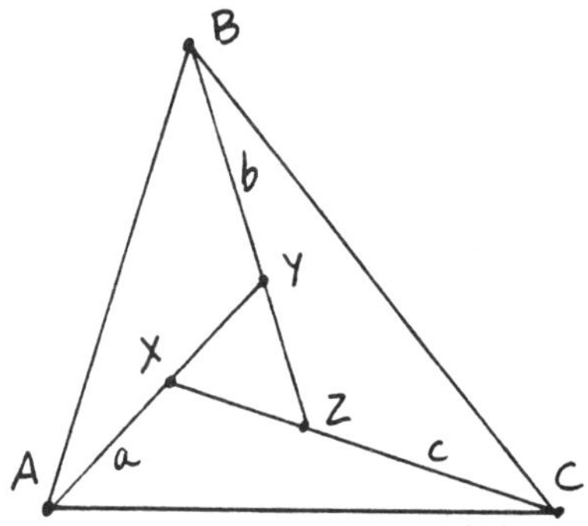

(i) $A*X*Y$ and $B*Y*Z$ and $C*Z*X$ (shown in diagram), or

(ii) $A*Y*X$ and $B*Z*Y$ and $C*X*Z$.

8 Axioms of Congruence for Line Segments

To the earlier undefined notions of point, line, and betweenness, and to the earlier axioms (I1)–(I3), (B1)–(B4), we now add an undefined notion of congruence for line segments, and further axioms (C1)–(C3) regarding this notion. This congruence is what Euclid called equality of segments. We postulate an undefined notion of *congruence*, which is a relation between two line segments $\overline{AB}$ and $\overline{CD}$, written $AB \cong CD$. For simplicity we will drop the bars over AB in the notation for a line segment, so long as no confusion can result. This undefined notion is subject to the following three axioms

C1. Given a line segment AB, and given a ray r originating at a point C, there exists a unique point D on the ray r, such that $AB \cong CD$.

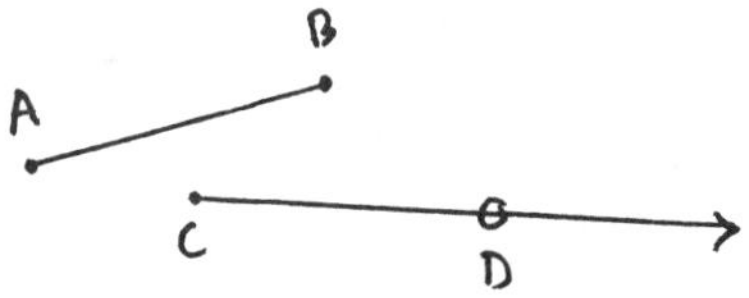

C2. If $AB \cong CD$ and $AB \cong EF$, then $CD \cong EF$. Every line segment is congruent to itself.

C3. (Addition). Given three points A, B, C on a line satisfying $A * B * C$, and three further points D, E, F on a line satisfying $D * E * F$, if $AB \cong DE$ and $BC \cong EF$, then $AC \cong DF$.

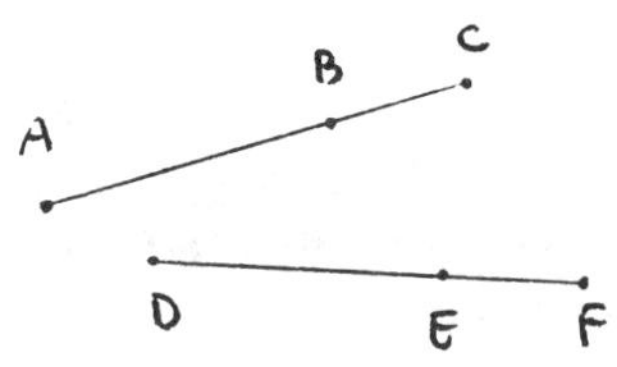

Let us observe how these axioms are similar to Euclid's postulates and how they are different. First of all, while Euclid phrases some of his postulates in terms of constructions ("to draw a line through any two given points", and "to draw a circle with any given center and radius"), Hilbert's axioms are existential. (I1) says for any two distinct points, there exists a unique line containing them. And here, in axiom (C1), it is the existence of the point D (corresponding to Euclid's construction (I.3)) which is taken as an axiom. Hilbert does not make use of ruler and compass constructions. In their place he puts the axiom (C1) of the existence of line segments and later (C4) the existence of angles. If you like, you can think of (C1) and (C4) as being tools, a "transporter of segments" and a "transporter of angles", and consider some of Hilbert's theorems as constructions with these tools.

The second congruence axiom (C2) corresponds to Euclid's common notion that "things equal to the same thing are equal to each other". This is one part of the modern notion of an equivalence relation, so to be comfortable in using congruence, let us show that it is indeed an equivalence relation.

Proposition 8.1 *Congruence is an equivalence relation on the set of line segments.*

Proof. To be an equivalence relation, congruence must satisfy three properties.

1) *Reflexivity*: every segment is congruent to itself. This is explicitly stated in (C2). And by the way, this corresponds to Euclid's fourth common notion that "things which coincide with each other are equal to each other".

2) *Symmetry*: if $AB \cong CD$, then $CD \cong AB$. This is a consequence of (C2): given $AB \cong CD$, and writing $AB \cong AB$ by reflexivity, we conclude from (C2) that $CD \cong AB$.

3) *Transitivity*: if $AB \cong CD$ and $CD \cong EF$, then $AB \cong EF$. This follows by first using symmetry to show $CD \cong AB$, and then applying (C2). Notice that Hilbert's formulation of (C2) was a clever way of including symmetry and transitivity in a single statement.

The third axiom (C3) is the counterpart of Euclid's second common notion, that "equals added to equals are equal". Let us amplify this by making a precise definition of the sum of two segments, and then showing that sums of congruent segments are congruent.

Definition. Let AB and CD be two given segments. Let r be the ray on the line AB consisting of B and all the points of l on the other side of B from A. Let E be the unique point on the ray r (whose existence is given by (C1)) such that $CD \cong BE$.

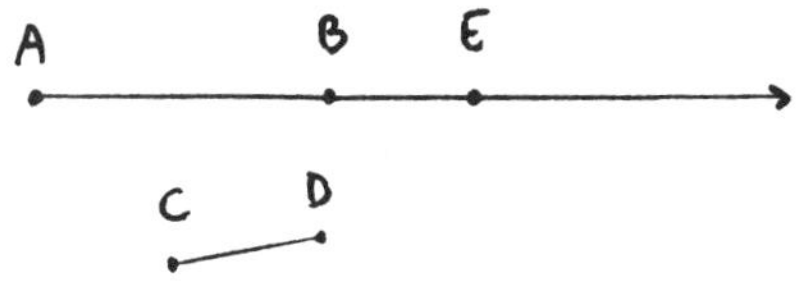

We then define the segment AE to be the *sum* of the segments AB and CD, and we will write $AE = AB + CD$.

Proposition 8.2 (Congruence of sums) *Suppose given segments $AB \cong A'B'$ and $CD \cong C'D'$. Then $AB + CD \cong A'B' + C'D'$.*

Proof. Let E' be the point on the line $A'B'$ defining the sum $A'E' = A'B' + C'D'$. Then $A * B * E$ by construction of the sum $AB + CD$, because E is on the ray from B opposite A. Similarly $A' * B' * E'$. We have $AB \cong A'B'$ by hypothesis. Furthermore we have $CD \cong C'D'$ by hypothesis, and $CD \cong BE$ and $C'D' \cong B'E'$ by construction of E and E'. From (8.1) we know that congruence is an equivalence relation, so $BE \cong B'E'$. Now by (C3) it follows that $AE \cong A'E'$ as required.

It follows from this proposition that addition is well-defined on congruence equivalence classes of line segments. So we can speak of addition of line segments or congruent segments without any danger (cf. also Exercise 8.1 which show that addition of line segments is associative and commutative, up to congruence). Euclid's third common notion is that "equals subtracted from equals are equal". Bearing in mind that subtraction does not always make sense, we can interpret this common notion as follows.

Proposition 8.3 *Given three points A, B, C on a line such that $A * B * C$, and given points E, F on a ray originating from a point D, suppose that $AB \cong DE$ and $AC \cong DF$. Then E will be between D and F, and $BC \cong EF$. (We regard BC as the* difference *of AC and BC.)*

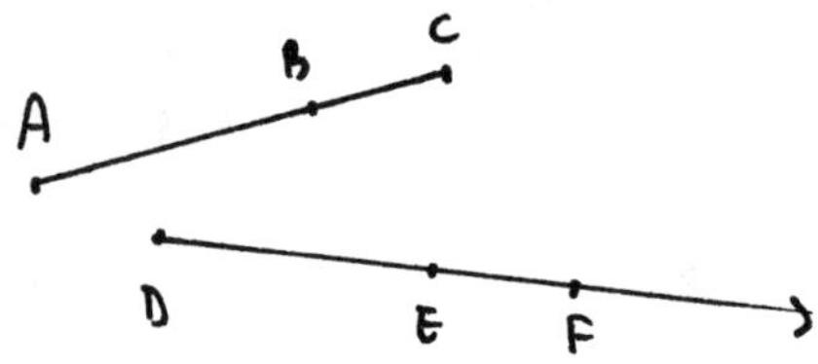

Proof. Let F' be the unique point on the ray originating at E, opposite to D, such that $BC \cong EF'$. Then from $AB \cong DE$ and $BC \cong EF'$ we conclude by (C3) that $AC \cong DF'$. But F and F' are on the same ray from D (check!) and also $AC \cong DF$, so by (C2) and the uniqueness part of (C1), we conclude that $F = F'$. It follows that $D * E * F$ and $BC \cong EF$, as required.

Note the role played by the uniqueness part of (C1) in the above proof. We can regard this uniqueness as corresponding to Euclid's fifth common notion, "the whole is greater than the part". Indeed, this statement could be interpreted as meaning, if $A * B * C$, then AB cannot be congruent to AC. And indeed, this follows from (C1), because B and C are on the same ray from A, and if $AB \cong AC$, then B and C would have to be equal by (C1).

So we see that Euclid's common notions, at least in the case of congruence of line segments, can be deduced as consequences of the new axioms (C1)–(C3). Another notion used by Euclid without definition is the notion of inequality of line segments. Let us see how we can define the notions of greater and lesser also using our axioms.

Definition. Let AB and CD be given line segments. We will say AB is *less than* CD, written $AB < CD$, if there exists a point E in between C and D, such that $AB \cong CE$. In this case we say also CD is *greater than* AB, written $CD > AB$.

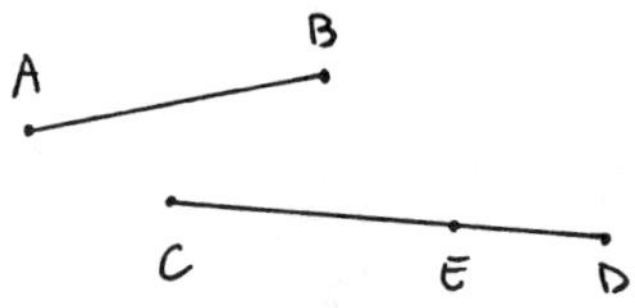

In the next proposition, we will see that this notion of less than is compatible with congruence, and gives an order relation on congruence equivalence classes of line segments.

Proposition 8.4 a) *Given line segments $AB \cong A'B'$ and $CD \cong C'D'$, then $AB < CD$ if and only if $A'B' < C'D'$.*

b) *The relation $<$ gives an order relation on line segments up to congruence, in the following sense*

(i) *If $AB < CD$, and $CD < EF$, then $AB < EF$.*

(ii) *Given two line segments AB, CD, one and only one of the three following conditions holds: $AB < CD$ or $AB \cong CD$, or $AB > CD$.*

Proof. a) Given $AB \cong A'B'$ and $CD \cong C'D'$, suppose that $AB < CD$. Then there is a point E such that $AB \cong CE$ and $C * E * D$. Let E' be the unique point on the ray $\overrightarrow{C'D'}$ such that $CE \cong C'E'$. It follows from (8.3) that $C' * E' * D'$. Furthermore, by transitivity of congruence, $A'B' \cong C'E'$, so $A'B' < C'D'$ as required. The "if and only if" statement follows by applying the same argument starting with $A'B' < C'D'$.

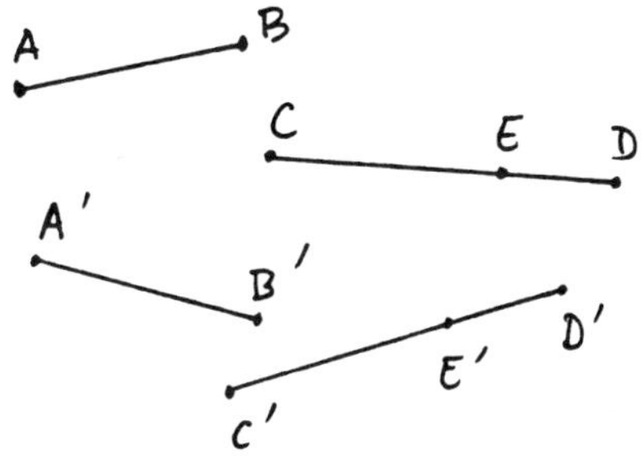

b) (i) Suppose given $AB < CD$ and $CD < EF$. Then by definition, there is a point $X \in \overline{CD}$ such that $AB \cong CX$, and there is a point $Y \in \overline{EF}$ such that $CD \cong EY$. Let $Z \in \overrightarrow{EF}$ be such that $CX \cong EZ$. Then by (8.3) we have $E * Z * Y$. It follows that $E * Z * F$ (Exercise 7.1) and that $AB \cong EZ$. Hence $AB < EF$ as required.

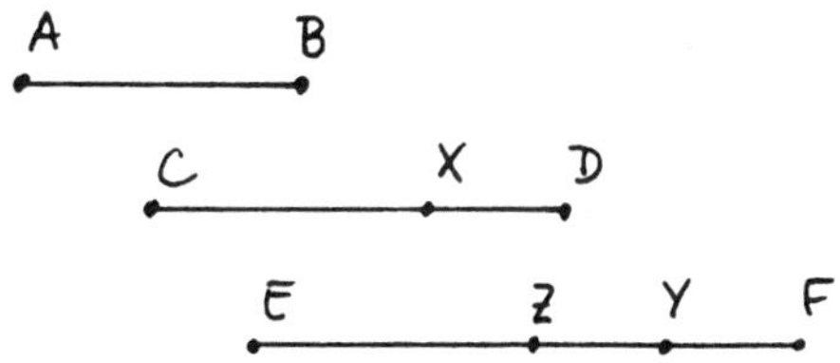

(ii) Given line segments AB and CD, let E be the unique point on the ray $\overrightarrow{CD}$ for which $AB \cong CE$. Then either $D = E$ or $C * E * D$ or $C * D * E$. We cannot have $D * C * E$ because D and E are on the same side of C. These conditions are equivalent to $AB \cong CD$, or $AB < CD$, or $AB > CD$ respectively, and one and only one of them must hold.

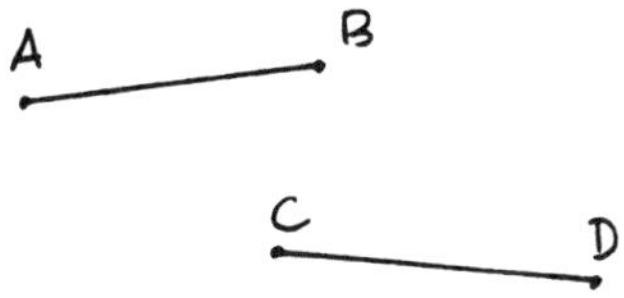

Example 8.4.1 Let us show how to define congruence for line segments in the real Cartesian plane $\mathbb{R}^2$, so that it becomes a model for the axioms (I1)–(I3), (B1)–(B4), and (C1)–(C3) which we have introduced so far. We have already seen how to define lines and betweenness (7.3.1). Given two points $A = (a_1, a_2)$ and $B = (b_1, b_2)$, we define the *distance* $d(A, B)$ by

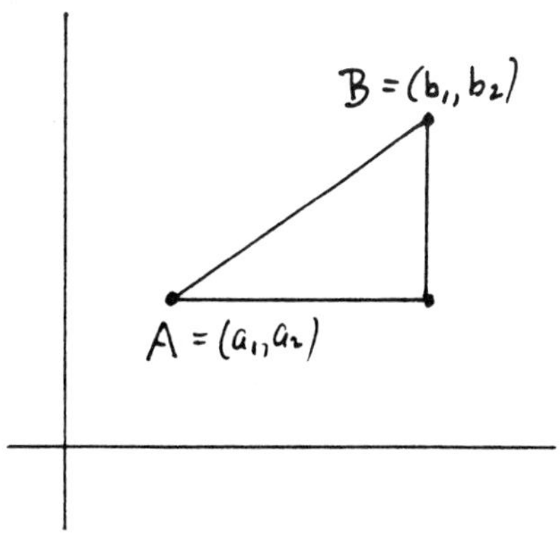

$$d(A, B) = \sqrt{(a_1 - b_1)^2 + (a_2 - b_2)^2}.$$

This is sometimes called the *Euclidean distance* or the *Euclidean metric* on $\mathbb{R}^2$. Note that $d(A, B) \geq 0$, and is $= 0$ only if $A = B$.

Now we can give an interpretation of the undefined notion of congruence in this model by *defining* $AB \cong CD$ if $d(A,B) = d(C,D)$. Let us verify that the axioms (C1), (C2), (C3) are satisfied.

For (C1), we suppose given a segment AB, and let $d = d(A,B)$. We also suppose given a point $C = (c_1, c_2)$ and a ray emanating from C. For simplicity we will assume that the ray has slope $m > 0$ and that it is going in the direction of increasing x-coordinate (we leave the other cases to the reader). Then any point D on this ray has coordinates $D = (c_1 + h, c_2 + mh)$ for some $h \geq 0$. The corresponding distance is

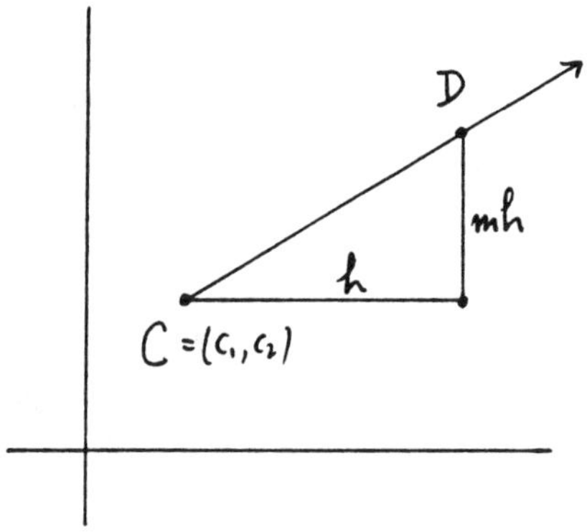

$$d(C,D) = h\sqrt{1+m^2}.$$

To find a point D with $AB \cong CD$ is then equivalent to solving the equation (in a variable $h > 0$)

$$h\sqrt{1+m^2} = d,$$

where m and $d > 0$ are given. Clearly there is a unique solution $h \in \mathbb{R}$, $h > 0$, for given d, m. This proves (C1).

The second axiom (C2) is trivial from the definition of congruence using a distance function.

To prove (C3), it will be sufficient to prove that the distance function is additive for points in a line: if $A * B * C$, then

$$d(A,B) + d(B,C) = d(A,C).$$

Suppose the line is $y = mx + b$, and $A = (a_1, a_2)$ is the point with smallest x-coordinate.

Then there are $h, k > 0$ so that

$$B = (a_1 + h, a_2 + mh)$$
$$C = (a_1 + h + k, a_2 + m(h+k)).$$

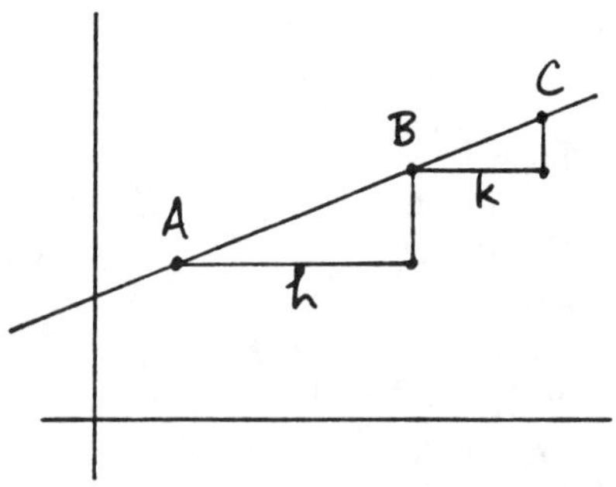

In this case

$$d(A,B) = h\sqrt{1+m^2}$$
$$d(B,C) = k\sqrt{1+m^2}$$
$$d(A,C) = (h+k)\sqrt{1+m^2},$$

so the additivity of the distance function follows.

We will sometimes call this model, the real Cartesian plane with congruence of segments defined by the Euclidean distance function, the *standard model* of our axiom system.

Exercises

8.1 a) Show that addition of line segments is associative: given segments AB, CD, EF, then $(AB+CD)+EF = AB+(CD+EF)$. (This means we obtain the same segment as the sum, not just congruent segments.)

b) Show that addition of line segments is commutative up to congruence: given segments AB, CD, then $AB + CD \cong CD + AB$.

8.2 Show that "halves of equals are equal" in the following sense: if $AB \cong CD$, and if E is a *midpoint* of AB in the sense that $A * E * B$ and $AE \cong EB$, and if F is a midpoint of CD, then $AE \cong CF$. (Note that we have not yet said anything about the existence of a midpoint: that will come later.) Conclude that a midpoint of AB, if it exists, is unique.

8.3 Show that addition preserves inequalities: if $AB < CD$ and if EF is any other segment, then $AB + EF < CD + EF$.

8.4 Let r be a ray originating at a point A, and let s be a ray originating at a point B. Show that there is a 1–1 mapping $\varphi : r \to s$ of the set r onto the set s, which preserves congruence and betweenness. In other words, if for any $X \in r$, we let $X' = \varphi(X) \in s$, then for any $X, Y, Z \in r$, $XY \cong X'Y'$, and $X * Y * Z \Leftrightarrow X' * Y' * Z'$.

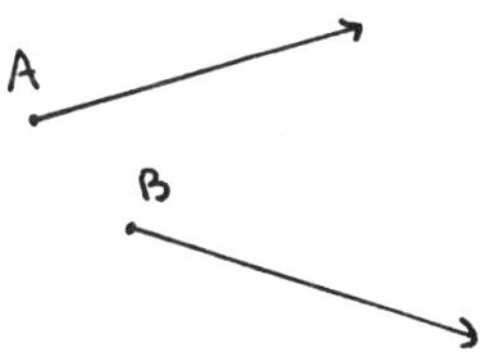

8.5 Given two distinct points O, A, we define the *circle* with *center* O and *radius* OA to be the set Γ of all points B such that $OA \cong OB$.

a) Show that any line through O meets the circle in exactly two points.

b) Show that a circle contains infinitely many points. (Warning: It is not obvious from this definition whether the center O is uniquely determined by the set of points Γ which form the circle. We will prove that later.)

8.6 Consider the *rational Cartesian plane* $\mathbb{Q}^2$ whose points are ordered pairs of rational numbers, where lines are defined by linear equations with rational coefficients and betweenness and congruence are defined as in the standard model (7.3.1) and (8.4.1). Verify that (I1)–(I3) and (B1)–(B4) are satisfied in this model. Then show that (C2) and (C3) hold in this model, but (C1) fails.

8.7 Consider the real Cartesian plane $\mathbb{R}^2$, with lines and betweenness as before (7.3.1), but define a different notion of congruence of line segments using the distance function given by the sum of the absolute values:

$$d(A, B) = |a_1 - b_1| + |a_2 - b_2|,$$

where $A = (a_1, a_2)$ and $B = (b_1, b_2)$. Some people call this "taxicab geometry" because it is similar to the distance by taxi from one point to another, in a city where all streets run east-west or north-south. Show that the axioms (C1), (C2), (C3) hold, so that this is another model of the axioms introduced so far. What does the circle with center $(0, 0)$ and radius 1 look like in this model?

8.8 Again consider the real Cartesian plane $\mathbb{R}^2$, and define a third notion of congruence for line segments using the sup of absolute values for distance function:

$$d(A, B) = \sup\{|a_1 - b_1|, |a_2 - b_2|\}.$$

Show that (C1), (C2), (C3) are also satisfied in this model. What does the circle with center $(0, 0)$ and radius 1 look like in this case?

8.9 Following our general principles, we say two models M, M' of our geometry are *isomorphic* if there exists a 1–1 mapping $\varphi : M \to M'$ of the set of points of M onto the set of points of M', denoted $\varphi(A) = A'$, which sends lines into lines, preserves betweenness, i.e. $A * B * C$ in $M \Leftrightarrow A' * B' * C'$ in M', and preserves congruence of line segments, i.e. $AB \cong CD$ in $M \Leftrightarrow A'B' \cong C'D'$ in M'.

Show that the models of Exercise 8.7 and Exercise 8.8 above are isomorphic to each other, but they are not isomorphic to the standard model (8.4.1). Note: To show the two models of Exercise 8.7 and Exercise 8.8 are isomorphic, you do not necessarily need to make the distance functions correspond. It is only the notion of congruence of line segments which must be preserved. To show that two models are not isomorphic, one method is to find some statement which is true in one model but not true in the other model.

9 Axioms of Congruence for Angles

Recall that we have defined an *angle* to be the union of two rays originating at the same point, and not lying on the same line. We postulate an undefined notion of *congruence* for angles, written $\cong$, which is subject to the following three axioms

C4. Given an angle $\angle BAC$ and given a ray $\overrightarrow{DF}$, there exists a unique ray $\overrightarrow{DE}$, on a given side of the line DF, such that $\angle BAC \cong \angle EDF$.

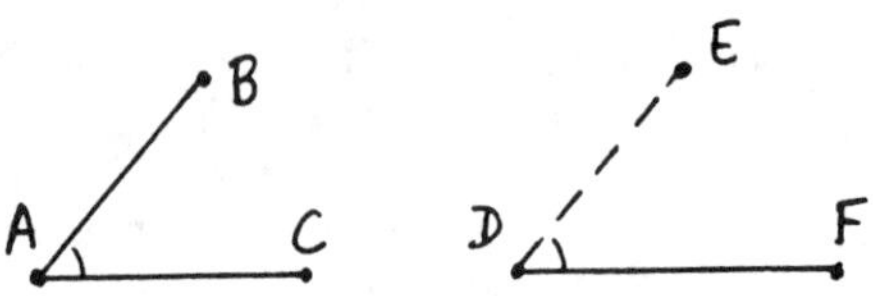

C5. For any three angles α, β, γ, if $\alpha \cong \beta$ and $\alpha \cong \gamma$, then $\beta \cong \gamma$. Every angle is congruent to itself.

C6. (SAS) Given triangles ABC and DEF, suppose that $AB \cong DE$ and $AC \cong DF$, and $\angle BAC \cong \angle EDF$. Then the two triangles are congruent, namely, $BC \cong EF$, $\angle ABC \cong \angle DEF$ and $\angle ACB \cong \angle DFE$.

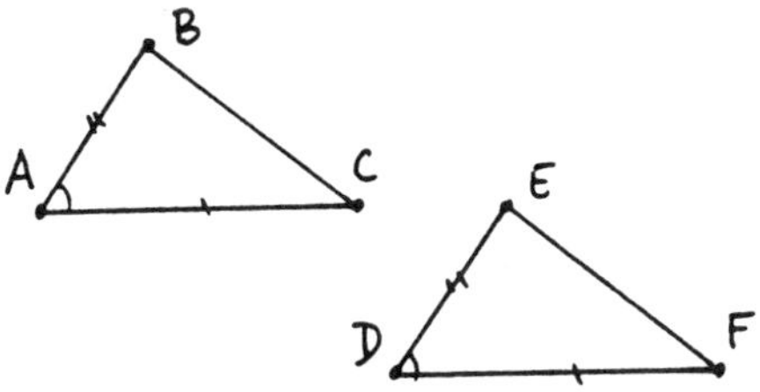

Note that Hilbert takes the existence of an angle congruent to a given one (C4) as an axiom, while Euclid proves this by a ruler and compass construction (I.23). Since Hilbert does not make use of the compass, we may regard this axiom as a tool, the "transporter of angles", which acts as a substitute for the compass.

As with (C2), we can use (C5) to show that congruence is an equivalence relation.

Proposition 9.1 *Congruence of angles is an equivalence relation.*

Proof. The proof is identical to the proof of (8.1), using (C5) in place of (C2).

As in the case of congruence of line segments, we would like to make sense of Euclid's common notions in the context of congruence of angles. This proposition (9.1) is the analogue of the first common notion, that "things equal to the same thing are equal to each other". The second common notion, that "equals added to equals are equal" becomes problematic in the case of angles, because in general we cannot define the sum of two angles.

If $\angle BAC$ is an angle, and if a ray $\overrightarrow{AD}$ lies in the interior of the angle $\angle BAC$, then we will say that the angle $\angle BAC$ is the *sum* of the angles $\angle DAC$ and $\angle BAD$.

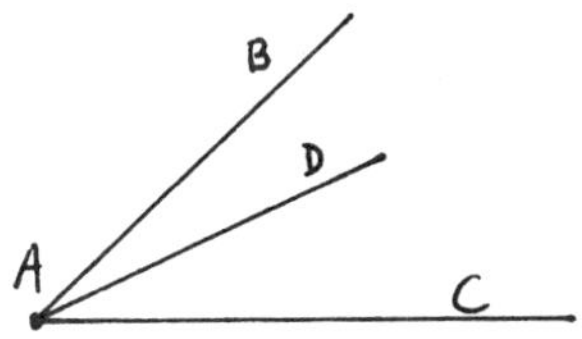

However, if we start with the two given angles, there may not be an angle which is their sum in this sense. For one thing, they may add up to a straight line, or "two right angles" as Euclid says, but this is not an angle. Or, their sum may be greater than 180°, in which case we get an angle, but the two original angles will not be in the interior of the new angle. So we must be careful how we state results having to do with sums of angles.

Note that we do not have an axiom about congruence of sums of angles analogous to the axiom (C3) about addition of line segments. That is because we can prove the corresponding result for angles. But in order to do so, we will need (C6).

Hilbert's use of (C6) = (SAS) as an axiom is a recognition of the difficulties in Euclid's proof of that result (I.4) using the method of superposition. To justify the method of superposition by introducing axioms allowing motion of figures in the plane would be foreign to Euclid's approach to geometry, so it seems prudent to take (C6) as an axiom. On the other hand, one can show that this axiom is necessary, since it is independent of the other axioms (Exercise 9.3). This axiom is essentially what tells us that our plane is homogeneous: geometry is the same at different places in the plane.

So now let us show how to deal with sums of angles and inequalities among angles, based on these axioms.

Definition. If $\angle BAC$ is an angle, and if D is a point on the line AC on the other side of A from C, then the angles $\angle BAC$ and $\angle BAD$ are *supplementary*.

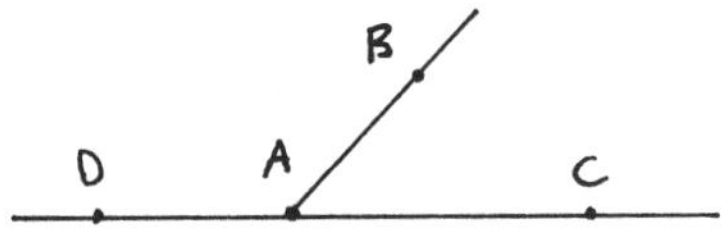

Proposition 9.2 *If $\angle BAC$ and $\angle BAD$ are supplementary angles, and if $\angle B'A'C'$ and $\angle B'A'D'$ are supplementary angles, and if $\angle BAC \cong \angle B'A'C'$, then also $\angle BAD \cong \angle B'A'D'$.*

Proof. Replacing B', C', D' by other points on the same rays, we may assume that $AB \cong A'B'$, $AC \cong A'C'$ and $AD \cong A'D'$. Draw the lines BC, BD, $B'C'$, and $B'D'$.

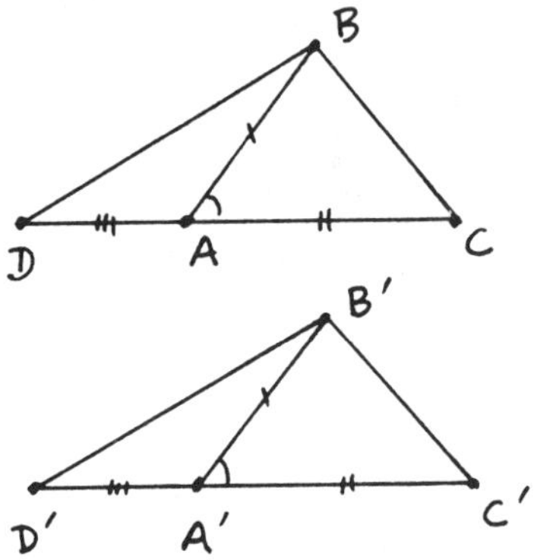

First we consider the triangles ABC and $A'B'C'$. By hypothesis we have $AB \cong A'B'$ and $AC \cong A'C'$ and $\angle BAC \cong \angle B'A'C'$. So by (C6) we conclude that the triangles are congruent. In particular, $BC \cong B'C'$ and $\angle BCA \cong \angle B'C'A'$.

Next we consider the triangles BCD and $B'C'D'$. Since $AC \cong A'C'$ and $AD \cong A'D'$, and $C*A*D$ and $C'*A'*D'$, we conclude from (C3) that $CD \cong C'D'$. Using $BC \cong B'C'$ and $\angle BCA \cong \angle B'C'A'$ proved above, we can apply (C6) again to see that the triangles BCD and $B'C'D'$ are congruent. In particular, $BD \cong B'D'$ and $\angle BDA \cong \angle B'D'A'$.

Now we consider the triangles BDA and $B'D'A'$. From the previous step we have $BD \cong B'D'$ and $\angle BDA \cong \angle B'D'A'$. But by hypothesis we have $DA \cong D'A'$. So a third application of (C6) shows that the triangles BDA and $B'D'A'$ are congruent. In particular, $\angle BAD \cong \angle B'A'D'$, which was to be proved.

Note: We may think of this result as a replacement for (I.13) which says that the angles made by a ray standing on a line are either right angles or are equal to two right angles. We have difficulty making sense of Euclid's statement, since the sum of two right angles is not an angle. However, the way Euclid uses (I.13) in applications can be replaced by an application of our result (9.2). So for example we have the

Corollary 9.3 *Vertical angles are congruent.*

Proof. The vertical angles α and α' are each supplementary to β, and β is congruent to itself, so by the proposition, α and α' are congruent.

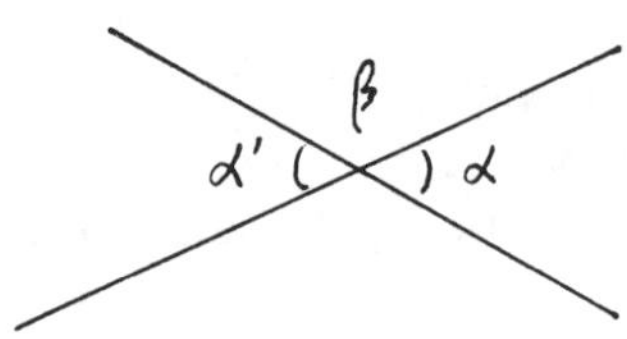

Proposition 9.4 (Addition of angles) *Suppose $\angle BAC$ is an angle, and the ray $\overrightarrow{AD}$ is in the interior of the angle $\angle BAC$. Suppose $\angle D'A'C' \cong \angle DAC$, and $\angle B'A'D' \cong \angle BAD$, and the rays $\overrightarrow{A'B'}$ and $\overrightarrow{A'C'}$ are on opposite sides of the line $A'D'$. Then*

the rays $\overrightarrow{A'B'}$ and $\overrightarrow{A'C'}$ form an angle, and $\angle B'A'C' \cong \angle BAC$, and the ray $\overrightarrow{A'D'}$ is in the interior of the angle $\angle B'A'C'$. For short, we say "sums of congruent angles are congruent".

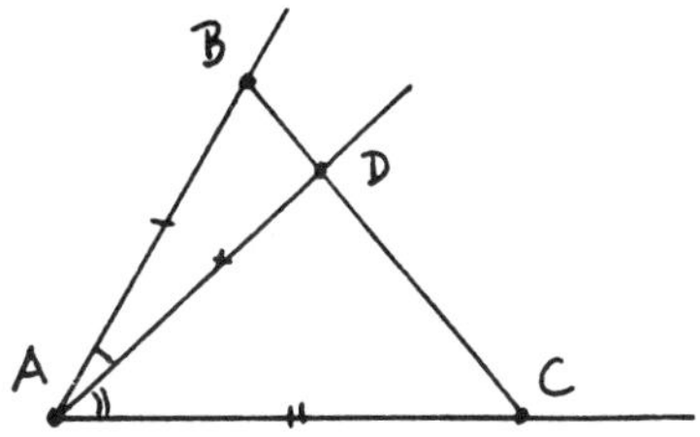

Proof. Draw the line BC. Then the ray $\overrightarrow{AD}$ must meet the segment $\overline{BC}$, by the crossbar theorem (7.3). Replacing the original D by this intersection point, we may assume that B, D, C lie on a line and $B * D * C$. On the other hand, replacing B', C', D' by other points on the same rays, we may assume that $AB \cong A'B'$, and $AC \cong A'C'$, and $AD \cong A'D'$. We also have $\angle BAD \cong \angle B'A'D'$ and $\angle DAC \cong \angle D'A'C'$ by hypothesis.

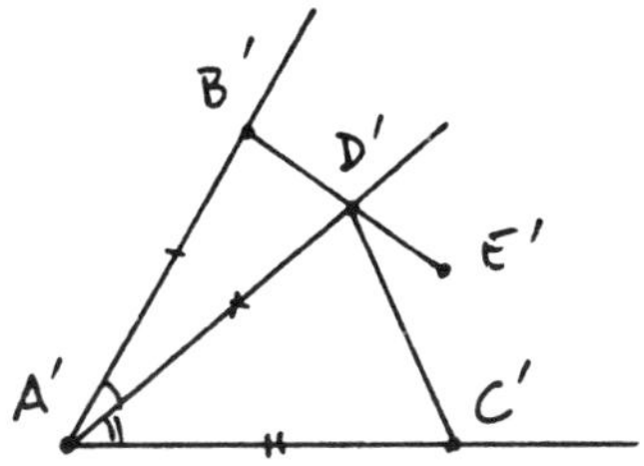

By (C6) we conclude that the triangles ΔBAD and $\Delta B'A'D'$ are congruent. In particular, $BD \cong B'D'$ and $\angle BDA \cong \angle B'D'A'$.

Again by (C6) we conclude that the triangle ΔDAC and $\Delta D'A'C'$ are congruent. In particular, $DC \cong D'C'$, and $\angle ADC \cong \angle A'D'C'$.

Let E' be a point on the line $B'D'$ with $B' * D' * E'$. Then $\angle A'D'E'$ is supplementary to $\angle A'D'B'$, which is congruent to $\angle ADB$. So by (9.2) and transitivity of congruence, we find that $\angle A'D'E' \cong \angle A'D'C'$. Since these angles are on the same side of the line $A'D'$, we conclude from the uniqueness part of (C4) that they are the same angle. In other words, the three points B', D', and C' lie on a line.

Then from (C3) we conclude that $BC \cong B'C'$. Since $\angle ABD \cong \angle A'B'D'$ by the first congruence of triangles used in the earlier part of the proof, we can apply (C6) once more to the triangles ABC and $A'B'C'$. The congruence of these triangles implies $\angle BAC \cong \angle B'A'C'$ as required. We have already used the fact that A', B', C' are not collinear. This is because we showed $B'D'C'$ are collinear. Thus $B'C'$ meets the line $A'D'$ at D', so it cannot contain A'. Also by construction, we see that $B' * D' * C'$, so that the ray $A'D'$ will be in the interior of the angle $\angle B'A'C'$ as required.

Next, we will define a notion of inequality for angles analogous to the inequality for line segments in §8.

Definition. Suppose given angles $\angle BAC$ and $\angle EDF$. We say $\angle BAC$ is *less than* $\angle EDF$, written $\angle BAC < \angle EDF$, if there exists a ray $\overrightarrow{DG}$ in the interior of the angle $\angle EDF$, such that $\angle BAC \cong \angle GDF$. In this case we will also say $\angle EDF$ is *greater than* $\angle BAC$.

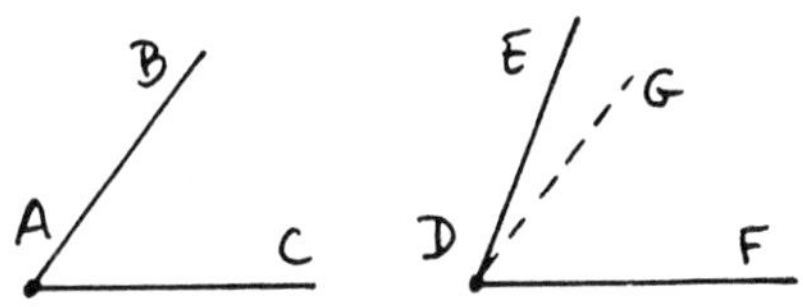

Proposition 9.5 a) *If $\alpha \cong \alpha'$ and $\beta \cong \beta'$, then $\alpha < \beta \Leftrightarrow \alpha' < \beta'$.*

b) *Inequality gives an order relation on angles, up to congruence. In other words,*

(i) *If* $\alpha < \beta$ *and* $\beta < \gamma$, *then* $\alpha < \gamma$.

(ii) *For any two angles* α *and* β, *one and only one of the following holds:* $\alpha < \beta$ *or* $\alpha \cong \beta$ *or* $\alpha > \beta$.

Proof. The proofs of these statements are essentially the same as the corresponding statements for line segments (8.4) so we will leave them to the reader.

Definition. A *right angle* is an angle α which is congruent to one of its supplementary angles β.

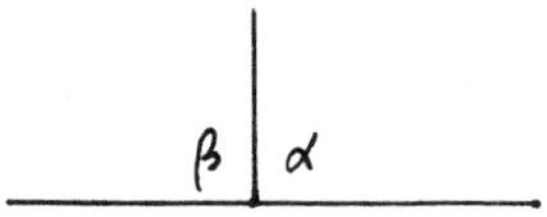

Proposition 9.6 *Any two right angles are congruent to each other.*

Proof. Suppose that $\alpha = \angle CAB$ and $\alpha' = \angle C'A'B'$ are right angles. Then they will be congruent to their supplementary angles β, β', by definition. Suppose α and α' are not congruent. Then by (9.5) either $\alpha < \alpha'$ or $\alpha' < \alpha$. Suppose for example $\alpha < \alpha'$. Then by definition of inequality there is a ray $\overrightarrow{A'E'}$ in the interior of angle α' so that $\alpha \cong \angle E'A'B'$.

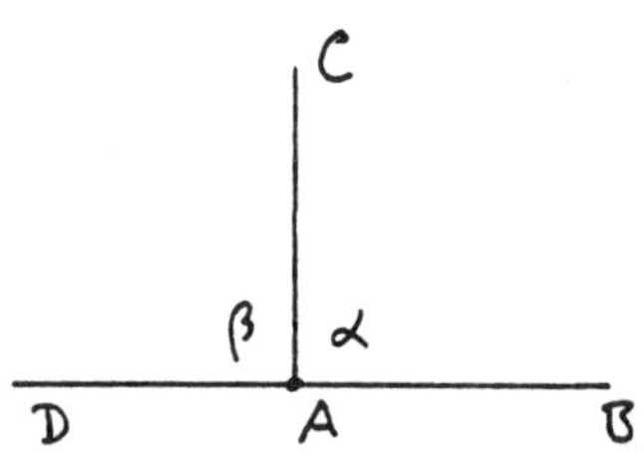

It follows (check!) that the ray $\overrightarrow{A'C'}$ is in the interior of $\angle E'A'D'$, so that $\beta' < \angle E'A'D'$. But $\angle E'A'D'$ is supplementary to $\angle E'A'B'$ which is congruent to α, so by (9.2), $\angle E'A'D' \cong \beta$. Therefore $\beta' < \beta$. But $\alpha \cong \beta$ and $\alpha' \cong \beta'$, so we conclude that $\alpha' < \alpha$, which is a contradiction, since we earlier supposed $\alpha < \alpha'$.

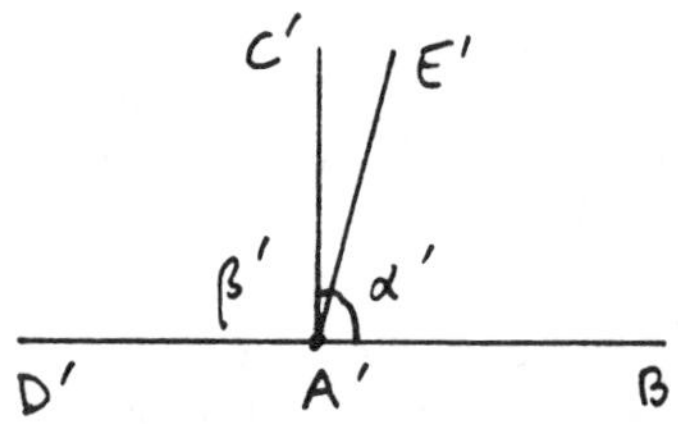

Note: Thus the equality of all right angles is a consequence of the axioms and does not need to be taken as an axiom as Euclid did (Postulate 4). The idea of this proof already appears in Proclus.

Example 9.6.1 We will show later (§17) that the real Cartesian plane $\mathbb{R}^2$ provides a model of all the axioms listed so far. You are probably willing to believe this, but the precise definition of what we mean by congruence of angles in this model, and the proof that axioms (C4)–(C6) hold, requires some work. We will postpone this work until we make a systematic study of Cartesian planes over arbitrary fields, and investigate just what properties of a field are needed. Ironically, to prove that (C6) holds in the Cartesian plane, we will go back to Euclid's method of superposition, though of course we will have to prove the existence of enough rigid motions for this to make sense.

Exercises

9.1 (Difference of angles). Suppose given congruent angles $\angle BAC \cong \angle B'A'C'$. Suppose also given a ray $\overrightarrow{AD}$ in the interior of $\angle BAC$. Then there exists a ray $\overrightarrow{A'D'}$ in the interior of $\angle B'A'C'$ such that $\angle DAC \cong \angle D'A'C'$ and $\angle BAD \cong \angle B'A'D'$. This statement corresponds to Euclid's Common Notion 3: "equals subtracted from equals are equal" where "equal" in this case means congruence of angles.

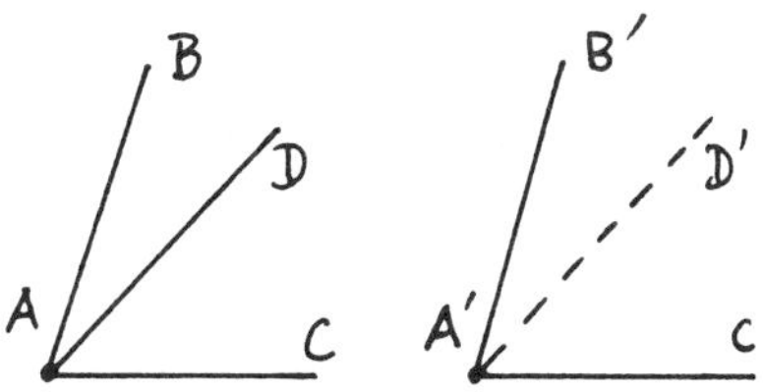

9.2 Suppose the ray $\overrightarrow{AD}$ is in the interior of the angle $\angle BAC$, and the ray $\overrightarrow{AE}$ is in the interior of the angle $\angle DAC$. Show that $\overrightarrow{AE}$ is also in the interior of $\angle BAC$.

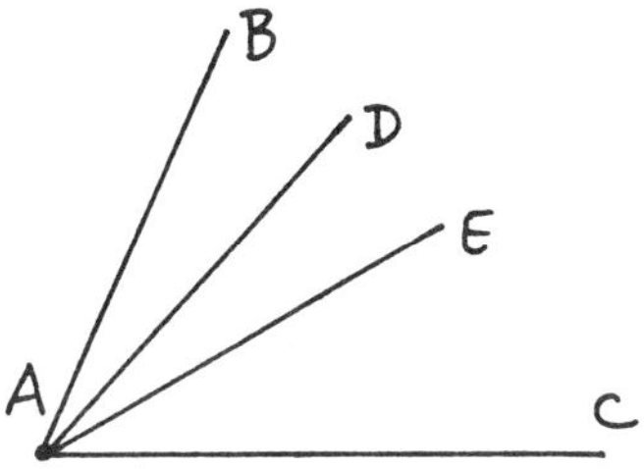

9.3 Consider the real Cartesian plane where congruence of line segments is given by the absolute value distance function (Exercise 8.7). Using the usual congruence of angles you know from analytic geometry, show that (C4) and (C5) hold in this model, but that (C6) fails. (Give a counterexample.)

9.4 Provide the missing betweenness arguments to complete Euclid's proof of (I.7) in the case he considers. Namely, assuming that the ray $\overrightarrow{AD}$ is in the interior of the angle $\angle CAB$, and assuming that D is outside the triangle ABC, prove that $\overrightarrow{CB}$ is in the interior of the angle $\angle ACD$ and $\overrightarrow{DA}$ is the interior of the angle $\angle CDB$.

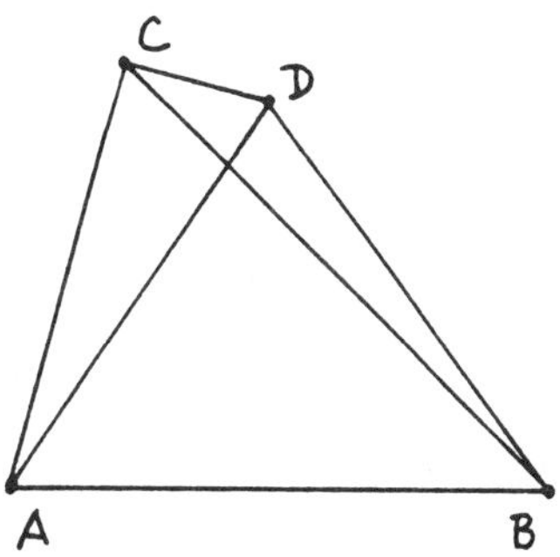

10 Hilbert Planes

We have now introduced the minimum of basic notions and axioms on which to found our study of geometry. To be precise, we will define a *Hilbert plane* to be a given set (of *points*) together with certain subsets called *lines*, and undefined notions of *betweenness*, *congruence* for line segments, and *congruence* for angles (as explained in the preceding sections), which satisfy the axioms (I1)–(I3), (B1)–(B4), and (C1)–(C6).

We could go on immediately and introduce the parallel axiom and axioms of intersection of lines and circles, so as to recover all of Euclid's *Elements*, but it seems worthwhile to pause at this point and see how much of the geometry we can develop on this minimal set of axioms. The main reason for doing this is that the axioms of a Hilbert plane form the basis for non-Euclidean as well as Euclidean geometry. In fact, some people call the Hilbert plane *neutral geometry*, because it neither affirms nor denies the parallel axiom.

In this section we will see how much of Euclid's Book I we can recover in a Hilbert plane. With two notable exceptions, we can recover everything which does not make use of the parallel postulate.

Let us work in a given Hilbert plane. Euclid's definitions, postulates, and common notions have been replaced by the undefined notions, definitions and axioms which we have discussed so far (excluding Playfair's axiom). We will now discuss the propositions of Euclid, Book I.

The first proposition (I.1) is our first exception! Without some additional axiom, it is not clear that the two circles in Euclid's construction will actually meet. There may be some other way to prove the existence of an equilateral triangle with a given side AB, using only the axioms of a Hilbert plane, but I do not know if that is possible. We will partially fill this gap by showing below (10.2) that there do exist isosceles triangles on a given segment.

Euclid's Propositions (I.2) and (I.3), about transporting line segments, are effectively replaced by axiom (C1). Proposition (I.4), (SAS), has been replaced by axiom (C6).

Proposition (I.5) and its proof are ok as they stand. In other words, every step of Euclid's proof can be justified in a straightforward manner within the framework of a Hilbert plane. To illustrate this process of reinterpreting one of Euclid's proofs within our new axiom system, let us look at Euclid's proof step by step.

Proof of (I.5). Let ABC be the given isosceles triangle, with $AB \cong AC$ (congruent line segments). We must prove that the base angles $\angle ABC$ and $\angle ACB$ are congruent. "In BD take any point F". This is possible by axiom (B2). "On AE cut off AG equal to AF". This is possible by (C1). Now $AC \cong AB$ and $AF \cong AG$ and the enclosed angle $\angle BAC$ is the same, so the triangles ΔAFC and ΔAGB are congruent by a direct application of (C6). So $FC \cong GB$ and $\angle AFC \cong \angle AGB$ and $\angle ACF \cong \angle ABG$.

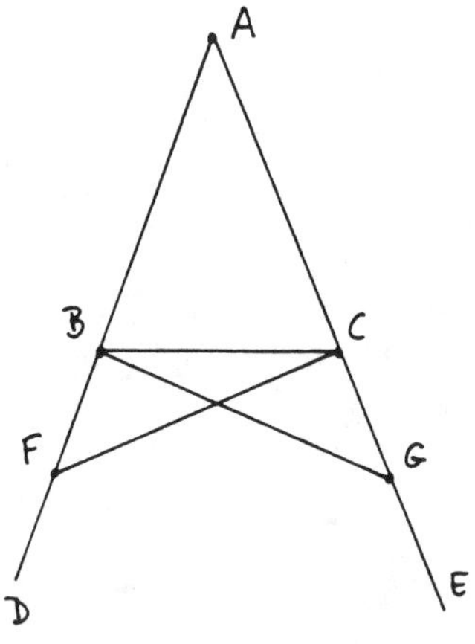

Since "equals subtracted from equals are equal", referring in this case to congruence of line segments, we conclude from (8.3) that $BF \cong CG$. Then by another application of (C6), the triangles ΔFBC and ΔGCB are congruent. It follows that $\angle CBG \cong \angle BCF$. Now by subtraction of congruent angles (Exercise 9.1), the base angles $\angle ABC$ and $\angle ACB$ are congruent, as required. (We omit the proof of the second assertion which follows similarly.)

At certain steps in this proof we need to know something about betweenness, which can also be formally proved from our axioms. For example, in order to subtract the line segment AB from AF, we need to know that B is between A and F. This follows from our choice of F. At the last step, subtracting angles, we need to know that the ray $\overrightarrow{BC}$ is in the interior of the angle $\angle ABG$. This follows from the fact that C is between A and G.

So in the following, when we say Euclid's proof is ok as is, we mean that each step can be justified in a natural way, without having to invent additional steps of proof, from Hilbert's axioms and the preliminary results we established in the previous sections.

Looking at (I.6), the converse of (I.5), everything is ok except for one doubtful step at the end. Euclid says "the triangle DBC is equal to the triangle ACB, the less to the greater; which is absurd". It is not clear what this means, since we have not defined a notion of inequality for triangles. However, a very slight change will give a satisfactory proof. Namely, from the congruence of the triangles $\Delta DBC \cong \Delta ACB$, it follows that $\angle DCB \cong \angle ABC$. But also $\angle ABC \cong \angle ACB$ by hypothesis. So $\angle DCB \cong \angle ACB$, "the less to the greater" as Euclid would say. For us, this is a contradiction of the uniqueness part of axiom (C4), since there can only be one angle on the same side of the ray $\overrightarrow{CB}$ congruent to the angle $\angle ACB$. We conclude that the rays $\overrightarrow{CA}$ and $\overrightarrow{CD}$ are equal, so $A = D$, and the triangle is isosceles, as required.

Proposition (I.7), as we have mentioned before, needs some additional justification regarding the relative positions of the lines, which can be supplied from our axioms of betweenness (Exercise 9.4).

For (I.8), (SSS), we will need a new proof, since Euclid's method of superposition cannot be justified from our axioms. The following proof is due to Hilbert.

Proposition 10.1 (SSS) *If two triangles ABC and $A'B'C'$ have their respective sides equal, namely $AB \cong A'B'$, $AC \cong A'C'$, and $BC \cong B'C'$, then the two triangles are congruent.*

Proof. Using (C4) and (C1), construct an angle $\angle C'A'B''$ on the other side of the ray $\overrightarrow{A'C'}$ from B', which is congruent to $\angle BAC$, and make $A'B''$ congruent to AB. Then $AB \cong A'B''$ by construction, $AC \cong A'C'$ by hypothesis, and $\angle BAC \cong \angle B''A'C'$ by construction, so by (C6), the triangle ΔABC is congruent to the triangle $\Delta A'B''C'$. It follows that $BC \cong B''C'$.

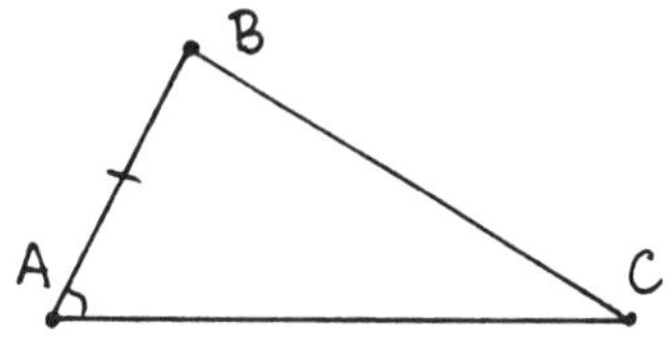

Draw the line $B'B''$. Now $A'B' \cong AB \cong A'B''$, so by transitivity, $A'B' \cong A'B''$. Thus the triangle $A'B'B''$ is isosceles, and so by (I.5) its base angles $\angle A'B'B''$ and $\angle A'B''B'$ are congruent. Similarly, $B'C' \cong B''C'$, so the triangle $C'B'B''$ is isosceles, and its base angles $\angle B''B'C'$ and $\angle B'B''C'$ are congruent. By addition of congruent angles (9.4) it follows that $\angle A'B'C' \cong \angle A'B''C'$.

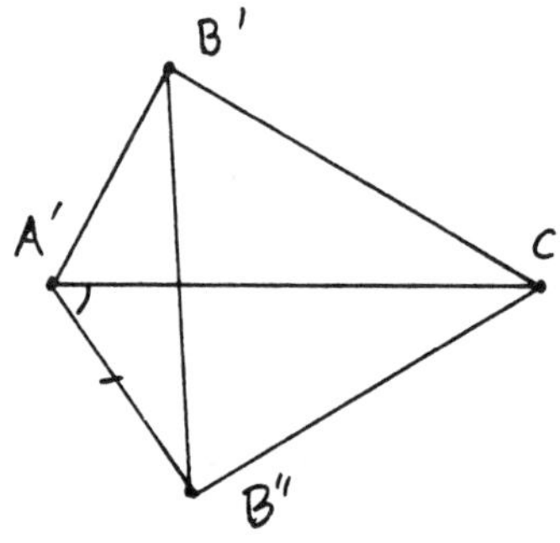

This latter triangle was shown congruent to ΔABC, so $\angle A'B''C' \cong \angle ABC$. Now by transitivity of congruence, $\angle ABC \cong \angle A'B'C'$, so we can apply (C6) again to conclude that the two triangles are congruent.

Note: This proof and the accompanying figure are for the case when A' and C' are on opposite sides of the line $B'B''$. The case when they are on the same side is analogous, and the case when one of A' or C' lies on the line $B'B''$ is easier, and left to the reader.

Starting with the next proposition (I.9) we have a series of constructions with ruler and compass. We cannot carry out these constructions in a Hilbert plane, because we have not yet added axioms to ensure that lines and circles will meet when they ought to. However, we can reinterpret these propositions as existence theorems, and these we can prove from Hilbert's axioms. Since we do not have the equilateral triangles which Euclid constructed in (I.1), we will prove the existence of isosceles triangles, and we will use them as a substitute for equilateral triangles in the following existence proofs.

Proposition 10.2 (Existence of isosceles triangles) *Given a line segment AB, there exists an isosceles triangle with base AB.*

Proof. Let AB be the given line segment. Let C be any point not on the line AB (axiom (I3)). Consider the triangle ΔABC. If the angles at A and B are equal, then ΔABC is isosceles (I.6). If not, then one angle is less than the other. Suppose $\angle CAB < \angle CBA$. Then there is a ray $\overrightarrow{BE}$ in the interior of the angle $\angle CBA$ such that $\angle CAB \cong \angle EBA$.

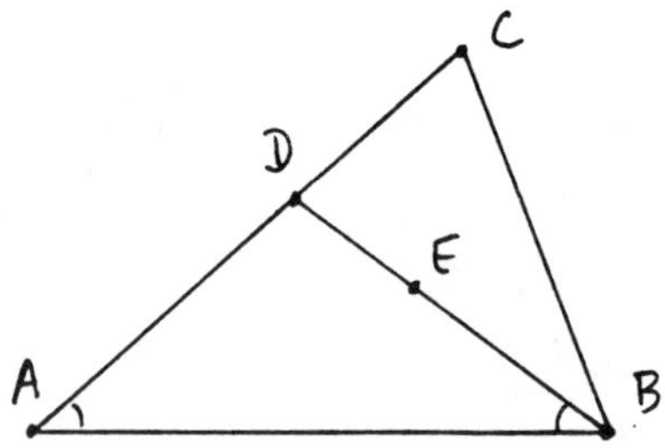

By the crossbar theorem (7.3) this ray must meet the opposite side AC in a point D. Now the base angles of the triangle DAB are equal, so by (I.6) it is isosceles.

Note: It would not suffice to construct equal angles at the two ends of the interval, because without the parallel axiom, even if the angles are small, there is no guarantee that the two rays would meet.

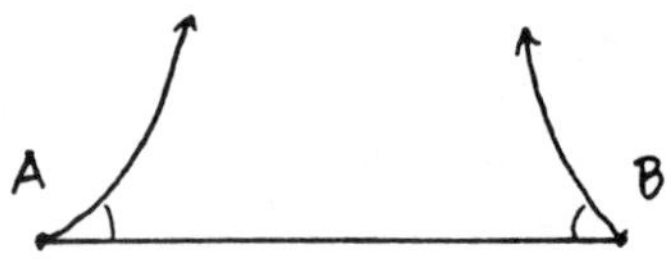

Now let us return to Euclid. We interpret (I.9) as asserting the existence of an angle bisector. We use the same method as Euclid, except that we use (10.2) to give the existence of an isosceles triangle ΔDEF where Euclid used an equilateral triangle. We may clearly assume that the equilateral triangle was constructed on the opposite side of DE from A. Then Euclid's proof, using (SSS) shows that $\angle DAF \cong \angle EAF$. It is not obvious from the construction that the ray $\overrightarrow{AF}$ is in the interior of the angle $\angle DAE$, but it does follow from the conclusion: for if $\overrightarrow{AF}$ was not in the interior of the angle, then $\overrightarrow{AD}$ and $\overrightarrow{AE}$ would be on the same side of $\overrightarrow{AF}$, and in that case the congruence of the angles $\angle DAF \cong \angle EAF$ would contradict the uniqueness in axiom (C4).

For (I.10) to bisect a given line segment, we again use (10.2) to construct an isosceles triangle instead of an equilateral triangle. The rest of Euclid's proof then works to show that a midpoint of the segment exists.

For (I.11) we again use (10.2) and thus show how to construct a line perpendicular to a line at a point. By the way, this also proves the existence of right angles, which is not obvious a priori.

For (I.12), to drop a perpendicular from a point C to a line not containing C, Euclid's method using the compass does not work in a Hilbert plane. We need a new existence proof (see Exercise 10.1).

Proposition (I.13) has been replaced by the result on congruence of supplementary angles (9.2) and (I.14) is an easy consequence (Exercise 10.3). The congruence of vertical angles (I.15) has already been mentioned above (9.3). The theorem on exterior angles (I.16) is sufficiently important that we will reproduce Euclid's proof here, with the extra justifications necessary to make it work.

Proposition 10.3 (Exterior angle theorem (I.16)) *In any triangle, the exterior angle is greater than either of the opposite interior angles.*

Proof. Let ABC be the given triangle. We will show that the exterior angle $\angle ACD$ is greater than the opposite interior angle at A. Let E be the midpoint of AC (I.10), and extend BE to F so that $BE \cong EF$ (axiom (C1)). Draw the line CF. Now the vertical angles at E are equal (I.15) so by SAS (C6), the triangles ΔABE and ΔCFE are congruent. Hence $\angle A \cong \angle ECF$.

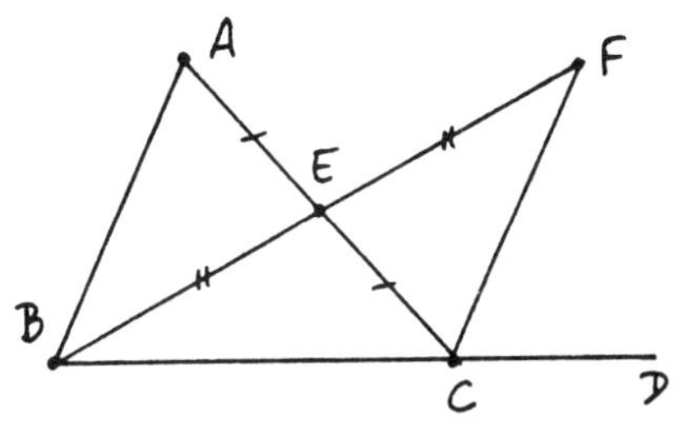

To finish the proof, that is to show that $\angle ECF$ is less than $\angle ACD$, we need to know that the ray $\overrightarrow{CF}$ is in the interior of the angle $\angle ACD$. This we can prove based on our axioms of betweenness. Since D is on the side BC of the triangle extended, B and D are on opposite sides of the line AC. Also, by construction of F, we have B and F on opposite sides of AC. So from the plane separation property (7.1) it follows that D and F are on the same side of the line AC.

Now consider sides of the line BC. Since $B * E * F$, it follows that E and F are on the same side of BC. Since $A * E * C$, it follows that A and E are on the same side of AC. By transitivity (7.1) it follows that A and F are on the same side of the line $BC = CD$. So by definition, F is in the interior of the angle $\angle ACD$, and hence the ray $\overrightarrow{CF}$ is also. Therefore, by definition of inequality for angles, $\angle BAC$ is less than $\angle ACD$, as required.

Propositions (I.17)–(I.21) are all ok as is, except that we should reinterpret the statement of (I.17). Instead of saying "any two angles of a triangle are less than two right angles", which does not make sense in our system, since "two right angles" is not an angle, we simply say: if α and β are any two angles of a triangle, then α is less than the supplementary angle of β.

Proposition (I.22) is our other exception. Without knowing that two circles intersect when they ought to, we cannot prove the existence of the triangle required in this proposition. In fact, we will see later (Exercise 16.9) that there are Hilbert planes in which a triangle with certain given sides satisfying the hypotheses of this proposition does not exist!

The next proposition (I.23), which Euclid proved using (I.22), is replaced by Hilbert's axiom (C4), the "transporter of angles".

The remaining results which Euclid proved without using the parallel postulate are ok as is in the Hilbert plane: (I.24), (I.25), (I.26) = (ASA) and (AAS), (I.27) "alternate interior angles equal implies parallel", and even the existence of parallel lines (I.31).

Thus using the axioms for a Hilbert plane we have been able to recover almost everything in the first part of Euclid's Book I.

Exercises

10.1 Given a line l and a point C not on l, prove the existence of a line through C, perpendicular to l, using only Hilbert's axioms and their consequences.

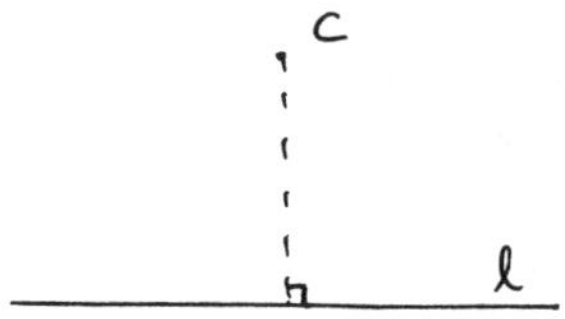

10.2 Write out a careful proof of Euclid (I.18), justifying every step in the context of a Hilbert plane, and paying especial attention to questions of betweenness and inequalities.

10.3 Rewrite the statement (I.14) so that it makes sense in a Hilbert plane, and then give a careful proof.

10.4 Write a careful proof of (I.20) in a Hilbert plane.

10.5 Show that the right-angle-side-side congruence theorem (RASS) holds in a Hilbert plane: If ABC and $A'B'C'$ are triangles with right angles at B and B', and if $AB \cong A'B'$ and $AC \cong A'C'$, then the triangles are congruent.

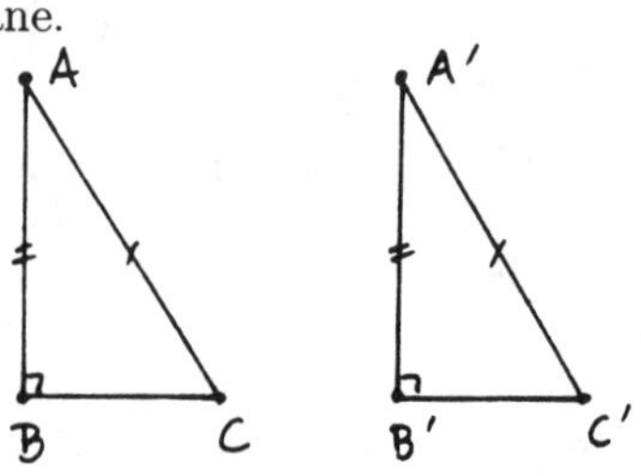

10.6 Given a finite set of points $A_1, \dots, A_n$ in a Hilbert plane, prove that there exists a line l for which all the points are on the same side of l.

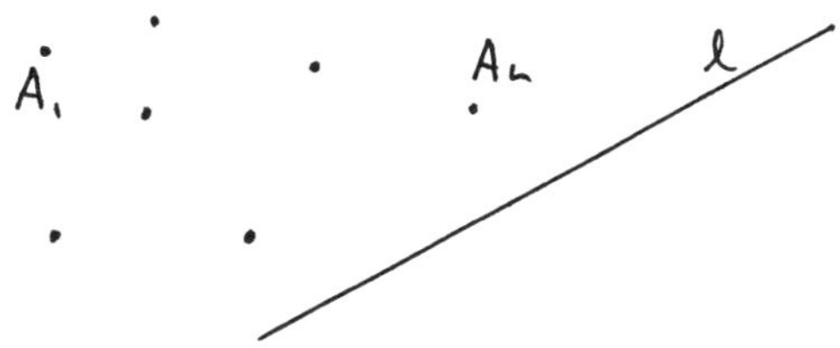

11 Intersections of Lines and Circles

In this section we will discuss the intersections of lines and circles in the Hilbert plane, and we will introduce the further axiom (E) which will guarantee that lines and circles will intersect when they "ought" to. With this axiom we can justify Euclid's ruler and compass constructions in Book I and Book III.

Definition. Given distinct points O, A, the *circle* Γ with center O and radius OA is the set of all points B such that $OA \cong OB$. The point O is the *center* of the circle. The segment OA is a *radius*.

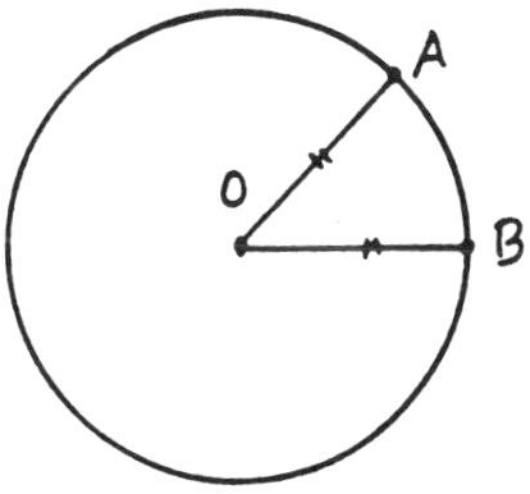

From this definition it is clear that a circle always has points. The point A is on the circle. Moreover, if l is any line through O, then by axiom (C1) there will be exactly two points on the line l, one on each side of O, lying on the circle. It is not obvious from the definition that the center is uniquely determined by the set of points of the circle.

Proposition 11.1 *Let Γ be a circle with center O and radius OA, and let Γ' be a circle with center O' and radius $O'A'$. Suppose $\Gamma = \Gamma'$ as point sets. Then $O = O'$. In other words the center of a circle is uniquely determined.*

Proof. Suppose $O \neq O'$. Then we consider the line l through O and O'. Since it passes through the center O of Γ, it must meet Γ in two points C, D, satisfying $C * O * D$, and $OC \cong OD$.

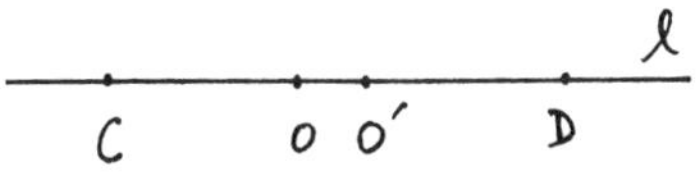

Since $\Gamma = \Gamma'$, the points C, D are also on Γ', so we have $O'C \cong O'D$ and $C * O' * D$. We do not know which of O or O' is close to C, but the two cases are symmetric, so let us assume $C * O * O'$. In this case we must have $O * O' * D$ by the properties of betweenness(!). Then $OC < O'C \cong O'D < OD$, which is impossible since $OC \cong OD$. Hence $O = O'$.

Now that we know the center of a circle is uniquely determined, it makes sense to define the inside and the outside of a circle.

Definition. Let Γ be a circle with center O and radius OA. A point B is *inside* Γ (or in the *interior* of Γ) if $B = O$ or if $OB < OA$. A point C is *outside* Γ (or *exterior* to Γ) if $OA < OC$.

Definition. We say a line l is *tangent* to a circle Γ if l and Γ meet in just one point A. We say a circle Γ is *tangent* to another circle Δ if Γ and Δ have just one point in common.

This definition of tangent circles is a little different from Euclid's: his definition of two circles touching is that they meet in a point but do not cut each other. Since it is not clear what he means by "cut", we prefer the definition above, and we will prove that these notions of tangency have the usual properties.

Proposition 11.2 *Let Γ be a circle with center O and radius OA. The line perpendicular to the radius OA at the point A is tangent to the circle, and (except for the point A) lies entirely outside the circle. Conversely, if a line l is tangent to Γ at A, then it is perpendicular to OA. In particular, for any point A of a circle, there exists a unique tangent line to the circle at that point.*

Proof. First, let l be the line perpendicular to OA at A. Let B be any other point on the line l. Then in the triangle OAB, the exterior angle at A is a right angle, so the angles at O and at B are less than a right angle (I.16). It follows (I.19) that $OB > OA$, so B is outside the circle. So we see that every other point of l lies outside the circle. In particular, l meets Γ only at the point A, so it is a tangent line.

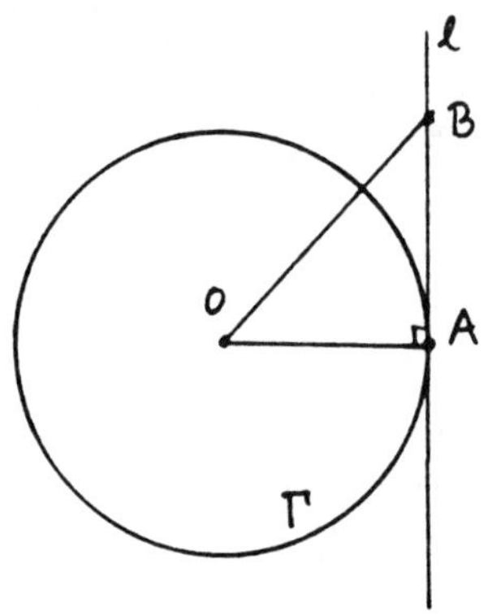

Now suppose that l is a line tangent to Γ at A, we must show l is perpendicular to OA. It cannot be equal to OA, because that line meets Γ in another point opposite A. So consider the line from O, perpendicular to l, meeting l at B. If $B \neq A$, take a point C on the other side of B from A, so that $AB \cong BC$ (axiom (C1)). Then $\Delta OBA \cong \Delta OBC$ by SAS, so we find $OA \cong OC$, and hence C is also on Γ. Since $C \neq A$, this is a contradiction. We conclude $B = A$ and so l is perpendicular to OA.

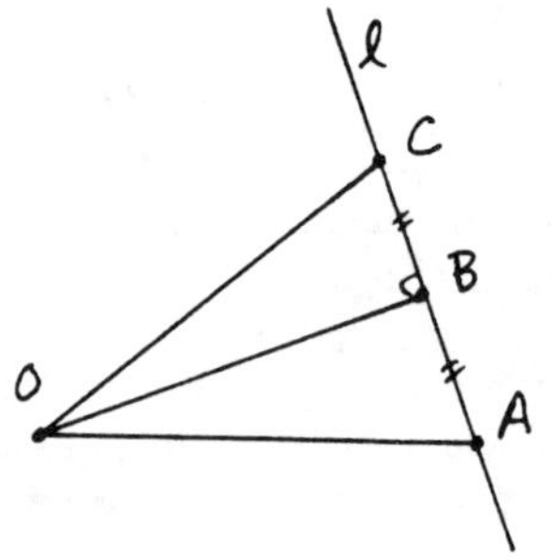

Corollary 11.3 *If a line l contains a point A of a circle Γ, but is not tangent to Γ, then it meets Γ in exactly two points.*

Proof. If l is not tangent to Γ at A, then it is not perpendicular to OA, in which case, as we saw in the previous proof, it meets Γ in another point C. We must show that l cannot contain any further points of Γ. For if D were another point of l on Γ, then $OD \cong OA$, OB is congruent to itself, so by (RASS) (Exercise 10.5) we would have $\Delta ODB \cong OAB$. Then $AB \cong BD$, so by axiom (C1) D must be equal to A or C.

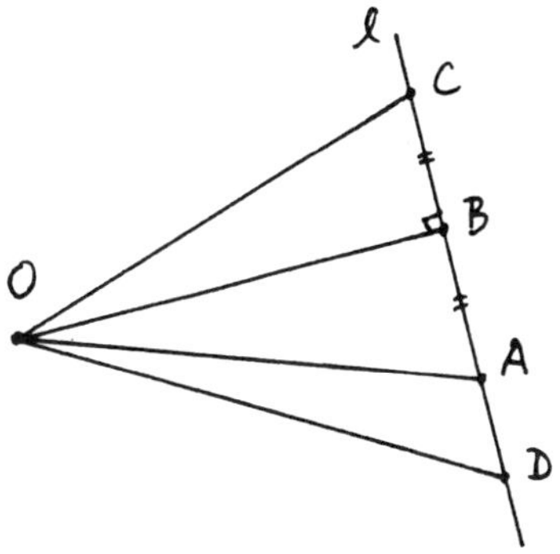

Proposition 11.4 *Let O, O', A be three distinct collinear points. Then the circle Γ with center O and radius OA is tangent to the circle Γ' with center O' and radius $O'A$. Conversely, if two circles Γ, Γ' are tangent at a point A, then their centers O, O' are collinear with A.*

Proof. Let O, O', A be collinear. We must show that the circles Γ and Γ' have no further points in common besides A. The argument of (11.1) shows that there is no other point on the line OO' which lies on both Γ and Γ'. So suppose there is a point B not on OO' lying on both Γ and Γ'. We divide into two cases depending on the relative position of O, O', and A.

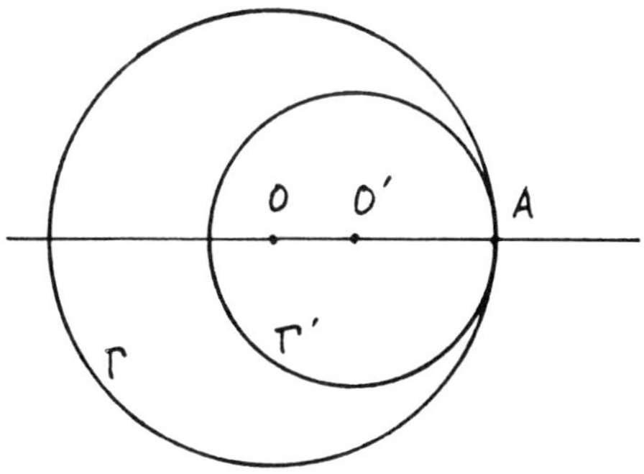

Case 1. $O * O' * A$. Since $OA = OB$, $\angle OAB \cong \angle OBA$. Also since $O'A = O'B$, $\angle O'AB \cong \angle O'BA$, using (I.5). It follows that $\angle OBA \cong \angle O'BA$ which contradicts axiom (C4). (This argument also applies if $O' * O * A$.)

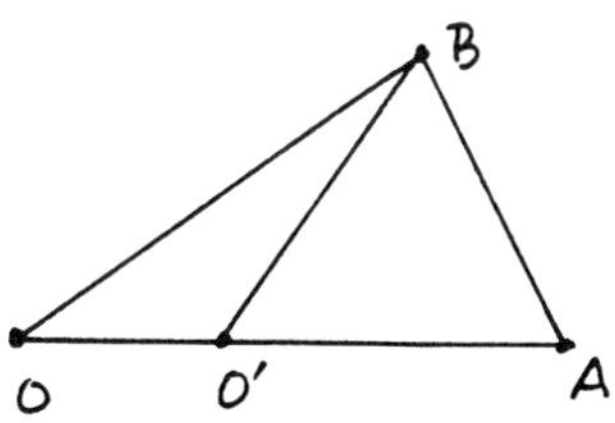

Case 2. $O * A * O'$. Again using (I.5) we find $\angle OAB \cong \angle OBA$ and $\angle O'AB \cong \angle O'BA$. But the two angles at A are supplementary, so it follows that the two angles at B are supplementary (9.2). But then O, B, and O' would be collinear, which is a contradiction.

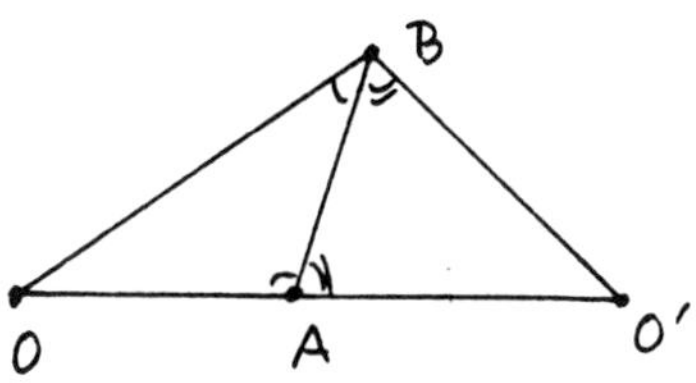

Conversely, suppose that Γ and Γ' are tangent at A, and suppose that O, O', A are not collinear. Then we let AC be perpendicular to the line OO', and choose B on the line AC on the other side of OO' with $AC \cong BC$. It follows by congruent triangles that $OA \cong OB$ and $O'A \cong O'B$, so B also lies on Γ and Γ', contradicting the hypothesis Γ tangent to Γ'. We conclude that O, O', A are collinear.

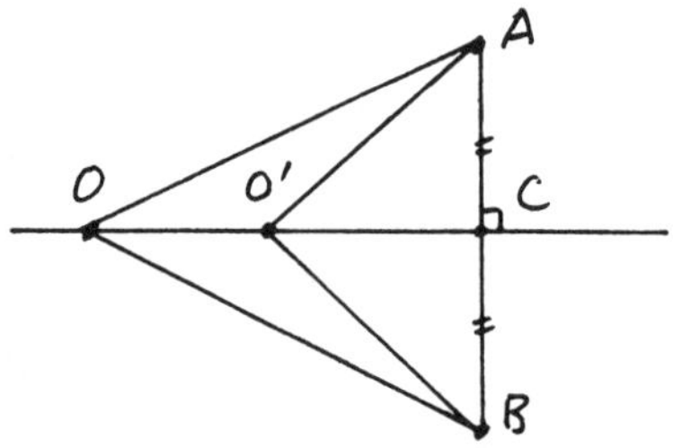

Corollary 11.5 *If two circles meet at a point A but are not tangent, then they have exactly two points in common.*

Proof. We have seen above that if they are not tangent, then O, O', A are not collinear, and they meet in an additional point B. We must show there are no further intersection points. If D is a third point on Γ and Γ', then $OD \cong OA$ and $O'D \cong O'A$, so by (I.7), D must be equal to A or B.

In the above discussion of lines and circles meeting, we have seen that a line and a circle, or two circles, can be tangent (meeting in just one point), or if they meet but are not tangent, they will meet in exactly two points. There is nothing here to guarantee that a line and a circle, or two circles, will actually meet if they are in a position so that they "ought" to meet according to the usual intuition. For this we need an additional axiom (and we will see later (§16) that this axiom is independent of the axioms of a Hilbert plane).

E. (Circle-circle intersection property). Given two circles Γ, Δ, if Δ contains at least one point inside Γ, and Δ contains at least one point outside Γ, then Γ and Δ will meet. (*Note*: It follows from Exercise 11.3 and (11.5) that they will then meet in exactly two points.)

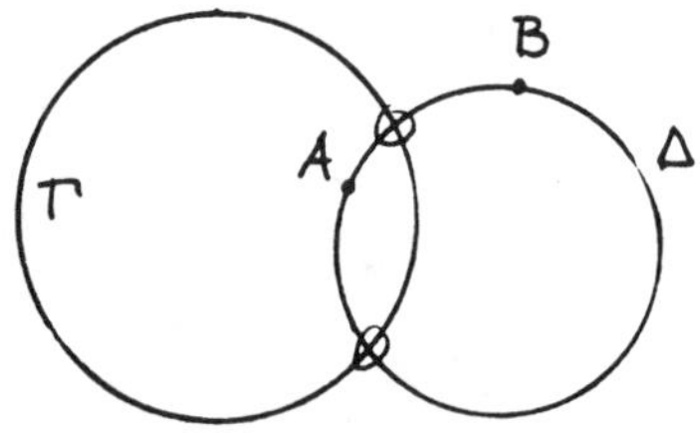

Proposition 11.6 (Line-circle intersection property LCI) *In a Hilbert plane with the extra axiom* (E), *if a line l contains a point A inside a circle* Γ, *then l will meet* Γ *(necessarily in two points, because of* (11.2) *and* (11.3)*).*

Proof. Suppose given the line L with a point A inside the circle Γ. Our strategy is to construct another circle Δ, show that Δ meets Γ, and then show the intersection point also lies on l. Let OB be the perpendicular from O to l (if O is on the line l, we already know l meets Γ by (C1)). Find a point O' on the other side of l from O, on the line OB, with $O'B \cong OB$. Let Δ be the circle with center O' and radius r = radius of Γ.

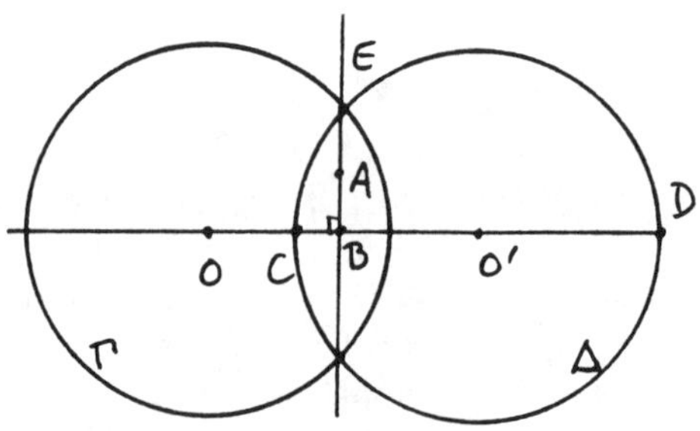

N O T E S.

PROP. XXII. B. I.

Some Authors blame Euclid becaufe he does not demonftrate that the two circles made ufe of in the conftruction of this Problem fhall cut one another. but this is very plain from the determination he has given, viz. that any two of the ftraight lines DF, FG, GH muft be greater than the third. for who is fo dull, tho' only beginning to learn the Elements, as not to perceive that the circle defcribed from the centre F, at the diftance FD, muft meet FH betwixt F and H, becaufe FD is leffer than FH; and that, for the like reafon, the circle defcribed from the centre G, at the diftance GH or GM muft meet DG betwixt D and G; and that thefe circles muft meet one another, becaufe FD and GH are together greater than FG? and this determination is eafier to be underftood than that which Mr. Thomas Simpfon derives from it, and puts inftead of Euclid's, in the 49. page of his Elements of Geometry, that he may fupply the omiffion he blames Euclid for; which determination is, that any of the three ftraight lines muft be leffer than the fum, but greater than the difference of the other two. from this he fhews the circles muft meet one another, in one cafe; and fays that it may be proved after the fame manner in any other cafe. but the ftraight line GM which he bids take from GF may be greater than it, as in the figure here annexed, in which cafe his demonftration muft be changed into another.

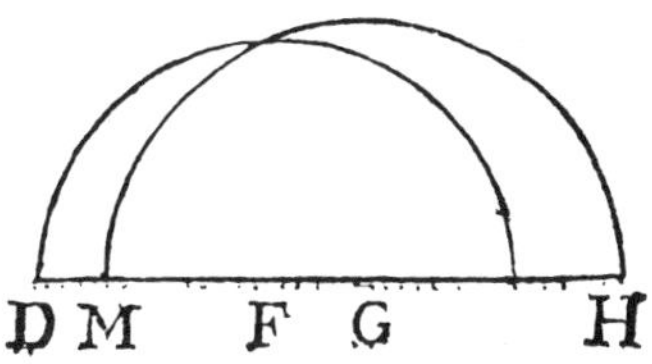

Plate IV. A note on (I.22) from Simson's Euclid (1756).

Now the line OO' meets Δ in two points C, D, where we choose them such that O, C are on the same side of O', and D on the opposite side.

By hypothesis, A is a point on l, inside Γ. Hence $OA < r$. In the right triangle OAB, using (I.19) we see $OB < OA$, so $OB < r$. It follows that $O'B < r = O'C$, so O' and C are on opposite sides of l. Hence O, C are on the same side of l. We wish to show C is inside Γ. There are two cases.

Case 1. If $O * C * B$ then $OC < OB < r$, so C is inside Γ.

Case 2. If $C * O * B$, then also $C * O * O'$, so $OC < O'C = r$, and again we see C is inside Γ. On the other hand, the point D satisfies $O * O' * D$, so $OD > O'D = r$, so D is outside Γ.

Now we can apply the axiom (E) to conclude that Γ meets Δ at a point E. We must show that E lies on l. We know $OE \cong r \cong O'E$ and $OB \cong O'B$ by construction, and BE is equal to itself, so by (SSS) $\Delta OEB \cong \Delta O'EB$. It follows that the angles at B are equal, so they are right angles, so BE is equal to the line l, and so E lies on l and Γ, as required.

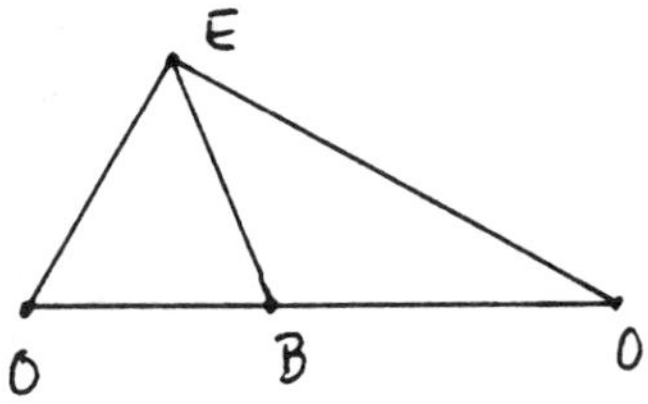

Remark 11.6.1. We will see later (§16) that in the Cartesian plane over a field, the circle-circle intersection property is equivalent to the line-circle intersection property. It seems likely that these two statements may be equivalent in any Hilbert plane, but I do not know how to prove it.

Using the new axiom (E) we can now justify Euclid's first construction (I.1), the equilateral triangle. Given the segment AB, let Γ be the circle with center A and radius AB. Let Δ be the circle with center B and radius BA. Then A is on the circle Δ, and it is inside Γ because it is the center of Γ. The line AB meets Δ in another point D, such that $A * B * D$. Hence $AD > AB$, so D is outside Γ.

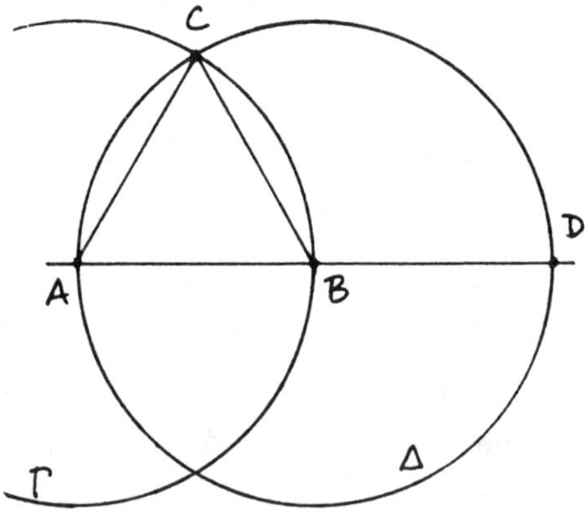

Thus Δ contains a point inside Γ and a point outside Γ, so it must meet Γ in a point C. From here, Euclid's proof shows that ΔABC is an equilateral triangle.

In a similar way one can justify Euclid's other ruler and compass constructions in Book I. Several of them depend only on using the equilateral triangle constructed in (I.1). For (I.12) and (I.22) see the exercises.

We can also justify the results of Euclid, Book III up through (III.19) (since (III.20) and beyond need the parallel axiom). The statements (III.10), (III.11), (III.12) about circles meeting, and (III.16), (III.18), (III.19) about tangent lines can be replaced by the propositions of this section. In (III.14) Euclid uses (I.47) to prove (RASS), but that is not necessary: one can prove it with only the axioms of a Hilbert plane (Exercise 10.5). For (III.17), to draw a tangent to a circle from a point outside the circle, we need the line-circle intersection property (11.6) and hence the axiom (E). (Note that the other popular construction of the tangent line using (III.31) requires the parallel axiom!) The other results of Book III, up to (III.19) (except (III.17)) are valid in any Hilbert plane, and their proofs are ok as is, except as noted.

Exercises

11.1 a) The interior of a circle Γ is a *convex* set: namely, if B, C are in the interior of Γ, and if D is a point such that $B * D * C$, then D is also in the interior of Γ.

b) If B, C are two points outside a circle Γ, then there exists a third point D such that the segments BD and DC are entirely outside Γ. (This implies that the exterior of Γ is a *segment-connected* set.)

11.2 Two circles Γ, Γ' which meet at a point A are tangent if and only if the tangent line to Γ at A is equal to the tangent line to Γ' at A.

11.3 If two circles Γ and Δ are tangent to each other at a point A, show that (except for the point A), Δ lies either entirely inside Γ or entirely outside Γ.

11.4 Use the line-circle intersection property (11.6) to give a careful justification of Euclid's construction (I.12) of a line from a point perpendicular to a given line.

11.5 Given three line segments such that any two taken together are greater than the third, use (E) to justify Euclid's construction (I.22) of a triangle with sides congruent to the three given segments.

11.6 Show that Euclid's construction of the circle inscribed in a triangle (IV.4) is valid in any Hilbert plane. Conclude that the three angle bisectors of a triangle meet in a point.

11.7 Using (E), show that Euclid's construction of a hexagon inscribed in a circle (IV.15) makes sense. Without using (P) or results depending on it, which sides can you show are equal to each other?

12 Euclidean Planes

Let us look back at this point and see how well Hilbert's axioms have fulfilled their goal of providing a new solid base for developing Euclid's geometry. The major problems we found with Euclid's method have been settled: questions of relative position of figures have been clarified by the axioms of betweenness; the problematic use of the method of superposition has been replaced by the device of taking SAS as an axiom; the existence of points needed in ruler and compass constructions is guaranteed by the circle-circle intersection property stated as axiom (E). Also in the process of rewriting the foundations of geometry we have disengaged a new notion, the *Hilbert plane*, which provides a minimum context in which to develop the beginnings of a geometry, free from the parallel axiom, which serves as a basis for Euclidean geometry, and also later, for the non-Euclidean geometries.

In this section we will complete the earlier sections by showing how adding the parallel axiom allows us to recover almost all of the first four books of Euclid's *Elements.* We will also mention two more axioms, those of Archimedes and of Dedekind, which will be used in some parts of later chapters.

We define a *Euclidean plane* to be a Hilbert plane satisfying the additional axioms (E), the circle-circle intersection property, and (P), Playfair's axiom, also called the parallel axiom. In other words, a Euclidean plane is a set of points with subsets called lines, and undefined notions of betweenness and congruence satisfying the axioms (I1)–(I3), (B1)–(B4), (C1)–(C6), (E), and (P). The Euclidean plane represents our modern formulation of the axiomatic basis for developing the geometry of Euclid's *Elements.*

We have already seen in the earlier sections how to recover those results of Euclid's Books I and III which do not depend on the parallel axiom. The first use of the parallel axiom is in (I.29). Since we have replaced Euclid's fifth postulate by Playfair's axiom, we need to modify Euclid's proofs of a few early results in the theory of parallels.

So for example, to prove (I.29) we proceed as follows. Given two parallel lines l, m, and a transversal line n, we must show that the alternate interior angles α and β are equal. If not, construct a line l' through A making an angle α with n (axiom (C4)). By (I.27), l' will be parallel to m. But then l and l' are two lines through A, parallel to m, so by (P), we must have $l = l'$, hence $\alpha = \beta$.

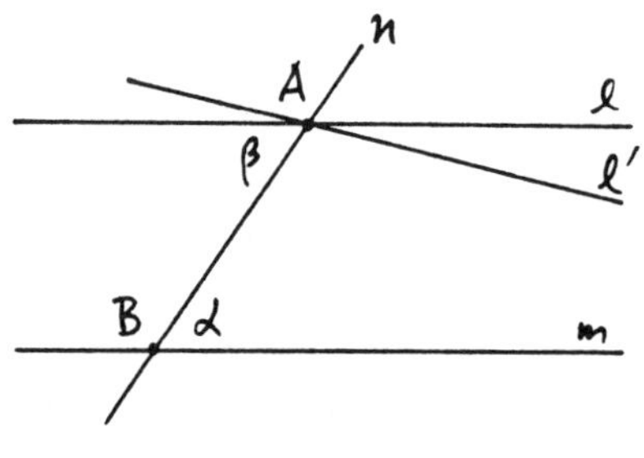

Proposition (I.30) is essentially equivalent to (P). The existence of parallel lines (I.31) follows from (C4) and (I.27) as mentioned before, so now we can reinterpret (I.31) in the stronger form that given a point A not on a line l, there exists a unique parallel to l passing through A. The remaining propositions using (P), namely (I.32)–(I.34) follow without difficulty. In particular we have the famous (I.32), that "the sum of the angles of a triangle is equal to two right angles", though if we want to be scrupulous, we would have to say that sum is not defined, and rephrase the theorem by saying that the sum of any two angles of a triangle is supplementary to the third angle.

Starting with (I.35), and continuing to the end of Book I and through Book II is Euclid's theory of area. Since Euclid does not define what he means by this new equality, we must presume that he takes it as another undefined notion, which we call *equal content*, just as the notion of congruence for line segments and angles were taken as undefined notions. Since Euclid freely applies the common notions to this concept, we may say that he has taken the common notions applied to equal content as further axioms, for example "figures having equal content to a third figure have equal content to each other", or "halves of figures of equal content have equal content".

Hilbert showed in his Grundlagen der Geometrie that it is not necessary to take the notion of equal content as a further undefined notion subject to further axioms. He shows instead that it is possible to *define* the notion of equal content for figures (by cutting them up, rearranging, and adding and subtracting), and that one can *prove* the properties suggested by Euclid's common notions. We will present this work in Chapter 5. For the moment we simply acknowledge that it is possible to provide the necessary definitions and preliminary results to justify Euclid's treatment of area within the axiomatic framework of a Euclidean plane. We will refer to this work roughly as the "theory of area", and note which results of Euclid's Books I–IV depend on it. The principal ways this theory is used is for the Pythagorean theorem (I.47), the construction of the golden ratio (II.11), the results on chords and tangents of circles (III.35)–(III.37), and in the construction of the regular pentagon (IV.10), (IV.11).

In Book III, the first use of the parallel axiom is in (III.20), that the angle at the center of a circle subtending a given arc is twice the angle on the circumference subtending the same arc. This result uses (I.32), that the exterior angle of a triangle is equal to the sum of the two opposite interior angles, and thus depends on the parallel axiom (P). The remaining results of Book III, which depend on this one, namely (III.21)–(III.34) follow without significant modification.

Most of the results of Book IV require the parallel axiom (P), some need circle-circle intersection (E), and some, notably (IV.10), (IV.11), require (P), (E), and the theory of area. Thus we may regard the construction of the regular pentagon as the crowning result of the first four books of the *Elements*, making use of all the results developed so far.

We end this section with a discussion of two further axioms which are not needed for Books I–IV, but will be used later. The first is *Archimedes' axiom.*

A. Given line segments AB and CD, there is a natural number n so that n copies of AB added together will be greater than CD.

This axiom is used implicitly in the theory of proportion developed in Book V, for example in Definition 4, where he says quantities have a ratio, when one can be multiplied to exceed the other. It appears explicitly in Book X, Proposition 1, in the form reminiscent of the ε-arguments of calculus, saying that given two quantities, AB and CD, if we remove from AB more than its half, and again from the remainder remove more than its half, and continue in this fashion, then eventually we will have a quantity less than CD. In modern texts this would appear as the statement "given any $\varepsilon > 0$, there is an integer n sufficiently large so that $1/2^n < \varepsilon$". Euclid applies this method to the study of the volume of three-dimensional figures in Book XII with what is called the method of exhaustion. When he cannot compare solids by cutting into a finite number of pieces and reassembling, he uses a limiting process where the solid is represented as a union of a sequence of sub-solids so that the remainder can be made as small as you like.

This axiom is independent of the others, so we will see examples of *archimedean* geometries which satisfy (A) and *non-archimedean* geometries which do not (§18).

The other axiom we would like to consider is *Dedekind's axiom*, based on Dedekind's definition of the real numbers in the late 19th century:

D. Suppose the points of a line l are divided into two non-empty subsets S, T in such a way that no point of S is between two points of T, and no point of T is between two points of S. Then there exists a unique point (P) such that for any $A \in S$ and any $B \in T$, either $A = P$ or $B = P$ or the point (P) is between A and B.

This axiom is very strong. It implies (A) and (E), and a Euclidean plane with (D) is essentially forced to be isomorphic to the Cartesian plane over the real numbers. (See (Exercise 12.2), (Exercise 12.3), (15.5), and (21.1).) So if you want a categorical axiom system, just add (D) to the axioms of a Euclidean plane. From the point of view of this book, however, there are two reasons to avoid using Dedekind's axiom. First of all, it belongs to the modern development of the real numbers and notions of continuity, which is not in the spirit of Euclid's geometry. Second, it is too strong. By essentially introducing the real numbers into our geometry, it masks many of the more subtle distinctions and obscures questions such as constructibility which we will discuss in Chapter 6. So we include this axiom only to acknowledge that it is there, but with no intention of using it.

Exercises

12.1 Show that in a Euclidean plane, the perpendicular bisectors of the sides of a triangle will meet in a point, and thus justify Euclid's construction of the circumscribed circle of a triangle (IV.5). *Note*: In a non-Euclidean geometry, there may be triangles having no circumscribed circle (41.1).

12.2 Show that in a Hilbert plane Dedekind's axiom (D) implies Archimedes' axiom (A). *Hint*: Given segments AB and CD, let T be the set of all points E on the ray $\overrightarrow{CD}$ for which there is no integer n with $n \cdot AB > CE$. Let S be the set of points of the line CD not in T, and apply (D).

12.3 Show that in Hilbert plane (D) implies (E). *Hint*: Follow the discussion in Heath, vol. I, p. 238.

CHAPTER 3

Geometry over Fields

Beginning with the familiar example of the real Cartesian plane, we show how to construct a geometry satisfying Hilbert's axioms over an abstract field. The axioms of incidence are valid over any field (§14). For the notion of betweenness we need an ordered field (§15). For the axiom (C1) on transferring a line segment to a given ray, we need a property (∗) on the existence of certain square roots in the field F. To carry out Euclidean constructions, we need a slightly stronger property (∗∗)—see §16.

To prove the (SAS) axiom over a field F, we revert to Euclid's method of superposition. In the case of the geometry over a field this can be justified by showing the existence of sufficiently many rigid motions (§17).

We end the chapter with some non-intuitive examples of geometries which do not satisfy Archimedes' axiom (§18).

13 The Real Cartesian Plane

We have seen that the geometry developed in Euclid's *Elements* does not make use of numbers to measure lengths or angles or areas. It is purely geometric in that it deals with points, lines, circles, triangles, and the relationships among these.

In the centuries after Euclid, geometers began using numbers more and more. As techniques of algebra and solving equations become more widespread, a connection between geometry and algebra arose. The first systematic exploitation of this connection occurs in *La Géométrie* of René Descartes (1596–1650), who represented points in the plane by a pair of numbers, and curves in the plane by equations relating these two numbers. This method has grown into the subject now taught as *analytic geometry*, where methods of abstract algebra are used to handle the equations of lines, circles, and other curves in the plane. We refer to the plane with the identification of its points as pairs of numbers as the *Cartesian plane*.

At first there was not much critical discussion of exactly what sort of numbers were allowed, but by the end of the 19th century considerations of limits and continuity made the *real numbers* $\mathbb{R}$ into the standard to be used in geometry, analytic geometry, calculus, and topology.

The geometry taught today in high schools and colleges has thus become a sort of hybrid between the purely geometric methods of Euclid and the analytic geometry of Descartes, with occasional notions of continuity thrown in. One of the purposes of this course is to clarify the blurred distinctions between these different approaches. So we will pursue two different logical tracks. One is the axiomatic approach of Euclid, starting with geometrical postulates and proving results in logical sequence from them. The other is a geometry over the real numbers, using algebra to give proofs of geometrical results. Properties of the real numbers and algebra form the logical starting point of this second track. In the following sections of this book, we will see that the real numbers can be replaced by any abstract ordered field, and this will give rise to a number of different geometries.

In this section we will make clear what we mean by the *real Cartesian plane*, which is the plane geometry over the real numbers.

We accept as given the field of real numbers $\mathbb{R}$. We call a *point* an ordered pair (a, b) of real numbers, and the set of all such ordered pairs is the Cartesian plane. As usual we call the set of points $(a, 0)$ the *x-axis*, and the set of points $(0, b)$ the *y-axis*, and their intersection $(0, 0)$ the *origin*.

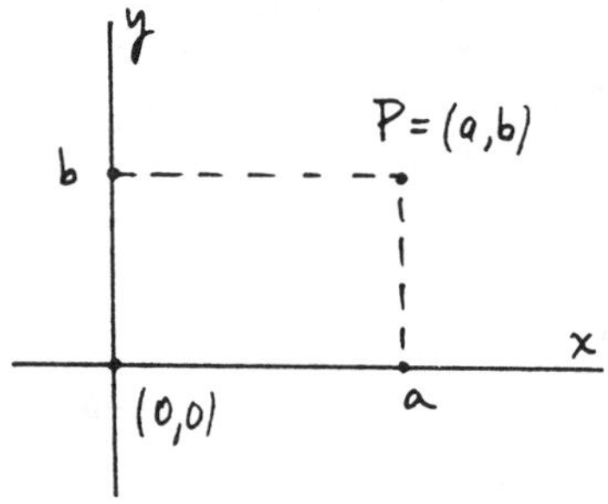

A *line* in this plane is the subset defined by a linear equation $ax + by + c = 0$, with a, b not both zero. Among these are the *vertical* lines which can be written as $x = a$, and every other line can be written in the form $y = mx + b$. In this case we call m the *slope* of the line, and b its y-intercept. For completeness, we will say that a vertical line has slope ∞.

Two lines are called *parallel* if they have no points in common. By looking at the equations of two lines, and solving those equations simultaneously, we see that the lines are parallel if and only if they have the same slope, but are not the same line. It follows immediately that if l_1, l_2, l_3 are three distinct lines, and $l_1 \| l_2$ and $l_2 \| l_3$, then $l_1 \| l_3$. Indeed, all three must have the same slope. In Euclid's *Elements*, this result appears as (I.30) and is proved there using the parallel postulate plus earlier results from Book I. Here in the Cartesian plane, we have a trivial proof just by looking at the equations of the lines.

Let us give another less trivial example of how useful the analytic method can be for proving geometric results. We will show that the three altitudes of a triangle meet at a point. (Compare this with the geometric proofs given in Chapter 1.)

Proposition 13.1 *In the real Cartesian plane, the three altitudes of any triangle all meet at a single point.*

Proof. Recall that an *altitude* of a triangle is the line through one vertex which is perpendicular to the opposite side. First let us move the triangle so that one edge lies along the x-axis, and the opposite vertex lies on the y-axis.

Then we can call the vertices $A = (a, 0)$, $B = (0, b)$, and $C = (c, 0)$. The y-axis is by construction one of the altitudes of the triangle. Our strategy is to find the equations of the other two altitudes, see where they meet the y-axis, and verify that they meet it at the same point.

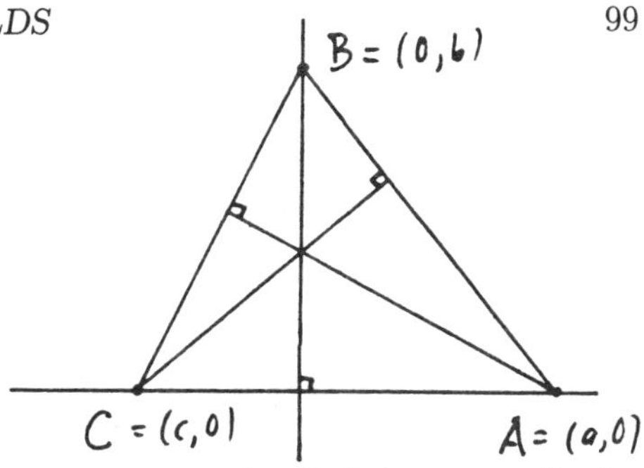

The line AB has slope $-\frac{b}{a}$, so the altitude through C, which is perpendicular to this line, will have slope $\frac{a}{b}$. (Here we use the fact that if two perpendicular lines have slopes m_1 and m_2, then $m_1 m_2 = -1$.) So the equation of the altitude through C, using the point-slope formula, is

$$y = \frac{a}{b}(x - c).$$

To intersect this with the y-axis, we set $x = 0$ and find $y = -\frac{ac}{b}$.

Now consider the line BC. It has slope $-\frac{b}{c}$, so the altitude through A will have slope $\frac{c}{b}$. Its equation becomes

$$y = \frac{c}{b}(x - a).$$

Setting $x = 0$ we find $y = -\frac{ac}{b}$. Since this is the same point as the previous calculation, we find the three altitudes meet at a point. □

Let us reflect a moment on the significance of this proof.

First of all, the reader may object that we have used some facts without proof, such as the result about the slopes of perpendicular lines, or the possibility of moving the triangle into the special position of the proof. However, I am assuming that anyone who has studied some analytic geometry could fill in those missing arguments satisfactorily.

The more serious question is, how do we respond to someone who says, well, with a simple analytic proof like that, why bother with abstract geometric proofs from axioms? If you believe there is only one true geometry, then indeed this proof would be sufficient. But modern mathematics has abandoned the naive position that there is only one truth. Instead it asks, what can be proved within each logical framework, within each separate mathematical theory? This proof shows that the result is true within the logical framework of the real Cartesian plane, using algebra of the real numbers as a logical base. Having found the result is true in this framework, we certainly expect it to be true in the framework of axiomatic Euclidean geometry. However, this proof gives no hint at all about how to find a proof in the abstract axiomatic geometry. In fact, it is even conceivable that while the axioms we have accepted are sufficient to prove the results we know so far, some further axiom might be necessary to prove this result. In other words, the present proof shows that the result is true in the geometry of the real Cartesian plane, but it does not imply a proof, or even guarantee the existence of a proof, in the abstract axiomatic geometry.

For these reasons we will preserve two separate logical tracks, the abstract axiomatic approach, and the analytic geometry approach, until such time as we can prove that the two tracks converge again, using abstract ordered fields.

Next we turn to one of the great insights provided by the algebraic perspective, namely Descartes' discovery that the ruler and compass constructions of Euclid's geometry correspond to the solution of quadratic equations in algebra. To be more precise, let us regard a construction problem as giving certain points in the plane, and requiring the construction of certain other points.

Theorem 13.2 (Descartes) *Suppose given points $P_1 = (a_1, b_1), \ldots, P_n = (a_n, b_n)$ in the real Cartesian plane. (We also assume given the points $(0,0)$ and $(1,0)$.) Then it is possible to construct a point $Q = (\alpha, \beta)$ with ruler and compass if and only if α and β can be obtained from $a_1, \ldots, a_n, b_1, \ldots, b_n$ by field operations $+$, $-$, $\cdot$, $\div$ and the solution of a finite number of successive linear and quadratic equations, involving the square roots of positive real numbers.*

Proof. A ruler and compass construction consists of drawing lines through given points, constructing circles with given center and radius, and finding intersections of lines and circles.

Given two points $P_1 = (a_1, b_1)$ and $P_2 = (a_2, b_2)$, the line passing through them has equation

$$y - b_1 = \frac{b_2 - b_1}{a_2 - a_1}(x - a_1).$$

Its coefficients are obtained by field operations from the initial data a_1, a_2, b_1, b_2.

A circle with center (a, b) and radius r has equation

$$(x - a)^2 + (y - b)^2 = r^2.$$

This is a quadratic equation whose coefficients depend on a, b, and r^2. Note that r may be determined as the distance between two points $P_1 = (a_1, b_1)$ and $P_2 = (a_2, b_2)$ in which case

$$r^2 = (a_1 - a_2)^2 + (b_1 - b_2)^2.$$

To find the intersection of two lines, we solve two linear equations, and get an answer which requires solving a linear equation in x.

To intersect a line with a circle, we solve the equations simultaneously, which requires solving a quadratic equation in x. Assuming the line meets the circle, we will need to take square roots of positive numbers only.

To intersect two circles, we first subtract the two equations, which eliminates the x^2 and y^2 terms. Then we must solve a quadratic with a linear equation, leading to another quadratic equation in x.

In other words, to find the coordinates of a point $Q = (\alpha, \beta)$ obtained by a ruler and compass construction from the initial data $P_1, \ldots, P_n$, requires solving a finite number of linear and quadratic equations whose coefficients depend on the coordinates a_i, b_i and on quantities constructed in earlier steps.

Conversely, the roots of any linear or quadratic equation can be constructed by ruler and compass, given lengths corresponding to the coefficients of the equations, and given a standard length 1. Indeed, such equations can be solved (using the quadratic formula) by a finite number of applications of field operations $+$, $-$, $\cdot$, $\div$ and extractions of square roots of positive numbers, and each of these five operations can be accomplished using ruler and compass. For $+$, $-$ this is obvious, and for $\cdot$, $\div$, $\sqrt{\ }$ see Exercises 13.1–13.3. $\square$

So here we have an algebraic criterion for deciding the possibility of a ruler and compass construction. The method of proof may not lead to an elegant construction, but at least one can determine the possibility of such a construction in a systematic manner. This theorem is a striking example of the insight into geometrical questions given by the algebraic point of view. As Descartes says[1]

> "One can construct all the problems of ordinary geometry without doing anything more than what little is contained in the four figures which I have explained; which is something I do not believe the ancients had noticed: for otherwise they would not have taken the trouble to write so many fat books, where already the order of their propositions makes it clear that they did not have the true method for finding them all, but merely collected those which they happened to come across."

As a practical application of this result, we will find expressions using nested square roots for some lengths which are constructible with ruler and compass, such as the sides of regular polygons inscribed in a circle. Also note that if a particular angle α is constructible, then its trigonometric functions, in particular $\sin\alpha$ and $\cos\alpha$, can be expressed using square roots.

For example, from the right isosceles triangle with sides $1, 1, \sqrt{2}$ we find

$$\sin 45^\circ = \cos 45^\circ = \frac{1}{2}\sqrt{2}.$$

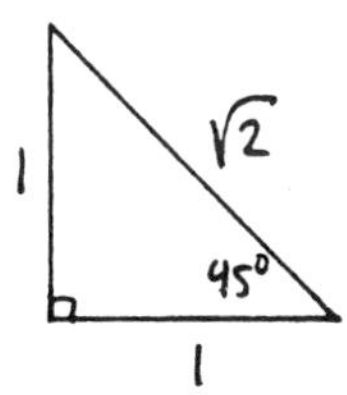

From the 30°–60°–90° triangle with sides 1, $\sqrt{3}$, and 2 we find

$$\begin{aligned} \cos 60^\circ &= \sin 30^\circ = \tfrac{1}{2} \\ \sin 60^\circ &= \cos 30^\circ = \tfrac{1}{2}\sqrt{3}. \end{aligned}$$

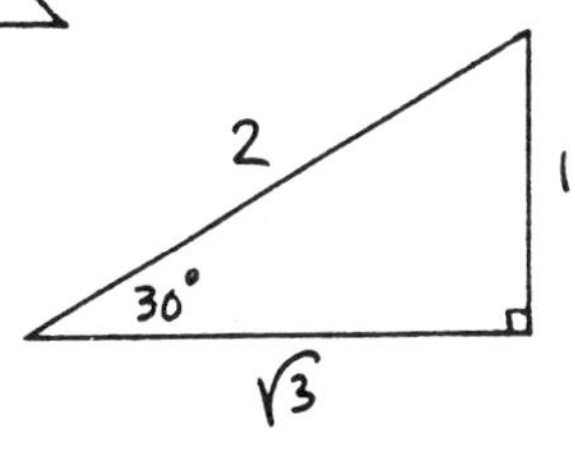

Proposition 13.3 *The length of the chord d of a circle of radius 1 subtending an angle α at the center of the circle is given by*

$$d = \sqrt{2 - 2\cos\alpha}.$$

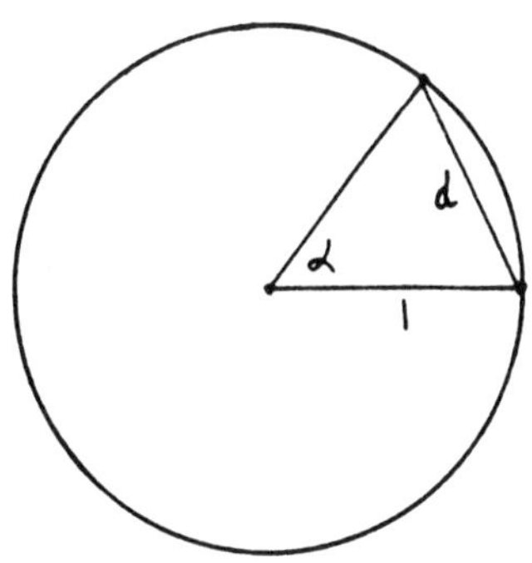

Proof. Draw an altitude of the triangle, and label a, b, c as shown. Then

$$\begin{aligned} a &= \cos\alpha \\ c &= \sin\alpha \\ b &= 1 - a \\ d &= \sqrt{b^2 + c^2}. \end{aligned}$$

[1]Descartes (1886) p. 7 (my translation).

Substituting we find

$$d = \sqrt{(1-\cos\alpha)^2 + \sin^2\alpha}.$$

Expanding and simplifying (using $\sin^2\alpha + \cos^2\alpha = 1$) we get

$$d = \sqrt{2 - 2\cos\alpha}.$$

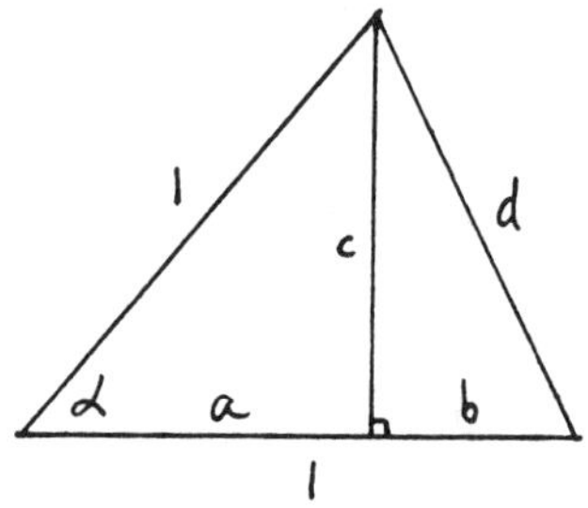

□

So for example, the side of the regular octagon inscribed in the unit circle will be

$$\begin{aligned} d &= \sqrt{2 - 2\cos 45^\circ} \\ &= \sqrt{2 - \sqrt{2}}. \end{aligned}$$

Proposition 13.4 *In a circle of radius* 1, *the length of the side of the inscribed regular pentagon is* $\frac{1}{2}\sqrt{10 - 2\sqrt{5}}$.

Proof. Recall that the key step in the construction of the regular pentagon is to find an isosceles triangle ABC whose base angles β are equal to twice the top angle α (4.1). This is done using (II.11) by dividing a given segment AB in extreme and mean ratio, that is, finding a point D such that $AB \times BD = AD^2$. Then we take AD to be the base of the isosceles triangle (IV.10).

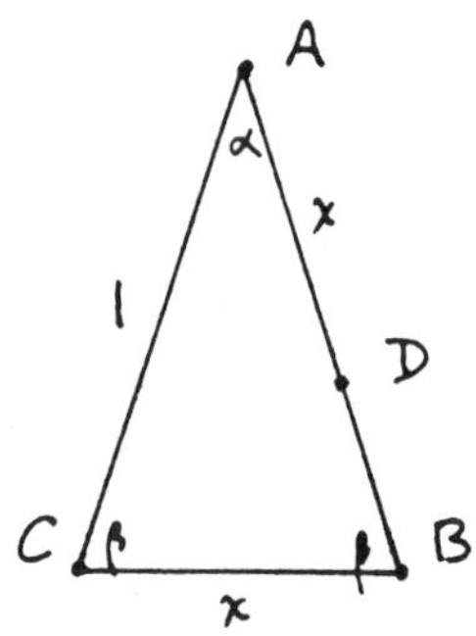

So let AB have length 1, and let AD have length x. Then, converting areas into numbers, we have $1 - x = x^2$, or $x^2 + x - 1 = 0$. Solving by the quadratic formula gives $x = \frac{1}{2}(\sqrt{5} - 1)$.

If we drop an altitude from A to BC, then it is clear that $\cos\beta = \frac{1}{2}x = \frac{1}{4}(\sqrt{5} - 1)$. Since $\beta = 2\alpha$ and $\alpha + 2\beta = 180^\circ$, we find $\beta = 72^\circ$.

On the other hand, the angle subtended at the center of the circle by a side of the regular pentagon is 72°, so by (13.3), the side of the pentagon is

$$\begin{aligned} d &= \sqrt{2 - 2\cos 72^\circ} \\ &= \sqrt{2 - \tfrac{1}{2}(\sqrt{5} - 1)} \\ &= \tfrac{1}{2}\sqrt{10 - 2\sqrt{5}}. \end{aligned}$$

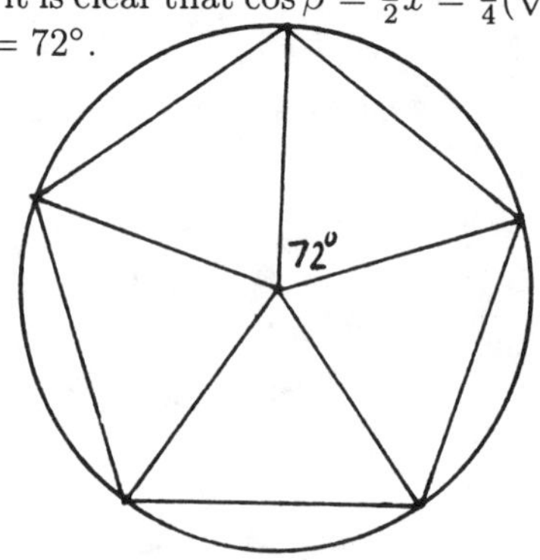

Exercises

13.1 Given segments of lengths a and b, and given a segment of length 1, construct with ruler and compass a segment of length ab.

13.2 With the same data as Exercise 13.1, construct a segment of length a/b.

13.3 Given segments of lengths a and 1, construct a segment of length $\sqrt{a}$.

13.4 Show that any quantity obtainable from the rational numbers by a finite number of operations $+, -, \cdot, \div, \sqrt{\ }$, can be written in a *standard form* $r \cdot A$, where $r \in \mathbb{Q}$ is a rational number, and A is an expression involving only integers, $+, -, \cdot$, and $\sqrt{\ }$. In the following problems, please express your answers always in standard form. (Unfortunately this standard form is not unique — see Exercises 13.9, 13.14 below.)

13.5 Express $\frac{\sqrt{5}+1}{\sqrt{10+2\sqrt{5}}}$ in standard form.

13.6 Find $\sin 22\frac{1}{2}^\circ$ and $\cos 22\frac{1}{2}^\circ$ as expressions involving square roots in standard form. Check your result by finding the decimal equivalent with a calculator.

13.7 Find the side of a regular 16-gon inscribed in the unit circle.

13.8 Find $\sin 11\frac{1}{4}^\circ$ and $\cos 11\frac{1}{4}^\circ$ in standard form.

13.9 Three students working on the same problem came up with the following answers.

a. $-\sqrt{5} + \sqrt{11 + 6\sqrt{2}}$

b. $3 + \sqrt{7 - 2\sqrt{10}}$

c. $\sqrt{2} + \sqrt{14 - 6\sqrt{5}}$.

Two answers were correct, and one false. Find which two were correct, and prove they are equal. Can you express the correct answer in a simpler form? How can you modify the third student's answer so that it becomes correct?

13.10 Find the length of the edge of a regular tetrahedron inscribed in a unit sphere.

13.11 Find the area of the largest equilateral triangle which is contained in a square of side 1.

13.12 If $a, b \in \mathbb{Z}$, and if $a + b\sqrt{2}$ has a square root in $\mathbb{Q}(\sqrt{2})$, then the square root is actually in $\mathbb{Z}[\sqrt{2}]$.

13.13 If $a, b \in \mathbb{Z}$, give a method for deciding whether $\sqrt{a+b\sqrt{2}} \in \mathbb{Z}[\sqrt{2}]$. Are the following squares in $\mathbb{Z}[\sqrt{2}]$? If so, find the square root.

a. $627 + 442\sqrt{2}$

b. $1507 + 1024\sqrt{2}$

c. $2107 + 1470\sqrt{2}$.

13.14 Verify

$$\sqrt{5+2\sqrt{5}} - \sqrt{5-2\sqrt{5}} = \sqrt{10-2\sqrt{5}}.$$

Also show that none of these three nested radicals is in $\mathbb{Q}(\sqrt{5})$. This is another example of non-uniqueness of the standard form.

13.15 Express $\cos 72°$ and $\sin 72°$ as nested radicals in standard form. Check by computing decimal equivalents with a calculator.

13.16 Find $\cos 36°$, $\sin 36°$, and the side of the regular decagon inscribed in the unit circle.

13.17 Find $\cos 24°$, $\sin 24°$, $\cos 12°$, $\sin 12°$, and the side of the regular 15-sided polygon inscribed in the unit circle. Express in standard form, and check decimal equivalents with the calculator.

13.18 Find the side of a regular pentagon circumscribed around a unit circle.

13.19 Given a regular pentagon of side 1, find the distance from the center to a vertex, in standard form.

13.20 Given a right angle at O, a point B on one arm, and a point A, construct with ruler and compass a circle with center O, meeting the arms of the right angle at C, D, such that AD is parallel to BC. (par = 9 steps, not counting lines AD, BC).

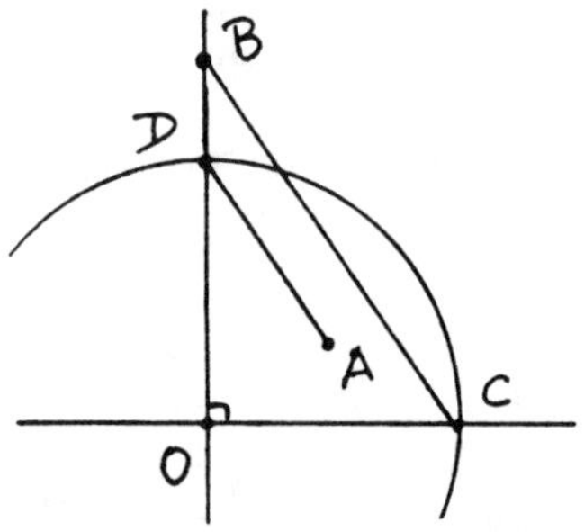

13.21 Given segments of lengths $1, a, b$ in the plane, construct with ruler and compass a length x satisfying $x^2 - ax - b = 0$. (If you use the quadratic formula, par = 21; using geometrical ideas from Exercise 13.20, par = 14.)

14 Abstract Fields and Incidence

In this section we start with the algebraic structure of a field, and based on this field we will obtain a geometry. In this way, using a field, we obtain a *model* of the abstract geometry determined by Hilbert's axioms. Different fields will give different geometries, and in this way we will obtain many different models, many different Euclidean geometries. We will investigate what properties of the field are needed to make the different axioms in Hilbert's axiom system hold. This will help us understand the independence of the axioms.

Hilbert's axiom system is based on the undefined notions of *point*, *line*, *betweenness*, and *congruence* for line segments and for angles. These undefined notions are limited only by having to satisfy all the axioms.

To make a model of the geometry within another mathematical framework, in this case algebra, we must say what the *interpretation* of the undefined notions is to be in our model, and then we must *prove* that the axioms hold in this interpretation.

We start then, with a field, and to fix the ideas we recall the definition of a field.

Definition. A *field* is a set F, together with two operations, $+$, $\cdot$, i.e. for each $a, b \in F$ there are given $a + b \in F$ and $a \cdot b \in F$, subject to the following conditions:

1) The set F, together with the operation $+$ forms an abelian group, namely

   (i) $(a + b) + c = a + (b + c)$ for any $a, b, c \in F$

   (ii) $a + b = b + a$ for any $a, b \in F$

   (iii) there is an element $0 \in F$ such that $a + 0 = a$ for all $a \in F$

   (iv) for each $a \in F$ there is an element $-a \in F$ such that $a + (-a) = 0$.

2) The set $F^* = F - \{0\}$, together with the operation $\cdot$ forms an abelian group, namely

   (i) $(ab)c = a(bc)$ for all $a, b, c \in F^*$

   (ii) $ab = ba$ for all $a, b \in F^*$

   (iii) there is an element $1 \in F^*$ such that $a \cdot 1 = a$ for all $a \in F^*$

   (iv) for all $a \in F^*$, there is an $a^{-1} \in F^*$ such that $a \cdot a^{-1} = 1$.

3) The operations $+$ and $\cdot$ are related by the distributive law

$$a(b + c) = ab + ac.$$

Note in particular that in our definition of a field $0 \neq 1$, and multiplication is always commutative. We leave to the reader to verify the other properties sometimes mentioned in the definition of a field, such as $0 \cdot a = 0$ for all $a \in F$.

Our first step in making a geometry is to say what we mean by points and lines. Of course we take our cue from the "standard" model of Euclidean geometry, the real Cartesian plane, given by ordered pairs of real numbers. This is the geometry we might call "high school geometry", where the abstract geometry of axioms is confused with the Descartes model $\mathbb{R}^2$, and we assume everything is over the real numbers. In that model we suppose everyone already "knows" what points, lines, angles, betweenness, and congruence mean.

But now, since we are starting with an arbitrary field F, which may not be the real numbers, we need to make our definitions precise.

Definition. The plane Π (or Π_F if we want to indicate the field), called the *Cartesian plane over the field* F, is the set F^2 of ordered pairs of elements of the field F, and we call these the *points* of Π. A *line* will be the subset defined by a linear equation

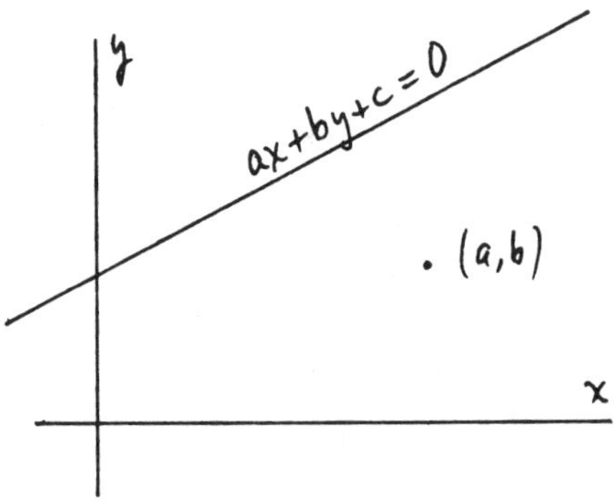

$$ax + by + c = 0$$

for some $a, b, c \in F$, with a, b not both zero.

Any line can also be written in either the form $x = c$, in which case we call it *vertical*, or in the form $y = mx + b$. In the latter case we say the line has *slope* m, and for the line $x = c$, we say it has slope ∞. Here ∞ is just a symbol (it is not an element of the field F).

Example 14.0.1 Let F be the field of two elements $F = \{0, 1\}$ with addition and multiplication (mod 2). Then the plane Π over F has exactly four points and six lines, shown schematically in the diagram. Note in particular that the two "diagonal" lines do not meet in this geometry.

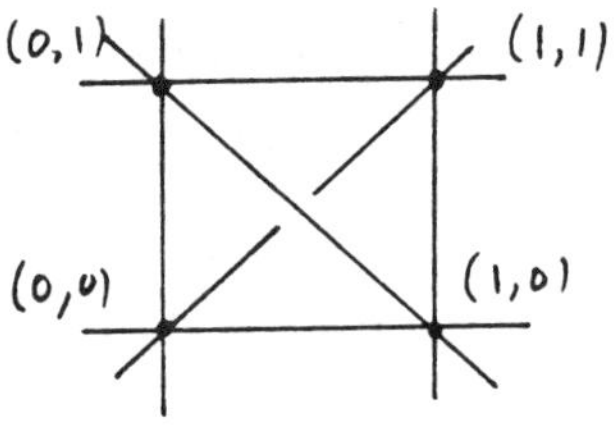

Proposition 14.1 *If F is any field, the Cartesian plane Π_F satisfies Hilbert's incidence axioms* (I1), (I2), (I3), *and the parallel axiom* (P).

Proof. (I1) says any two points lie on a line. Since we can perform rational operations $+$, $-$, $\cdot$, $\div$ in the field F, the usual "two point formula" of analytic geometry shows that we can find a linear equation, with coefficients in the field F, which determines a line containing the two given points.

(I2) says every line has at least two points. Since any field F has at least the two distinct elements $0, 1$, by putting $x = 0, 1$ if the line has the form $y = mx + b$, or by putting $y = 0, 1$ if the line is $x = c$, we obtain two points on any line.

(I3) says there exist three noncollinear points. Indeed, we can always take $(0, 0)$, $(0, 1)$, $(1, 0)$ and we can see easily these do not lie on any line.

(P) says there is at most one parallel to a given line l through a given point $P \notin l$. In fact the stronger statement holds, that there is *exactly* one line parallel to l through P. Recall that parallel means two lines do not meet. In the plane Π_F, we see immediately that two lines are parallel if and only if they have the same slope and are not equal to each other. So given a line l, let its slope be m. Then the familiar "point-slope" formula of analytic geometry shows that there is a unique line of slope m passing through the point P. This will be the parallel to l.

Before introducing the further notions of betweenness, and congruence into our Cartesian plane over a field F, there are already some interesting connections between algebraic properties of the field F and incidence properties in the plane Π_F. To investigate these, it is useful to be able to change coordinates.

Proposition 14.2 *In the Cartesian plane* Π *over a field* F, *we can make a linear change of variables*

$$\begin{cases} x' = ax + by + c \\ y' = dx + ey + f \end{cases}$$

so that the new coordinate axes are any two intersecting lines, and the new unit points are any given points P, Q *on them, not equal to their intersection point* E.

Proof. Since a composition of linear changes of variables is again one, we can proceed one step at a time. First, a change of the form

$$\begin{cases} x' = x - a \\ y' = y - b \end{cases}$$

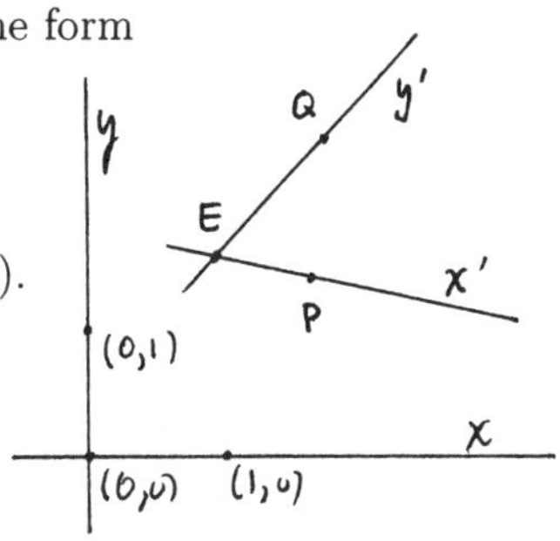

will move the origin $(0,0)$ to the point $E = (a, b)$.

Next, a transformation of the form

$$\begin{cases} x' = ax \\ y' = by \end{cases}$$

will move the unit points to any other points on the same axes.

Then, a change of the form

$$\begin{cases} x' = x - ay \\ y' = y \end{cases}$$

will keep the x-axis fixed, but replace the y-axis by another line through the origin. The x-axis may be moved by another such transformation, interchanging the roles of x and y.

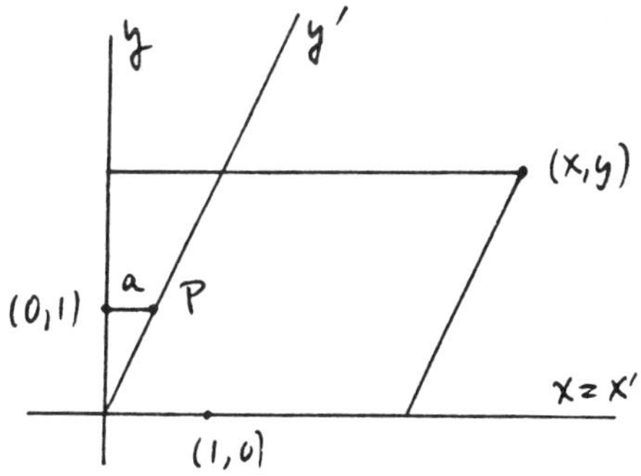

Combining all these gives a transformation which moves the original axes and unit points to any other desired axes and unit points.

Remark. Since the change of coordinates is linear, lines in the new coordinate system are still given by linear equations, so it is equivalent to describe the geometry of the plane Π_F using either the old or the new coordinates.

Now we give some applications.

Proposition 14.3 *There exists a configuration in the plane* Π_F *of four points* A, B, C, D *such that* $AB \| CD$, $AC \| BD$ *and* $AD \| BC$, *if and only if the characteristic of* F *is* 2.

Proof. We have already seen the existence (14.0.1), since any field F of char. 2 contains the subfield $\{0, 1\}$ of two elements with addition and multiplication (mod 2).

For the converse, suppose that such a configuration exists in Π_F. Then make a linear change of coordinates such that C becomes the new origin, and A, D are the unit points. Then B will be the point $(1, 1)$; BC will be the line $x = y$, and AD

will be the line $x + y = 1$. In this configuration, $AD \| BC$, so the equations $x = y$ and $x + y = 1$ must have no common solution. Solving, we find $2x = 1$, which has a solution in F as long as $2 \neq 0$. We conclude that this configuration exists only if $2 = 0$, i.e. char. $F = 2$.

Proposition 14.4 (Pappus' theorem) *In the Cartesian plane over a field F, suppose given lines l, m, and points $A, B, C \in l$ and $A', B', C' \in m$ such that $AC' \| A'C$ and $BC' \| B'C$. Then also $AB' \| A'B$.*

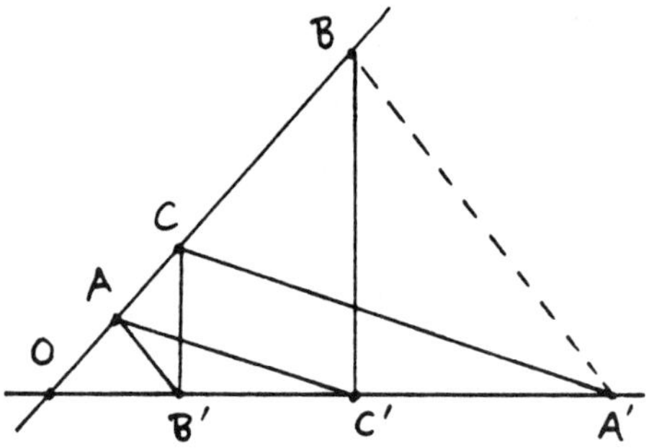

Proof. Suppose that l, m meet at a point O (we leave the case $l \| m$ as exercise). Choose coordinates so that O is the origin, and A, B' are the unit points. Let C be the point $(0, a)$ and let C' be the point $(b, 0)$. Then writing the equations of the lines involved, we find $B = (0, ab)$ and $A' = (ab, 0)$. Thus the line BA' has slope -1, hence is parallel to AB'.

Remark 14.4.1. It is possible to define a Cartesian plane over a *skew field* F (which is an algebraic structure the same as a field, except that the multiplication need not be commutative). Then Hilbert has shown [Foundations] that the skew field F is commutative, if and only if Pappus' theorem holds in the associated plane Π_F.

Example 14.4.2 In the Cartesian plane over the field F, assuming characteristic 0, there is a configuration such as the one shown (where all lines which appear parallel are assumed to be parallel, namely $DE \| BC$, $DF \| AC$, $EF \| AB$, $GH \| BC$, and $BH \| GE$) if and only if $\sqrt{2} \in F$.

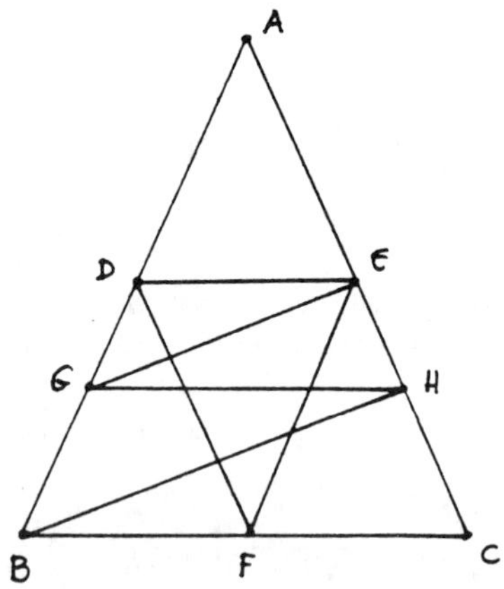

Indeed, to analyze this situation, take B to be the origin, BC and BA to be the axes, and D, F the unit points. Then $A = (0, 2)$, $E = (1, 1)$, $C = (2, 0)$. Let G have coordinates $(0, a)$. Then $H = (2 - a, a)$. The line BH will have slope $a/(2-a)$, and the line GE will have slope $1 - a$. The parallelism $BH \| GE$ then requires $a/(2 - a) = 1 - a$, or equivalently $a^2 - 4a + 2 = 0$. Solving with the quadratic formula gives $a = 2 \pm \sqrt{2}$.

For this configuration to exist, it is necessary and sufficient that $a \in F$, and this is clearly equivalent to $\sqrt{2} \in F$, as required.

Exercises

14.1 Show that Pappus' theorem (14.4) still holds in Π_F in the case that $l \| m$.

14.2 Show that Desargues' theorem holds in the Cartesian plane over a field F: Given a configuration as shown, with $AC \| A'C'$ and $AB \| A'B'$, prove that $BC \| B'C'$.

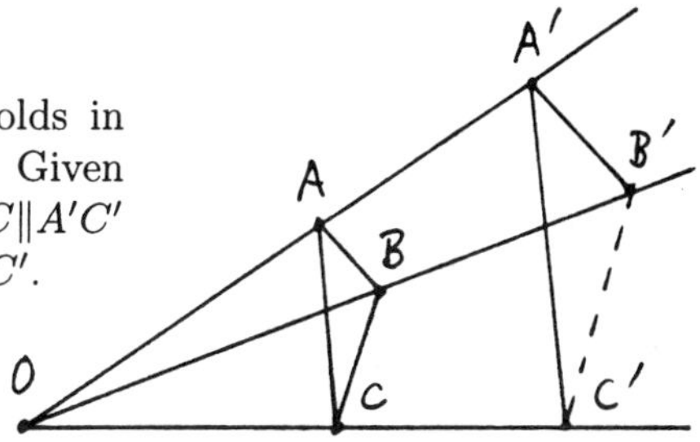

For each of the following problems, assume you are working in the Cartesian plane Π over a field F of characteristic 0. Give necessary and sufficient conditions on the field F for the given configuration to exist. Assume that all lines which appear to be parallel are parallel.

14.3 Ans: $\sqrt{3} \in F$.

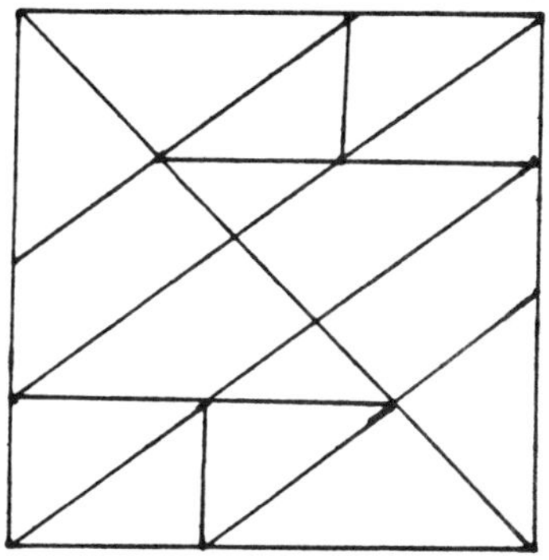

14.4 Ans: $\sqrt{13} \in F$.

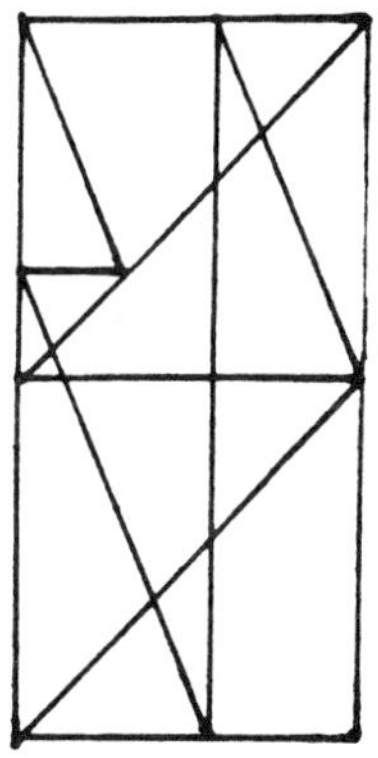

14.5

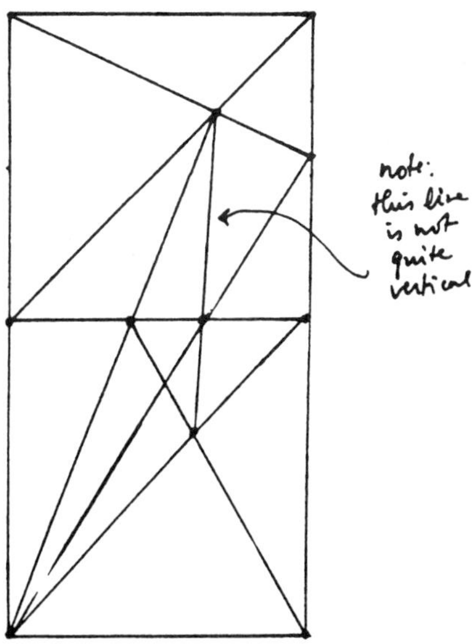

14.6

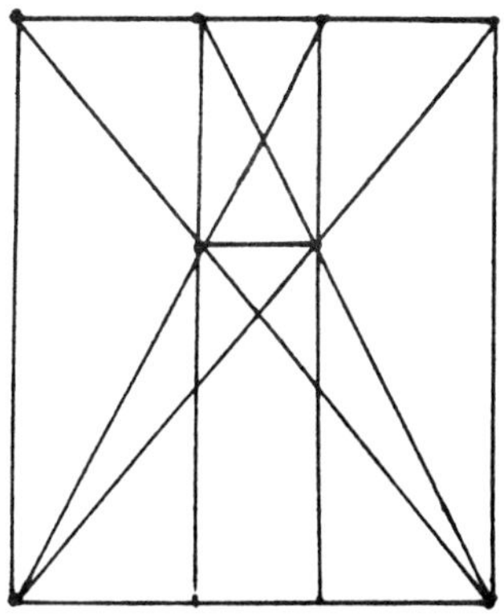

14.7

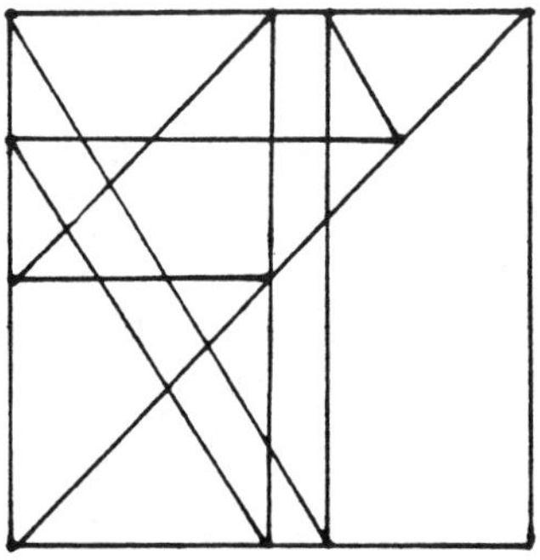

14.8

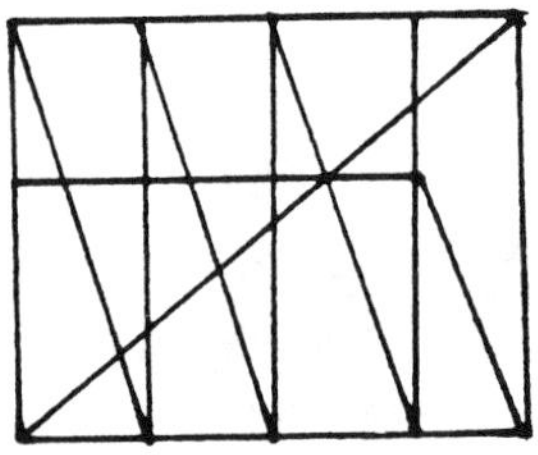

In each of the following four problems, suppose given the triangle ABC, and make a ruler and compass construction of the diagram shown. In the first three, D, E, F are the midpoints of the sides. In the last, they are one third of the way along each side. (Par = 20 to 25 steps each.)

14.9

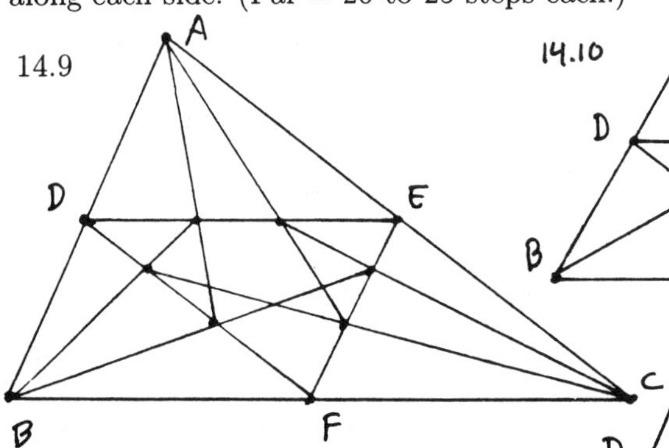

14.10

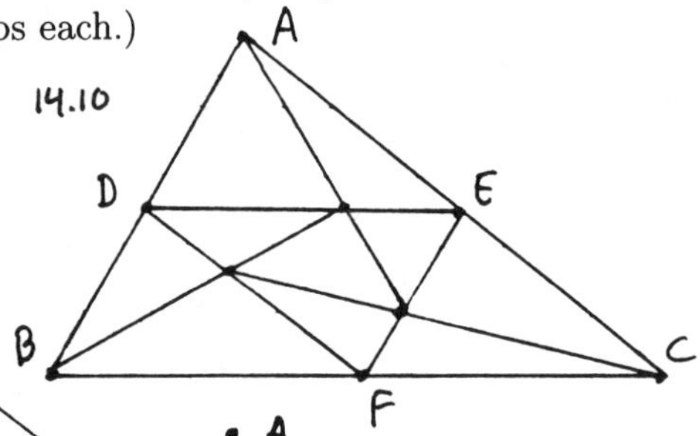

14.11

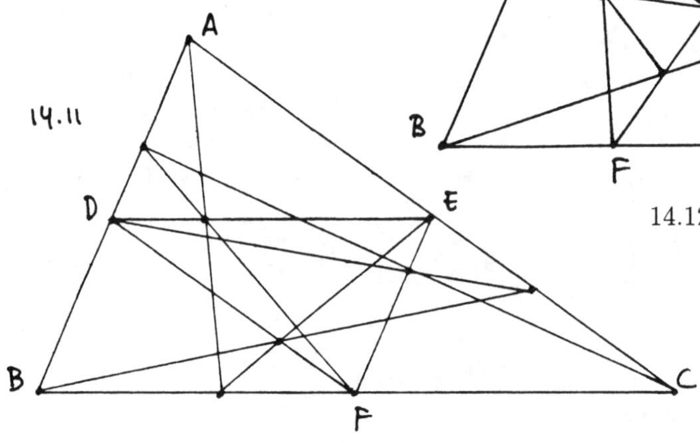

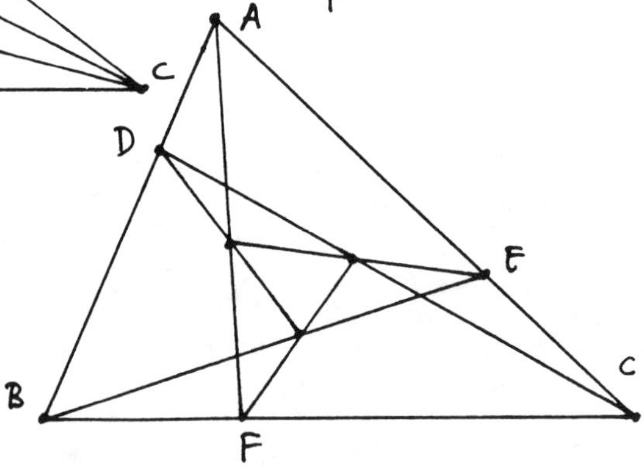

14.12

15 Ordered Fields and Betweenness

The next undefined notion we need to interpret in the Cartesian plane over a field is *betweenness.* It turns out that this is not possible over an arbitrary field.

We will have to impose some additional structure on the field to make a definition of betweenness possible. To see why this is so, suppose we had a notion of betweenness in our geometry which satisfies Hilbert's axioms. Then the x-axis (whose points are in one-to-one correspondence with the elements of our field F) could be divided into subsets consisting of the "positive x-axis" meaning all points on the same side of 0 as 1, the origin 0, and the "negative x-axis" consisting of all points on the other side of 0 from 1. In this way we can define a notion of "positive" elements of the field F, analogous to the usual notion of positive real numbers.

To make sense of this notion in an abstract field, we introduce the notion of an ordered field.

Definition. An *ordered field* is a field F, together with a subset P, whose elements are called *positive*, satisfying

(i) if $a, b \in P$, then $a + b \in P$ and $ab \in P$.

(ii) For any $a \in F$, one and only one of the following holds: $a \in P$ or $a = 0$ or $-a \in P$.

Here are a few elementary properties of an ordered field.

Proposition 15.1 *Let F, P be an ordered field. Then*

a) *$1 \in P$, i.e. 1 is a positive element*

b) *F has characteristic 0*

c) *the smallest subfield of F containing 1 is isomorphic to the rational numbers $\mathbb{Q}$*

d) *for any $a \neq 0 \in F$, $a^2 \in P$.*

Proof. a) In any field, $1 \neq 0$, so either $1 \in P$ or $-1 \in P$. If $1 \in P$ we are done. If $-1 \in P$, then by (i) also $(-1) \cdot (-1) = 1 \in P$ which contradicts (ii). Hence $1 \in P$.

b) Since $1 \in P$, $1+1+1+\cdots+1$ any number of times is also in P. In particular, it is never 0, so F has characteristic 0.

c) The natural map of the positive integers $\mathbb{N}$ to F given by n goes to $1+1+\cdots+1$ (n times) is injective, by b), and extends to an injective map of $\mathbb{Q}$ to F whose image is isomorphic to $\mathbb{Q}$ and which is the smallest subfield of F containing 1. Whenever no confusion can arise, we will identify $\mathbb{Q}$ with its image in F. So for example if $n \in \mathbb{Z}$, then n will also denote the corresponding element of F.

d) If $a \neq 0$, then either $a \in P$ or $-a \in P$. If $a \in P$, then $a^2 \in P$ by (i). If $-a \in P$, then $(-a)(-a) = a^2 \in P$.

Proposition 15.2 *In an ordered field F, P, we define $a > b$ if $a - b \in P$, and $a < b$ if $b - a \in P$. This notion of inequality satisfies the usual properties, namely*

(i) *If $a > b$ and $c \in F$, then $a + c > b + c$*

(ii) *If $a > b$ and $b > c$, then $a > c$*

(iii) *If $a > b$ and $c > 0$, then $ac > bc$*

(iv) *Given $a, b \in F$, one and only one of the following holds: $a > b$ or $a = b$ or $a < b$.*

Examples 15.2.1 The rational numbers $\mathbb{Q}$ form an ordered field, where we take for P the positive rational numbers, in the usual sense.

15.2.2 The field of real numbers $\mathbb{R}$ is an ordered field with the usual notion of positive elements.

15.2.3 The field of complex numbers $\mathbb{C}$ cannot be an ordered field (i.e. there is no subset P of $\mathbb{C}$ satisfying the definition) because $i^2 = -1 < 0$, which contradicts (15.1d).

15.2.4 Since an ordering on a field is extra structure, in general there may be more than one way to make a given abstract field into an ordered field. For example, let $F = \mathbb{Q}(\sqrt{2})$. Then F is a subfield of $\mathbb{R}$, so we can make it into an ordered field by taking P to be the subset of elements of F which are positive in $\mathbb{R}$. But there is another embedding $\varphi : F \to \mathbb{R}$ given by $\varphi(a + b\sqrt{2}) = a - b\sqrt{2}$ for all $a, b \in \mathbb{Q}$, and we can put another ordering on F by taking P to be the set of elements $x \in F$ for which $\varphi(x) > 0$ in $\mathbb{R}$.

Proposition 15.3 *If F is a field, and if there is a notion of betweenness in the Cartesian plane Π_F satisfying Hilbert's axioms* (B1)–(B4), *then F must be an ordered field. Conversely, if F, P is an ordered field, we can define betweenness in Π_F so as to satisfy* (B1)–(B4).

Proof. First suppose that F is a field and that there is a notion of betweenness in the plane Π_F satisfying (B1)–(B4). We define the subset $P \subseteq F$ to consist of all $a \in F$, $a \neq 0$, such that the point $(a, 0)$ of the x-axis is on the same side of 0 as 1. Since addition in the field corresponds to laying out line segments consecutively on the x-axis, it is clear that $a, b \in P \Rightarrow a + b \in P$.

For multiplication, given $a, b \in P$, put a on the x-axis, put $1, b$ on the y-axis, draw the line from $(0, 1)$ to $(a, 0)$ and draw the line parallel to this one through $(0, b)$. It will meet the x-axis in the point $(ab, 0)$. Now clearly $1, a, b \in P \Rightarrow ab \in P$ (we leave to the reader to see exactly how this follows from (B1)–(B4)!), so P satisfies the first property of the definition of an ordered field. By construction, F is the disjoint union of $P \cup \{0\} \cup -P$, so that F, P is an ordered field.

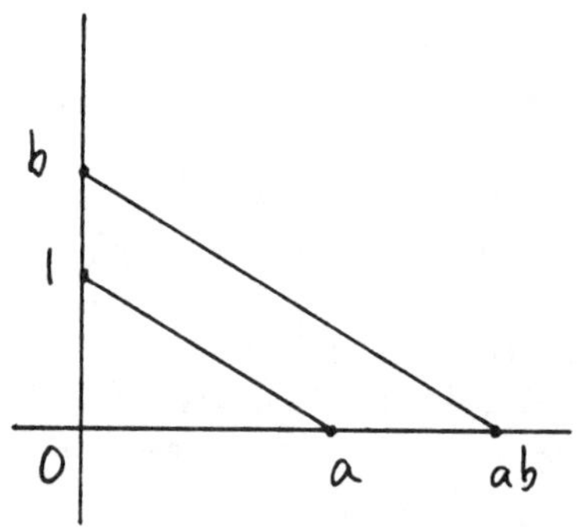

Now suppose conversely that F, P is a given ordered field. We define *betweenness* for points on a line as follows: let $A = (a_1, a_2)$, $B = (b_1, b_2)$, $C = (c_1, c_2)$ be three distinct points on a line $y = mx + b$. We say B is *between* A and C ($A * B * C$) if

$$\text{either } a_1 < b_1 < c_1 \text{ or } a_1 > b_1 > c_1.$$

If the line is vertical, we use instead the second coordinates in the same way.

We must verify the axioms (B1)–(B4).

(B1) is obvious from our definition.

(B2) This follows from the corresponding fact, true in any ordered field, that given $b < d \in F$, there exist $a, c, e \in F$ such that $a < b < c < d < e$. Indeed, we can always take, for example, $a = b - 1$, $c = \frac{1}{2}(b + d)$ and $e = d + 1$. Note that since F has characteristic 0, by (15.1), $\frac{1}{2} \in F$.

(B3) follows from the fact that in an ordered field F, if a, b, c are three distinct elements, then one and only one of the following six possibilities can occur:

$$\begin{gathered} a < b < c \\ a < c < b \\ b < a < c \\ b < c < a \\ c < a < b \\ c < b < a. \end{gathered}$$

(B4). Suppose given a triangle ABC and a line l which meets the side AB. Assuming $A, B, C \notin l$, we must show that l also meets either AC or BC, but not both.

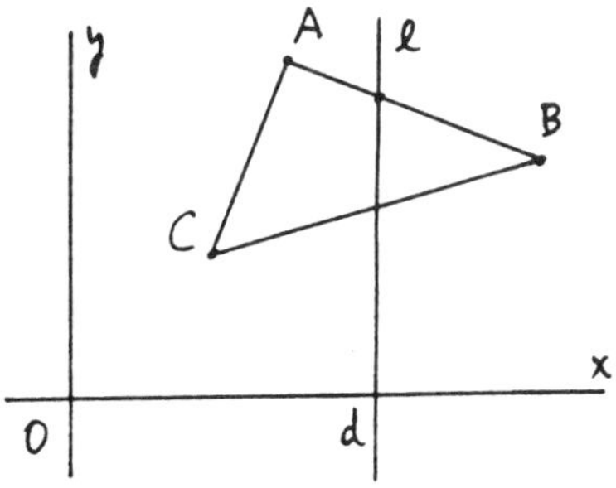

First suppose the line l is vertical, with equation say $x = d$. Let a, b, c be the x-coordinates of A, B, C. By hypothesis, either $a < d < b$ or $b < d < a$. By symmetry, let us assume $a < d < b$. Then it is clear if $c < d$, (as in the picture), then l will meet BC but not AC. If $c > d$, then l will meet AC and not BC, as required.

If l is not vertical, we make a change of coordinates (14.2) so that l becomes vertical. Since linear changes of variables either preserve or reverse inequalities, this does not affect the notion of betweenness, and so we reduce to the previous case.

To complete this section, we will discuss Archimedes' axiom (A) and Dedekind's axiom (D)—cf. §12.

Proposition 15.4 *Let F, P be an ordered field. Then the Cartesian plane Π_F will satisfy* (A) *or* (D) *if and only if the field F satisfies the corresponding property for a field, namely*

(A′) (Archimedes' axiom for a field). *For any $a > 0$ in F, there is an integer n such that $n > a$.*

(D′) (Dedekind's axiom for a field). *Suppose we can write the field F as the disjoint union of two non-empty subsets $F = S \cup T$, and assume that for all $a \in S$ and all $b \in T$ we have $a < b$. Then there exists a unique element $c \in F$ such that for all $a \in S$ and all $b \in T$ we have $a \leq c \leq b$.*

Planche IX

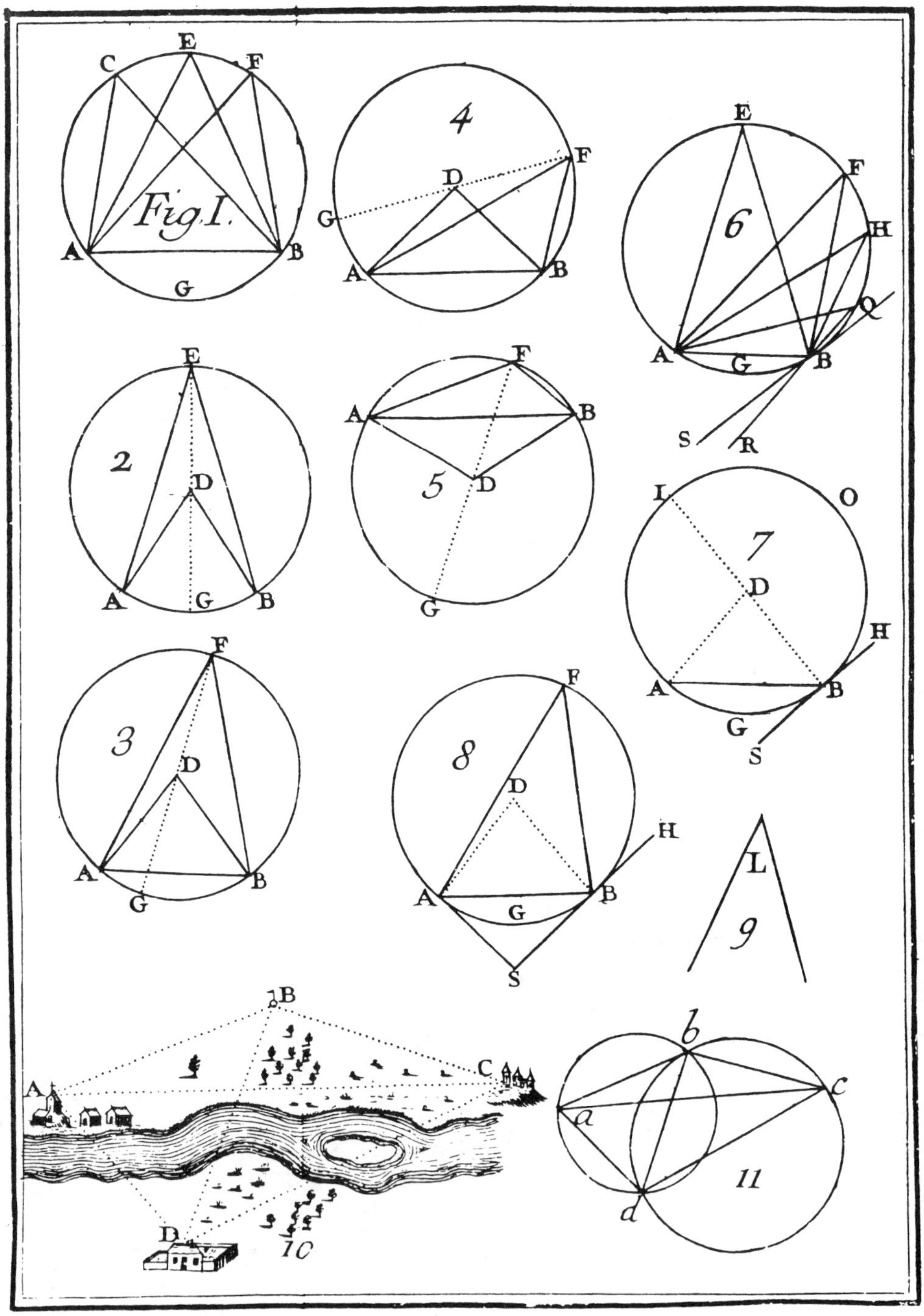

Plate V. Diagrams from Clairaut (1775).

Proof. For (A), we can choose coordinates such that the first segment AB is a unit segment. If C and D on the same line correspond to elements $c < d \in F$, then n copies of AB will exceed CD if and only if $n > d - c$.

For (D), choose coordinates so that the line in question is the x-axis, and identify its points with elements of F. Then the statements are the same.

Proposition 15.5 *Let F be an ordered field satisfying Archimedes' axiom* (A′). *Then F is isomorphic, with its ordering, to a subfield of $\mathbb{R}$. Furthermore, in this case, F satisfies Dedekind's axiom* (D′) *if and only if this subfield is equal to $\mathbb{R}$.*

Proof. We saw earlier (15.1) that F contains a subfield F_0 isomorphic to $\mathbb{Q}$. This gives us a unique isomorphism $\varphi_0 : F_0 \to \mathbb{Q} \subseteq \mathbb{R}$. We will extend φ_0 to an isomorphism of F into $\mathbb{R}$. Let $\alpha \in F$. Because of Archimedes axiom, there are integers both smaller and bigger than α. So let a_0 be the unique integer n such that $n \leq \alpha < n + 1$. Next define $a_1 \in \frac{1}{10}\mathbb{Z}$ to be the unique one-tenth integer such that $10a_1 \leq \alpha < 10a_1 + 1$. Similarly define $a_2 \in (1/100)\mathbb{Z}$ such that $100a_2 \leq \alpha < 100a_2 + 1$. Continuing in this way we obtain a sequence $a_0 \leq a_1 \leq a_2 \leq \ldots$ of rational numbers with the property that for each n, $a_n \leq \alpha < a_n + 10^{-n}$. In the field of real numbers $\mathbb{R}$, these converge to a certain real number which we call $\varphi(\alpha)$. This defines a map $\varphi : F \to \mathbb{R}$. It is easy to verify that $\varphi(\alpha + \beta) = \varphi(\alpha) + \varphi(\beta)$ and $\varphi(\alpha\beta) = \varphi(\alpha) \cdot \varphi(\beta)$. So φ is a homomorphism of fields, which is necessarily an isomorphism onto its image. One checks also $\alpha < \beta \Rightarrow \varphi(\alpha) < \varphi(\beta)$, so it is an order isomorphism of F onto $\varphi(F) \subseteq \mathbb{R}$.

Now condition (D′) on F is equivalent to (D′) on $\varphi(F)$ since the fields are order-isomorphic. Each real number $r \in \mathbb{R}$ is characterized by the sets $\Sigma_1 = \{a \in \mathbb{R} \mid a \leq r\}$ and $\Sigma_2 = \{a \in \mathbb{R} \mid a > r\}$, so clearly (D′) holds in $\varphi(F)$ if and only if $\varphi(F) = \mathbb{R}$.

Remark 15.5.1. As a converse to this result, note that any subfield F of $\mathbb{R}$ becomes an ordered field if we take for $P \subseteq F$ those elements of F which are positive (in the usual sense) in $\mathbb{R}$. Thus the study of Archimedean ordered fields is equivalent to the study of subfields of $\mathbb{R}$.

See §18 for some examples of non-Archimedean ordered fields.

Exercises

15.1 Let F be an ordered field, and let $a > 0$. Show that if a has a square root in F, i.e. an element $b \in F$ such that $b^2 = a$, then a has exactly two square roots in F, one of which is positive, and the other negative. We use the notation $\sqrt{a}$ to denote the positive square root.

15.2 Let F be an ordered field, let $d > 0$, and suppose that d does not have a square root in F. Let $F(\sqrt{d})$ denote the set of all $a + b\sqrt{d}$, with $a, b \in F$, where $\sqrt{d}$ is a square root in some extension field of F.

a) Show that $F(\sqrt{d})$ is a field.

b) Show how to define an ordering on $F(\sqrt{d})$, with $\sqrt{d} > 0$, so that it becomes an ordered field.

15.3 In an ordered field F, show that Dedekind's axiom (D$'$) implies Archimedes axiom (A$'$). *Hint*: If F did not satisfy (A$'$), let $S = \{\alpha \in F \mid \exists n \in \mathbb{Z}, \text{ with } \alpha < n\}$, and let $T = F - S$. Then apply (D$'$).

15.4 In the proof of (15.5), verify that $\varphi(\alpha\beta) = \varphi(\alpha) \cdot \varphi(\beta)$.

16 Congruence of Segments and Angles

Next we need to define the notion of congruence for line segments and for angles. We assume from now on that we are starting from an *ordered* field F, P, so that we have betweenness as studied above. Then we can define the *line segment* AB to be the set of all points on the line AB which are between A and B, plus the endpoints A, B. We would like to define congruence of line segments using the usual *Euclidean distance* function for two points $A = (a_1, a_2)$, $B = (b_1, b_2)$, namely

$$\text{dist}(A, B) = \sqrt{(a_1 - b_1)^2 + (a_2 - b_2)^2}.$$

However, our field F may not have square roots in it, so we will use instead the "distance-squared" function

$$\text{dist}^2(A, B) = (a_1 - b_1)^2 + (a_2 - b_2)^2.$$

This will give the same notion of congruence.

Definition. Two line segments AB and CD in the Cartesian plane over an ordered field F are *congruent* if

$$\text{dist}^2(A, B) = \text{dist}^2(C, D).$$

Since congruence is defined using the function dist^2 from line segments to the field, the axiom (C2), transitivity of congruence will be obvious. Notice because of the ordering on F, if A, B are distinct points, then $\text{dist}^2(A, B) > 0$.

Next we will define congruence for angles, by defining a function $\tan \alpha \in F$ for any angle α. This is motivated by the usual tangent function of trigonometry, but since we are working over an abstract field, you should not assume any properties of this function until they have been proved.

Recall that a *ray* is a subset of a line consisting of a point plus all the points of the line on one side of that point. An *angle* is a set of two rays emanating from the same point, which do not lie on the same line. The *interior* of the angle α consists of all points of the plane which are on the same side of l as r' and on the same side of l' as r.

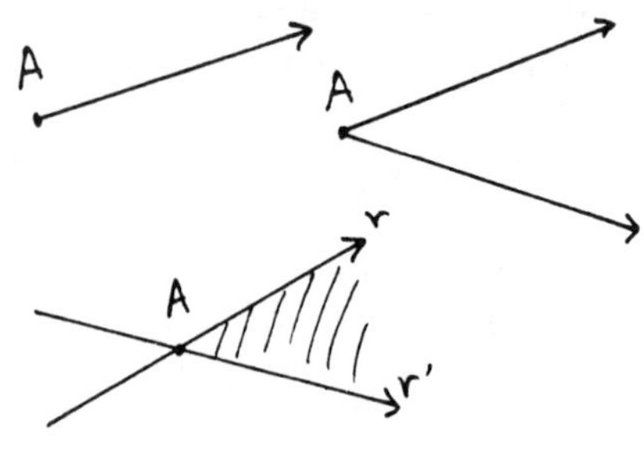

We say that an angle is a *right angle* if the slopes of the lines its rays lie on satisfy $mm' = -1$. Then we say an angle is *acute* if it is contained in the interior of a right angle; it is *obtuse* if it contains a right angle in its interior.

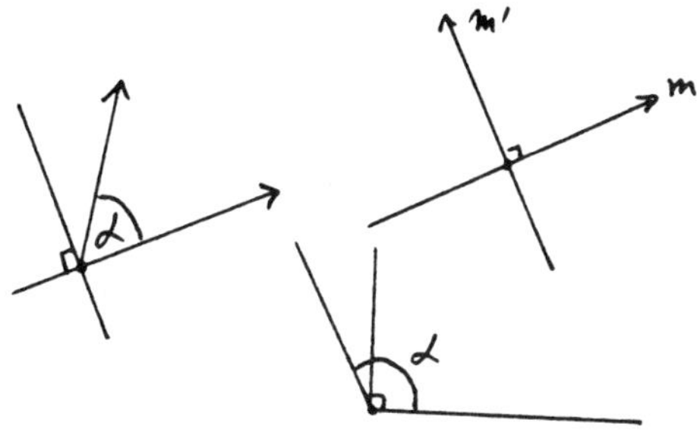

Definition. If α is an angle formed by two rays r, r' lying on lines of slopes m, m', we define the *tangent* of α to be

$$\tan \alpha = \pm \left| \frac{m' - m}{1 + mm'} \right|$$

where we take $+$ if the angle is acute, and $-$ if the angle is obtuse.

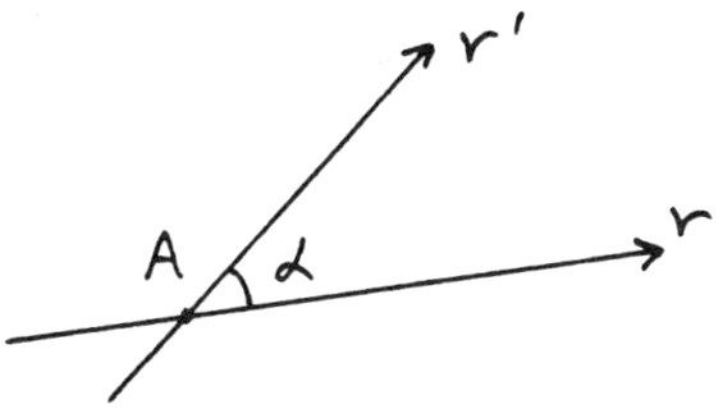

Note that the awkwardness of this definition is due to the fact that the slopes depend only on the lines, and do not distinguish the rays on those lines. So any formula using slopes cannot distinguish an angle from its supplement.

Note also that by this definition, the tangent of a strictly acute or strictly obtuse angle is an element of the field F, while the tangent of a right angle we take to be the symbol ∞. In case one of the slopes m or m' in the definition is ∞, we can still make sense of the formula by using rules such as

$$\frac{\infty - m}{1 + m \cdot \infty} = \frac{1}{m}.$$

Definition. Two angles in the Cartesian plane over an ordered field F are *congruent* if they have the same tangent, considered as an element of the set $F \cup \{\infty\}$.

Because congruence is defined by a function with values in $F \cup \{\infty\}$, axiom (C5), transitivity of congruence, becomes obvious.

Proposition 16.1 *Let F be an ordered field, and let Π_F be the associated Cartesian plane. Then Π_F satisfies axioms* (C2)–(C5). *Furthermore, Π_F satisfies* (C1) *if and only if F satisfies the condition*

(∗) *For any element $a \in F$, the element $1 + a^2$ has a square root in F (in which case we say the field F is* Pythagorean*).*

Proof. (C2) is transitivity of congruence of segments, which follows immediately from our definition of congruence using the dist^2 function.

(C3) is left as an exercise (Exercise 16.1).

(C4) is the axiom about laying off angles. So suppose given an angle α, and a ray emanating from a point A, with slope m. We must find a line passing through A with slope m' such that

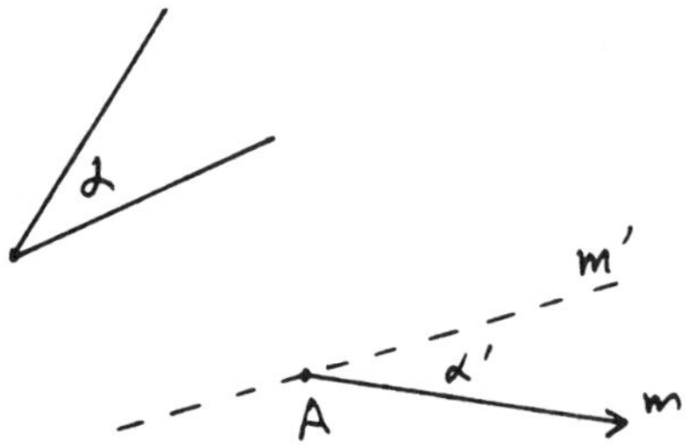

$$\tan\alpha = \pm\left|\frac{m'-m}{1+mm'}\right|$$

where the sign is adjusted according to whether α is acute or obtuse. This gives equations which are linear in m', so can be solved in F. We find

$$m' = \frac{m \pm \tan\alpha}{1 \mp m\tan\alpha}.$$

The two solutions give angles on either side of the given line through A, so that we can construct the new angle α' on the desired side of the line.

(C5) is the transitivity of congruence of angles, which is immediate from our definition of congruence using the tangent function.

Now let us consider the axiom (C1) about laying off line segments. This does not hold over an arbitrary field. For example let $\mathbb{Q}$ be the field of rational numbers. Then the segment from $(0,0)$ to $(1,1)$ cannot be laid off on the x-axis, because its length, $\sqrt{2}$, is not in the field.

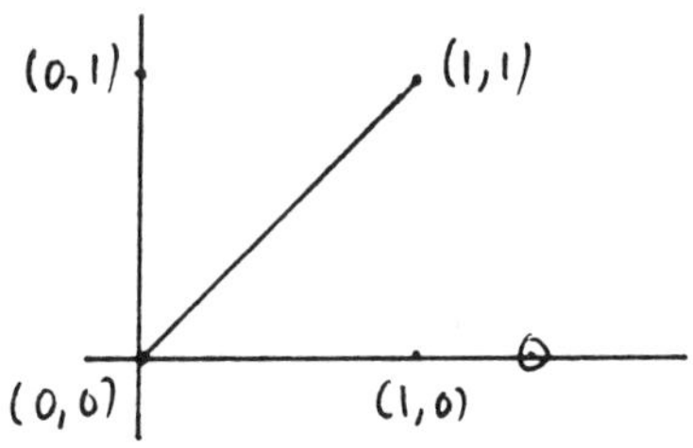

More generally, over an arbitrary field F, if $a \in F$ is any element, let us consider the segment from $(0,0)$ to $(a,1)$. There will be a segment congruent to this one, laid off on the x-axis starting from 0, if and only if there is an element $b \in F$ such that

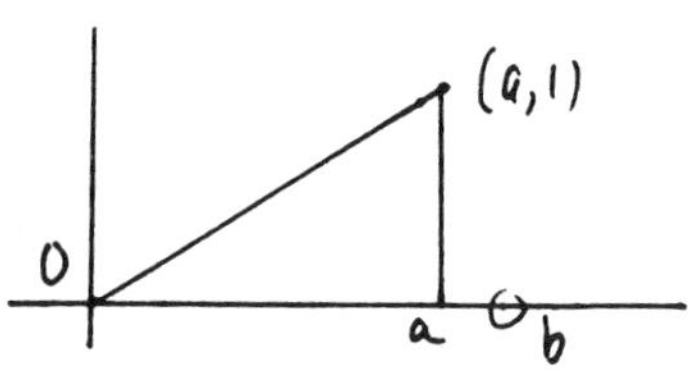

$$\text{dist}^2((0,0),(a,1)) = \text{dist}^2((0,0),(b,0)).$$

This says

$$1 + a^2 = b^2.$$

Thus we need $b \in F$ which is a square root of $1+a^2$. So we see that in order for (C1) to hold in Π_F, we must have the condition $(*)$ on the field F.

Conversely, suppose that F satisfies $(*)$, namely for any $c \in F$, we have $\sqrt{1+c^2} \in F$. Notice first that for any $a, b \in F$, with $a \neq 0$, we can write

$$a^2 + b^2 = a^2\left(1 + \left(\frac{b}{a}\right)^2\right).$$

Now letting $c = b/a$ we see that

$$\sqrt{a^2 + b^2} = |a| \cdot \sqrt{1 + c^2}$$

is also in F. From this it follows that for any two points $A, B \in \Pi_F$, the *distance* between A, B is also in F, so we have the distance function

$$\text{dist}(A, B) = \sqrt{(a_1 - b_1)^2 + (a_2 - b_2)^2} \in F.$$

Now suppose given a line $y = mx + b$ and a point A on the line, and suppose we wish to lay off a segment of length d. We can write $A = (a, ma + b)$, and we are looking for a point $C = (c, mc + b)$ on the same line such that

$$\text{dist}(A, C) = d.$$

This says

$$\sqrt{(a - c)^2 + (ma + b - (mc + b))^2} = d$$

which becomes

$$|a - c| \cdot \sqrt{1 + m^2} = d.$$

Since F satisfies $(*)$, the quantity $\sqrt{1 + m^2} \in F$, so we can solve this equation for c. Note that there will be two solutions, corresponding to the two directions from A along the line l.

Remark. We defer consideration of (C6), the (SAS) axiom, to the next section where we discuss rigid motions and Euclid's method of superposition.

To complete this section, we discuss the intersections of lines and circles. Recall from §11 the circle-circle intersection property which we called axiom (E), and the line-circle intersection property (LCI) which was proved in (11.6) as a consequence of (E).

Proposition 16.2 *Let Π be the Cartesian plane over an ordered field F. Then the following conditions are equivalent:*

(i) Π *satisfies the circle-circle intersection property* (E).

(ii) Π *satisfies the line-circle intersection property* (LCI).

(iii) *the field F satisfies $(**)$: for any $a \in F$, $a > 0$, there is a square root of a in F (in which case we say the field F is* Euclidean*).*

Proof. (i) $\Rightarrow$ (ii) is proved in (11.6), using Hilbert's axioms for Π which we have already established. Note that the proof given there uses (C1) and (C6), but in fact that construction and proof are valid over any ordered field F.

(ii) $\Rightarrow$ (iii). Now we assume (LCI) and we must prove that F has square roots of positive element. Given an element $a \in F$, $a > 0$, consider the points $O = (0,0)$, $A = (a,0)$ and $A' = (a+1,0)$. Let Γ be the circle with center $(\frac{1}{2}(a+1),0)$ and radius $\frac{1}{2}(a+1)$. Consider the vertical line l through the point A. Clearly A is inside the circle, so by (LCI), the line l must meet the circle Γ in a point B. Solving the equations, we find $B = (a, \sqrt{a})$, so $\sqrt{a} \in F$.

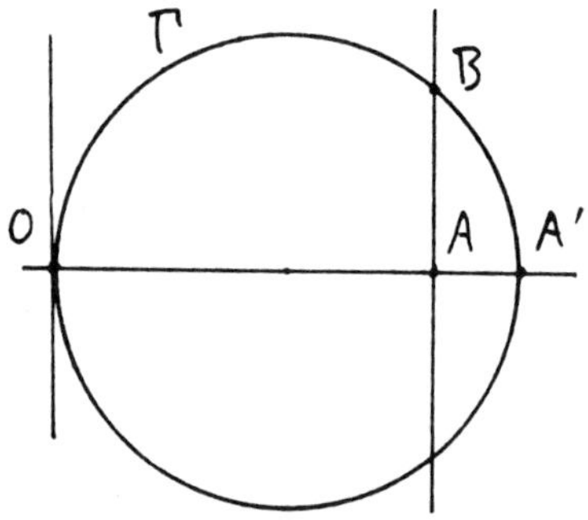

(iii) $\Rightarrow$ (i). This time we assume the existence of square roots of positive elements in F, and must prove (E). If Γ and Γ' are circles in Π, their equations can be written

$$\begin{aligned}(x-a)^2 + (y-b)^2 &= r^2 \\ (x-c)^2 + (y-d)^2 &= s^2\end{aligned}$$

where (a,b) and (c,d) are the centers of the two circles, and r, s their radii (which are elements of F because of our hypothesis $(**)$ on existence of square roots).

Normally, solving two quadratic equations simultaneously would lead to a fourth degree equation, but in this case, the coefficients of x^2 and y^2 in both equations are 1. Thus we can subtract one equation from the other to get a linear equation. This can be solved simultaneously with one of the quadratic equations using only square roots, and so, using $(**)$, the intersection points of the circles have coordinates in F, so they exist in the plane Π. We leave to the reader the troublesome verification that if Γ and Γ' satisfy the hypothesis of (E), then the square roots we need will be square roots of *positive* elements of F, so will exist by $(**)$, cf. Exercise 16.5.

Remark 16.2.1. This shows that (E) and (LCI) are equivalent in the Cartesian plane over any ordered field F. We do not know whether they are equivalent in an abstract Hilbert plane.

Proposition 16.3 *Let Ω be the set of all real numbers which can be expressed starting from the rational numbers and using a finite number of operations $+$, $-$, $\cdot$, $\div$ and $c \mapsto \sqrt{1+c^2}$. (Note that for any $c \in \mathbb{R}$, $1+c^2 > 0$, so $\sqrt{1+c^2} \in \mathbb{R}$.) Then Ω is an ordered field satisfying $(*)$.*

Proof. To show that Ω is a field, let $a, b \in \Omega$. Then each of a, b can be expressed in a finite number of steps using rational numbers and operations $+$, $-$, $\cdot$, $\div$, $c \mapsto \sqrt{1+c^2}$. Hence the same is true of $a \pm b$, $a \cdot b$, and a/b provided $b \neq 0$. If c is any element of Ω, then c can be expressed in a finite number of such steps, so $\sqrt{1+c^2}$ can also, so $\sqrt{1+c^2} \in \Omega$. Hence Ω satisfies $(*)$. Ω is an ordered field, because it is a subfield of $\mathbb{R}$, so we can take P to be those elements of Ω, which are positive as real numbers.

Remark 16.3.1. Clearly Ω is the smallest ordered field satisfying $(*)$. We call it *Hilbert's field*, since Hilbert studied this field in his "Foundations of Geometry".

It is also the smallest field over which all of Hilbert's axioms of betweenness and congruence will hold (17.3.1).

Proposition 16.4 *Let K be the set of all real numbers which can be obtained from the rational numbers by a finite number of operations $+$, $-$, $\cdot$, $\div$, and $a > 0 \mapsto \sqrt{a}$. Then K is an ordered field satisfying* (**).

Proof. Similar to proof of (16.3). Note that we may only take square roots of positive elements. Since K is also a subfield of $\mathbb{R}$, we get the ordering on K from $\mathbb{R}$ as above.

Remark 16.4.1. We call this the *constructible field* because it is the smallest field over which we can carry out ruler and compass constructions. Note also that $\Omega \subseteq K$ since $1 + c^2 > 0$ for any $c \in \Omega$. To show $\Omega \neq K$, see Exercise 16.8.

Exercises

16.1 If Π is the plane over an ordered field F, show that (C3), congruence of added line segments, holds. Do not assume condition (*) for the field F.

16.2 Make up a set of rules for dealing with ∞ so that we can do arithmetic in $F \cup \{\infty\}$ and get the results we want with slopes and tangents of angles.

16.3 If F is an ordered field satisfying (*), prove the *triangle inequality* in the corresponding plane Π, namely, if A, B, C are three points in Π, then

$$\text{dist}(A, C) \leq \text{dist}(A, B) + \text{dist}(B, C),$$

and equality holds if and only if A, B, C are collinear and B is between A and C.

16.4 Using the definition of the tangent of an angle given in the text, verify that for any two acute angles α, β,

$$\tan(\alpha + \beta) = \frac{\tan \alpha + \tan \beta}{1 - \tan \alpha \tan \beta}.$$

16.5 Let Π be the plane over an ordered field F satisfying (**). Verify that a circle Γ meets a line l in two points if and only if l has a point inside Γ. *Hint*: Compute the shortest distance from the center O of the circle to the line l. Show that this is less than the radius of the circle if and only if the square roots needed to solve the equations are square roots of *positive* elements.

16.6 Let F be an ordered field (without assuming (*) or (**)). Let A, B be points of the associated plane Π. Show that the circle Γ with center A and passing through B has infinitely many points on it. *Hint*: First do the case of the circle of radius 1 and center $(0, 0)$ over $\mathbb{Q}$.

16.7 Prove, using high school geometry and trigonometry, that if l and l' are lines in $\mathbb{R}^2$, which are not perpendicular to each other, and which have slopes m, m', then the (usual) tangent of the acute angle α between the two lines is given by

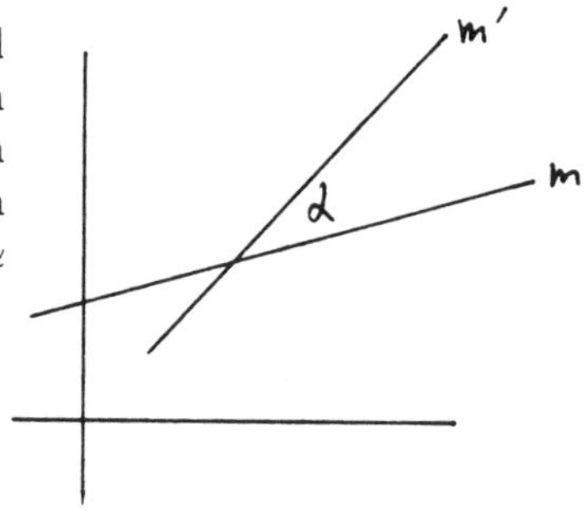

$$\tan \alpha = \left| \frac{m' - m}{1 + mm'} \right| .$$

(In other words, you are checking that what we took as a *definition* of the tangent is equal to the usual tangent in $\mathbb{R}^2$.)

16.8 *Totally real field extensions* (this exercise requires some knowledge of field theory). We consider *algebraic numbers*, which are complex numbers satisfying some polynomial equation with rational coefficients. We denote the set of algebraic numbers by $\bar{\mathbb{Q}}$. An algebraic number $a \in \bar{\mathbb{Q}}$ is *totally real* if it and all its conjugates are real $\in \mathbb{R}$. We say a is *totally positive* if it and all its conjugates are real and positive. A subfield $F \subseteq \bar{\mathbb{Q}}$ is *totally real* if all its elements are totally real. Show the following:

a) If $F \subseteq \bar{\mathbb{Q}}$ is a totally real subfield, and if $a \in F$ is a totally positive element, then the extension field $F' = F(\sqrt{a})$ is totally real.

b) If $a_1, \ldots, a_r$ are elements of a totally real field F, then Σa_i^2 is a totally positive element of F.

c) Hilbert's field Ω (cf. 16.3) is a totally real field.

d) The number $a = \sqrt{1 + \sqrt{2}}$ is in the constructible field K (cf. 16.4) but not in Ω. Thus $\Omega < K$.

e) In fact, Ω is equal to the set of totally real elements in the field K.

f) The converse of b) is a theorem of Emil Artin: If b is a totally positive element of a totally real field F, then there exist $a_1, \ldots, a_r \in F$ such that $b = \Sigma a_i^2$. Verify directly that this is true in the field $\mathbb{Q}(\sqrt{2})$: every totally positive element is a sum of squares.

16.9 Use ideas from (Exercise 16.8) to give an example of three line segments in the Cartesian plane over Ω, any two exceeding the third, but so that the triangle with sides equal to those segments does not exist. Thus (I.22) fails in this plane.

17 Rigid Motions and SAS

Our first goal in this section is to show that Hilbert's axiom (C6), the "side-angle-side" criterion for congruence of triangles, holds in the geometry over an ordered field F. This will complete the proof that all of Hilbert's axioms hold in the geometry over a field (with suitable conditions on the field). After that we will study the properties of rigid motions in any plane, not necessarily one over a field.

One could criticize Hilbert for taking a statement as complicated as (SAS) for an axiom, just as one could criticize Euclid for his fifth postulate which is so much less elementary than his others. The response in both cases is the same: there is no avoidance of including a statement as an axiom if one cannot succeed in proving it from the other axioms. Now Euclid did not include (SAS) as an axiom, but "proved" it as (I.4). However his method of proof has been justly criticized, because he used the "method of superposition" which involves moving one triangle and placing it on the other. This cannot be justified on the basis of Euclid's postulates and common notions, because nothing in them speaks of the possibility of moving one figure onto another. In fact, if you think about it, the possibility of moving figures around, without distorting their shapes, is a rather strong statement about the homogeneity of space: the geometry is similar in different parts of the space. As such this is a deep fact about the geometry, and not to be taken lightly or assumed without proof.

Curiously enough, in order to show that (SAS) holds in the geometry over a field, we will use Euclid's method of superposition, but we will first prove that this is possible in this context. We will define the notion of *rigid motion* of a plane, and we will prove the existence of enough rigid motions in the geometry over a field, so as to be able to use Euclid's method rigorously.

Conversely, we will see that in an arbitrary Hilbert plane, we can prove the existence of rigid motions. Thus the existence of enough rigid motions is essentially equivalent to the statement (SAS), in the presence of the other axioms. This gives a satisfactory modern understanding of the meaning of Euclid's method of superposition. It also introduces us to the group of rigid motions of the plane and validates Felix Klein's point of view, expressed in his "Erlanger Program" in the late 19th century, that one should classify different geometries according to the groups of motions that act on them.

To start with we define the notion of a rigid motion.

Definition. If Π is a geometry, consisting of the undefined notions of point, line, betweenness, and congruence of line segments and angles, which may or may not satisfy various of Hilbert's axioms, we define a *rigid motion* of Π to be a mapping $\varphi : \Pi \to \Pi$ defined on all points, such that

1) φ is a 1-1 mapping of the points of Π onto itself,

2) φ sends lines into lines,

3) φ preserves betweenness of collinear points,

4) for any two points A, B, we have $AB \cong \varphi(A)\varphi(B)$,

5) for any angle α, we have $\angle\alpha \cong \angle\varphi(\alpha)$.

In other words φ preserves the structures determined by the undefined notions in our geometry.

Remark 17.0.1. For example, the *identity* transformation of Π to itself, which leaves every point fixed, is a (trivial) rigid motion. However, it is not obvious in general that there are any other rigid motions besides the identity. It is clear, on the other hand, that the set G of all rigid motions forms a *group*, because the composition of any two is another one. We will use functional notion for composition: $\varphi\psi(A) = \varphi(\psi(A))$.

Now we can express what is needed to justify Euclid's method of superposition in the following principle

ERM. (Existence of Rigid Motions)

1) For any two points $A, A' \in \Pi$, there is a rigid motion $\varphi \in G$ such that $\varphi(A) = A'$.

2) For any three points O, A, A' satisfying $OA \cong OA'$, there is a rigid motion $\varphi \in G$ such that $\varphi(O) = O$ and $\varphi(A) = A'$.

3) For any line l, there is a rigid motion $\varphi \in G$ such that $\varphi(P) = P$ for all $P \in l$ and φ interchanges the two sides of l.

Proposition 17.1 *In a plane satisfying the incidence and betweenness axioms, and assuming* (C2), (C5), *and the uniqueness portions of* (C1) *and* (C4) *only,* (ERM) *implies* (C6) = (SAS).

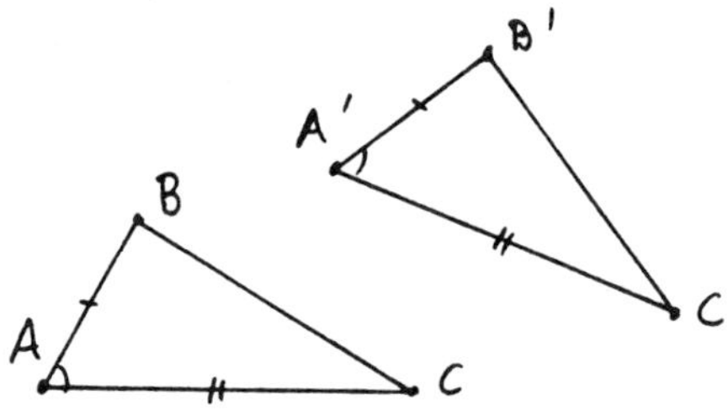

Proof. We are now assuming the existence of rigid motions (ERM) and will prove (SAS) by Euclid's method. So suppose given two triangles ABC and $A'B'C'$, and suppose $AB \cong A'B'$, $AC \cong A'C'$, and the angle $\angle BAC \cong \angle B'A'C'$. Then we must show that $\Delta ABC \cong \Delta A'B'C'$, namely that $BC \cong B'C'$ and the angles at B, C are congruent to the angles at B' and C' respectively.

By (ERM) 1), there is a rigid motion φ which takes A to A'. Let $B'' = \varphi(B)$. Then $AB \cong A'B''$ since φ is a rigid motion and $AB \cong A'B'$ by hypothesis, so $A'B' \cong A'B''$ by (C2).

Next, by (ERM) 2), there is a rigid motion ψ which leaves A' fixed and sends B'' to B'. Let $C'' = \psi\varphi(C)$.

Then we consider the line $l = A'B'$, and the two rays $A'C'$ and $A'C''$. If they are on the same side of l we do nothing, but if on opposite sides, by (ERM) 3) there is a rigid motion σ leaving the points of l fixed, and interchanging the sides.

Let us denote by $\theta \in G$ the composition $\psi\varphi$ or $\sigma\psi\varphi$ if we used σ. Then θ has the following properties: $\theta(A) = A'$, $\theta(B) = B'$, and $C''' = \theta(C)$ is on the same side of $A'B'$ as C'.

Since θ is a rigid motion, $\angle BAC \cong \angle B'A'C''$. But also $\angle BAC \cong \angle B'A'C'$ by hypothesis, so by (C5), $\angle B'A'C' \cong \angle B'A'C'''$. Furthermore, C' and C''' are on the same side of $A'B'$. So by the uniqueness portion of (C4) we conclude that the rays $A'C'$ and $A'C'''$ are equal.

Now $A'C' \cong AC$ by hypothesis, and $AC \cong A'C'''$ since θ is a rigid motion, so by (C2) $A'C' \cong A'C'''$. Furthermore C' and C''' are on the same side of A'. So by the uniqueness portion of (C1) we conclude that $C' = C'''$.

Thus $\theta(B) = B'$ and $\theta(C) = C'$. Since θ is a rigid motion, $BC \cong B'C'$ as required. Similarly, for the angles, θ takes $\angle ABC$ to $\angle A'B'C'$. So being a rigid motion we conclude $\angle ABC \cong \angle A'B'C'$. The same method shows $\angle ACB \cong \angle A'B'C'$. This concludes the proof of (SAS).

Next we will show that (ERM) holds in the Cartesian plane over a field.

Theorem 17.2 *Let F be an ordered field, and let Π be the associated Cartesian plane. Then* (ERM) *holds in Π.*

Proof. We think of Π as having coordinates (x, y). We will consider certain transformations of Π defined by functions of x and y, we will show that these are rigid motions, and then we will see that there are enough of them to prove (ERM).

First of all consider a point $A = (a, b)$ and the transformation τ (called a *translation*) given by

$$\begin{cases} x' = x + a \\ y' = y + b. \end{cases}$$

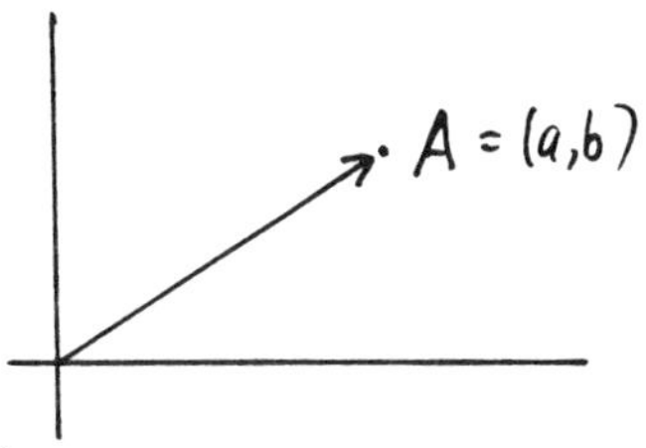

Clearly τ is 1-1 and onto, because it has an inverse

$$\begin{cases} x = x' - a \\ y = y' - b. \end{cases}$$

A line $y = mx + k$ under this transformation becomes

$$y' - b = m(x' - a) + k$$

or

$$y' = mx' - ma + k + b.$$

In particular its image is a line, so we see that τ takes lines into lines. Second we notice that the slope of the new line is the same as the slope of the old line, so τ preserves angles. Clearly τ preserves betweenness, because this reduces to questions of inequalities in the field F, which are unchanged by adding constants.

Finally we must check that τ preserves the dist^2 function to get congruence of segments. This is obvious, since we add the same constant to the coordinates of two points A, B, so in computing the dist^2 function we get the same thing.

Thus the mapping τ is a rigid motion. Given two points B, C, we can take a, b to be the difference of their x- and y-coordinates, so $\tau(B) = C$, and we have satisfied condition 1) of (ERM).

To prove condition 2) of (ERM), we will first consider another kind of transformation (which is not a rigid motion) called a *stretch-rotation.* Given any point $A = (a, b)$ in the plane, not equal to $(0, 0)$, we consider the transformation $\theta = \theta_A$ defined by

$$\begin{cases} x' = ax - by \\ y' = bx + ay. \end{cases}$$

This transformation has an inverse

$$\begin{cases} x = \frac{1}{a^2+b^2}(ax' + by') \\ y = \frac{1}{a^2+b^2}(-bx' + ay') \end{cases}$$

so it is 1-1 and onto. If $y = mx + f$ is a line, it is transformed by θ (as a short calculation shows) into the line

$$y' = \frac{b+ma}{a-mb}x' + \frac{f(a^2+b^2)}{a-mb}.$$

Thus θ takes lines to lines. In particular, the slope of the new line is

$$m' = \frac{b+ma}{a-mb}.$$

Since linear transformations either preserve or reverse inequalities, θ preserves betweenness.

Next we show that θ preserves angles. Given two lines with slopes m_1 and m_2, let m_1' and m_2' be the new slopes. Since congruence of angles is determined by their tangents, it will be enough to show that

$$(1) \qquad \frac{m_1' - m_2'}{1 + m_1' m_2'} = \frac{m_1 - m_2}{1 + m_1 m_2}.$$

This is an elementary calculation (left to the reader).

Finally, let us see what happens to the distance function. Let $C = (c_1, c_2)$ and $D = (d_1, d_2)$ be two points. Then another amusing little calculation, left to the reader, shows that

$$(2) \qquad \text{dist}^2(\theta(C), \theta(D)) = (a^2 + b^2)\ \text{dist}^2(C, D).$$

Now we can verify the condition 2) of (ERM) in the Cartesian plane Π_F. Given points O, A, A' satisfying $OA \cong OA'$, we must show there is a rigid motion leaving O fixed and taking A to A'. By using a translation, we can reduce to the case $O =$ origin. Let φ be the composition of stretch-rotations $\varphi = \theta_{A'}\theta_A^{-1}$. Then φ is 1-1 and onto, preserves lines, betweenness, and angles, because the stretch rotations do. Furthermore φ sends A to $(1, 0)$ and then to A', and leaves O fixed, as required. It remains only to show that φ preserves distances. Indeed, θ_A and $\theta_{A'}$ multiply the dist^2 function by factors equal to $\text{dist}^2(0, A)$ and $\text{dist}^2(0, A')$, which are the same, so φ preserves distances, hence is a rigid motion, as required.

To complete the proof of (ERM), we must verify condition 3), that for every line l, there is a rigid motion (called a *reflection*) leaving l pointwise fixed, and interchanging the two sides of l. Using a translation from a point of l to the origin O, we may assume that $O \in l$. Let A be any other point of l, and let θ_A be the stretch-rotation which sends $(1, 0)$ to A. Let σ be the *reflection* in the x-axis, defined by

$$\begin{cases} x' = x \\ y' = -y. \end{cases}$$

Clearly this is a rigid motion which leaves the x-axis pointwise fixed and interchanges the two sides. Now $\varphi = \theta_A \sigma \theta_A^{-1}$ is the required reflection in the line l. □

Remark 17.2.1. A stretch-rotation with scale factor $a^2+b^2 = 1$ is called a *rotation*. Note that we did not prove the existence of a rotation sending any line through O

to any other line through O. In fact these may not exist. For example, over the rational numbers $\mathbb{Q}$, there is no rotation taking the x-axis to the line $y = x$. One can see easily that the group of rigid motions will be transitive on the set of lines through O if and only if the field F is Pythagorean (condition $(*)$ of (16.1)).

Corollary 17.3 *If F is any ordered field, then axiom* (C6) *holds in the Cartesian plane over F.*

Proof. Indeed, by the theorem (ERM) holds in Π. Since we have previously verified that (C2)–(C5) hold in Π (16.1), and since the uniqueness part of (C1) holds in Π even without condition $(*)$ of (16.1), it follows from (17.1) that (C6) holds in Π.

Remark 17.3.1. This completes the proof that Hilbert's axioms hold in the Cartesian plane over a field. To be more precise, for any ordered field F, the axioms of incidence (I1)–(I3), betweenness (B1)–(B4), congruence (C2)–(C6) and Playfair's axiom (P) all hold. In order for (C1) to hold, we need condition $(*)$ that F is Pythagorean, and then Π_F will be a Hilbert plane. If we assume the stronger condition $(**)$, that F is Euclidean, then Π will be Euclidean.

Next we will prove a sort of converse to (17.1), namely that (ERM) holds in any Hilbert plane.

Proposition 17.4 *In any Hilbert plane (cf. §10), there are enough rigid motions:* (ERM) *holds.*

Proof. First we will show the existence of reflections. Then we will build other rigid motions out of these.

Suppose given a line l. We will construct a rigid motion σ, called the *reflection* in l, which leaves the points of l fixed and interchanges the two sides of l. For any point $P \in l$ we define $\sigma(P) = P$. For any point $A \notin l$, drop the perpendicular AA_0 to l, and extend on the far side of l so that $AA_0 \cong A_0A'$. Then we set $\sigma(A) = A'$. Clearly $\sigma^2 =$ id, so σ is 1-1 and onto.

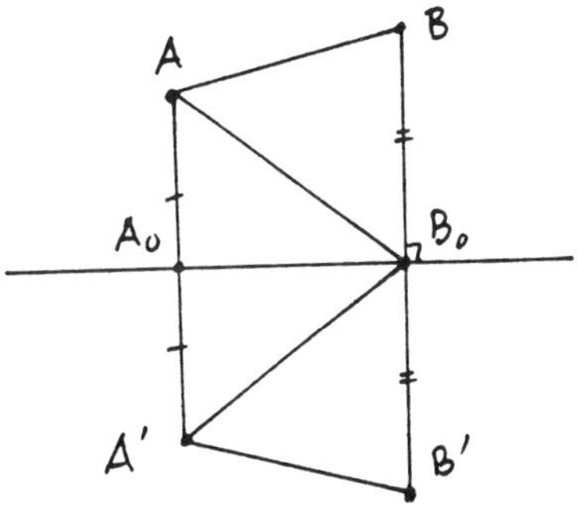

Let A, B be any two points not on l. We will show that $AB \cong A'B'$, where $\sigma(A) = A'$, and $\sigma(B) = B'$. If A, B lie on the same line perpendicular to l, this is immediate from subtracting congruent line segments. If A, B are on different perpendiculars, as in the figure, let A_0, B_0 be the feet of those perpendiculars. Then ΔA_0AB_0 is congruent to $\Delta A_0A'B_0$, using the right angles at A_0, by (SAS). Therefore the angles $\angle AB_0A_0$ and $\angle A'B_0A_0$ are congruent. Subtracting from the right angles at B_0, we find $\angle AB_0B \cong \angle A'B_0B'$. On the other hand, $AB_0 \cong A'B_0$ from the first triangles. Now we can apply (SAS) again to conclude that $\Delta AB_0B \cong \Delta A'B_0B'$. In particular, $AB \cong A'B'$ as claimed.

Now suppose A, B, C are three non-collinear points, whose images by σ are A', B', C'. Then by (SSS) we conclude that $\Delta ABC \cong \Delta A'B'C'$, and in particular $\angle BAC \cong \angle B'A'C'$. Thus σ preserves angles. From here it is easy to verify that σ preserves lines and betweenness, so in fact σ is a rigid motion (details left to reader).

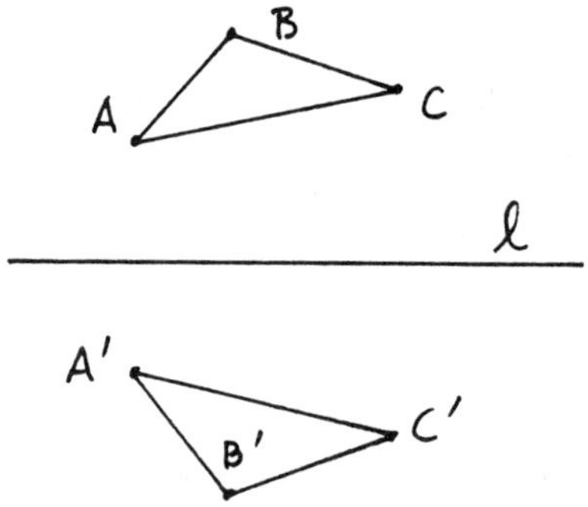

To verify that (ERM) holds, we have just established property 3) by the existence of reflections. If A, A' are any two points, let l be the perpendicular bisector of the segment AA'. Then σ_l will send A to A'. Thus condition 1) of (ERM) holds. For condition 2), let O, A, A' be three points with $OA \cong OA'$. Let l be the bisector of the angle $\angle AOA'$. Then the reflection σ_l will leave O fixed and send A to A'. Thus (ERM) holds.

Corollary 17.5 *In the presence of all the axioms of a Hilbert plane except* (C6), *the axiom* (C6) *is equivalent to* (ERM).

Proof. Combine (17.1) and (17.4).

Exercises

17.1 In the proof of theorem (17.2) verify

a) angles are preserved by a stretch-rotation, formula (1)

b) dist^2 is multiplied by $(a^2 + b^2)$, formula (2).

17.2 Let $\varphi : \Pi \to \Pi$ be a map of a Hilbert plane into itself. For any point A, denote $\varphi(A)$ by A'. Assume $AB \cong A'B'$ for any two points A, B.

a) Prove that φ is 1-1 and onto.

b) Show in fact φ is a rigid motion.

17.3 In a Hilbert plane Π, show:

a) Any rigid motion with at least three non-collinear fixed points must be the identity.

b) Any rigid motion is equal to the product of at most three reflections.

17.4 In a Hilbert plane Π, define a *rotation* around a point O to be a rigid motion ρ leaving O fixed, and such that for any two points A, B, the angles $\angle AOA'$ and $\angle BOB'$ are equal, where $\rho(A) = A'$, $\rho(B) = B'$. Show

a) For any two points A, A' with $OA \cong OA'$, there exists a unique rotation around O sending A to A'.

b) The set of rotations around a fixed point O, together with the identity, is an abelian subgroup of the group of all rigid motions.

c) Any rotation can be written as the product of two reflections.

d) A rigid motion having exactly one fixed point must be a rotation.

17.5 In a Euclidean plane Π, define a *translation* to be a rigid motion τ such that for any two points A, B, we have $AA' \cong BB'$, where $\tau(A) = A'$, $\tau(B) = B'$. Show

a) For any two points A, A', there exists a unique translation τ such that $\tau(A) = A'$.

b) If τ is a translation, then for any two points A, B, $AB \| A'B'$ and $AA' \| BB'$.

c) The set of translations forms an abelian subgroup T of the group of all rigid motions.

d) Any translation is a product of two reflections.

e) Is the group T of translations a normal subgroup of the group G of all rigid motions? Prove yes or no.

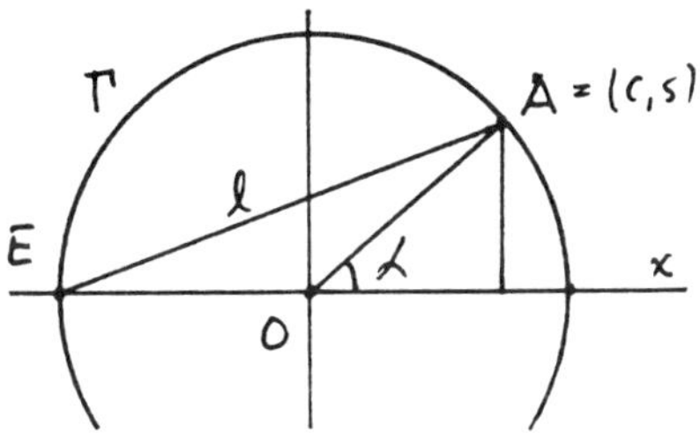

17.6 In this exercise we establish an algebraic interpretation of the group of rotations around a point in a Cartesian plane. Let F be an ordered field. In the Cartesian plane Π over F, let Γ be the unit circle, and let $E = (-1, 0)$. Let a line l through E meet the circle at a point A.

a) If the line l has slope t, show that the coordinates of the point A are (c, s), where

$$c = \frac{1 - t^2}{1 + t^2}, \qquad s = \frac{2t}{1 + t^2}.$$

(*Note*: We use this notation, because if $F = \mathbb{R}$, and if α is the angle that OA makes with the positive x-axis, then by trigonometry, we find $t = \tan \frac{1}{2}\alpha$, $c = \cos \alpha$, $s = \sin \alpha$, and these are the usual formulas for expressing c, s in terms of t. Do you remember those substitutions used in calculus classes for rationalizing trigonometric integrals?)

b) Let ρ_t be the corresponding stretch-rotation

$$\rho_t : \begin{cases} x' = cx - sy \\ y' = sx + cy \end{cases}$$

as defined in the text. Show that ρ_t is a rotation, and show that the mapping $t \mapsto \rho_t$ gives a 1-1 correspondence between the set $F \cup \{\infty\}$ and the group R of rotations of Π with center O.

c) Under the correspondence in b), show that the group operation in R corresponds to the operation

$$t \circ t' = \frac{t + t'}{1 - tt'}$$

in the set $F \cup \{\infty\}$. We call this set $F \cup \{\infty\}$ with the operation $\circ$ the *circle group* of the field F.

d) If $F = \mathbb{R}$, show that the circle group of F is isomorphic to the abstract group $(\mathbb{R}/\mathbb{Z}, +)$.

17.7 Let F be a field (not necessarily an ordered field) which does not contain a square root of -1. In analogy to the situation above, we define the *circle group* of F to be the set $C(F) = F \cup \{\infty\}$ with the operation

$$a \circ b = \frac{a+b}{1-ab} \text{ for } a, b \in C(F).$$

a) Verify directly that $C(F), \circ$ is an abelian group by verifying the group axioms. Make clear your rules for treating ∞, and point out where you use the hypothesis $\sqrt{-1} \notin F$.

b) Show that $+1$, -1 are elements of finite order in $C(F)$.

c) If $F = \mathbb{Q}$, show that $\pm 1, \infty$ are the *only* elements of finite order different from the identity.

d) If $F = \mathbb{R}$, find explicitly four elements of order 5 in $C(F)$.

17.8 With F a field as in (Exercise 17.7), consider an extension field $F(i)$ with $i = \sqrt{-1}$. Elements of $F(i)$ can then be written as $\alpha = a + bi$, $a, b \in F$. Define the *norm* of an element α by $N(\alpha) = a^2 + b^2$.

a) For any $\alpha, \beta \in F(i)$ verify that $N(\alpha\beta) = N(\alpha)N(\beta)$.

b) Let $S = \{\alpha \in F(i) \mid N(\alpha) = 1\}$. Then S is a group under multiplication. Show that the map $S \to C(F)$ defined by

$$\alpha = a + bi \mapsto \frac{b}{a+1}$$

is an isomorphism of $(S, \cdot)$ with the circle group $(C(F), \circ)$.

Note: This mysterious isomorphism is motivated by the figure in (Exercise 17.6), in which $t = \tan \frac{1}{2}\alpha = s/(c+1)$.

17.9 Let F be a finite field of p elements, $p \equiv 3 \pmod 4$.

a) Show that -1 does not have a square root in F.

b) Show that the circle group $C(F)$ is *cyclic* of order $p + 1$.

17.10 Let ABC and $A'B'C'$ be two congruent triangles in a Hilbert plane. Show that there exists a rigid motion φ of the plane with $\varphi(A) = A'$, $\varphi(B) = B'$, and $\varphi(C) = C'$.

17.11 In a Euclidean plane, show that the product of two rotations around different points is equal to either a rotation around a third point or a translation. *Hint*: Show that it has at most one fixed point.

17.12 In a Euclidean plane, show that the product of an odd number of reflections cannot be equal to the identity. *Hint*: Use Exercise 17.11 to reduce products of four reflections to products of two reflections, and go by induction.

18 Non-Archimedean Geometry

The Archimedean principle, that given two line segments, some multiple of the first will exceed the second, is so embedded in our experience of the world that it is hard to imagine a geometry in which this would not hold. Even the farthest star has a distance from the earth that can be measured in light years, and even if we take the inch as our standard unit of length, some number of inches, albeit a very large number, will exceed the distance to that farthest star. As long as we retain the notion that geometry somehow represents the real world, we are bound to accept Archimedes principle as a truth.

In abstract mathematics, on the other hand, a geometry is anything which satisfies a certain set of axioms. In this chapter we have seen how to construct a geometry over an abstract ordered field. The elements of the field need not be numbers or distances. Any abstract field will do.

In this section we will take advantage of this abstraction to construct some non-Archimedean geometries. These examples will serve two functions. One is to show the independence of Archimedes axiom (A) and Playfair's axiom (P) from the axioms of a Hilbert plane. The other is to free our minds from the constraints of habit by studying the properties of a logically constructed geometry in which Archimedes axiom (A) does not hold. Such geometries are called non-Archimedean geometries.

Proposition 18.1 *Let $\mathbb{R}$ be the field of real numbers, let t be an indeterminate, and let $\mathbb{R}(t)$ be the field of all* rational functions *of t, that is, all quotients $f(t)/g(t)$ where f and g are polynomial functions of t with real coefficients, and $g(t)$ is not identically zero. Then the field $F = \mathbb{R}(t)$ has a natural ordering which makes it into a non-Archimedean ordered field.*

Proof. If $\varphi \in F$, we think of φ as a function $\varphi(t) = f(t)/g(t)$ from $\mathbb{R}$ to $\mathbb{R}$, defined everywhere except at the finite number of points where $g(t) = 0$. We define the set P of "positive" elements of F to be the set of those functions φ which are positive for all large enough values:

$$P = \{\varphi \in F \mid \exists a_0 \in \mathbb{R} \text{ such that } \varphi(b) > 0 \text{ for all } b > a_0\}.$$

Now F is a field, because any sum, difference, product, or quotient of rational functions is again a rational function. The set P is closed under sums and products, because the sum and product of two eventually positive functions is again eventually positive. To show that (F, P) is an ordered field, it remains to show that if $\varphi \in F$, $\varphi \neq 0$, then either $\varphi \in P$ or $-\varphi \in P$, but not both. Indeed, if $\varphi \neq 0$, then it is the quotient of two non-zero polynomials $\varphi = f(t)/g(t)$. Each of these has a finite number of zeros. If we take $a_0 \in \mathbb{R}$ larger than all the zeros of $f(t)$ and $g(t)$, then φ is continuous and never 0 for all $b > a_0$. Thus by the intermediate value theorem, φ is either always positive for $b > a_0$, or always negative for $b > a_0$. In the first case $\varphi \in P$, in the second case $-\varphi \in P$.

Finally, we observe that F is non-Archimedean. Consider the element $t \in F$. For any integer $n > 0$, we have $t > n$ as elements of F. Indeed, for $b > n$, the function $\varphi(t) = t - n$ is positive. Note that in this field we have

$$0 < 1 < 2 < \cdots < t < t + 1 < t + 2 < \cdots < t^2 < t^3 < \ldots$$

Definition. Let F be a non-Archimedean ordered field. We will say an element $a \in F$ is *finite* if there exists a positive integer n for which $-n < a < n$. Otherwise we say that a is *infinite*. We say an element $a \in F$ is *infinitesimal* if for every positive integer n, we have $-1/n < a < 1/n$.

Next we will construct non-Archimedean fields satisfying the properties $(*)$ of (16.1) and $(**)$ of (16.2).

Proposition 18.2 *There is a (non-Archimedean) ordered field Ω' containing the field $\mathbb{R}(t)$, and satisfying condition $(*)$.*

Proof. We start with the field $\mathbb{R}(t)$ of rational functions in an indeterminate t, described above, and we consider $\mathbb{R}(t)$ as a subset of the set $\mathcal{C}$ of all continuous real-valued functions from $\mathbb{R}$ to $\mathbb{R}$, defined at all except a finite number of points, and having only a finite number of zeros (except for the identically 0 function). Beware that $\mathcal{C}$ is not a field(!) because for example the functions $2 \in \mathcal{C}$ and $2 + \sin t \in \mathcal{C}$, but their difference $\sin t \notin \mathcal{C}$, because it has infinitely many zeros. Nevertheless, $\mathcal{C}$ has a nice order, because we can define the subset $P_{\mathcal{C}}$ of positive functions as before: $\varphi(t) \in \mathcal{C}$ is *positive* if $\exists a_0 \in \mathbb{R}$ for which $\varphi(b) > 0$ for all $b > a_0$. Then $P_{\mathcal{C}}$ satisfies properties (i) and (ii) of the definition of an ordered field, even though $\mathcal{C}$ is not a field. We use the fact that a continuous function on an interval (a_0, ∞) which has no zeros is either always positive or always negative.

Now let Ω' be the set of all elements of $\mathcal{C}$ which can be obtained from $\mathbb{R}(t)$ by a finite number of operations $+, -, \cdot, \div$ and $c \mapsto \sqrt{1+c^2}$. The hard part is to show Ω' is a field. Once we know Ω' is a field, the property $(*)$ is easy, because for any $c \in \mathcal{C}$, $1 + c^2$ is a function which is strictly positive whenever it is defined (at all except the finite number of points where c is not defined) so $\sqrt{1+c^2}$ is another such function, hence also in $\mathcal{C}$. Thus if $c \in \Omega'$, $\sqrt{1+c^2} \in \Omega'$ also. We make Ω' into an ordered field by taking as the positive elements $P' = P_{\mathcal{C}} \cap \Omega'$, and P' satisfies (i) and (ii) because $P_{\mathcal{C}}$ does.

Lemma 18.3 *Let F be a subset of Ω' which is a field, and let $\omega \in F$, $\sqrt{1+\omega^2} \notin F$. Then*

$$F' = \{\alpha + \beta\sqrt{1+\omega^2} \mid \alpha, \beta \in F\}$$

is also a subset of Ω' which is a field.

Proof. First we show every element of F' is in Ω'. Since α, β, ω are obtained from $\mathbb{R}(t)$ by a finite number of operations $+, -, \cdot, \div, c \mapsto \sqrt{1+c^2}$, so are the elements of F'. The elements of F' are defined except at the finite number of points where α, β, ω may fail to be defined. They are continuous because α, β, ω are. The only problem is to show that $\alpha + \beta\sqrt{1+\omega^2}$ has only finitely many zeros. Any zero t_0 of this function satisfies

$$\alpha(t_0) + \beta(t_0)\sqrt{1+\omega(t_0)^2} = 0.$$

Separating the two pieces, squaring, and combining again we find

$$\alpha(t_0)^2 - \beta(t_0)^2(1+\omega(t_0)^2) = 0.$$

In other words, t_0 is a zero of the function

$$\alpha^2 - \beta^2(1+\omega^2) \in F.$$

Hence there are only finitely many such zeros, since $F \subseteq \Omega'$. Note that $\alpha+\beta\sqrt{1+\omega^2}$ is not identically zero because then $\sqrt{1+\omega^2} \in F$. Thus $\alpha + \beta\sqrt{1+\omega^2}$ has only finitely many zeros, and so $F' \subseteq \Omega'$.

To show that F' is a field is standard. It is clearly closed under $+$, $-$, $\cdot$. And to show closed under $\div$ one rationalizes the denominator by multiplying by its conjugate:

$$\frac{a+b\sqrt{f}}{c+d\sqrt{f}} \cdot \frac{c-d\sqrt{f}}{c-d\sqrt{f}} = \frac{(a+b\sqrt{f})(c-d\sqrt{f})}{c^2-d^2 f}.$$

Proof of 18.2, Continued. To show that Ω' is a field, suppose $\alpha, \beta \in \Omega'$. We must show that $\alpha \pm \beta$, $\alpha \cdot \beta$, $\alpha/\beta \in \Omega'$ (provided $\beta \neq 0$). Since α is obtained from $\mathbb{R}(t)$ by a finite number of operations $+$, $-$, $\cdot$, $\div$, $\omega \mapsto \sqrt{1+\omega^2}$, by applying the lemma each time we take a square root, we obtain a subfield $F \subseteq \Omega'$ which contains α. Now starting from F, and applying the lemma again each time we use a square root in the description of β, we find a field $F \subseteq G \subseteq \Omega'$, with $\alpha, \beta \in G$. Then clearly $\alpha \pm \beta$, $\alpha \cdot \beta$, $\alpha/\beta \in G \subseteq \Omega'$.

Proposition 18.4 *There is a (non-Archimedean) field K' containing $\mathbb{R}(t)$ which satisfies the condition* $(**)$.

Proof. We follow the same plan of proof as for (18.2), except that now we consider the space $\mathcal{C}'$ as follows: $\mathcal{C}'$ consists of continuous real-valued functions defined on some interval (a_0, ∞) of $\mathbb{R}$, and which are never 0. Two functions f on (a_0, ∞) and g on (a_1, ∞) are *equivalent* if $\exists a_2 > a_0, a_1$ such that $f = g$ on (a_2, ∞). We say f is *positive* if $f(b) > 0$ for all $b > a_0$. The set $P_{\mathcal{C}'}$ of positive functions clearly satisfies (i) and (ii) of the definition of ordered field. Note again $\mathcal{C}'$ is not a field. But if $\varphi \in \mathcal{C}'$, $\varphi > 0$ then $\sqrt{\varphi} \in \mathcal{C}'$ also.

Now we take K' to be the set of all elements of $\mathcal{C}'$ which can be obtained from $\mathbb{R}(t)$ by a finite number of operations $+$, $-$, $\cdot$, $\div$ and $\varphi > 0 \mapsto \sqrt{\varphi}$.

The proof that K' is a field can be carried out exactly as in the proof of (18.2). Clearly K' satisfies $(**)$, and taking $P' = K' \cap P_{\mathcal{C}'}$ makes K' into an ordered field.

Example 18.4.1 Let Π be the Cartesian plane over the field Ω' of (18.2). Then Π is a Hilbert plane satisfying (P) but not (A). In particular, this shows that (A) is independent of the axioms of a Hilbert plane.

Example 18.4.2 Let Π be the Cartesian plane over the field K' of (18.4). Then Π is a Euclidean plane which does not satisfy (A).

Example 18.4.3 Let Π be the non-Archimedean geometry described in (18.4.2). Let Π^0 be the subset consisting of all points of Π whose distance from the origin is finite. A *line* of Π^0 will be the intersection of a line of Π with Π^0, whenever that intersection is nonempty. Take betweenness and congruence to have the same meaning as in Π. Then Π^0 is a Hilbert plane satisfying neither (A) nor (P). In particular, this shows that (P) is independent of the axioms of a Hilbert plane.

Exercises

18.1 In the ordered field $\mathbb{R}(t)$, arrange the following elements in increasing order: 0, 1, 5, t, $1/t$, $t+1$, $1/(t+1)$, $t-1$, $\frac{1}{2}t$, t^2-t, t^2-1, $t+\frac{1}{t}$, $(t-1)/(t+1)$.

18.2 Show that the field Ω' of (18.2) does not satisfy (**), by showing that $\sqrt{t} \notin \Omega'$.

18.3 Show that the plane Π^0 of (18.4.3) satisfies the axioms for a Hilbert plane. Pay special attention to (I2), (B2), and (C1).

18.4 Again let Π^0 be the plane of (18.4.3).

a) Show that Π^0 does not satisfy (P).

b) Show that Π^0 does satisfy (I.32): the angle sum of every triangle is two right angles.

c) Show that Π^0 does not satisfy (IV.5), by giving an example of a triangle which has no circumscribed circle.

But when the sceptre devolved to Almamon, the seventh of the Abbassides, he completed the designs of his grandfather, and invited the Muses from their ancient seats. His ambassadors at Constantinople, his agents in Armenia, Syria, and Egypt, collected the volumes of Grecian science: at his command they were translated by the most skillful interpreters into the Arabic language: his subjects were exhorted assiduously to peruse these instructive writings; and the successor of Mahomet assisted with pleasure and modesty at the assemblies and disputations of the learned...

The sages of Greece were translated and illustrated in the Arabic language, and some treatises, now lost in the original, have been recovered in the version of the East, which possessed and studied the writings of Aristotle and Plato, of Euclid and Apollonius, of Ptolemy, Hippocrates, and Galen.

– from *The History of the Decline and Fall of the Roman Empire*
by Edward Gibbon, vol V, ch 52
Bigelow, NY (1845)

CHAPTER 4

Segment Arithmetic

Segment arithmetic allows us to complete the chain of logical connections between an abstract geometry satisfying axioms studied in Chapter 2, with the geometries over fields studied in Chapter 3. We will show how to define addition and multiplication of line segments in a Hilbert plane satisfying the parallel axiom (P). In this way, the congruence equivalence classes of line segments become the positive elements of an ordered field F (§19). Using this field F we can recover the usual theory of similar triangles (§20).

To complete the circle, we show that if you start with a Hilbert plane Π satisfying (P), and if F is the associated field of segment arithmetic, then Π is isomorphic to the Cartesian plane over the field F (§21).

19 Addition and Multiplication of Line Segments

In studying Euclid's *Elements*, we have noted the absence of numbers in his development. There is no notion of the *length* of a line segment, for example. There is an undefined notion of congruence of segments, which we can think of as their being the same. This is in contrast to ordinary high school geometry where each segment has a length, which is thought of as a real number, and two segments are congruent if they have the same length.

Similarly in the case of angles, there is no degree measure attached to an angle, although there is a notion of congruence of angles.

In the study of area, Euclid does not assign a number to a plane figure, as for example in high school one takes a triangle and assigns to it the number $\frac{1}{2}bh$ as its area, where b is the length of the base and h is the length of the height. Instead, Euclid treats area by adding and subtracting congruent figures.

For the material of Books I–IV of the *Elements*, we have also seen that Euclid succeeds remarkably well in developing a "pure" geometry with no numbers. They really are not necessary and one can develop a beautiful theory without them. Hilbert has reinforced this by providing a set of purely geometric axioms on which to base Euclid's geometry in a way which will satisfy modern criteria of rigor.

Just for contrast, you might look at some other 20-th century proposals for a set of axioms on which to base the study of geometry, where the real numbers are presupposed from the beginning in the axioms. (See for example [Birkhoff], or the SMSG postulates (1966). Both can be found as appendices to [Cederberg].)

For me this is unsatisfactory because a) it is not purely geometric, and b) the concept of a real number is a rather sophisticated modern notion, dating from the 19-th century, and is not in the elementary spirit of Euclid's geometry. While Euclid was clearly aware of irrational numbers, and indeed includes in his *Elements* a proof of the fact that $\sqrt{2}$ is irrational (X.117), I find it difficult to support any argument that Euclid had a concept of the totality of real numbers.

While Euclid was able to develop the material of Books I–IV without any notion of number, it is a different matter when we come to the concept of *similar* triangles as taught in high school. These are triangles whose sides are not equal, but all in some common *ratio* to each other. If that ratio is an integer, for example 2, it is not difficult to develop a theory of triangles which are doubles of each other, as we did in §5. With a little more effort, one could extend this theory to triangles whose sides were integer multiples of each other, or (with even a little more effort), rational number multiples of each other. But if the ratio is not rational, as for example when comparing an isosceles right triangle to its half formed by drawing an altitude, how can one even express the notion of sides being proportional to each other without having numbers? One would like to say that the ratios of the lengths of the sides are equal, but this is difficult if one has no notion of length as a number and does not have the ability to divide one such number by another.

Euclid handles this difficulty with the theory of *proportion* developed in Book V of the *Elements*. The key concept is in Book V, definition 5, where he says that magnitudes (which could be line segments, areas, or whatever) are in the *same ratio* (in symbols $a : b = c : d$) if whenever equal integer multiples (say n times) be taken of a and c, and whenever equal integer multiples (say m times) be taken of b and d, then $na > mb$ or $na = mb$ or $na < mb$ if and only if $nc > md$, or $nc = md$, or $nc < md$, respectively. If a, b, c, d are numbers, this is equivalent to saying that a rational number m/n is less than, equal to, or greater than a/b if and only if that same rational number is less than, equal to, or greater than c/d. If furthermore a, b, c, d are real numbers, this is equivalent, as we know, to saying that a/b and c/d are equal as real numbers, since the rational numbers are dense in the set of real numbers. In fact, this is word for word the same notion used by Dedekind in constructing the real numbers by his so-called Dedekind cuts (cf. [Dedekind]).

Aha!, you may say, so Euclid did know about the real numbers, and wrote their definition 2000 years before Dedekind! But here is the difference. Euclid used this criterion to distinguish between ratios which existed in his geometry, such as the ratios of line segments which might be obtained by ruler and compass constructions, and which might be irrational. But I see no evidence that he conceived of the existence of other real numbers (such as e for example) which did not arise naturally in his geometry. Whereas, Dedekind could conceive of the totality of all Dedekind cuts of rational numbers, and take this set to be a new mathematical object called the set of real numbers. It is this process of creating a new mathematical object as a set of all subsets of another set with certain properties, which seems very modern to me.

Even in the classical problem of the trisection of the angle it seems that the emphasis was on finding a construction which would produce an angle equal to one-third of a given angle, and there is no evidence that they believed in the existence of such an angle before it was constructed.

Note also that for Euclid's theory of proportion to work, we implicitly need Archimedes axiom. This is clear from Book V, Definition 4, which says that magnitudes have a ratio to each other if each, when multiplied, is capable of exceeding the other. Without Archimedes axiom, some quantities would be incomparable. Also, one would fail to distinguish unequal quantities. For example, if F is a non-Archimedean ordered field with an infinite element t, then Euclid's test would fail to distinguish between $\sqrt{2}$ and $\sqrt{2} + 1/t$, which are distinct elements of the field.

Having developed the theory of proportion abstractly in Book V, Euclid proceeds to apply his theory to geometry in Book VI, and develops what we recognize as the familiar theory of similar triangles. The key result here, which forms the basis of the subsequent development, is (VI.2), which says that a line parallel to the base of a triangle cuts the sides proportionately, and conversely. Euclid's proof is a tour de force, using the theory of area previously developed in Book I to establish this result.

So there are two reasons for us to seek an alternative development of the theory of similar triangles: one is to free ourselves from dependence on Archimedes axiom, and the other is to avoid Euclid's use of the theory of area, which we have not yet treated satisfactorily (cf. Chapter 5).

So now, after this rather lengthy introduction, we come to the main point of this section. That is to create an arithmetic of line segments. We will define notions of *addition* and *multiplication* for line segments up to congruence, that is to say the sum or product of congruent segments will be congruent. Or, if you prefer, the operations $+$ and $\cdot$ will be defined on the set P of equivalence classes of line segments modulo congruence. We will show that these operations obey all the usual rules of arithmetic for positive numbers. And then, by a natural construction which introduces also an element 0 and negatives of line segments, we will construct an ordered *field* whose positive elements are the congruence classes of line segments. Here is where the concepts of modern abstract algebra play an essential role, because instead of using some preexisting notion of number, such as the rational numbers or the real numbers, we create a set which occurs naturally in our geometry and give this set the structure of a field in the sense of abstract algebra.

Using this field we will then in the next section be able to define the notion of length of a segment (as an element of this field) and to develop the theory of similar triangles, where ratios are quotients of elements of this field.

We will now define the arithmetic operations on congruence equivalence classes of line segments, following the ideas of [Hilbert], with simplifications suggested by material in the supplements to that book, apparently due to Enriques. We will work in a Hilbert plane satisfying the parallel axiom (P).

Definition. Given congruence equivalence classes of line segments a, b, we define their *sum* as follows. Choose points A, B such that the segment AB represents the class a.

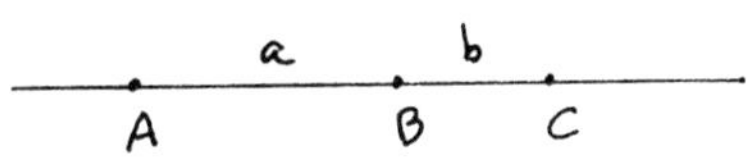

Then on the line AB choose a point C with $A * B * C$, such that the segment

BC represents the class b. Then we define $a + b$ to be represented by the segment AC.

Proposition 19.1 *In any Hilbert plane, addition of line segment classes has the following properties.*

1) $a + b$ *is well-defined, i.e. different choices of* A, B, C *in the definition will give rise to congruent segments.*

2) $a + b = b + a$, *i.e. the corresponding line segments are congruent.*

3) $(a + b) + c = a + (b + c)$.

4) *Given any two classes* a, b, *one and only one of the following holds:*

i) $a = b$.

ii) *there is a class* c *such that* $a + c = b$.

iii) *there is a class* d *such that* $a = b + d$.

Proof. 1) If we choose a different representative $A'B'$ of the class a, and lay off C' on the line $A'B'$ so that $B'C'$ represents b, then $AC \cong A'C'$ by axiom (C3).

2) Let AB represent a, and choose C so that $A*B*C$ and BC represents b, as in the definition. Then AC represents $a + b$. Now take DE to represent b, and lay off F so that $D * E * F$ and EF represents a. Then DF represents $b+a$. But $AB \cong FE$ and $BC \cong ED$, so $AC \cong FD$ by (C3). This shows that $a + b = b + a$.

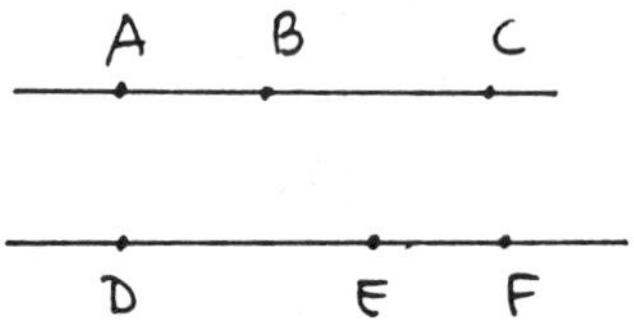

3) To get $(a + b) + c$ we first choose $AB \in a$, then find C such that $A * B * C$ and $BC \in b$, then find D such that $A * C * D$ and $CD \in c$. Then AD represents $(a + b) + c$.

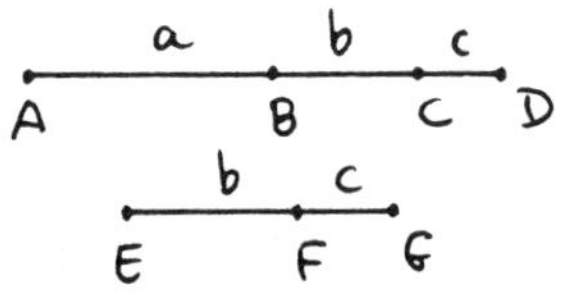

On the other hand, let $EF \in b$ and choose G such that $FG \in c$. Then EG represents $b + c$. To get $a + (b + c)$ we need to find a point H with $A * B * H$ and $BH \cong EG$. But $BD \cong EG$ by (C3), so $H = D$ by the uniqueness part of (C1). Therefore $(a + b) + c = a + (b + c)$.

4) Given two classes a, b, on a ray from a point A lay off points B, C so that $AB \in a$ and $AC \in b$. If $B = C$, then $a = b$. If $A * B * C$, then $a + [BC] = b$. If $A * C * B$, then $a = b + [CB]$. By (B3) these are the only possibilities, and this proves 4).

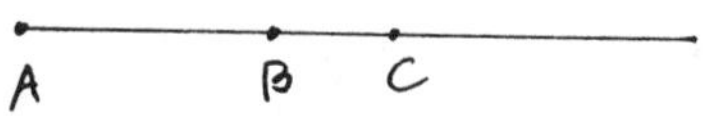

Before we define multiplication, we need a standard unit segment. So choose arbitrarily, and then fix once and for all, a segment class we call the *unit segment*, and denote it by 1. We also need the parallel axiom (P), even for the definition of the product.

Definition. Given two segment classes a, b, we define their *product* ab as follows. First make a right triangle ABC with $AB \in 1$ and $BC \in a$, where the right angle is at B. Let α be the angle $\angle BAC$. Now make a new right triangle DEF with $DE \in b$ and having the same angle α at D. Then we define ab to be the class of side EF of this new triangle.

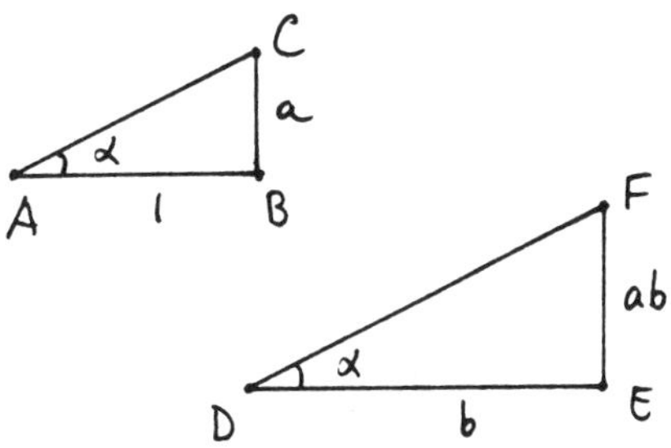

Proposition 19.2 *In any Hilbert plane with* (P), *multiplication of segment classes has the following properties.*

1) *ab is well-defined.*
2) *$a \cdot 1 = a$ for all a.*
3) *$ab = ba$ for all a, b.*
4) *$a(bc) = (ab)c$ for all a, b, c.*
5) *For any a, there is a unique b such that $ab = 1$.*
6) *$a(b + c) = ab + ac$ for all a, b, c.*

Proof. 1) The product is well-defined. If $A'B'C'$ is another right triangle with sides $1, a$, then it is congruent to ABC by (SAS). Hence we get a congruent angle α. If $D'E'F'$ is another right triangle with angle α and side b, then it is congruent to DEF by (ASA). So we get a congruent segment $E'F'$.

2) To compute $a \cdot 1$, we take the triangle DEF to have side $b = 1$ and angle α. Then $DEF \cong ABC$ by (ASA), so $a \cdot 1 = a$.

3) Given a, b, first make a right triangle ABC with sides $1, a$. This determines the angle $\alpha = \angle BAC$. Now extend CB on the other side of AB to D, so that $BD \in b$, and draw a line through D making an angle α with BD, on the far side of BD from A. Let this line meet AB extended to E. Then DBE is a right triangle with side b and angle α, so the segment BE represents ab by definition.

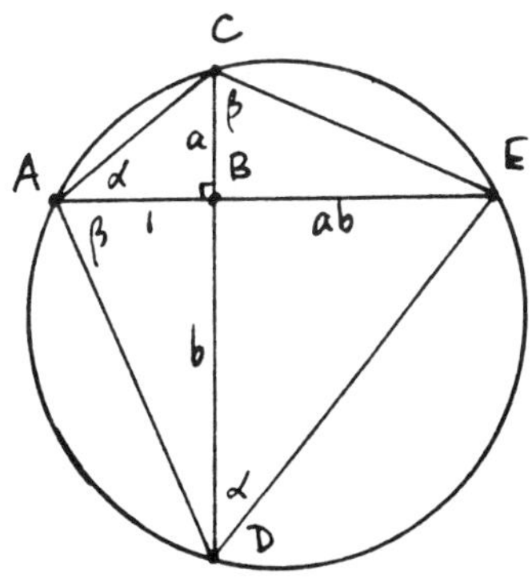

Now consider the four points $ACDE$. We will use the method of cyclic quadrilaterals developed in §5. Because the angles $\angle CAE$ and $\angle CDE$ are both equal to α, they satisfy the hypotheses of (5.8), so the points $ACDE$ form a cyclic quadrilateral. Applying (5.8) to the same points in a different order, it follows that the angles $\angle DAE$ and $\angle DCE$ are equal, say β. To compute the product ba, and then use the triangle CBE which has angle β and side a. This shows that BE represents the product ba. Thus $ab = ba$.

4) For the associative law, we proceed as follows. Make right triangles with $1, a$ to define the angle α, and with $1, c$ to define the angle γ. Make a right triangle ABC with angle α and side b to determine ab.

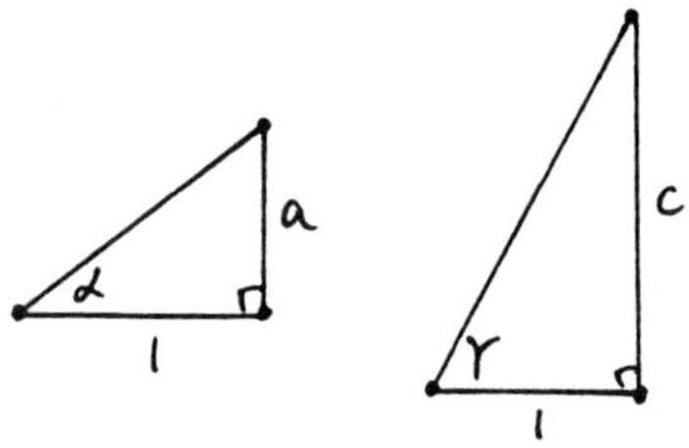

Extend CB on the other side of AB to meet a line from A making an angle γ with AB. Then BD represents cb. Now make a line at D with angle α to meet AB extended at E. Then BE will represent $a(cb)$.

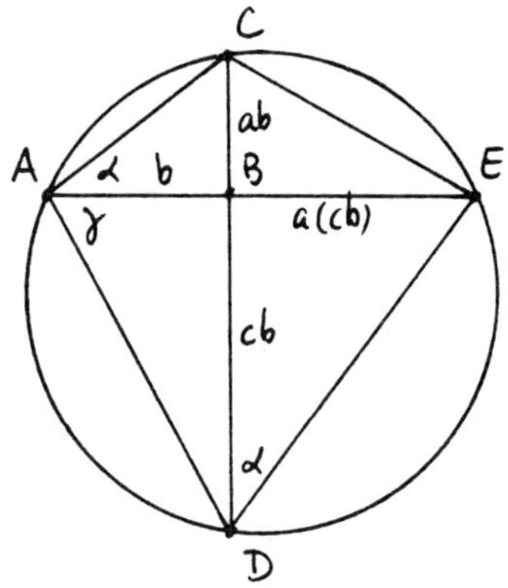

As in the previous proof, the angles α at A and D show that $ACDE$ is a cyclic quadrilateral, by (5.8). Then by (5.8) again we conclude that $\angle BCE = \gamma$. It follows that BE also represents the segment $c(ab)$. Thus $a(cb) = c(ab)$. Then using the commutative law already proved, this shows that $a(bc) = (ab)c$.

5) Given a, make a right triangle with sides $1, a$ to define α, and let β be the other acute angle in that triangle. Then make a right triangle with angle β and side 1 to determine a new segment b. Since the other angle in this triangle is α (I.32) this second triangle shows that $ab = 1$.

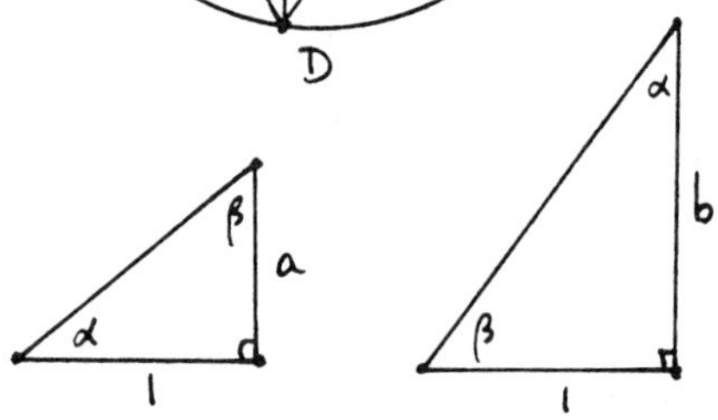

6) Given a, b, c, let α be determined by the right triangle with sides $1, a$. Make a right triangle ABC with side b and angle α to determine $BC \in ab$. Choose D on the line AB such that $A * B * D$ and $BD \in c$. Draw CE parallel to AB, and DEF perpendicular to AB. Then $\angle ECF = \alpha$, and $CE \in c$, so EF represents ac.

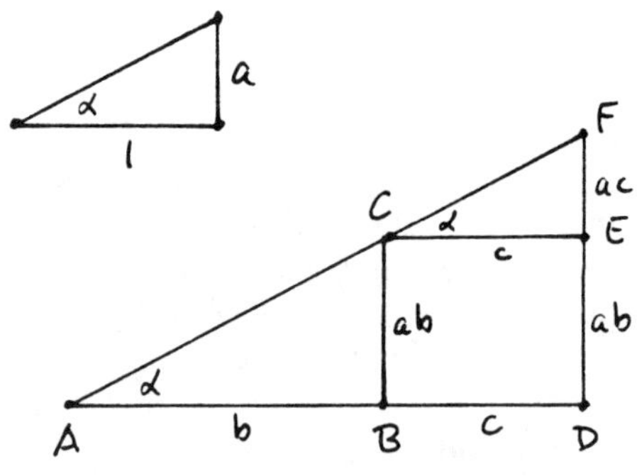

Because $BCDE$ is a rectangle, $DE \in ab$. Now by definition of sum, AD represents $b + c$ and DF represents $ab + ac$. On the other hand, the triangle ADF has side $b + c$ and angle α, so DF also represents $a(b + c)$. Hence $a(b + c) = ab + ac$.

Remark 19.2.1 Let us examine carefully the hypotheses needed for the validity of these two results (19.1) and (19.2). The first, concerning addition of line segments, is valid in any Hilbert plane (cf. §8). On the other hand, even for the definition of the product, we need (P), or its equivalent, Euclid's fifth postulate, to guarantee that the point F exists. For the proof of (19.2), you might feel more comfortable assuming the hypotheses of a Euclidean plane, in which case we have justified the needed results from Book III (§12). But, if you look closely, we do not need the Euclidean axiom (E) on intersection of circles. For (5.8) we do need the existence of the circumscribed circle (IV.5), which holds in a Hilbert plane with (P). We also need to know that a line which meets a circle once will meet it again, and that is true without assuming (E)—cf. (11.3). We also need (I.32) which uses (P) but not (E). Thus these two results hold in a Hilbert plane with (P). We do not need (E), nor did we ever use Archimedes' axiom (A).

Proposition 19.3 *Given a Hilbert plane satisfying* (P), *there is a unique (up to isomorphism) ordered field F, whose set of positive elements P is the set of congruence equivalence classes of line segments with operations $+$, $\cdot$ defined above.*

Proof. This is a consequence of the purely algebraic lemma which follows.

Lemma 19.4 *Let P be a set, with two operations $+$, $\cdot$ defined on it, which satisfy the properties listed in* (19.1) *and* (19.2). *Then there is a unique ordered field F whose positive elements form the set P.*

Proof. One is tempted to define F to be the set $P \cup \{0\} \cup -P$, for intuitively, this is what is happening. F will consist of the original set P, plus a 0 element, plus another set of "negative" elements, which is in 1-1 correspondence with the set of positive elements. However, I believe we can obtain a cleaner proof by imitating the definition of the quotient field of an integral domain using ordered pairs. Except this time our ordered pairs will represent differences of elements of P.

So here is the formal construction. Let F be the set of equivalence classes (a, b) of ordered pairs (think of (a, b) as being $a - b$) of elements of P, where

$$(a, b) \sim (a', b') \text{ if } a + b' = a' + b.$$

Define addition by

$$(a, b) + (c, d) = (a + c, b + d)$$

and multiplication by

$$(a, b)(c, d) = (ac + bd, ad + bc).$$

We must verify that these operations are well-defined, i.e. if we replace an ordered pair by an equivalent ordered pair, the result is equivalent (!). (The symbol (!) means a trivial verification left to the reader. There will be lots of these, and all will result by using properties of $+$ and $\cdot$ in the set P.)

Then we let 0 denote the equivalence class of (a, a) for any $a \in P$, and note that 0 acts as an additive identity (!). Also note that addition is commutative (!) and associative (!). For any pair (a, b) note that (b, a) acts as an additive inverse (!). Thus the set F, together with the operation $+$ is an abelian group.

Next verify that multiplication is associative (!), commutative (!) and distributive over addition (!). Let 1 be the class of $(1 + a, a)$ for any $a \in P$. Thus 1 acts as a multiplicative identity (!) and there exist multiplicative inverses (!). Hence F together with $+$, $\cdot$ is a field.

We define a mapping $\varphi : P \to F$ by $a \in P$ goes to $(a+b, b)$ for any $b \in P$. This mapping is 1-1 onto its image (!) which we therefore identify with P. Also φ preserves $+, \cdot$ (!) so that P has already two of the three properties required for P to be the set of positive elements of an ordered field. It remains to verify the trichotomy, namely for any $x = (a, b)$ in F, either $x \in P$ or $x = 0$ or $-x \in P$. If $a = b$, then $x = 0$. We will use property 4) of (19.1). If there exists a c such that $a + c = b$, then $x = (a, b) = (a, a + c)$, and the negative of this element $-x = (a + c, a) \in P$. If on the other hand there is a d such that $a = b + d$, then $x = (a, b) = (b + d, b) \in P$.

This concludes the proof modulo a million tedious verifications (!) left to the reader!

Exercises

19.1 Supply the missing verifications in the proof of (19.4).

19.2 If we start with the Cartesian plane over a field F_0, show that the field F of segment arithmetic constructed in (19.3) is naturally isomorphic to the original field F_0.

20 Similar Triangles

We continue to work in a Hilbert plane satisfying (P).

Now that we have defined the arithmetic of line segments, and have constructed a field F whose positive elements correspond to congruence classes of line segments, we can establish a theory of proportion and similar triangles. The results are the same as Euclid's in Book VI, but our methods are different.

For any line segment AB, its congruence equivalence class a is an element of the field F. We will call a the *length* of AB, to conform with the usual terminology. If AB and CD are two segments, with lengths a, b, we can speak of their *ratio* as the quotient $a/b \in F$. We say four segments, with lengths a, b, c, d, are *proportional* if $a/b = c/d$ as elements of the field F.

Definition (VI, Definition 1). Two triangles ABC and $A'B'C'$ are *similar* if the three angles of one are respectively equal to the three angles of the other, and the corresponding sides are proportional, i.e.

$$a/a' = b/b' = c/c'.$$

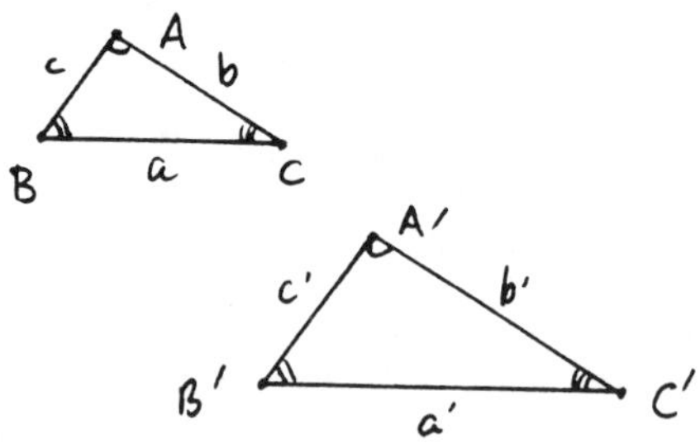

Proposition 20.1 (Sim AAA) (VI.4) *If two triangles ABC and DEF have their three angles respectively equal, then the two triangles are similar.*

Proof. Our definition of multiplication in the field of segment arithmetic was based on a special case of the notion of similar triangles, namely, comparing the legs of equal-angled right triangles. So we will prove this result, following [Hilbert] by reducing to this case.

77

ΕΥΚΛΕΙΔΟΥ ΣΤΟΙ-
ΧΕΙΟΝ ΠΡΩΤΟΝ.

EVCLIDIS ELEMENTORVM GEO-
metricorum liber primus.

ISt hic liber primus totus ferè elementarius, non tantum ad reliquos sequentes huius Operis libros, sed etiam ad aliorum Geometrarum scripta intelligenda necessarius. Nam in hoc libro communium uocabulorum, quę subinde in geometria uersanti occurrunt, definitiones continentur. Pręceptiones deinde ducendi perpendicularem, quomodo item Trilaterę figurę, secundum latera uel angulos diuersæ, & Quadrilateræ, formari debeant. Figura item aliqua proposita, quomodo illa in alterius formæ figuram permutanda sit, præceptiones, ut diximus, traduntur. Cum igitur talia doceantur, & plura etiam alia, quàm hoc loco commemorare uoluimus, facile erit cuiuis, non solum quàm sit necessarius, sed etiam ad reliqua perdiscenda liber iste quàm utilis, perspicere.

ΟΡΟΙ.

Σημεῖόν ἐστιν, οὗ μέρος οὐθέν. Γραμμὴ δέ, μῆκος ἀπλατές. Γραμμῆς δὲ πέρατα, σημεῖα. Εὐθεῖα γραμμή ἐστιν, ἥτις ἐξ ἴσου τοῖς ἐφ᾽ ἑαυτῆς σημείοις κεῖται.

DEFINITIONES.

Punctum est, cuius pars nulla. 2. Linea uerò, longitudo latitudinis expers. Lineę autem termini puncta. 3. Recta linea est, quę ęqualiter inter sua puncta iacet.

Prima defini.

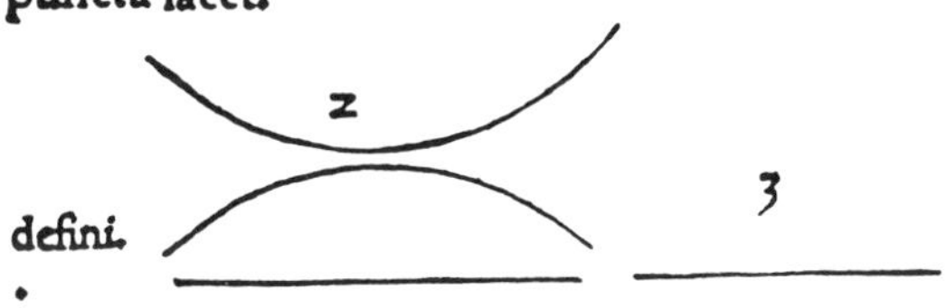

Ἐπιφάνειά ἐστιν, ὃ μῆκος καὶ πλάτος μόνον ἔχει. Ἐπιφανείας δὲ πέρατα, γραμμαί. Ἐπίπεδος ἐπιφάνειά ἐστιν, ἥτις ἐξ ἴσου ταῖς ἐφ᾽ ἑαυτῆς εὐθείαις κεῖται.

4. Superficies est, quę longitudinem & latitudinem tantum habet.

K 3 Super-

Plate VI. The beginning of Book I from Scheubel's Euclid (1550) showing the Greek text of the first few definitions.

In the first triangle, draw the angle bisectors of the three angles, and let them meet at the point I (cf. (Exercise 1.8) or (IV.4)). Recall from the proof of (IV.4) that I is equidistant from the three sides of the triangle: if we drop perpendiculars from I to the three sides, we obtain three congruent segments h. Also in the course of the proof we found congruent triangles about each vertex: $AFI \cong AEI$, etc. Thus we get congruent segments $AE \cong AF$, which we call x, $BD \cong BF = y$ and $CD \cong DE = z$.

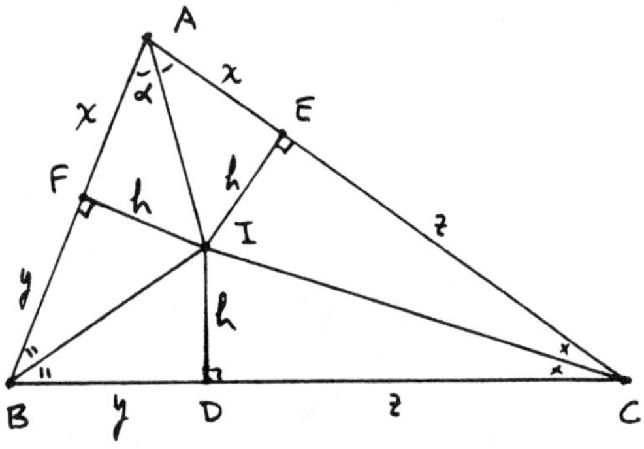

Make a similar construction in the second triangle $A'B'C'$, obtaining points D', E', F', I' and segments x', y', z', h'.

Let α be one half of the angle at A, and draw a right triangle with one leg equal to 1, and let r be the other leg. Then by the definition of segment multiplication, $h = rx$. In the second triangle, the angle at A' is equal to the angle at A by hypothesis, so one half of it is also α, so we find similarly that $h' = rx'$. Dividing one equation by the other, we find that $x/x' = h/h'$.

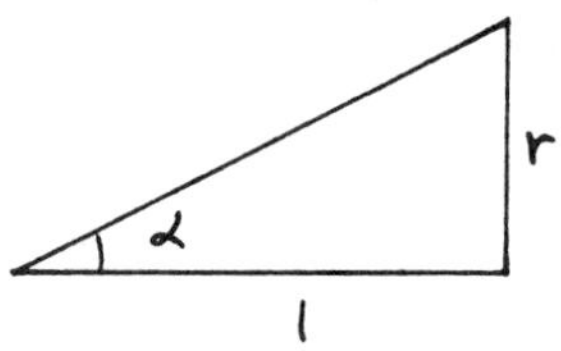

In the same way, working from the other two vertices of the triangle, we find $y/y' = h/h'$ and $z/z' = h/h'$. If we let $h/h' = k$, then we can write these results as

$$x = kx'$$
$$y = ky'$$
$$z = kz'.$$

The sides of the original triangle are formed of sums of these. Thus $a = y + z$ and $a' = y' + z'$. It follows from the distributive law that

$$a = ka',$$

and by the same reasoning also

$$b = kb'$$
$$c = kc'.$$

Then $a/a' = b/b' = c/c'$, so the two triangles are similar.

While this proof is different from anything in Euclid's Elements, the other basic results of Book VI will follow by methods similar to Euclid's, but perhaps not in the same order. We give three such results, after which the other results of Book VI will follow as in Euclid.

Proposition 20.2 (VI.2) *In any triangle ABC, let $B'C'$ be drawn parallel to BC. Then the sides AB and AC are proportional to AB' and AC'. Conversely, if the sides are divided by points B', D such that AB, AC are proportional to AB', AD, then $B'D$ is parallel to BC.*

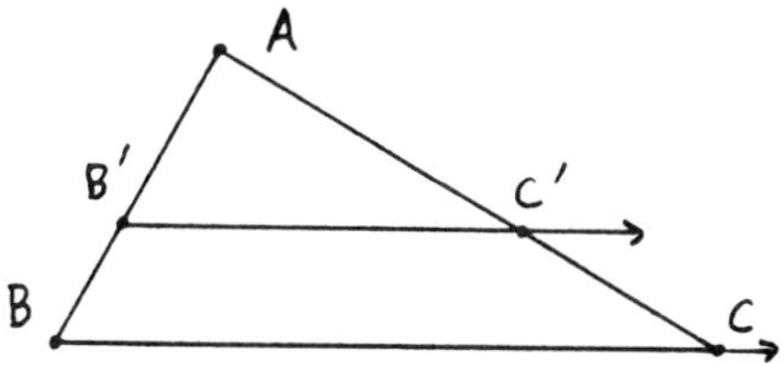

Proof. Since $B'C'$ is parallel to BC, the angles at B', C' are equal to the angles at B, C, respectively (I.29). Since the angle at A is common, the triangles ABC and $AB'C'$ have their three angles equal, and so they are similar (20.1). It follows that the sides are proportional.

Conversely, suppose given B', D such that AB, AC are proportional to AB', AD. Draw $B'C'$ parallel to BC. Then also AB, AC are proportional to AB', AC'. Since we are working in a field F, the fourth proportional to three given quantities is uniquely determined. Hence $AD \cong AC'$. Since the points D, C' lie on the same ray from A, the points D, C' are equal (axiom (C1)). Hence $B'D$ is parallel to BC.

Proposition 20.3 (Sim SSS) (VI.5) *Suppose two triangles ABC and $A'B'C'$ have their three sides respectively proportional to each other. Then the two triangles are similar.*

Proof. Supposing the second triangle to be larger, find a point D on the segment $B'A'$ such that $B'D \cong BA$. Then draw a line through D parallel to $A'C'$. It follows (20.2) that the triangles $A'B'C'$ and $DB'E$ are similar, and in particular, their sides are proportional. But the sides of ABC are also proportional to the sides of $A'B'C'$, so it follows (from field arithmetic) that the sides of ABC are proportional to the sides of $DB'E$.

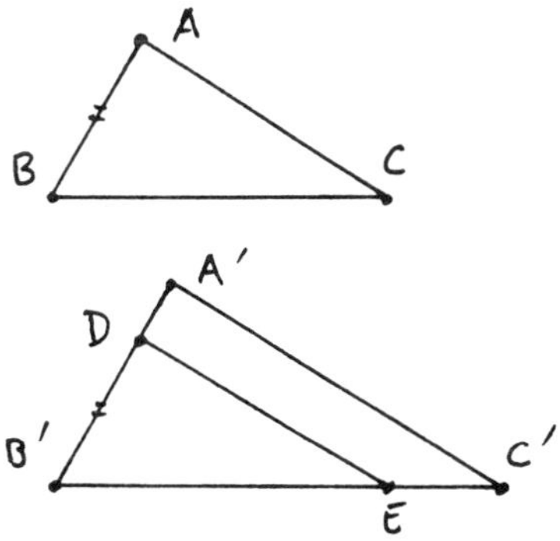

On the other hand, $B'D$ was chosen congruent to BA. So the proportionality factor is 1, and it follows that all three sides of ABC are congruent to all three sides of $DB'E$. Then by the congruence criterion (SSS), the triangle ABC is congruent to $DB'E$. In particular, the three angles of ABC are equal to the three angles of $DB'E$, which in turn are equal to the three angles of $A'B'C'$ since the latter two are similar. Thus we have proved that the angles of ABC are equal to the angles of $A'B'C'$, and so the two triangles are similar.

Proposition 20.4 (Sim SAS) (VI.6) *Suppose that two triangles ABC and $A'B'C'$ have the angles at A and A' equal, and the two sides AB, AC are proportional to the two sides $A'B'$, $A'C'$. Then the two triangles are similar.*

Proof (Exercise 20.1).

Next, using our segment arithmetic, and the theory of similar triangles, we can prove some analogues of theorems which Euclid stated in terms of area, but which we will state as equations in the field F.

Proposition 20.5 *If ABC is a right triangle with legs a, b and hypotenuse c, then*

$$a^2 + b^2 = c^2$$

in the field F of segment arithmetic.

Proof. This, of course, is another version of the Pythagorean theorem (I.47) which Euclid proved in terms of the areas of the squares built on the sides of the triangle. The present statement in terms of segment arithmetic, is of a totally different nature, and neither implies, nor is implied by, the previous statement, until we have made some connection between area and segment arithmetic (cf. Chapter 5).

To prove the current statement, drop a perpendicular CD from the vertex with the right angle to the hypotenuse. Then we find that the original triangle ABC has the same angles as the two new triangles ACD and CBD. Hence all three triangles are similar, by (20.1). (This statement is (VI.8) in Euclid.) Then corresponding sides are proportional, and we find

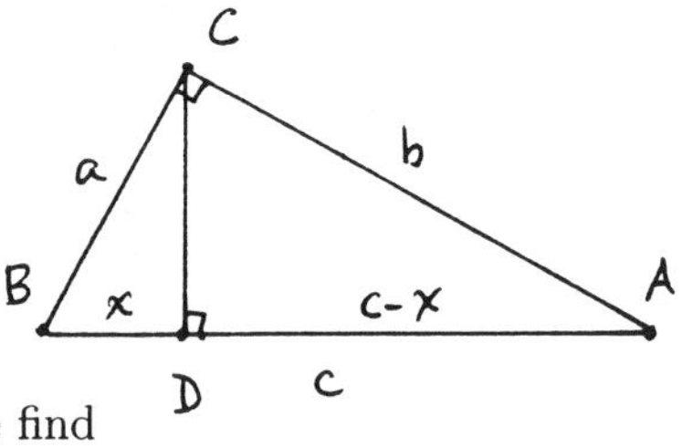

$$\frac{x}{a} = \frac{a}{c}$$

using CBD similar to ABC, and we find

$$\frac{c-x}{b} = \frac{b}{c}$$

using ACD similar to ABC. Cross multiplying we find

$$\begin{aligned} cx &= a^2 \\ c^2 - cx &= b^2 \end{aligned}$$

from which by substituting we find

$$a^2 + b^2 = c^2.$$

Proposition 20.6 (cf. (III.35)) *If two chords of a circle meet, cutting each other in segments of length a, b, c, d, then*

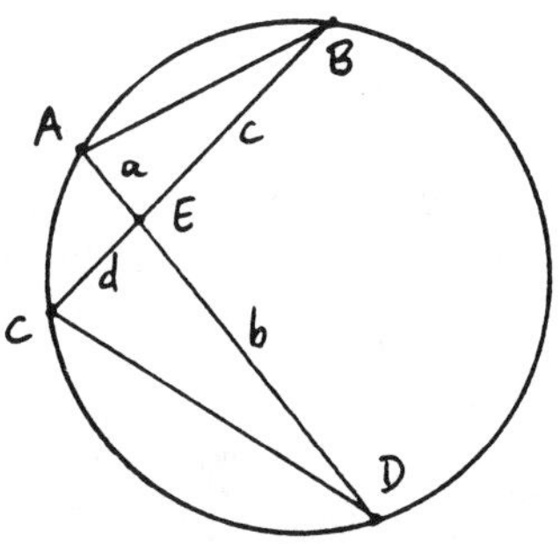

$$ab = cd$$

in the field F of segment arithmetic.

Proof. Draw the lines AB and CD. Then by (III.21) the angles at B and D are equal, and the angles at A and C are equal. Hence the triangles ABE and CDE have the same angles, so are similar (20.1). It follows that corresponding sides are proportional: $a/c = d/b$. Cross-multiplying, we find $ab = cd$.

Proposition 20.7 (cf. (III.36)) *Let A be a point outside a circle, let the line AB be tangent to the circle at B, and let the line ACD cut the circle at C and D. Then, in the field of segment arithmetic,*

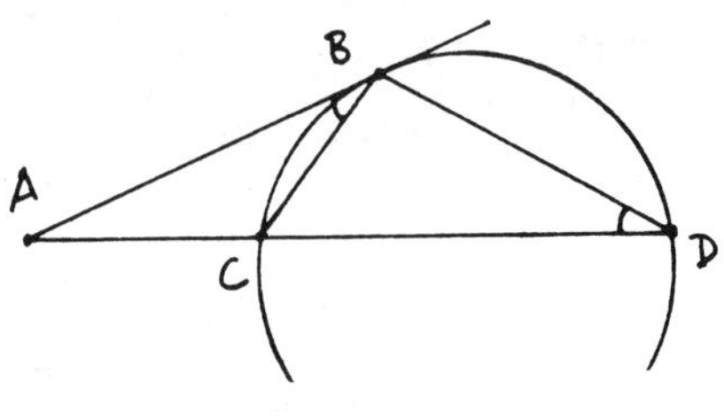

$$(AB)^2 = (AC) \cdot (AD).$$

Proof. Draw the lines BC and BD. Then $\angle ABC = \angle ADB$ by (III.32). Since the angle at A is common, the triangles ABC and ADB have two (and hence three) angles equal, so they are similar (20.1). It follows that corresponding sides are proportional, namely

$$AB/AC = AD/AB.$$

Cross-multiplying gives

$$(AB)^2 = (AC) \cdot (AD).$$

As applications of similar triangles, we give some other well-known theorems here and in the exercises.

Proposition 20.8 (Menelaus' theorem) *Let ABC be any triangle, and let a line l cut the sides of the triangle (extended if necessary) in points D, E, F. Then*

$$\frac{AD}{BD} \cdot \frac{BF}{CF} \cdot \frac{CE}{AE} = 1.$$

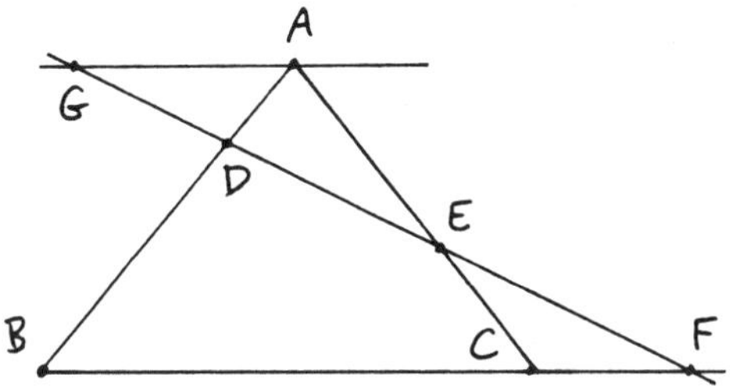

Proof. Draw a line through A parallel to BC, and let it meet l at G. Then the triangle ADG is similar to BDF, and the triangle AEG is similar to CEF. From this we obtain

$$\frac{AD}{AG} = \frac{BD}{BF} \quad \text{and} \quad \frac{AE}{AG} = \frac{CE}{CF}.$$

Eliminating AG from these equations, and rearranging, gives the result.

Exercises

20.1 Prove (Sim SAS) (20.4).

20.2 Let ABC be any triangle, and let AD be the angle bisector at A. Prove that AB and AC are proportional to BD and DC.

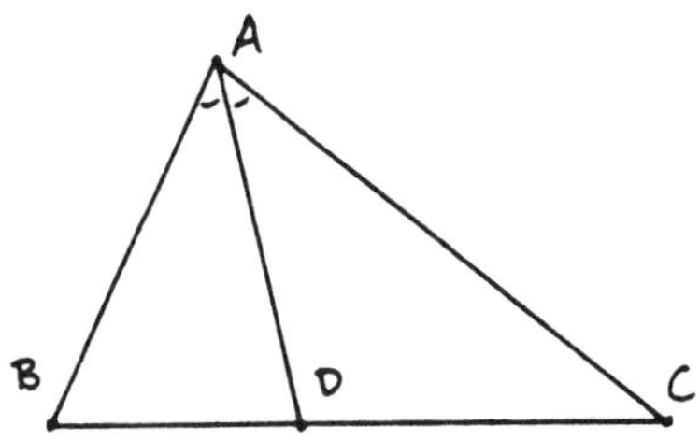

20.3 Let A be a point outside a circle, and draw any two lines through A cutting the circle at B, C and D, E. Then show that

$$(AB) \cdot (AC) = (AD) \cdot (AE).$$

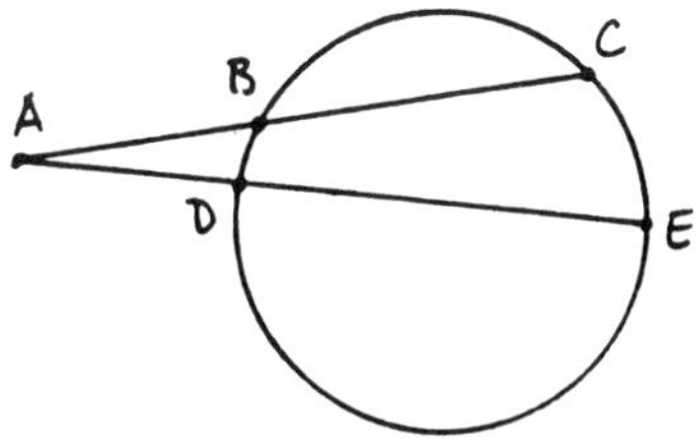

20.4 (Ceva's theorem). Let ABC be any triangle, and let P be any point inside the triangle. Draw lines from the vertices through P, meeting the opposite sides at D, E, F. Then show that

$$\frac{AD}{BD} \cdot \frac{BF}{CF} \cdot \frac{CE}{AE} = 1.$$

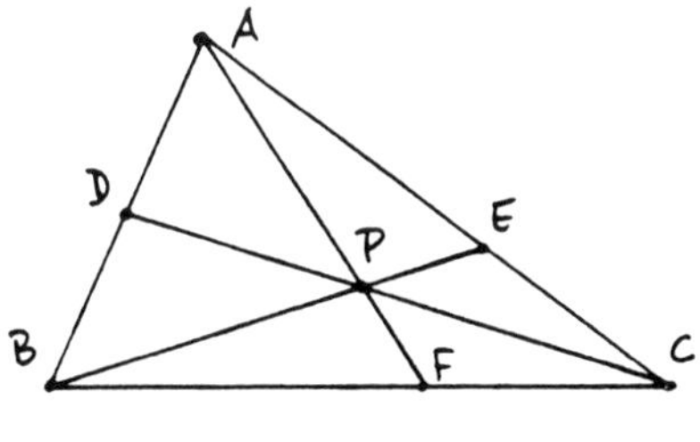

20.5 (Desargues' theorem). Let ABC, and $A'B'C'$ be two triangles. Assume that AA', BB', CC' all pass through a single point O (we can say ABC and $A'B'C'$ are *perspective* from O). Assume further that AB is parallel to $A'B'$ and BC is parallel to $B'C'$. Prove that AC is parallel to $A'C'$.

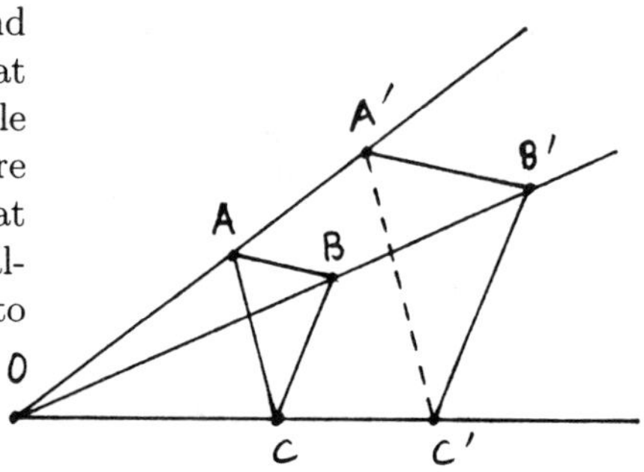

20.6 Prove the field analogue of (II.11), as follows. Let AB be a given line segment. Construct $AC \cong AB$, and perpendicular to it. Let D be the midpoint of AC. Then find E on AC so that $DE \cong DB$. Find F on AB so that $AE \cong AF$. Prove that

$$(AF)^2 = (AB)(BF),$$

in the field of segment arithmetic. *Hint*: Use (20.5). We say AB has been divided in *extreme and mean ratio*.

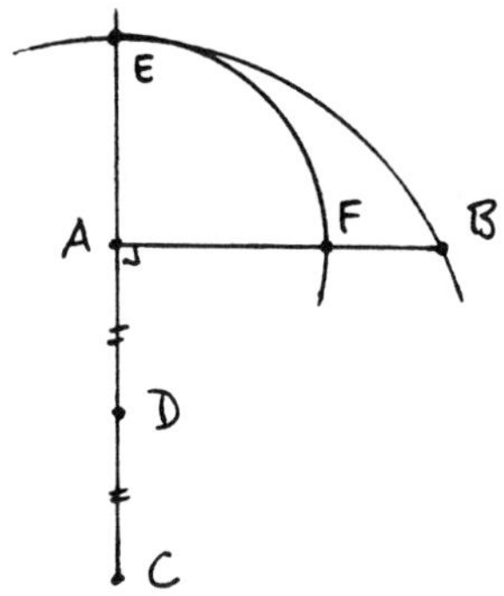

20.7 Give a new proof of (IV.10) as follows. Let the segment AB be divided in extreme and mean ratio as in (Exercise 20.6) above: $(AF)^2 = (AB) \cdot (BF)$. Construct a triangle ABC so that $AC \cong AB$ and $BC \cong AF$. Prove that the base angles of the isosceles triangle ABC are each equal to twice the vertex angle at A. *Hint*: Use (Sim SAS) (20.4) to find similar triangles.

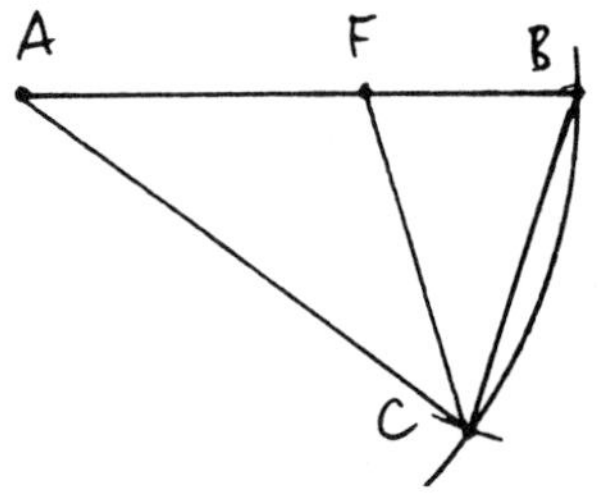

20.8 Prove that in a Hilbert plane with (P) (without assuming (E)), there exists an equilateral triangle with given side AB. *Hint*: First show the field F of segment arithmetic contains an element $\frac{1}{2}\sqrt{3}$.

20.9 Given a triangle ABC with acute angles at B and C, make a ruler and compass construction for a square with one edge along the side BC, and the other two vertices on the sides AB, AC. We call this an *inscribed square*.

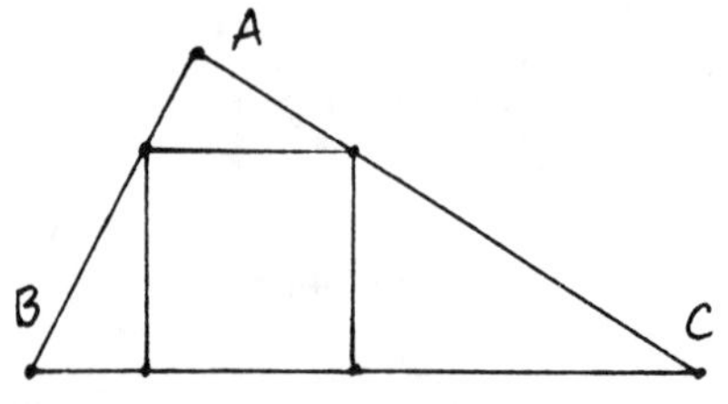

21 Introduction of Coordinates

In this section we will complete a logical circle by showing that if Π is an abstract geometry satisfying the axioms of a Hilbert plane plus (P) and if F is the field of segment arithmetic for Π (introduced in §19), then Π is isomorphic to the Cartesian plane over the field F.

Let me explain this in some more detail. We started our study of geometry from two different perspectives. On the one hand we considered a purely geometric development, where points, lines, congruence, etc., were undefined notions subject to certain axioms, from which we prove theorems. This is Euclid's approach, improved by Hilbert, who gave us a set of axioms including Euclid's unstated assumptions, so that we could develop his geometry on a rigorous basis.

On the other hand, we constructed examples, or *models* of this abstract geometry, based on the logical foundations of modern algebra, by starting with an ordered field F (for example the real numbers), and making a geometry whose points are ordered pairs of elements of the field F. This is the Cartesian approach (cf. §13). In this model, we defined lines and congruence, using linear equations and a distance function, and then proved, by algebraic methods, that the axioms of abstract geometry were true in this model.

For any particular field F, it may happen that certain things are true which do not hold in every geometry: for the plane F^2 is just one of many possible models of an abstract geometry. For example, if $F = \mathbb{R}$, then Dedekind's axiom (D) holds, but this would not be true if F was the field of *constructible* numbers (16.4).

Perhaps more interesting is that we can prove certain results which hold in the geometry over any field F, though we do not know how to prove the corresponding statement in abstract geometry. For example, over any field F, the line-circle intersection property (LCI) is equivalent to the circle-circle intersection property (E), because we have shown that both of these are equivalent to the condition (**) on the field F, (16.2). We do not know any purely geometric proof of this equivalence.

Of course it might be that the geometries constructed over fields were only some possible geometries, and that there were other abstract geometries, not corresponding to any field, with properties different from the geometries over fields. If we drop the parallel axiom, this is indeed the case, as we will see in the study of non-Euclidean geometries (Chapter 7). However, we will show that any abstract geometry with (P) is isomorphic to a geometry over a field.

To understand this, we need to be clear what we mean by an *isomorphism* of geometries. The Greek roots iso + morph mean the same form. Intuitively, both geometries behave the same way. Their outer structure might be different, but there is no way they can be distinguished internally. This is similar to the notion of a one-to-one correspondence between sets. If you have five apples in one bag, and five oranges in another bag, certainly apples are different from oranges, but from the point of view of set theory, both sets have five elements, so as abstract sets they are isomorphic.

The formal definition of isomorphism of geometries is as follows.

Definition. Let Π and Π' be two Hilbert planes. An *isomorphism* between Π and Π' is a one-to-one mapping $\varphi : \Pi \to \Pi'$ of Π onto Π', which is compatible with the undefined notions. This means

1) A subset $L \subseteq \Pi$ is a line if and only if $\varphi(L) \subseteq \Pi'$ is a line.

2) Three points $A, B, C \in \Pi$ satisfy the betweenness property $A * B * C$ if and only if $\varphi(A) * \varphi(B) * \varphi(C)$ in Π'.

3) Given four points $A, B, C, D \in \Pi$, the line segments AB and CD are congruent if and only if the line segments $\varphi(A)\varphi(B)$ and $\varphi(C)\varphi(D)$ are congruent in Π'.

4) If α is an angle formed by the rays AB and AC in Π, we denote by $\varphi(\alpha)$ the angle formed by the rays $\varphi(A)\varphi(B)$ and $\varphi(A)\varphi(C)$ in Π'. If α and β are two angles in Π, then α and β are congruent if and only if $\varphi(\alpha)$ and $\varphi(\beta)$ are congruent in Π'.

Theorem 21.1 (Introduction of Coordinates) *Let Π be a Hilbert plane satisfying the parallel axiom* (P). *Let F be the ordered field of segment arithmetic in* Π (19.7). *Then Π is isomorphic to the Cartesian plane F^2 over the field F.*

Proof. We start by fixing two perpendicular lines in the plane Π, which we call the *x-axis* and the *y-axis*. We call their intersection point O the *origin*. On each axis choose a point 1_x and 1_y so that the segments $O1_x$ and $O1_y$ both represent 1 in the field F. These then define the *positive rays* on the x-axis and the y-axis.

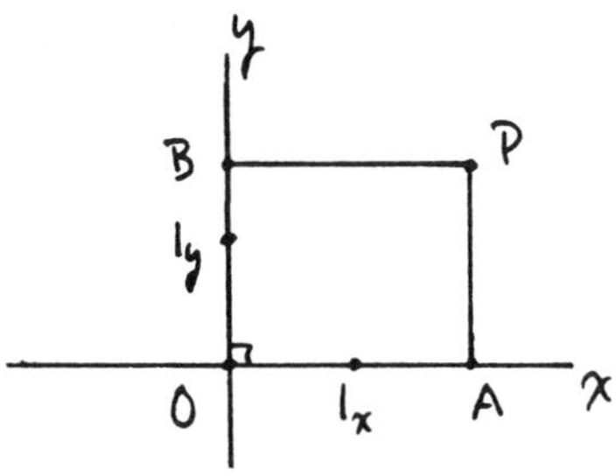

Now for any point P in the plane, we drop perpendicular PA to the x-axis and PB to the y-axis. Let the segment OA represent $a \in F$ and let OB represent $b \in F$.

Now we can define a mapping $\varphi : \Pi \to F^2$ by $\varphi(P) = (\pm a, \pm b)$, where we choose the $+$ sign if A (resp. B) is on the positive x-axis (resp. y-axis) and the $-$ sign if not. Clearly this construction gives a bijective correspondence between the set of points of Π and the set of ordered pairs of the field F, so φ is 1-1 and onto.

We must verify that φ is compatible with the notions of line, betweenness, congruence of segments, and congruence of angles. And remember that in Π these are undefined notions, whose properties are known only through the axioms and propositions of the geometry, while in F^2 they were defined in terms of algebraic conditions (Chapter 3).

Step 1. Let l be a line in Π. (For simplicity we will consider a general line, and let the reader check the special cases of horizontal or vertical lines(!).) Let l meet the x-axis at A. Measure off $AB \in 1$, let BC be a perpendicular, and let $m \in F$ be the class of BC. We call m the *slope* of the line.

Let l meet the y-axis at D, and let $b \in F$ represent that point (i.e. $b = OD$ if D is on the positive y-axis, otherwise $b = -OD$).

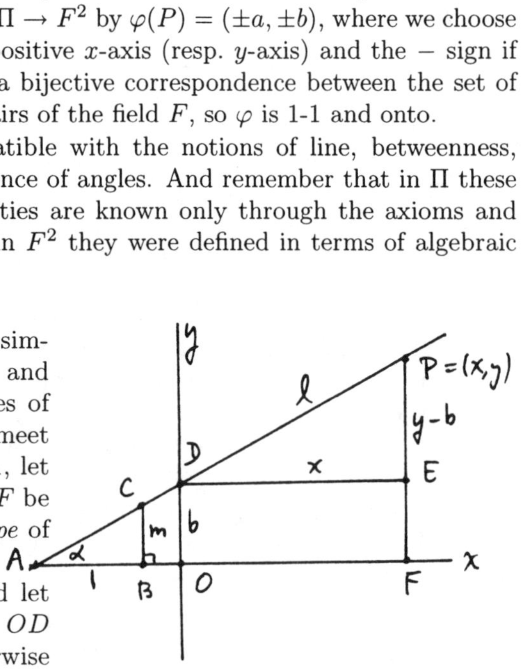

Now consider an arbitrary point $P = (x, y)$ in the plane. Make a triangle DPE using horizontal and vertical lines. Then $DE = x$ and $PE = y - b$ (in the case shown; otherwise adjust signs $\pm$ as needed (!)). This point P will lie on the line l if and only if the angle $PDE = \alpha$. Because of the definition of our segment arithmetic, this condition is equivalent to saying $y - b = mx$. In other words $P = (x, y)$ lies on the line l if and only if $y = mx+b$. Since lines in F^2 were defined by linear equations, this shows the first property of an isomorphism: $L \subseteq \Pi$ is a line $\Leftrightarrow \varphi(L) \subseteq \Pi'$ is a line.

Step 2. Let A, B, C be three collinear points in Π (which by Step 1 will guarantee that their images in Π' are collinear). Let A', B', C' be their projections on the x-axis (and again for simplicity we will treat the special case that A, B, C are in the first quadrant, leaving other cases to the reader (!)).

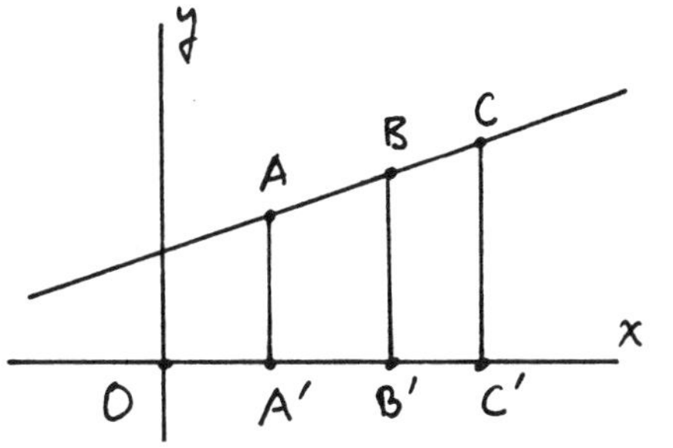

Since the lines AA', BB', CC' are parallel, A and C will be on opposite sides of the line BB' if and only if A' and C' are on opposite sides of the line BB' (axiom (B4)), so $A * B * C$ if and only if $A' * B' * C'$. Let the segments OA', OB', OC' represent $a, b, c \in F$. Then $A' * B' * C'$ means either the segments are related $OA' \le OB' \le OC'$ or vice versa $OC' \le OB' \le OA'$. This is equivalent to saying $a < b < c$ or $c < b < a$, by the way we defined inequality in the field F of segment arithmetic. And this in turn is equivalent to saying $\varphi(A) * \varphi(B) * \varphi(C)$ because of the definition of betweenness in F^2 (§15).

Step 3. Let A, B be two points in Π, and let the segment AB represent $d \in F$. On the other hand, let $\varphi(A) = (a_1, a_2)$ and $\varphi(B) = (b_1, b_2)$. Then if we draw the right triangle ABC with legs parallel to the axes, we find $AC = b_1 - a_1$ and $BC = b_2 - a_2$ by construction. We use the field version of the Pythagorean theorem (20.5) to conclude that

$$d^2 = (b_1 - a_1)^2 + (b_2 - a_2)^2$$

in F.

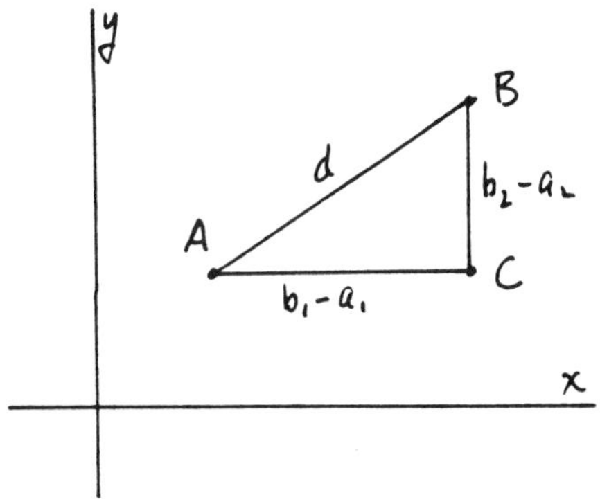

Now let $A'B'$ be another segment, with length $d' \in F$. Then similarly if $\varphi(A') = (a_1', a_2')$ and $\varphi(B') = (b_1', b_2')$, we have

$$d'^2 = (b_1' - a_1')^2 + (b_2' - a_2').$$

Now $AB \cong A'B'$ if and only if $d = d'$, because F was constructed from the set P of congruence equivalence classes of line segments. On the other hand $d = d'$ if and only if $d^2 = d'^2$, because both are positive elements of F. But the equations above show that d^2 and d'^2 are equal to the "distance squared" function which we used to define congruence of segments in F^2 (§16). Thus $AB \cong A'B'$ if and only if $\varphi(A)\varphi(B) \cong \varphi(A')\varphi(B')$.

Step 4. We show that two angles α, α' in Π are congruent if and only if $\varphi(\alpha)$ and $\varphi(\alpha')$ are congruent in Π'. For economy of exposition, we give an indirect proof though a direct proof is also possible (see Exercise 21.3).

Suppose given angles α and α' in Π. Let the vertices be A and A', and choose any two points B, C on the two rays of α. Then find B', C' on the rays of α' so that $AB \cong A'B'$ and $AC \cong A'C'$. Draw the lines BC and $B'C'$ to make triangles.

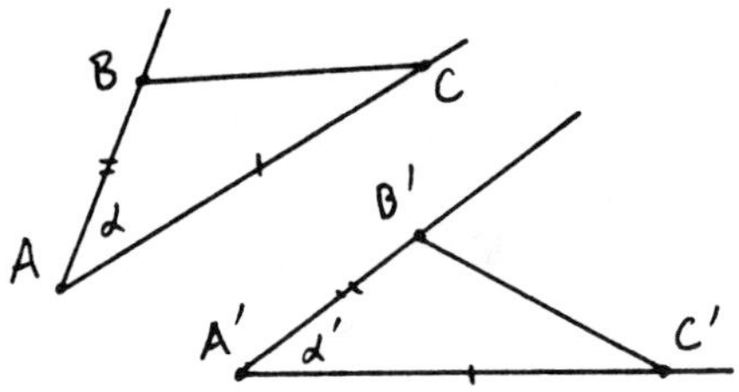

If $\alpha \cong \alpha'$ then by (SAS) it follows that the triangles ABC and $A'B'C'$ are congruent, and in particular $BC \cong B'C'$. Conversely, if $BC \cong B'C'$, then by (SSS) the two triangles are congruent, and so $\alpha \cong \alpha'$. Hence $\alpha \cong \alpha' \Leftrightarrow BC \cong B'C'$.

Apply φ to the six points A, B, C, A', B', C'. Then $\varphi(A)\varphi(B) \cong \varphi(A')\varphi(B')$ and $\varphi(A)\varphi(C) \cong \varphi(A')\varphi(C')$ by Step 3. Furthermore, we have shown that the geometry F^2 satisfies Hilbert's axioms (§17) and in particular (SAS) and (SSS) hold also in F^2. So by the same argument in F^2 we see that $\varphi(\alpha) \cong \varphi(\alpha')$ if and only if $\varphi(B)\varphi(C) \cong \varphi(B')\varphi(C')$.

Combining with Step 3 for the segments BC and $B'C'$, we see that $\alpha \cong \alpha' \Leftrightarrow BC \cong B'C' \Leftrightarrow \varphi(B)\varphi(C) \cong \varphi(B')\varphi(C') \Leftrightarrow \varphi(\alpha) \cong \varphi(\alpha')$. □

Corollary 21.2 *In any Hilbert plane* Π *satisfying* (P), (LCI) *is equivalent to* (E).

Proof. Indeed, we have shown this is true over a field (16.2), so by the theorem it is true in Π also.

Exercises

21.1 Given an angle α at a point A, choose any point B on one ray of α and drop a perpendicular BC to the other ray (extended if necessary). Define the tangent $\tan\alpha = b/a$ if α is acute, or $-b/a$ if α is obtuse, where $b = BC$ and $a = AC$.

Show that $\tan\alpha$, as an element of the field F of segment arithmetic of the abstract geometry Π, is independent of the choice of B, and the choice of ray AB. Show also $\alpha \cong \alpha' \Leftrightarrow \tan\alpha = \tan\alpha'$ in F.

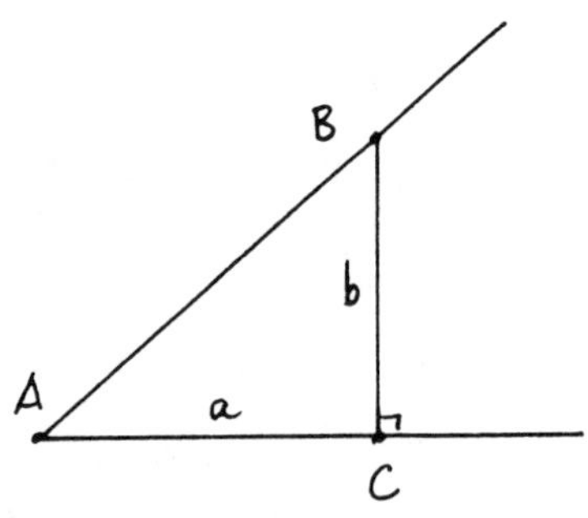

21.2 Given two adjacent nonoverlapping angles α, β at a point A, use the diagram shown, plus similar triangles, and (20.5), to show that

$$\tan(\alpha+\beta) = \frac{\tan\alpha + \tan\beta}{1 - \tan\alpha\tan\beta}$$

in F.

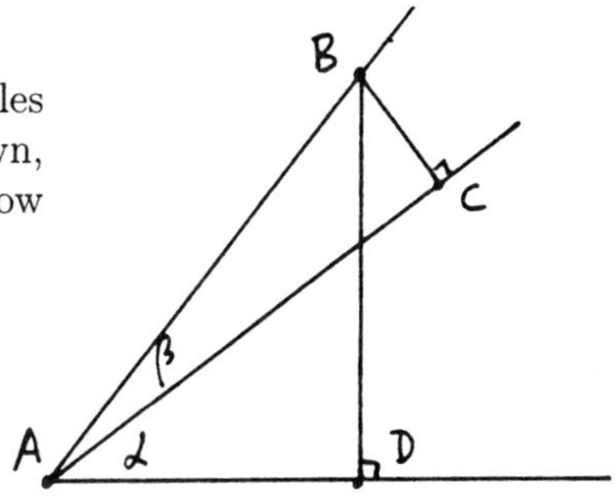

21.3 Use Exercises 21.1, 21.2 above to give a direct proof of part 4) of (21.1), namely that two angles α, α' in Π are congruent if and only if $\varphi(\alpha)$ and $\varphi(\alpha')$ are congruent in F^2, using the definition of congruence we gave for angles in F^2 (§16). In particular, show that the tangent just defined is equal to the tangent defined earlier in F^2.

Our two soules therefore, which are one,
Though I must goe, endure not yet
A breach, but an expansion,
Like gold to ayery thinnesse beate.

If they be two, they are two so
As stiffe twin compasses are two,
Thy soule the fixt foot, makes no show
To move, but doth, if the'other doe.

And though it in the center sit,
Yet when the other far doth rome,
It leanes, and hearkens after it,
And growes erect, as that comes home.

Such wilt thou be to mee, who must
Like th'other foot, obliquely runne;
Thy firmnes drawes my circle just,
And makes me end, where I begunne.

– from *A Valediction: Forbidding Mourning*
by John Donne (1572–1631)

CHAPTER 5

Area

Looking at Euclid's theory of area in Books I–IV, Hilbert saw how to give it a solid logical foundation. We define the notion of equal content if we can transform one figure into the other by adding and subtracting congruent triangles (§22). We can prove all the properties of area which Euclid uses, except that "the whole is greater than the part". This is established only when we relate the geometrical notion of equal content to the notion of an area function (§23).

In an archimedean Euclidean plane, we prove the theorem of Bolyai and Gerwien, that figures of equal area are equivalent by dissection (§24). We also investigate the practical problem of dissecting one figure into another.

We briefly discuss the classical problem of squaring the circle (§25) and its influence.

In comparing the volumes of three-dimensional figures, Euclid uses a limiting process, the "method of exhaustion" (§26). We give Dehn's solution of Hilbert's third problem, that solid figures of equal volume are not necessarily equivalent by dissection (§27), thus vindicating Euclid's use of an infinite limiting process in the study of volume.

22 Area in Euclid's Geometry

Starting with (I.35), Euclid introduces a new notion of equality between figures, which corresponds to what we would call "equal area". The area we are familiar with from high school attaches a number to each figure. So the area of a rectangle with sides a and b is the number ab; the area of a triangle with base b and height h is $\frac{1}{2}bh$. In Euclid's geometry, there are no numbers, so we cannot explain his concept of area this way.

Euclid does not define this new notion of equality, but we can infer from his proofs that he considers it another undefined notion, like congruence of segments or angles, which satisfies certain properties similar to the common notions. In particular, he assumes that

1. Congruent figures are "equal".

2. Sums of "equal" figures are "equal".

3. Differences of "equal" figures are "equal".

4. Halves of "equal" figures are "equal".
5. The whole is greater than the part.
6. If squares are "equal", then their sides are equal.

Properties 1, 2, and 3 are used in the proof of (I.35). Property 4 appears in the proof of (I.37), and Property 5 appears in the proof of (I.39). Property 6, which is actually a consequence of 5, is used in the proof of (I.48).

We could accept this notion of "equality" between figures as another undefined notion, with these properties as additional axioms. However, one is reluctant to encumber the foundations of geometry with unnecessary undefined notions and axioms. So instead, following Hilbert, we will show that one can define a suitable notion of "equal area" and prove its properties, thus providing a new foundation for Euclid's treatment. To avoid overuse of the word "equal", we introduce a new terminology and will say that certain figures have "equal content".

To begin with, let us be precise about our terminology. We presuppose the axioms of a Hilbert plane. When we speak of a *triangle* ABC in this chapter, we mean that the subset of the plane consisting of the three line segments AB, AC, BC, the *sides* of the triangle, plus all the points in the interior of the triangle.

Recall (§7) that the *interior* of a triangle ABC is the set of points which are on the same side of the line AB as C, on the same side of AC as B, and on the same side of BC as A. Two triangles are *non-overlapping* if they have no interior points in common. They may have common vertices or parts of edges.

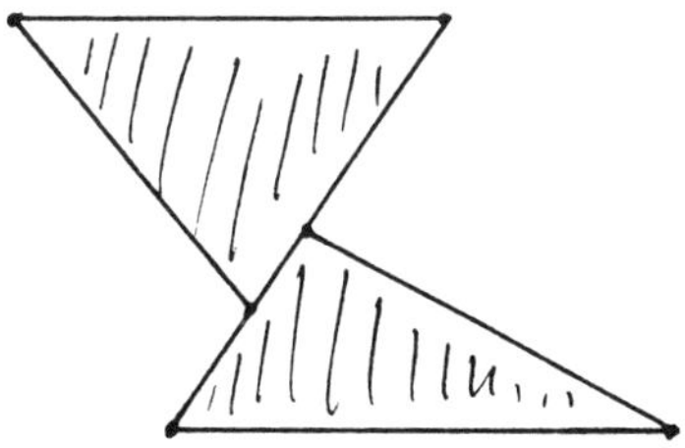

Definition. A *rectilineal figure* (or *figure* for short) is a subset of the plane which can be expressed as a finite non-overlapping union of triangles. A point D is in the *interior* of a figure P, if there is a triangle ABC entirely contained in P, such that D is in the interior of the triangle ABC. Two figures are non-overlapping, if they have no interior points in common.

Proposition 22.1 *The intersection of any two figures is a figure. The union of any two figures is a figure. The complement of one figure inside another figure (plus the line segments which form its sides) is a figure. In particular, any finite union of triangles is a figure.*

Proof. The basic idea is to deal with one triangle at a time. For example, if a triangle ABC is cut by a line l, then that portion of the triangle which lies on one side of the line is a figure. One side BDE in this example is a triangle. The other side is a union of two triangles, after we draw the line DC. We leave details to the reader (Exercises 22.1, 22.2, 22.3).

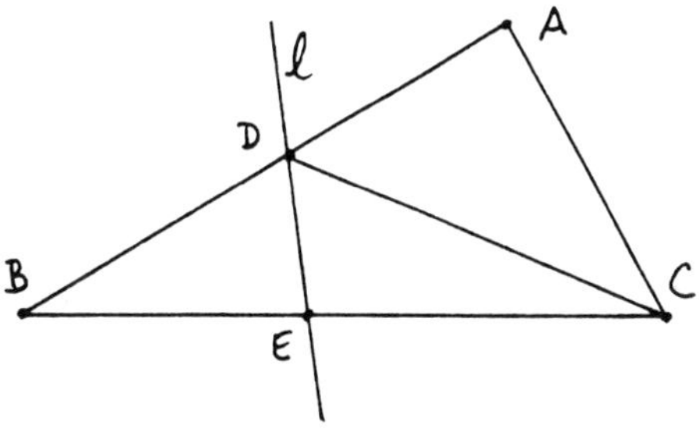

Definition. Two figures P, P' are *equidecomposable* if it is possible to write them as non-overlapping unions of triangles

$$\begin{aligned} P &= T_1 \cup \cdots \cup T_n \\ P' &= T_1' \cup \cdots \cup T_n' \end{aligned}$$

where for each i, the triangle T_i is congruent to the triangle T_i'.

Two figures P, P' have *equal content* if there are other figures Q, Q' such that

1) P and Q are non-overlapping,
2) P' and Q' are non-overlapping,
3) Q and Q' are equidecomposable, and
4) $P \cup Q$ and $P' \cup Q'$ are equidecomposable.

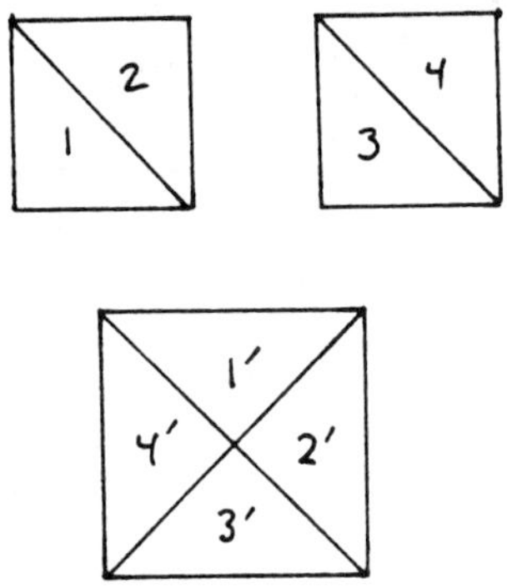

Example 22.1.1 If P is the union of two congruent squares in a Euclidean plane, and P' is a square built on the diagonal of one of the squares of P, then P and P' are equidecomposable. Indeed, we cut P and P' into four congruent triangles each, as shown in the diagram.

Example 22.1.2 Let $ABCD$ and $CDEF$ be two parallelograms on the same base CD and lying in the same parallels. Then Euclid's proof of (I.35) shows that $ABCD$ and $CDEF$ have equal content.

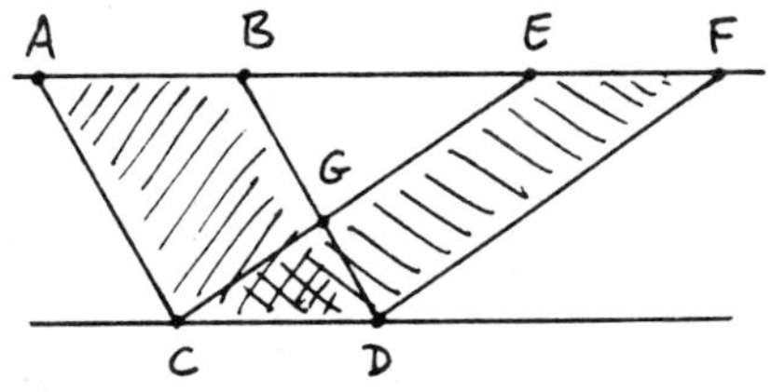

Indeed, if we let $P = ABCD$, and $P' = CDEF$, take $Q = Q' =$ triangle BGE. Then $P + Q$ and $P' + Q'$ are the unions of the congruent triangles ACE and BDF and the equal triangles CDG and CDG.

Example 22.1.3 If two figures P and P' are equidecomposable, then they have equal content, but the converse is not necessarily true. For example, consider the Cartesian plane over a non-Archimedean field F (cf. §18). Let t be an infinite element of the field F. Consider the unit square $ABCD$, and the parallelogram with base CD and top side EF, where $E = (t, 1)$ and $F = (t + 1, 1)$.

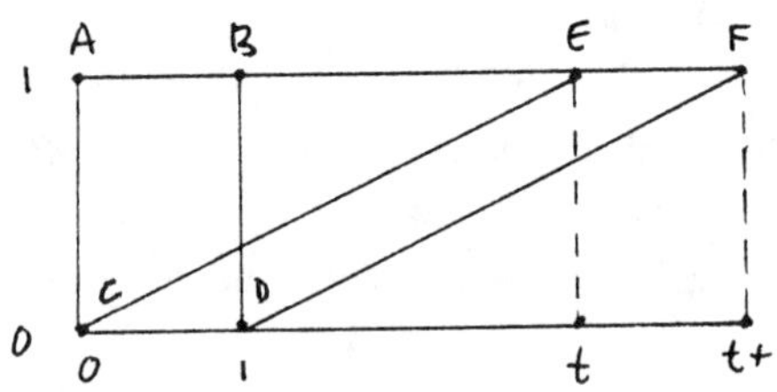

Then according to (I.35), $ABCD$ and $CDEF$ have equal content. However, they are not equidecomposable. Indeed, any triangle contained in the unit square has sides of length $\leq \sqrt{2}$. Any finite number of these sides, placed end to end still has finite length in the field F. But the side CE of the parallelogram has length $\sqrt{t^2 + 1} > t$ which is infinite. Thus no finite number of triangles contained in $ABCD$ can ever fill up the parallelogram $CDEF$.

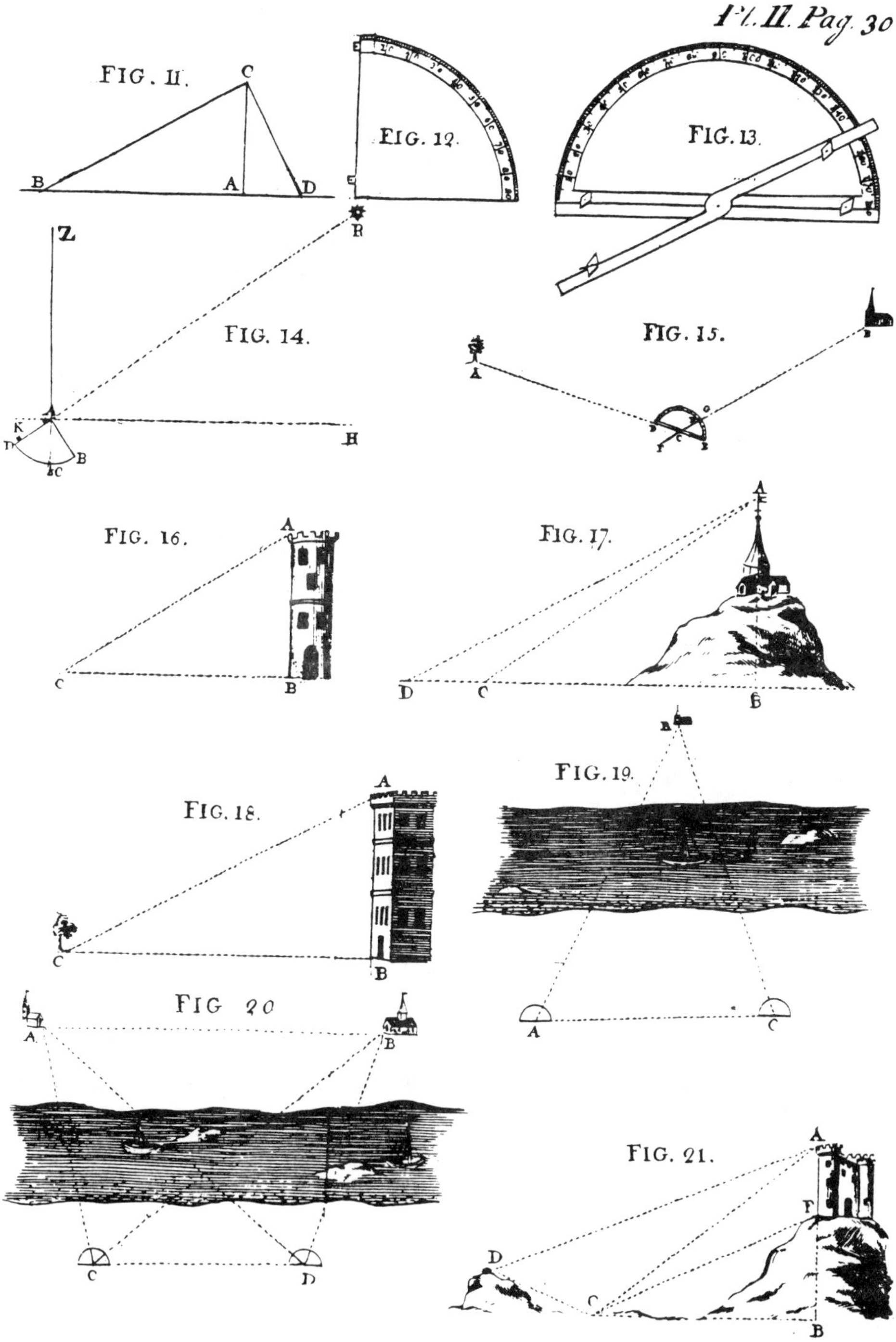

Plate VII. Figures from Gregory's "Treatise of Practical Geometry" (1751).

Remark 22.1.4. This example suggests the question, if we assume Archimedes axiom (A) in addition, are the notions of equidecomposable and equal content equivalent? We will see that the answer is yes in any Hilbert plane with (P) and (A), by using a measure of area function with values in the field of segment arithmetic (24.7.3). I do not know any purely geometric proof of this fact. In the non-Euclidean case, we obtain the same result using the defect of a triangle as a measure of area function (36.7.2).

Proposition 22.2 *In a Hilbert plane, the relation of two figures being equidecomposable is an equivalence relation. Non-overlapping unions of equidecomposable figures are equidecomposable.*

Proof. The relation is obviously symmetric ("P is equidecomposable with P") and reflexive ("if P and P' are equidecomposable, then P' and P are equidecomposable"). The non-trivial part is transitivity. So suppose that P and P' are equidecomposable, and P' and P'' are equidecomposable. Let

$$\begin{aligned} P &= T_1 \cup \cdots \cup T_n \\ P' &= T_1' \cup \cdots \cup T_n' \end{aligned}$$

where T_i and T_i' are congruent triangles, for each i. Also let

$$\begin{aligned} P' &= S_1' \cup \cdots \cup S_m' \\ P'' &= S_1'' \cup \cdots \cup S_m'' \end{aligned}$$

where S_j' and S_j'' are congruent triangles, for each j. We must show that P and P'' are equidecomposable.

To do this, we will refine the decompositions of P and P'' in order to express them both as unions of congruent triangles. For each i, j consider the intersection $T_i' \cap S_j'$ in P'. It may be empty, or may consist of points or line segments only. We ignore those. When it has a non-empty interior, it will be a figure (Exercise 22.1) which can be written as a union of triangles

$$T_i' \cap S_j' = \bigcup_{k=1}^{l} U_{ijk}'.$$

Now let $\varphi_i : T_i \to T_i'$ be a rigid motion (Exercise 17.10) taking the triangle T_i to the congruent triangle T_i'. We use φ_i to transport the triangles U_{ijk}' to new triangles $U_{ijk} = \varphi_i^{-1}(U_{ijk}')$ contained in T_i. Then

$$T_i = \bigcup_{j,k} U_{ijk},$$

and each U_{ijk} is congruent to U_{ijk}'.

Similarly, for each j, let ψ_j be a rigid motion taking S_j' to the congruent triangle S_j''. Let $U_{ijk}'' = \psi_j(U_{ijk}')$. Then

$$S_j'' = \bigcup_{i,k} U_{ijk}'',$$

and each U''_{ijk} is congruent to U'_{ijk}.

By construction, the U_{ijk} and the U''_{ijk} are non-overlapping triangles, and we can write

$$\begin{aligned} P &= \bigcup_{i,j,k} U_{ijk} \\ P'' &= \bigcup_{i,j,k} U''_{ijk} \end{aligned}$$

where U_{ijk} is congruent to U''_{ijk} for each i, j, k. Thus P and P'' are equidecomposable.

If P and P' are equidecomposable, and Q and Q' are equidecomposable, and if P does not overlap Q and P' does not overlap Q', then it is obvious that $P \cup Q$ and $P' \cup Q'$ are equidecomposable.

Proposition 22.3 *In a Hilbert plane, the relation of two figures having equal content has the following properties:*

a) *Equal content is an equivalence relation.*

b) *Equidecomposable figures have equal content.*

c) *Non-overlapping unions of figures of equal content have equal content.*

d) *If $Q \subseteq P$ and $Q' \subseteq P'$, and if Q and Q' have equal content, and P and P' have equal content, then $P - Q$ and $P' - Q'$ have equal content.*

Lemma 22.4 *Suppose P and P' are equidecomposable figures, and suppose P is expressed as a non-overlapping union of subfigures $P = P_1 \cup P_2$. Then there are subfigures P'_1, P'_2 of P' such that P' is the non-overlapping union of P'_1 and P'_2, and P_i and P'_i are equidecomposable for $i = 1, 2$.*

Proof. Suppose

$$\begin{aligned} P &= T_1 \cup \cdots \cup T_n \\ P' &= T'_1 \cup \cdots \cup T'_n \end{aligned}$$

where T_i and T'_i are congruent triangles for each i. As in the proof of (22.2) we will refine decompositions appropriately.

For each i, consider the intersections $T_i \cap P_1$ and $T_i \cap P_2$. We can write each as unions of triangles (22.1)

$$\begin{aligned} T_i \cap P_1 &= \bigcup_j S_{ij1} \\ T_i \cap P_2 &= \bigcup_j S_{ij2}. \end{aligned}$$

Use rigid motions $\varphi_i : T_i \to T'_i$ to transport these triangles and define

$$S'_{ijk} = \varphi_i(S_{ijk})$$

for each i, j, k. Let

$$\begin{aligned} P'_1 &= \bigcup S'_{ij1} \\ P'_2 &= \bigcup S'_{ij2}. \end{aligned}$$

Then P'_1, P'_2 satisfy the requirements of the lemma.

Proof of (22.3). a) The relation of equal content is obviously reflexive and symmetric. The non-trivial part is to show it is transitive. So suppose figures P and P' have equal content, and P' and P'' have equal content. Then there are equidecomposable figures Q and Q' such that $P \cup Q$ and $P' \cup Q'$ are equidecomposable, and there are further equidecomposable figures R' and R'' such that $P' \cup R'$ and $P'' \cup R''$ are equidecomposable.

The difficulty is that while the unions mentioned above are all non-overlapping, it may happen that Q' and R' overlap. To avoid this situation, we apply the lemma to the equidecomposable figures $P' \cup R'$ and $P'' \cup R''$ and the given decomposition of the first of these. Thus we may assume that the triangulation of $P' \cup R'$ arises from separate triangulations of P' and R'. Once this is so, we can move R' to some other position in the plane R^* and still have $P' \cup R^*$ equidecomposable with $P'' \cup R''$. In particular, we may choose R^* in such a way that Q' and R^* do not overlap (Exercise 22.6).

Now let R be a figure congruent to R^* which does not overlap P or Q, and let Q'' be a figure congruent to Q' which does not overlap P'' or R''. Then, by additivity of equidecomposability, we find $P \cup Q \cup R$ equidecomposable with $P'' \cup Q'' \cup R''$, and $Q \cup R$ equidecomposable with $Q'' \cup R''$, so by definition, P and P'' have equal content. This completes the proof of a).

Statement b) is trivial.

Statements c) and d) are not difficult to prove after using the lemma to avoid overlaps (Exercise 22.5).

Remark 22.4.1. Now we have defined a notion of equal content for rectilineal figures in the plane, and we have established enough properties to recover most of Euclid's results on area. In particular, this notion of equal content satisfies the properties 1, 2, 3 listed at the beginning of this section. However, our theory does not seem to be strong enough to establish properties 4, 5, 6 so we formulate what is missing as follows.

Z. (de Zolt's axiom). If Q is a figure contained in another figure P, and if $P - Q$ has a non-empty interior, then P and Q do not have the same content.

We can think of (Z) as a precise formulation of Euclid's Common Notion 5, "the whole is greater than the part", for the notion of content. However, we avoid the use of the words "greater" and "lesser", because these imply the existence of an order relation among figures, which we have not yet established. In fact, the existence of an order relation for content depends on (Z) (Exercise 22.7). We will also see in the exercises that (Z) implies the other two properties 4 and 6 listed at the beginning of this section (Exercises 22.6, 22.8).

I do not know of any purely geometric proof of (Z) from the definition of content we have given, and I do not know if (Z) holds in an arbitrary Hilbert plane. We will see in the next section, however, that (Z) holds whenever there is a measure of area function in the geometry. In particular, (Z) will hold in the Cartesian plane over any field F, and also in the Archimedean non-Euclidean geometries (§36).

Now let us review Euclid's results about area and their proofs, substituting everywhere his "equality" of figures by the notion of equal content developed in this section. We work in a Euclidean plane (§12), i.e. a Hilbert plane with (P) and (E).

We have already seen (22.1.2) that Euclid's proof of (I.35) shows that the two parallelograms have equal content. The next result (I.36) follows using transitivity of equal content. But in the proof of (I.37) Euclid uses the property that "halves of equals are equal", which depends on (Z) (Exercise 22.8). So we will give another proof which does not depend on (Z).

Proposition 22.5 (I.37) *Triangles on the same base and in the same parallels have equal content.*

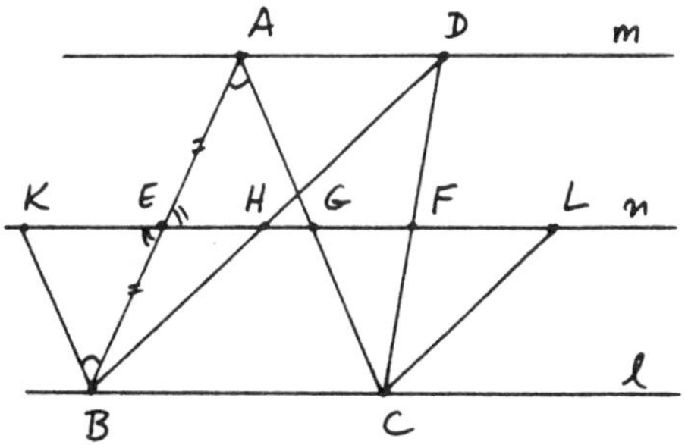

Proof. Let ABC and DBC be the given triangles, lying in the parallel lines l, m. Let E be the midpoint of AB, and draw a line n through E, parallel to l. Let this line meet DC at F. Then, from (5.1) (cf. Exercise 5.3) it follows that F is the midpoint of DC. Draw a line through B, parallel to AC, to meet n at K, and draw a line through C, parallel to BD, to meet n at L.

Then because of parallel lines, we get equal angles, which show that $\Delta AEG \cong \Delta BEK$ using (ASA). So the triangle ABC has equal content with the parallelogram $BCGK$. Similarly $\Delta DHF \cong \Delta CLF$, so that ΔBCD has equal content to the parallelogram $BCHL$.

Now by (I.35), these two parallelograms have equal content. So by transitivity it follows that the triangles ABC and DBC have equal content.

Continuing with our examination of Euclid's results, (I.38) follows by transitivity. Its converse (I.39) uses "the whole is greater than the part" in its proof, and so depends on (Z). Generally speaking, all of the results in which Euclid shows two figures are equal will be valid for the notion of equal content. However, when a hypothesis of equal content is used to conclude something involving congruence of segments or angles in the figure, then (Z) will be necessary. So (I.40) also depends on (Z). In (I.48) Euclid says if squares have equal content, then their sides are equal, so this result also depends on (Z) (Exercise 22.6). The remaining results of Book I, namely (I.41)–(I.45) and (I.47) are all valid for content. So for example, (I.47), the Pythagorean theorem, says that the square on the hypotenuse of a right triangle has equal content to the union of the squares on the two legs of the triangle.

In Book II, all of the results make statements about certain figures having equal content to certain others, and all of these are valid in our framework. Note that the line-circle intersection property (11.6) is used in (II.11), to divide a line such that the square on the larger piece has equal content to the rectangle on the whole and the smaller piece. It is also used in (II.14) to construct a square with equal content to any given rectilineal figure.

In Book III, Propositions (III.35) and (III.36) hold for equal content. The converse (III.37) of (III.36) requires property 6 above: if squares are equal, their sides are equal, and so depends on (Z).

In Book IV the only result needing Euclid's theory of area is (IV.10), to construct an isosceles triangle whose base angles are twice the angle at the vertex. The proof uses (III.37) and so depends on (Z). In particular, Euclid's proof of the construction of the regular pentagon (§4) depends on (Z). So we see again how the construction of the pentagon involves all the subtleties of Euclid's geometry!

At this point we could take (Z) as an additional axiom, and then we would have a satisfactory basis for Euclid's theory of area. However, we will see in the next section that (Z) holds in the Cartesian plane over a field, and hence, using the theorem of introduction of coordinates (21.1), it holds in any Euclidean plane.

Exercises

22.1 Show that the intersection of any two figures is a figure. *Hint*: First do the intersection of two triangles.

22.2 Show that the complement of one figure contained in another figure is a figure. *Hint*: First do the case when the smaller figure is a single triangle.

22.3 Show that the union of two figures is a figure.

22.4 In a Hilbert plane, given two figures P, Q, show that there is a rigid motion φ of the plane (cf. §17) such that P and $\varphi(Q)$ do not overlap.

22.5 Prove parts c) and d) of (22.3).

22.6 In a Euclidean plane, assuming (Z), suppose given segments AB and CD such that the squares on AB and CD have equal content. Show that AB and CD are congruent.

22.7 In a Euclidean plane with (Z), show that there is a total ordering on the set of figures, with the property that $P \leq Q$ whenever there exists a figure P' with same content as P, and P' is contained in Q. *Hint*: Use (I.44) to show that any figure P has equal content with a rectangle P' of given fixed side AB.

22.8 In a Euclidean plane with (Z) show that "halves of equals are equal", in the following sense: if P and Q are figures with equal content, and if $P = P_1 \cup P_2$ is a non-overlapping union, where P_1 and P_2 have equal content, and similarly $Q = Q_1 \cup Q_2$ with Q_1 and Q_2 having equal content, then P_1 and Q_1 have equal content.

22.9 In a Euclidean plane, suppose that (Z) fails in the sense that there exists a triangle ABC and a point D between B and C, such that ABC has equal content with the smaller triangle ADC. Then show that for any figure P, there exists a figure Q containing P, and Q has equal content with the empty set. (If you can find a contradiction resulting from this, you will have discovered a proof of (Z)!)

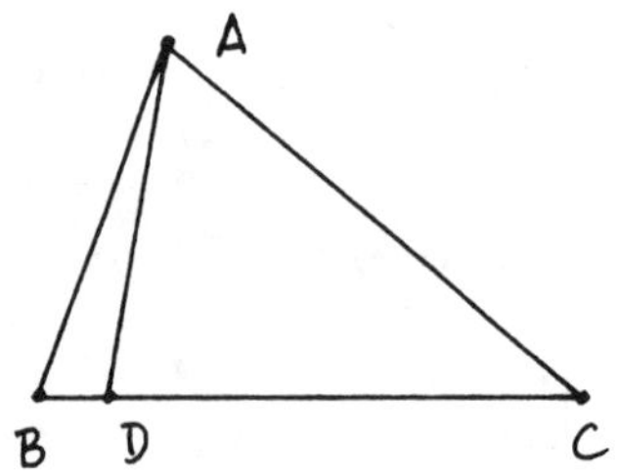

22.10 In a Euclidean plane with Archimedes' axiom (A), give a direct proof of the analogue of (I.35) for equidecomposability: two parallelograms on the same base and in the same parallels are equidecomposable.

22.11 *Simple closed polygons.* A *simple closed polygon* is a finite union of line segments $A_1A_2, A_2A_3, \ldots, A_iA_{i+1}, \ldots, A_nA_1$, where $A_1, \ldots, A_n$ are distinct points in the plane, and the line segments have no other points in common except their endpoints which each lie on two segments.

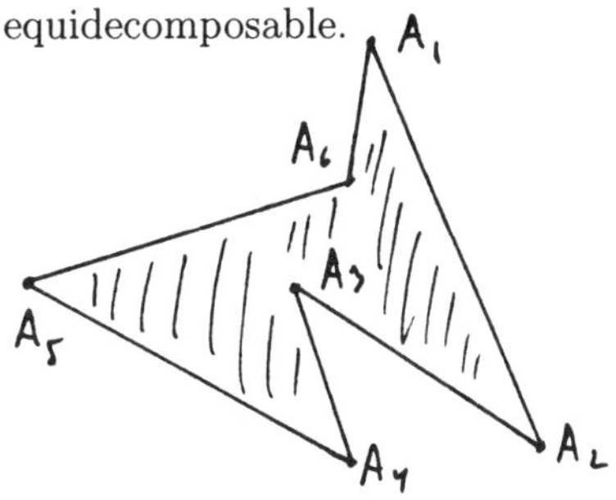

a) Show that a simple closed polygon divides the plane into two segment-connected subsets, its *interior* and its *exterior*. (This is the (much easier) polygonal analogue of the famous Jordan curve theorem for simple closed curves in $\mathbb{R}^2$.)

b) Show that the simple closed polygon, together with its interior, is a figure P (i.e. a finite union of triangles). Is it always possible to write an n-sided polygon P as a union of $n-2$ triangles?

23 Measure of Area Functions

In this section we will relate Euclid's theory of area discussed in the last section with the familiar notion of area as a number. In high school geometry, most likely you learned how to compute the area of various figures, but never saw a definition of area or a proof that it exists. So we will first define an area function by the properties we want it to have. Then we discuss the question of existence and uniqueness. We will see that an area function exists in a Hilbert plane with (P). We will also see that the existence of an area function implies de Zolt's axiom (Z) discussed in the last section. These two results then put Euclid's theory of area on a firm basis.

Definition. A *measure of area function* on a Hilbert plane is a function α, defined on the set $\mathcal{P}$ of all *figures* (see definition in §22), with values in an ordered abelian group G, such that

1) for any triangle T, we have $\alpha(T) > 0$ in G,
2) if T and T' are congruent triangles, then $\alpha(T) = \alpha(T')$,
3) if two figures P and Q do not overlap, then

$$\alpha(P \cup Q) = \alpha(P) + \alpha(Q).$$

We call $\alpha(P)$ the *area* of the figure P, with respect to the given measure of area function.

Proposition 23.1 *Suppose α is a measure of area function on a Hilbert plane.*

a) *If P is any figure with non-empty interior, then $\alpha(P) > 0$.*
b) *If P and P' are equidecomposable figures, then $\alpha(P) = \alpha(P')$.*
c) *If P and P' are figures with equal content, then $\alpha(P) = \alpha(P')$.*

d) *If a figure Q is contained in a figure P, and $P - Q$ has non-empty interior, then $\alpha(Q) < \alpha(P)$. In particular, P and Q cannot have equal content, so (Z) holds (cf. §22).*

Proof. a) Writing the figure P as a union of triangles T_i, it follows from the definition of a measure of area function that $\alpha(P) = \sum \alpha(T_i)$, and each $\alpha(T_i) > 0$, so $\alpha(P) > 0$.

b) This follows from the property that congruent triangles have equal area function.

c) If P and P' have equal content, then there are figures Q and Q' which are equidecomposable and do not overlap with P and P', respectively, such that $P \cup Q$ and $P' \cup Q'$ are equidecomposable. Hence $\alpha(Q) = \alpha(Q')$ and $\alpha(P \cup Q) = \alpha(P' \cup Q')$. Using the additivity property 3) and subtracting in the group G, we find $\alpha(P) = \alpha(P')$.

d) Write $P = Q \cup (P - Q)$. Since $P - Q$ has non-empty interior, $\alpha(P - Q) > 0$. Hence by additivity, $\alpha(Q) < \alpha(P)$. It follows from c) that P and Q cannot have equal content. In other words de Zolt's axiom (Z) stated in §22 holds.

Now that we know what a measure of area function is and have seen some of its properties, the main work of this section is to prove the existence of such a function in a Hilbert plane with (P).

Theorem 23.2 *In a Hilbert plane with* (P), *there is an area function α, which satisfies and is uniquely determined by the following additional condition: For any triangle ABC, whenever we choose one side AB to be the* base *and let it have length $b \in F$, and let h be the length of an altitude perpendicular to the base, then $\alpha(ABC) = \frac{1}{2}bh$.*

Proof. In this theorem, the uniqueness is obvious, because the additional condition tells us the value of α for any triangle, and any figure is a finite union of triangles.

For the existence of α, there is no choice: For any figure $P \in \mathcal{P}$, write P as a union of triangles $P = T_1 \cup \cdots \cup T_n$, for each triangle T_i choose one side to be the base b_i, let h_i be the corresponding altitude, and define

$$\alpha(P) = \sum \frac{1}{2} b_i h_i.$$

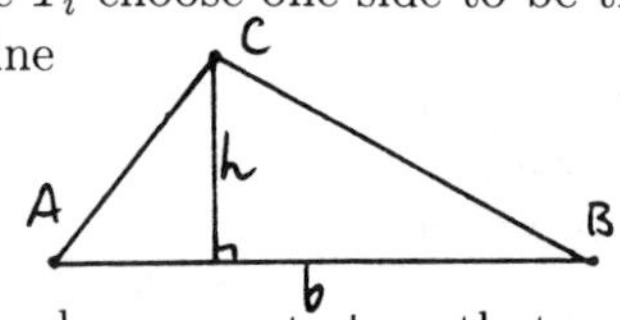

The problem is to show that this notion is well-defined: we must show that α of a triangle is independent of which side we choose as base, and then we must show that $\alpha(P)$ is independent of the triangulation chosen. After that we must verify that α satisfies the properties of an area function. We will deal with some of these questions as separate lemmas.

Lemma 23.3 *In a Hilbert plane with* (P), *let ABC be any triangle. Let b be one choice of base, with corresponding altitude h, and let b' be another choice of base, with altitude h'. Then $\frac{1}{2}bh = \frac{1}{2}b'h'$ in the field of segment arithmetic.*

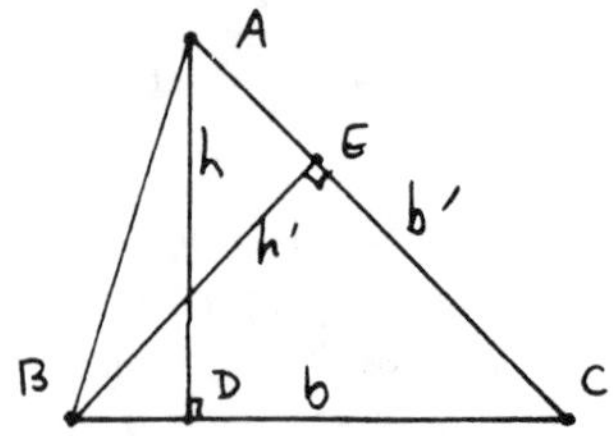

Proof. Let ABC be the triangle, with $b = BC$, $h = AD$, $b' = AC$, $h' = BE$. The two right triangles ADC and BEC have the angle at C in common, hence all three angles equal (I.32). So by (Sim AAA) (20.1) they are similar triangles. It follows that the ratios of corresponding sides are equal, so

$$h/b' = h'/b.$$

Cross-multiplying, $bh = b'h'$, and so $\frac{1}{2}bh = \frac{1}{2}b'h'$ as required.

Thus the function α is well-defined for triangles. The key point in studying arbitrary triangulations is to see what happens when a triangle is divided into smaller triangles.

Lemma 23.4 *If a triangle T be subdivided into smaller triangles T_i in any way whatsoever (but still a finite number) then the measure of area of the big triangle is equal to the sum of the measures of area of the small ones:* $\alpha(T) = \sum \alpha(T_i)$.

Proof. Step 1. We consider the special case where a triangle ABC is divided into two triangles by a single *transversal*, namely a line which goes from one vertex (say C) to a point D on the opposite side AB.

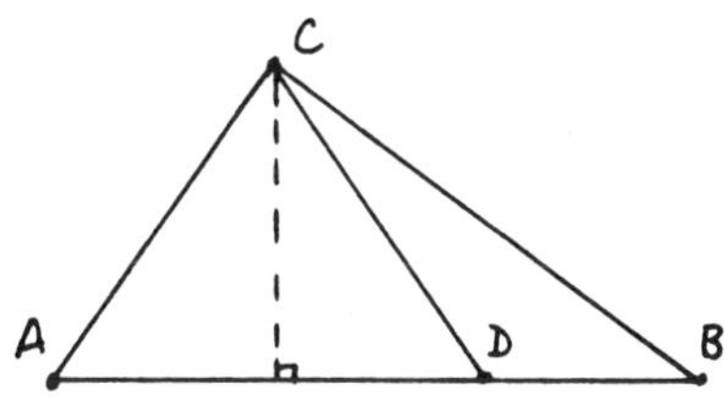

Then choosing side AB as a base for the big triangle, and sides AD and DB as bases for the two smaller triangles, all three triangles have the same height, and the bases AD, DB add up to the base AB, so clearly $\alpha(ABC) = \alpha(ACD) + \alpha(BCD)$.

Step 2. Next we do a slightly harder case. Suppose a triangle ABC is subdivided into triangles T_i so that there are no new vertices in the *interior* of the triangle, and at least one edge (AC in the picture) is free from new vertices on the sides of the triangle. Then $\alpha(ABC) = \sum \alpha(T_i)$.

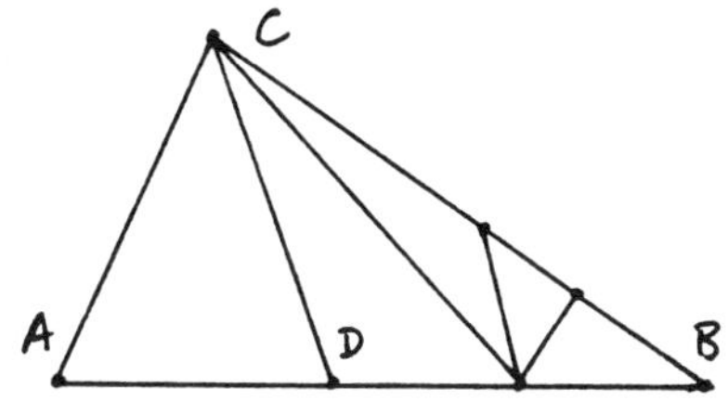

We prove this result by induction on the number of the smaller triangles T_i. If there are just two of them, then we are in the situation of Step 1 above. So suppose there are more than two. The free side (say AC) must belong to one of the small triangles, say T_1, and the third vertex D of T_1 must be on one of the sides AB or BC (suppose it is AB). Then by Step 1,

$$\alpha(ABC) = \alpha(T_1) + \alpha(BCD).$$

Notice that BCD has one fewer triangle in its subdivision than ABC. Furthermore, BCD satisfies the hypotheses of Step 2, because it has no interior vertices (since ABC didn't) and the side CD, being interior to ABC, has no vertices on it. Thus by the induction hypothesis, $\alpha(BCD) = \sum_{i=2}^{n} \alpha(T_i)$ and we are done.

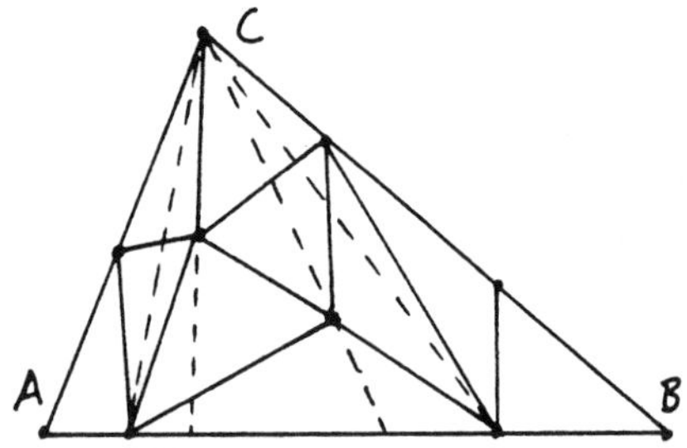

Step 3. The general case. Let the triangle ABC be divided into subtriangles T_i. Choose one vertex of ABC, say C, and draw lines (dotted in the diagram) from C to each of the vertices of the triangulation, including those on the opposite side AB, and extend these lines down to the side AB.

Then we have another subdivision of ABC, this time by triangles S_j. Note that the subdivision S_j satisfies the hypotheses of Step 2, so

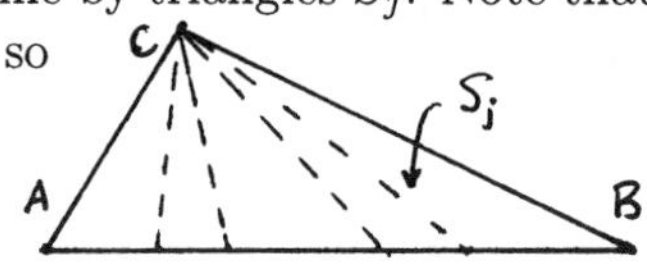

$$\alpha(ABC) = \sum_j \alpha(S_j). \tag{1}$$

Taking all the T_i and the S_j together gives another subdivision $ABC = \cup_{i,j} T_i \cap S_j$. These figures $T_i \cap S_j$ may be triangles or may have four sides. Add extra lines so that they become unions of triangles: $T_i \cap S_j = \cup_k U_{ijk}$. Thus we have a third triangulation of ABC into the triangles U_{ijk}.

Next, observe that each S_j is the union of triangles U_{ijk} as i and k vary, and this triangulation of S_j satisfies the conditions of Step 2! It has no interior vertices, because the lines forming the triangles S_j went through all the vertices of the original triangulation. Furthermore, the side of S_j along the base AB is free of vertices, for the same reason. Thus by Step 2, for each j we have

$$\alpha(S_j) = \sum_{i,k} \alpha(U_{ijk}). \tag{2}$$

Combining with the earlier result (1) about ABC as union of the S_j we find

$$\alpha(ABC) = \sum_{i,j,k} \alpha(U_{ijk}). \tag{3}$$

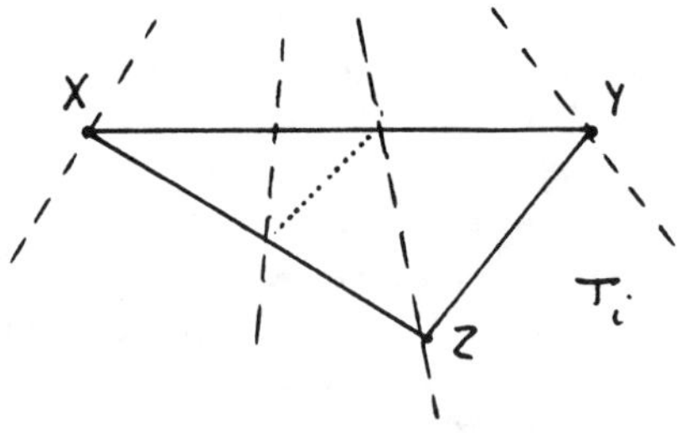

It remains to discuss the division of each T_i into the smaller triangles U_{ijk}. Typically one of the lines from C through the three vertices XYZ of T_i will cut T_i in halves, say T_i', T_i''. (In the drawing the line through Z cuts T_i in half.) Then by Step 1, $\alpha(T_i) = \alpha(T_i') + \alpha(T_i'')$.

Each of these halves is further subdivided by lines through C and additional lines we have added to cut quadrilaterals into two triangles. These triangulations of T_i' and T_i'' satisfy the conditions of Step 2: there are no interior vertices, and the side through Z, which we used to separate T_i into T_i' and T_i'', contains no vertices. Hence by Step 2 again, each of T_i' and T_i'' has measure equal to the sum of the measures of the U_{ijk} which comprise it, and hence

$$\alpha(T_i) = \sum_{j,k} \alpha(U_{ijk}). \tag{4}$$

Now, finally, combining this with the equation (3) we obtain

$$\alpha(ABC) = \sum_i \alpha(T_i) \tag{5}$$

as required.

Lemma 23.5 *The measure of area of a rectilineal figure is independent of the triangulation used to define it.*

Proof. If a figure P has two triangulations

$$P = T_1 \cup \cdots \cup T_n$$

and

$$P = T'_1 \cup \cdots \cup T'_m,$$

then the intersections of the T_i and T'_j can be further subdivided into triangles U_{ijk}, as in the proof of (22.2). Now applying the lemma (23.4) to each $T_i = \cup_{j,k} U_{ijk}$ and to each $T'_j = \cup_{i,k} U_{ijk}$, we find that

$$\sum_i \alpha(T_i) = \sum_{i,j,k} \alpha(U_{ijk}) = \sum_j \alpha(T'_j)$$

so that $\alpha(P)$ comes out the same either way.

Proof of (23.2), Continued. From the lemmas (23.3), (23.4) and (23.5) we know that the function α is well-defined. We need only verify that it has the properties required of an area function. Since segments give positive elements of F, $\alpha(T) > 0$ for any triangle. Congruent triangles have congruent sides and congruent altitudes, so $\alpha(T) = \alpha(T')$ if T and T' are congruent triangles.

If P and Q are two figures with non-overlapping interior, and if we write $P = T_1 \cup \cdots \cup T_n$ and $Q = T'_1 \cup \cdots \cup T'_m$, we can use all the T_i and T'_i to triangulate $P \cup Q$. In that case the additivity is obvious.

Corollary 23.6 *In a Euclidean plane, all of Euclid's theory of area, namely* (I.35)–(I.48), (II.1)–(II.14), (III.35)–(III.37), *and* (IV.10) *hold, where we interpret "equality" of figures to mean equal content in the sense of* §22.

Proof. We have already seen in §22 that all these results follow from the definition of equal content plus the statement (Z). In this section we saw that (Z) is a consequence of the existence of an area function (23.1) and that an area function exists in a Euclidean plane (23.2).

Remark 23.6.1. This proof is analytic in that it makes use of the field of segment arithmetic and similar triangles. We do not know any purely geometric proof for example of (I.39) that triangles on the same base with equal content have the same altitude.

Proposition 23.7 *In a Hilbert plane with* (P), *let* α *be the measure of area function of* (23.2). *Then two figures* P, Q *have equal content if and only if* $\alpha(P) = \alpha(Q)$.

Proof. If P and Q have equal content, then $\alpha(P) = \alpha(Q)$ by (23.1). Conversely, suppose $\alpha(P) = \alpha(Q)$. By (I.44) we can find rectangles P' and Q' with content equal to P, Q, respectively, and furthermore, we may assume that one side of these rectangles is the unit 1 in the field of segment arithmetic. Let a, b be the other sides of these two rectangles. Then cutting each into two triangles, we see that $\alpha(P') = 1 \cdot a = a$, and $\alpha(Q') = 1 \cdot b = b$. On the other hand, figures of equal content have equal α, so $\alpha(P) = a$ and $\alpha(Q) = b$. Our hypothesis now implies $a = b$. Thus the two rectangles are congruent, so P' and Q' have the same content. By transitivity, P and Q have the same content.

Remark 23.7.1. Thus in a Hilbert plane with (P), the theory of content is essentially equivalent to the theory of area given by the area function of (23.2), so we can also restate Euclid's theory of area in terms of the area function.

Exercises

23.1 In Theorem (23.2), the uniqueness of the area function was established by requiring it to have the expected value on every triangle. Suppose instead, in the Cartesian plane over a field F, we consider measure of area functions, with values in the additive group of the field, with the weaker requirement that α of the unit square should be equal to 1.

a) If the field is Archimedean, show that α is uniquely determined by the above condition.

b) If the field F is non-Archimedean, show that there can be more than one area function having value 1 on the unit square.

23.2 Use the measure of area function of (23.2) to show that if two squares have equal area, then their sides are congruent.

23.3 Let ABC be any triangle. Let D, E, F divide the sides in thirds. Draw AE, BD, CF. Show that the small triangle formed in the inside has area equal to 1/7 the area of the whole triangle. *Hint*: Review the proof of (5.4) and use similar triangles.

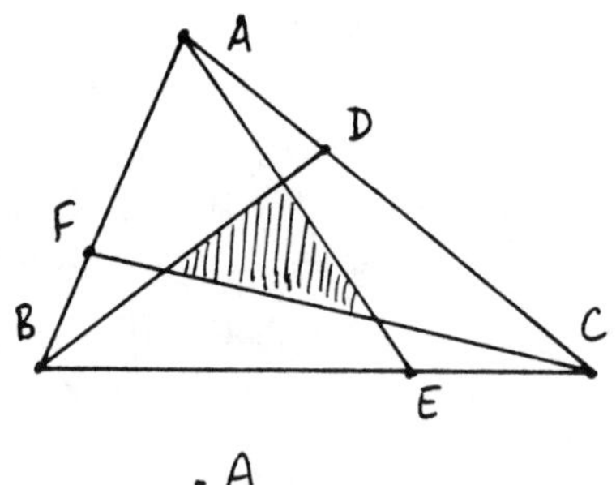

23.4 Let ABC be any triangle, let DE be a line parallel to the base, and let F be any point on DE. Show that the area of the union of the two triangles DBF and ECF is $\leq 1/4$ the area of the whole triangle, with equality if and only if D and E are the midpoints of AB and AC.

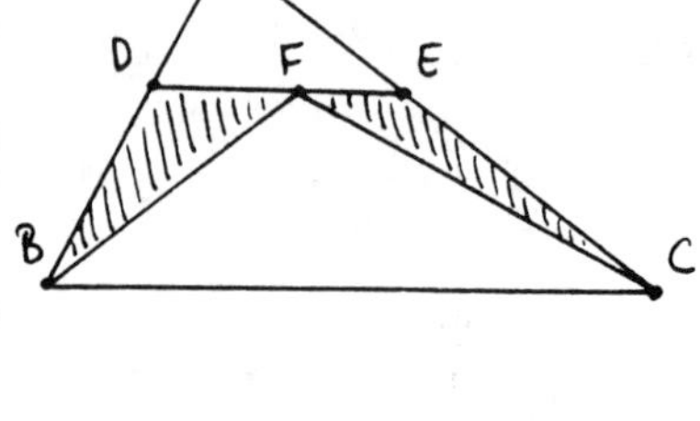

24 Dissection

As we have seen in the previous sections, Euclid bases his theory of area on adding and subtracting congruent figures. This notion is formalized by Hilbert's definition of equal content (§22), which can also be interpreted using a measure of area function (§23). While the notion of content gives a good general theory, one could complain that it is not practical, in the sense that if two figures have equal content, one does not know how much must be added in order to make them equidecomposable. In Example (22.1.3), the two figures being compared both have area 1, but to make them equidecomposable, one must add a triangle whose area is infinite in the non-Archimedean field F.

In this section we will investigate the stricter notion of when two figures are equidecomposable (§22). This leads to the practical problem of *dissection*: given two figures, to find, if possible, an efficient decomposition of the first as a non-overlapping union of smaller figures, not necessarily triangles, which can be reassembled into the second. A dissection exists if and only if the two figures are equidecomposable. In this case we will also say that one figure can be dissected into the other, or that they are *equivalent by dissection.*

We work in a Hilbert plane with (P), and in some cases use also (A) or (E). We will see that certain of the results, such as the Pythagorean theorem (I.47), which Euclid proved for content, are also true in the stronger sense of dissection. We will prove the theorem, due to Bolyai and Gerwien, that in an Archimedean plane, any two figures of equal area (§23) are equivalent by dissection.

The practical problem of finding dissections of one figure into another has received a certain amount of attention among recreational mathematicians, but at this point it seems to be more of an art than a science. Amateurs have discovered a number of clever dissections, but proofs that certain dissections are minimal, or effective bounds on the number of pieces required, seem to be lacking.

We begin this section with some general results on existence of dissections.

Proposition 24.1 *In a Hilbert plane with* (P), *any triangle can be dissected into a parallelogram.*

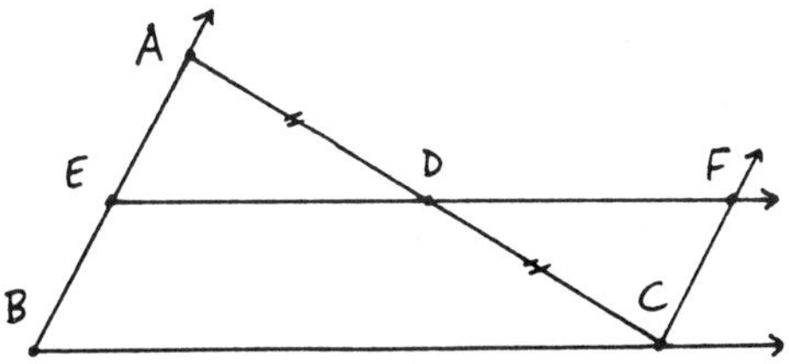

Proof (cf. 22.5). Let ABC be the triangle. Let D be the midpoint of AC. Draw lines through D parallel to BC and through C parallel to AB, meeting at F. Then $\Delta ADE \cong \Delta CDF$ by (ASA). So ΔABC can be dissected into the parallelogram $BCEF$.

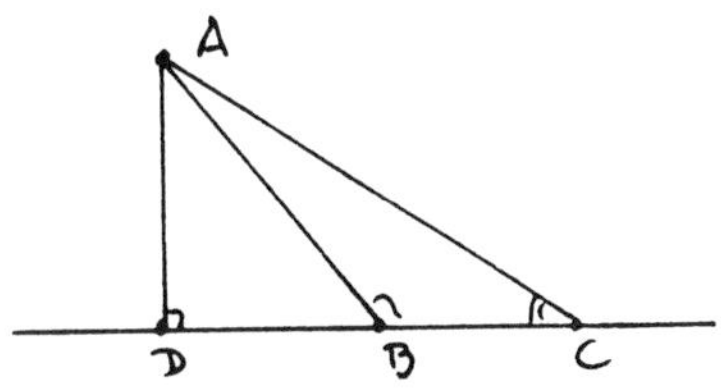

Lemma 24.2 *Let ABC be a triangle, and suppose that the foot D of the altitude from A to side BC lands outside the interval BC. Then one of the angles of triangle at B or C is obtuse.*

Proof. Suppose B is between D and C. Then the angle $\angle ABC$ is an exterior angle of the right triangle ADB, so $\angle ABC$ is greater than a right angle, by (I.16).

Proposition 24.3 *Any parallelogram can be dissected into a rectangle.*

Proof. Let $ABCD$ be the given parallelogram. Drop the altitudes from C and D to the side AB, with feet E, F. Suppose for example that E is inside the interval AB. Then the triangle ΔACE is congruent to the triangle BDF. So the parallelogram is dissected into the rectangle $CDEF$.

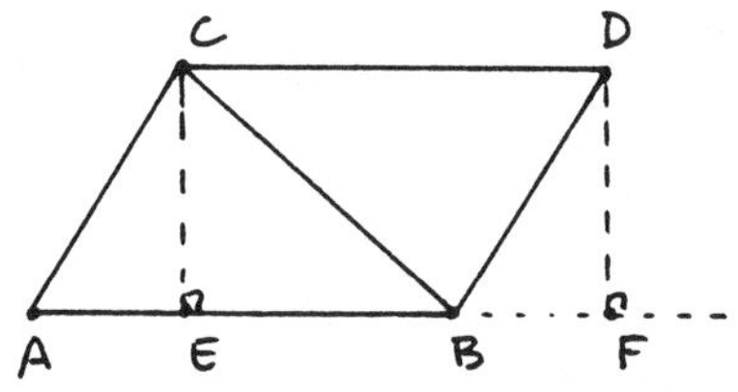

Now let us see that we can always apply this construction. In any parallelogram, the opposite angles are equal (I.34) and the sum of the angles is four right angles, so two of the opposite angles are acute. (If not, all four are right angles and there is nothing to prove.) So we may assume the angle at A is acute.

Now if the altitude from C lands outside AB, then by (24.2) the angle ABC must be obtuse. This forces the angle ACB to be acute (I.32), and in that case, exchanging the roles of B and C, the altitude from B to AC will land inside the segment AC (24.2), so we can apply the construction above.

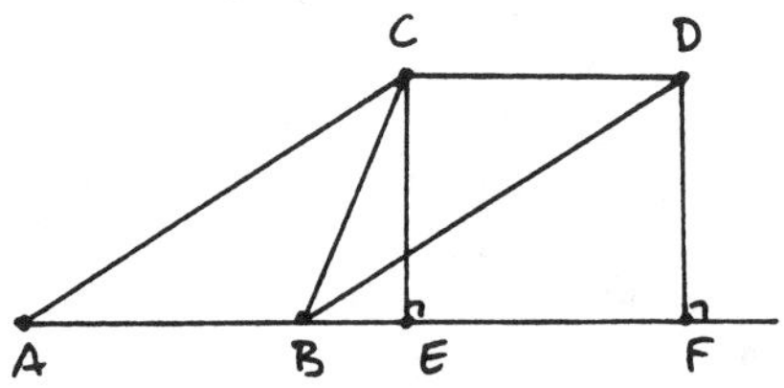

Proposition 24.4 *Given a rectangle $ABCD$, and given a segment AE such that $AB < AE < 2AB$, the rectangle $ABCD$ can be dissected into a rectangle having one side equal to AE.*

Proof. Suppose given the rectangle $ABCD$ and the point E. Let C and E be joined, meeting BD at F. Let G be chosen on AC so that $CG \cong BF$. Construct the rectangle $AEGH$, and let K, L be as shown. Because of parallel lines, the angles at C and F are equal. And CG was equal to BF by construction, so by (ASA), $\Delta CGK \cong \Delta FBE$. It follows that $GK \cong BE$, and so by subtraction, $CD \cong AB \cong KH$. Now by (ASA) again, $\Delta CDF \cong \Delta KHE$. This gives a dissection of the rectangle $ABCD$ into the rectangle $AEGH$, as required.

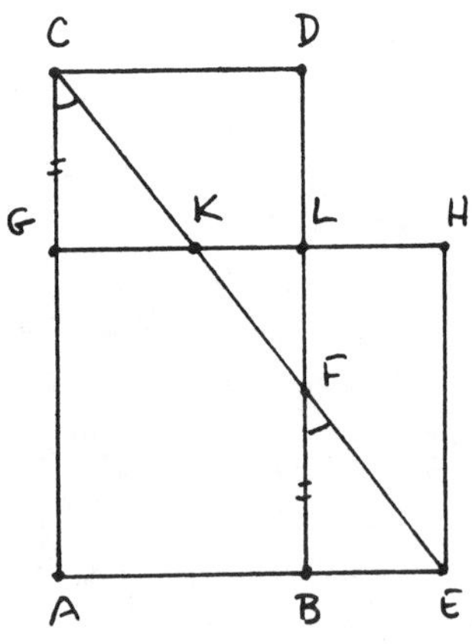

Note that in order for this dissection to work, we need to know that F is below the midpoint of BD, so that G is above the midpoint of AC, and so F is below L. This follows from the hypothesis $AB < AE < 2AB$, because the line from C to the midpoint of BD would meet the line AB at a point M with $AM = 2AB$.

Proposition 24.5 *Assume Archimedes axiom* (A). *Given any rectangle $ABCD$ and given any segment EF, there is a rectangle $EFGH$ with area equal to $ABCD$.*

Proof. Given any rectangle, by cutting it in half and reassembling the two halves along the other sides, we can dissect the original rectangle into a new rectangle with one half the height and twice the base of the original one. Now by Archimedes axiom, after doubling or halving AB a finite number of times, we may assume $AB \leq EF < 2AB$. Then we apply (24.4) to get a rectangle $EFGH$ with area equal to $ABCD$.

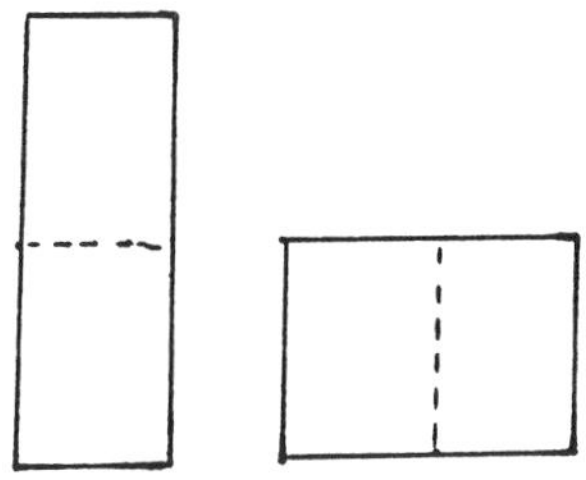

Corollary 24.6 *Assuming* (A) *and given a segment EF, any rectilineal figure* (§22) *can be dissected into a rectangle with one side EF.*

Proof. Divide the figure P into a finite number of triangles $T_1, T_2, \ldots, T_n$. For each i, first dissect the triangle T_i into a parallelogram (24.1), then to a rectangle (24.3), then to a rectangle R_i with base equal to EF (24.5). Now stacking the rectangles $R_1, R_2, \ldots, R_n$, all of which have base equal to EF, on top of each other, we obtain one big rectangle with base EF, as required.

Theorem 24.7 (Bolyai, Gerwien) *In a Hilbert plane with* (P) *and* (A), *let α be the area function of* (23.2). *Then two figures P and Q are equivalent by dissection if and only if they have the same area.*

Proof. One direction follows from (23.1). The new part, which requires the use of Archimedes axiom, is that figures of equal area are equivalent by dissection. The proof parallels (23.7) where the analogous result was proved for content.

So suppose figures P and Q have equal area: $\alpha(P) = \alpha(Q)$. By (24.6), they can each be dissected into rectangles with sides $1, a$, and $1, b$. Then $\alpha(P) = a$, $\alpha(Q) = b$, so $a = b$, and the rectangles are equal. Thus by transitivity of equidecomposability (22.2), P can be dissected into Q.

Remark 24.7.1. This result is false without (A), as is shown by Example (22.1.3).

Remark 24.7.2. This proof is effective, since given figures P and Q of equal area, and given a decomposition of each into triangles, the method of proof of the results of this section will lead to a dissection of one to the other. Of course the dissection found in this way may not be efficient in terms of the number of pieces required.

Remark 24.7.3. Combining this result with (23.7) we see that in a Hilbert plane with (P) and (A), two figures have equal content if and only if they are equidecomposable (cf. 22.1.4).

Proposition 24.8 *Assume* (E). *Let $ABCD$ be a rectangle with sides $a = AB$ and $b = AC$ satisfying $a \leq b \leq 4a$. Then there is a segment $c = AE$ such that the rectangle $ABCD$ can be dissected into a square with side AE.*

Proof. Form a segment of length $a + b$, let O be its midpoint, draw a circle with center O passing through its end points, and let c be the segment cut off on a perpendicular dividing the segment into $a+b$, to where it meets the circle. (Here we use (E) for the existence of the intersection point.)

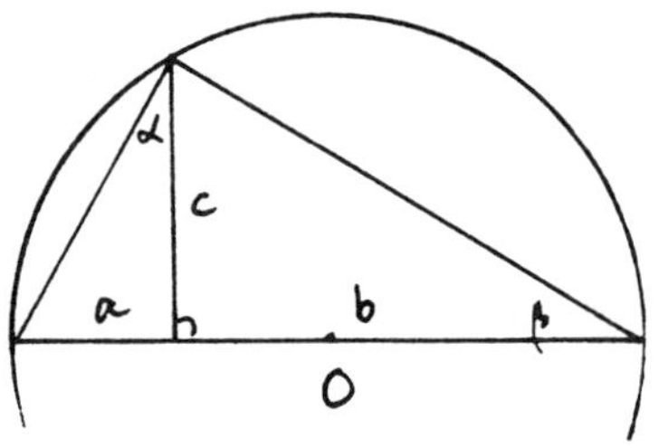

Now let $ABCD$ be the given rectangle, let AE be congruent to the segment c, and let $AEFG$ be the square on side AE. Since the angle in a semicircle is a right angle (III.31), the angles α, β in the circle diagram are equal. These are the same as the angles α, β of the second diagram, by congruent triangles (SAS). Hence the two diagonal lines FB and CE are parallel (I.28). It follows that $CF \cong MB$ and $FK \cong BE$ (I.33). From here the same argument as in the proof of (24.4) shows that $ABCD$ can be dissected into the square $AEFG$.

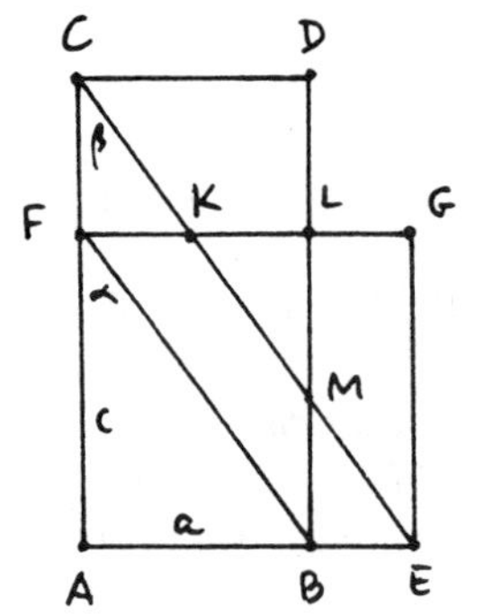

In order for this construction to work, we need to know that $AB \leq AE < 2AB$. Well, we assumed that $a \leq b$, so the angle $\alpha = \beta$ is less than one half right angle, and so $a \leq c \leq b$. On the other hand, if $c \geq 2a$, then by similar triangles it follows that $b \geq 2c$. This implies $b \geq 4a$, contrary to our hypothesis.

Corollary 24.9 *In a Euclidean plane with* (A), *any rectilineal figure can be dissected into a square.*

Proof. Choose a segment AB. By (24.6) the original figure can be dissected into a rectangle $ABCD$ with side AB. Cutting the rectangle in half and reassembling a finite number of times (again using (A)), as in the proof of (24.5), we may assume that the sides $a = AB$ and $b = AC$ satisfy $a \leq b \leq 4a$. Then by (24.8) this rectangle can be dissected into a square.

Next, let us look at the pythagorean theorem (I.47). Euclid's proof shows that the union of the squares on the legs of the right triangle has equal *content* to the square on the hypotenuse. Using Archimedes axiom (A) one can improve Euclid's results about parallelograms (I.35) and triangles (I.37) to hold also for dissection. Then Euclid's proof will show that (I.47) holds for dissection assuming (A). However, even better than this are several direct proofs of (I.47) by dissection which do not need Archimedes axiom.

Let us first consider the proof attributed to Thabit b. Qurra (826–901 AD).

Proposition 24.10 (I.47) *In a Hilbert plane with* (P), *the union of the squares on the legs of a right triangle can be dissected into the square on the hypotenuse.*

Proof. Let ABC be the original triangle, let $ABDE$ be the square on AB, let $ACFG$ be the square on AC, fill in the square $GHEK$ and the line LM so that $DM \cong BC$. Then the three triangles ADF, FHG, CKG are congruent to the original triangle ABC, and $GHEK$ is congruent to the square on BC.

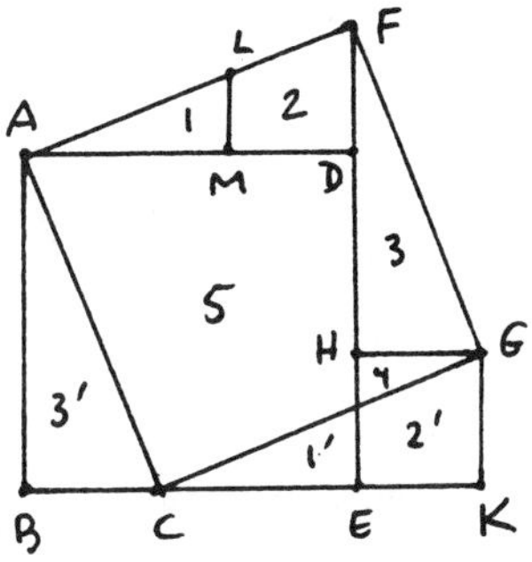

Now the square on the hypotenuse is dissected into the five pieces $1, 2, 3, 4, 5$. The first three are congruent to $1', 2', 3'$ by construction. The square on AB is dissected into $1', 3', 5$, while the square on BC is dissected into $2', 4$. So the result is proved.

Remark 24.10.1. Here is another proof of (I.47) by dissection. Let ABC be the original right triangle. Draw the squares on the three sides as shown. Let O be the center of the square $ACFG$, and draw lines through O parallel to the sides of the square $BCRS$.

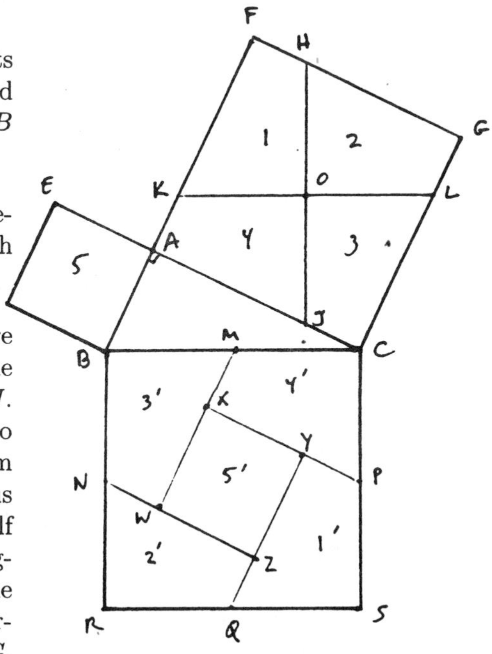

Now let M, N, P, Q be the midpoints of the sides of the square $BCRS$, and draw lines through them parallel to AB and AC, to form the figure shown.

We claim that $1', 2', 3', 4', 5'$ are respectively congruent to $1, 2, 3, 4, 5$, which will complete the proof by dissection.

Since O is the center of the square $ACFG$, it follows that O is also the midpoint of the segments KL and HJ. These two segments are also equal to each other, and from the parallelogram $BCKL$, they are equal to BC. Thus OK, OL, OH, OJ are all equal to one half of BC, hence are congruent to the segments BM, MC, DP, PS, etc. Since the lines drawn in the big square are parallel to the sides of the square $ACFG$, it follows that $1, 2, 3, 4$ are congruent to $1', 2', 3', 4'$.

It remains to show that $5 \cong 5'$. Note that $MW \cong LC$, since they are corresponding sides of 3 and $3'$. From the parallelogram $BCKL$ we see that $LC \cong BK$. But also $BK \cong MX$. So by subtraction $XW \cong AB$. The figure is symmetrical, so one can also show $AJ \cong LC$ and $JC \cong KA$, then a similar argument shows $XY \cong AB$. Thus $5'$ is a square congruent to 5.

Note that a proof by dissection requires

1) a diagram to show the pieces and how they correspond,

2) a careful construction stating how and in what order the diagram was drawn, and

3) a proof, based on the construction 2) showing that corresponding pieces are congruent.

There may be many ways of making the construction, and it is worth contemplating which will make the proof easier.

Exercises

The following exercises take place in a Hilbert plane with (P). Assume that it is a Euclidean plane, or the Cartesian plane over a field, as needed.

24.1 Using Archimedes axiom (A) give a direct proof of (I.37) by dissection.

24.2 Dissect three squares of side 1, to be reassembled into a single square of side $\sqrt{3}$.

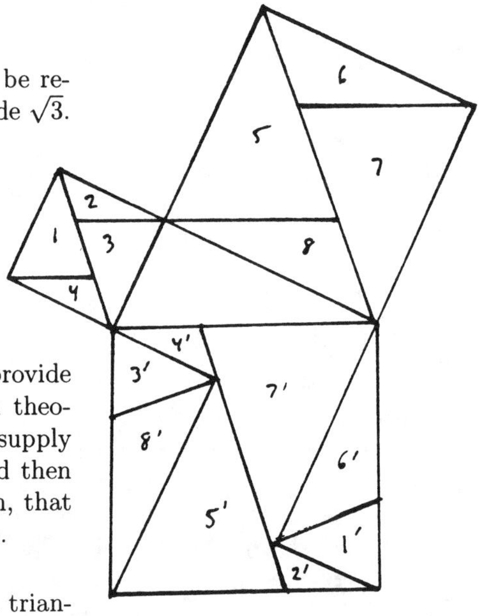

24.3 Use the accompanying diagram to provide another proof of the Pythagorean theorem (I.47) by dissection. You must supply a construction of the diagram, and then a proof based on your construction, that corresponding pieces are congruent.

24.4 Find a dissection of an equilateral triangle into a square. After writing your construction and proof, make a model with construction paper and numbered pieces to illustrate your dissection.

24.5 Prove or disprove the existence of a dissection of a square of side 21 into a rectangle of sides 13, 34 according to the diagram below.

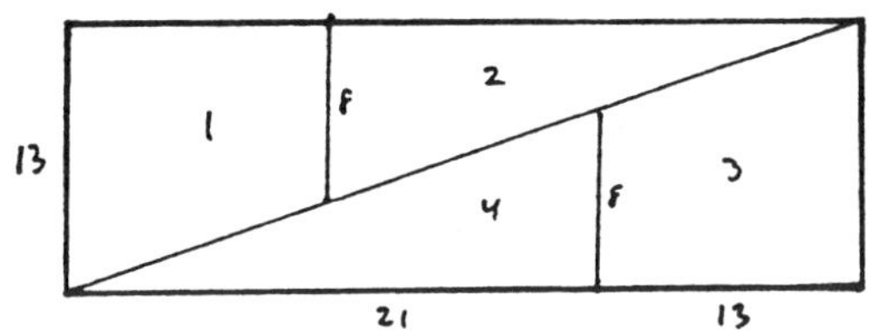

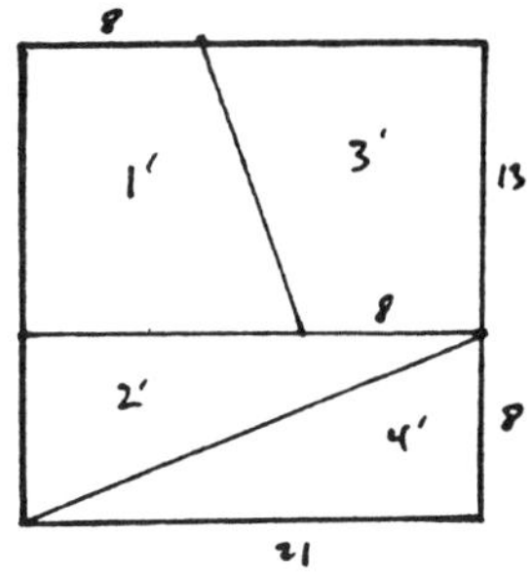

24.6 Let ABC be any triangle. Let D be a point on AC. Prove that there is a dissection of ABC into a trapezoid $PQRS$ as shown. What condition is necessary on the point D for this to work?

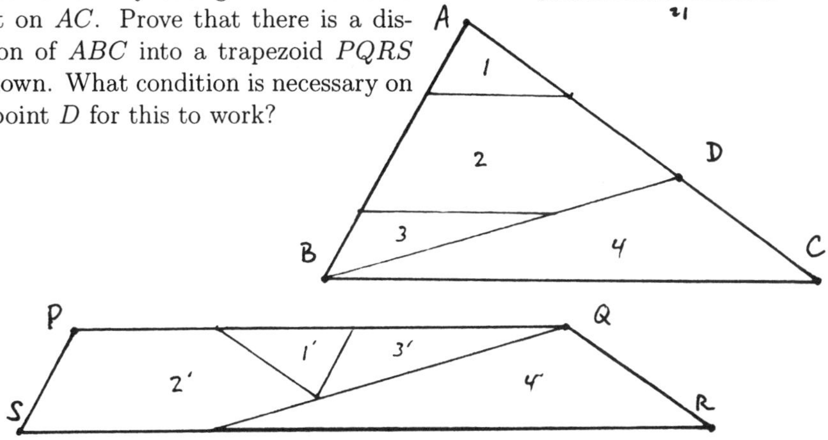

24.7 Use the previous exercise to find a dissection of two equilateral triangles of side 1 into a single equilateral triangle of side $\sqrt{2}$.

24.8 Prove the existence of Dudeney's dissection of an equilateral triangle into a square using four pieces, as suggested in the accompanying diagram.

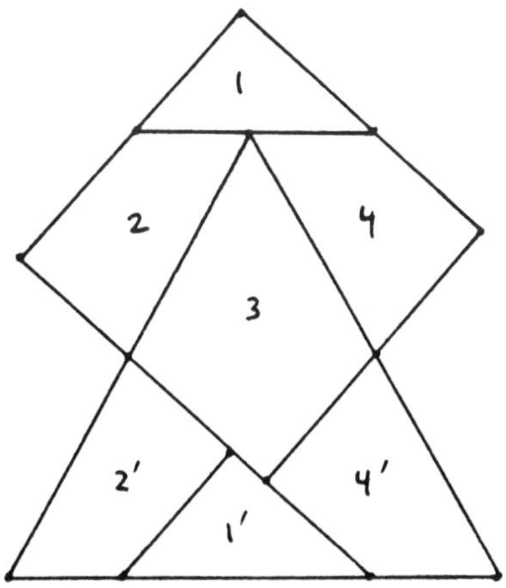

24.9 Dissect a regular pentagon into a square.

24.10 Prove or disprove: For any integer $n \geq 1$, the minimum number of triangles required to dissect a square of side 1 into a rectangle of sides $n \times \frac{1}{n}$ is $2n$.

24.11 Dissect a square into eight strictly acute triangles (all angles $< 90°$).

24.12 If a square is dissected into n triangles of equal area, then n must be an even number.

24.13 Consider a restricted form of dissection problem, where you can cut the figure in pieces, and move the pieces around in the plane, *using translations only* (i.e. no rotations, and no turning pieces over). Show that you can dissect the unit square, with translations only, into the unit square with any other given orientation.

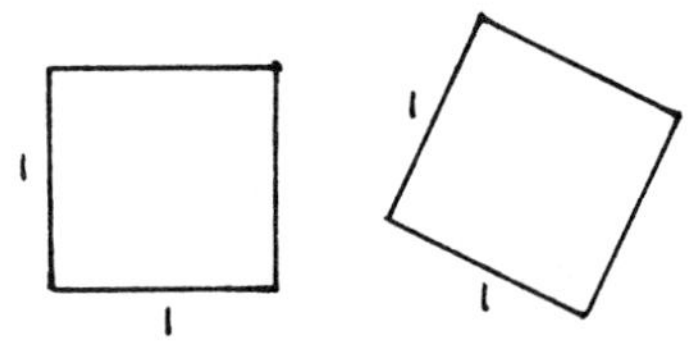

24.14 Divide the unit square by its diagonal into two (congruent) triangles T_1, T_2. Show that it is impossible to dissect T_1 into T_2 using translations only.

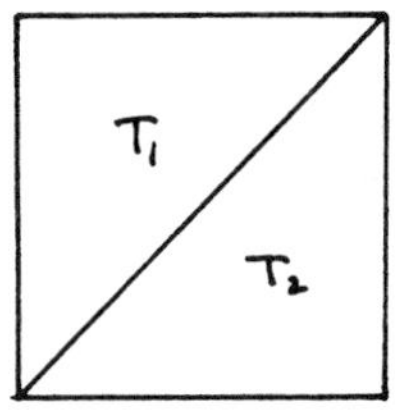

24.15 If P and P' are any two figures with the same area, and assuming Archimedes axiom, show that it is possible to dissect P into P' using translations and $180°$ rotations only.

24.16 Let T be a triangle whose smallest angle is $\geq 45°$. Then T can be dissected into a square using 5 or fewer pieces.

24.17 Let T be a triangle having an angle α with $\tan \alpha < \frac{1}{a}$ for some $a \in F$, $a > 1$. Then any dissection of T into a square requires $\geq \frac{1}{2}\sqrt{a}$ pieces. In particular, there exist triangles requiring arbitrarily many pieces to dissect into a square.

24.18 Let T be a triangle whose smallest angle α satisfies $\tan \alpha > \frac{1}{a}$, for some $a \in F$, $a \geq 1$. Then T can be dissected into a square using no more than $3\sqrt{a} + 4$ pieces.

25 Quadratura Circuli

No discussion of area would be complete without mention of the classical problem of squaring the circle. Through the ages this problem has had an influence way beyond the confines of mathematics as an archetype of the insoluble problem. In one of his religious meditations, John Donne (early 17th century) mentions squaring the circle as something only God can achieve. Perhaps the expression "to put a round peg in a square hole" is a modern vestige of this old problem. And yet what a natural problem it is! The circle is the perfect round form, while the square is complete with its four equal sides and equal angles. What could be more natural than to transform the one into the other? But the solution has been as elusive as the alchemist's search for a way to transform lead into gold.

The problem is to construct a square with area equal to a given circle. What does this mean, exactly? By "construction" of course is meant a geometrical construction with ruler and compass. But what is the area of a circle? In the theory of area presented so far, which corresponds to Euclid's usage in Books I–VI of the *Elements*, we have only discussed the area of rectilineal figures. So there is another problem implicit in the problem of squaring the circle, namely to give a good definition of the area of a circle. In the language of §23, we might ask for a measure of area function defined on all plane regions which can be bounded by straight line segments and arcs of circles. A modern approach to this (in the real Cartesian plane) is to use a definite integral to define the area.

Euclid never defines the area of a circle, and wisely avoids the problem of squaring the circle, but in Book XII, he does have a proposition saying that the ratio of the area of one circle to another is equal to the ratio of the squares of their radii. His proof makes use of the "method of exhaustion", which we would call a limiting process. The key result, which is remarkably similar to the treatment of the Riemann integral in modern calculus classes, is this.

Proposition 25.1 *In a Euclidean plane with* (A), *given a circle* Γ, *one can find an inscribed polygon* P *and a circumscribed polygon* P' *such that the difference in area* $\alpha(P') - \alpha(P)$ *is less than any preassigned quantity (as an element of the field* F *of segment arithmetic).*

Proof. Start with any inscribed polygon P and any circumscribed polygon P', where the vertices of P are the points of tangency of P'. If we double the number of sides of P and P' by bisecting all the central angles of P, then we get new polygons P_1 and P_1'. We compare the difference of P_1 and P_1' to the difference of P and P'. Let us look at the triangle ABC, which is one piece of $P' - P$. The new difference $P_1' - P_1$ intersected with ABC is represented by the shaded area.

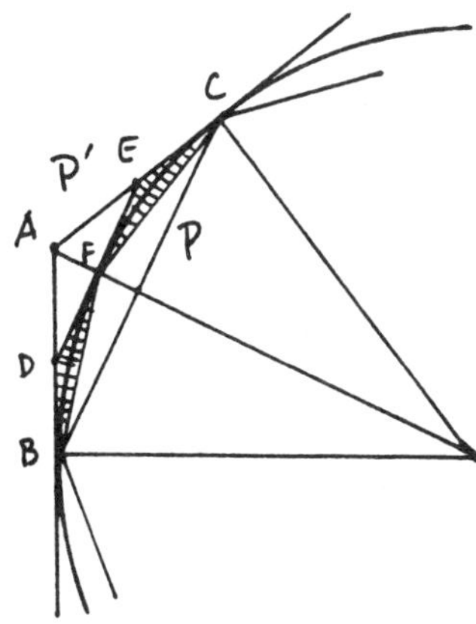

Euclid shows that the shaded area is less than one half of the area of the triangle ABC. In fact, its area is $\leq \frac{1}{4}$ of the area of that triangle (Exercise 23.4). So every time we double the number of sides, the new difference in areas of outside and inside polygons is less than half of the previous difference. Hence, using Archimedes axiom (A), it can be made less than any preassigned quantity.

Approximating the circle by its inscribed and circumscribed polygons, Euclid could show that the ratio of the area of a circle to the square of its radius was a constant independent of the size of the circle. This constant, of course, is what we now call π. Archimedes, who lived shortly after Euclid, used the method of approximating by polygons, to obtain the estimate $3\frac{10}{71} < \pi < 3\frac{1}{7}$, or in fractions, $\frac{223}{71} < \pi < \frac{22}{7}$. Thus began one facet of the work inspired by the problem of squaring the circle, namely to find increasingly accurate numerical approximations for π. Energetic human calculators worked hard on this problem, so that by 1600, the value of π was known to 15 decimal places, by 1700 to 71 places, and by 1873 to more than 500 places. Even today, the urge to compute is still strong, and with the help of electronic calculators, the value of π has been found to more than one billion decimal places.

Coming back to the original problem of squaring the circle, it has always been regarded as very difficult, if not impossible. Since Lindemann proved in the 1880's that π is transcendental, we know that the problem as originally stated, is mathematically impossible (cf. 13.2). Nevertheless, the problem has exerted a great fascination on the inquiring mind, and over the years, many people have come forward with solutions. The study of these solutions forms a colorful chapter in the history of mathematics.

The solutions can be divided into three general categories.

I. Honest ruler and compass constructions which however give only an approximate solution to the problem. One of the best of these depends on the accidental proximity of the number $\frac{1}{5}(3\sqrt{5}+9) = 3.14164\ldots$ to π.

II. Constructions using additional tools which give an exact solution to the problem. One such known from antiquity uses an auxiliary curve, the *quadratrix.*

Take a quadrant AOB of the unit circle. Let OA_1 bisect the right angle, and let B_1 bisect the interval OB. Let the radius OA_1 meet the horizontal line through B_1 at C_1. Bisect the two $45°$ angles at O with OA_2 and OA_3, and bisect the intervals OB_1 and B_1B with B_2, B_3. Let the radii meet the corresponding horizontal lines in C_2 and C_3. Continuing in this manner, bisecting angles and segments, one can construct as many points as one likes $C_1, C_2, C_3, \ldots$. The curve passing through these points is called the quadratrix.

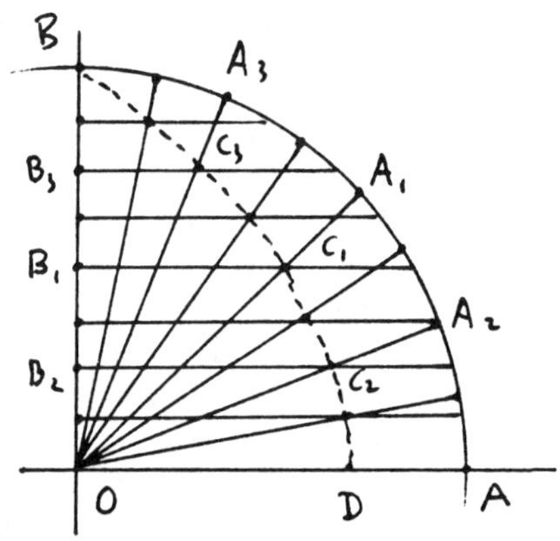

With a little help from modern trigonometry we can express the curve parametrically as

$$\begin{cases} x = \frac{2\theta}{\pi \tan\theta} \\ y = \frac{2\theta}{\pi} \end{cases}$$

where θ is the angle of the radius to the x-axis. As θ tends to O we obtain the point D where the curve meets the x-axis. Its x-coordinate is $2/\pi$ (Exercise 25.1). So clearly if we are given a copy of the quadratrix curve, we can obtain the quantity π, and hence can construct a square with area equal to a given circle. Note however, that this is not an honest ruler and compass construction: though we can obtain as many of the C_i as we like by ruler and compass, the point D is obtained only as a limit of the C_i as they approach the x-axis.

III. Solutions by people I would like to call fringe mathematicians, though they have been called by worse names, who have little mathematical training, but are attracted to famous problems like the moth to the flame. The hint that a problem may be insoluble inspires them to work even harder, as they have a conviction that they will succeed where others have failed. Their solutions are a jumble of facts and fallacies lacking in logical coherence. Pointing out their errors and quoting the results of established mathematicians only confirms their belief that they have discovered the real truth which all the scholars were too blind to see. Sometimes their solutions come from divine inspiration, which leaves little room for discussion, since one can't argue with God.

A case in point is Joseph Ignati Carl von Leistner, a knight in the service of Francis I (1708–1765), duke of Lorraine, and later Holy Roman emperor. Herr Leistner discovered that the exact value of π was $3844/1225$. When his discovery did not occasion the acclaim he expected, he used his influence with the duke to appoint a imperial commission to report on his work. This commission naturally pointed out his errors, upon which he published a lengthy rebuttal attacking his critics and defending his work. Apparently Herr Leistner believed that the true value of π would be a ratio of integers, but that no one had yet been clever enough to find the right ones, as he did. One of his arguments goes like this: "Archimedes proposed the values $22/7$ and $223/71$, but even he acknowledged that they were not correct, while mine is the correct value". Herr Leistner failed to notice that his value fell outside of the bounds proved by Archimedes. On the occasion of his lord's marriage to Maria Theresa of Austria in 1736, Herr Leistner published a treatise on the wonderful coincidence of the numbers 3844 and 1225 with various dates and other numbers from the life of the duke and his bride.

Exercises

25.1 a) Derive the parametric equations for the quadratrix given in the text.

b) Use a little first-year calculus to show that the point D has x-coordinate $2/\pi$.

25.2 a) Let Γ be the circle of radius 1. Find the area of the inscribed polygons of $6, 12, 24, 48, 96, 192$ sides, and express them in standard form (Exercise 13.4) as a rational number times an expression involving integers and square roots. Then verify that the area of the inscribed 192-gon is $> 223/71$, thus confirming Archimedes' estimate $223/71 < \pi$. (Check: To four decimals, area of inscribed 48-gon is 3.1326, 96-gon is 3.1394, 192-gon is 3.1410.)

b) Show that the area of the circumscribed 96-gon is $< 22/7$, confirming the inequality $\pi < 22/7$.

26 Euclid's Theory of Volume

Euclid discusses three-dimensional geometry in Book XI, and beginning with (XI.28) he treats the volume of solid figures. As with area in Book I, there is no definition

Unwiderrufflicher, Wohlgegründter
Und
Ohnendlicher Beweiß
Der Wahren
QUADRATUR
des Circuls,
Oder
Des Durchmessers zu seinem Umcreyß,
Wie 1225. zu 3844. oder 3844. zu 1225.
Zu
Abermaligen und noch mehreren Ehren-Ruf der gantzen Hoch-Edlen Teutschen Nation und Aufnahm der edlen Mathematischen Wissenschafften, auch zum Besten des gemeinen Weesens und Nutzens der studirenden Jugend,
Besonders aber
Zur Erkandtnuß der goldenen Wahrheit Wider die sich anmassende und ungegründte Widerlegung
Hn. J. J. MARINONII.
Verabfasset,
Und zu einem unpartheyischen Urtheil der gantzen Welt in offentlichen Druck befördert
Von
JOSEPHO IGNATIO CAROLO DE LEISTNER,
Der Römisch-Kayserlichen Majestät Rittmeistern,
Als
INVENTORE QUADRATURÆ CIRCULI.
Cum Gratia & Privilegio S.C.R.C. Majestatis, & Licentia Superiorum.

Wienn, gedruckt und zu finden bey Johann Ignatz Heyinger, Universitäts-Buchdruckern, 1737.

Plate VIII. Title page of Leistner (1737).

of volume, and no acknowledgment that we are dealing with a new sort of equality different from congruence. From his proofs it becomes apparent that he is using "equality" of solid figures to mean equality of volume. Exactly what his notion of volume is becomes more clear as we progress; to begin with, we can think of a notion similar to the notion of *content* for plane figures, achieved by adding and subtracting congruent pieces of a dissection of the figure.

To simplify our discussion of Euclid's text, let us suppose that we are working over a field F, and suppose given a *volume function* v which to each solid figure P associates a non-negative element $v(P) \in F$; which assigns the same volume to congruent figures; and which is additive for unions with non-overlapping interiors. Each of Euclid's results can be interpreted as saying certain figures have equal volume. Our attention will be focused on *how* Euclid proves this equality.

First we fix our terminology. A *pyramid* is a solid figure formed by taking a plane figure, say $ABCD$, and joining its vertices to a point O outside the plane of $ABCD$. If the base is a triangle, we speak of a *triangular pyramid*, if the base is a square, a *square pyramid*, etc.

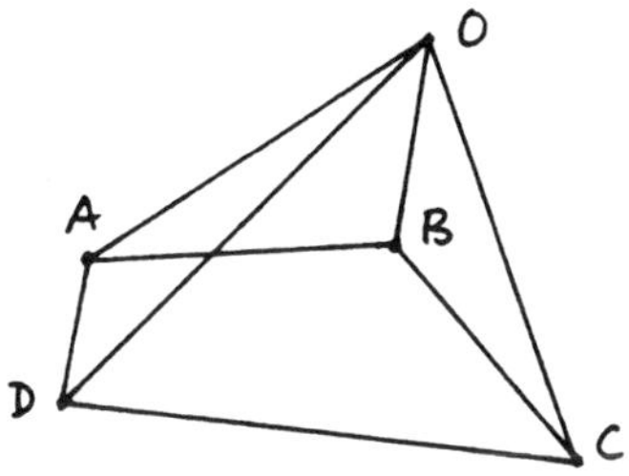

A *prism* is a solid figure formed by taking two congruent figures in parallel planes, with parallel edges, and joining their corresponding vertices. Its faces consist of the two original congruent figures, plus parallelograms joining the corresponding sides. If the base is a triangle, it is a *triangular prism* (as shown).

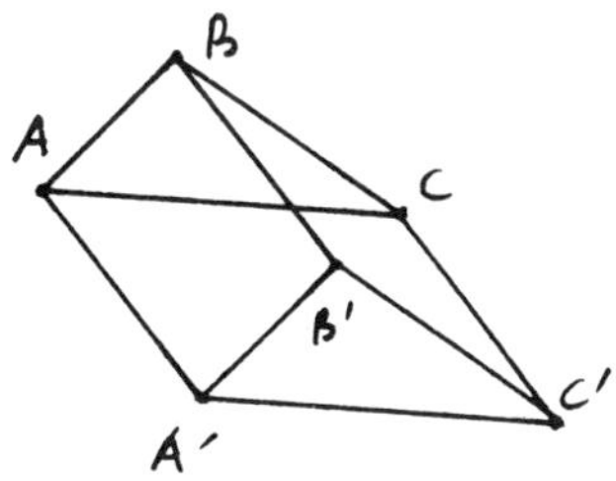

A *parallelepiped* is a prism whose base is a parallelogram. It is formed by three pairs of parallelograms in parallel planes.

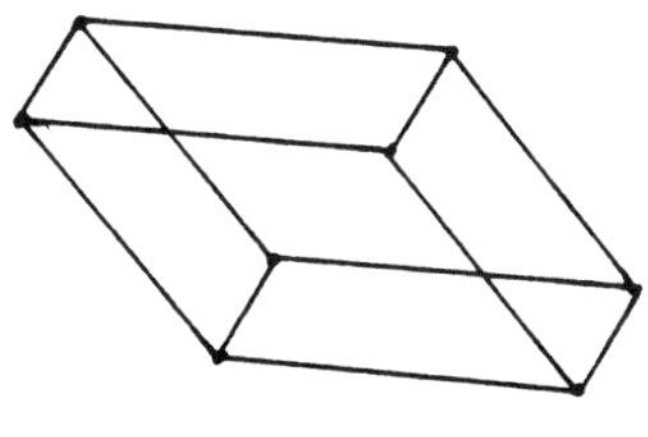

Euclid's first results concerning volume are proved by methods entirely analogous to the methods of Book I. So for example in (XI.28) he shows that a parallelepiped is bisected by the plane through two opposite edges. Note that Euclid proves the two halves have congruent faces and angles. But they cannot be superimposed on each other in 3-dimensional space, since they are mirror images. However, Legendre gives a proof that they are equivalent in the sense of dissection, and we will see later (Exercise 26.1) that any solid figure is equivalent by dissection to its mirror image.

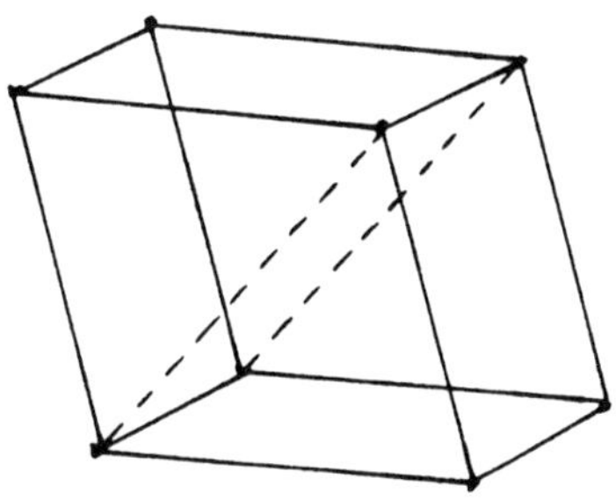

Using techniques exactly analogous to those of Book I, Euclid shows (XI.29) and (XI.30) that parallelepipeds with the same base and of the same height have the same volume. Then (XI.31) he shows, using the method of (I.44), that parallelepipeds an equal basis (in the sense of same content) and of the same height have equal volume. Using these results, one could show as a corollary, that any parallelepiped has the same volume as a rectangular parallelepiped with sides 1, 1, and a, for some $a \in F$. In fact, assuming the field to be Archimedean, one can accomplish this by *dissection*, that is, by cutting the solid into pieces and rearranging.

As an interesting application of these methods, Euclid shows (XI.39) that if given two triangular prisms, one lying on its side, and the other standing up, and if the parallelogram base of the one is equal to twice (in the sense of content) the triangular base of the other, and if they have the same height, then they have equal volume.

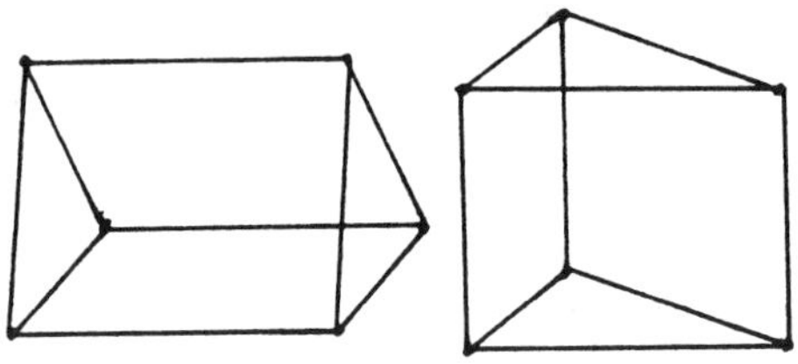

The method is to double each to get parallelepipeds, which have equal volume by the earlier results, and then appeal to the principle that "halves of equals are equal". When we think of volume as a function with values in a field, this principle poses no problem. However, if we sought to develop a purely geometric theory of volume by dissection or by content, as we did for area, it would require justification.

So we see that the theory of volume of parallelepipeds and prisms can be handled by methods familiar from the theory of area, and involves really nothing new beyond translating ideas from the plane into three-dimensional space.

When we come to the study of the volume of pyramids in Book XII, however, the situation is quite different. The key result here is (XII.5), which, in a special case, states

Proposition 26.1 (XII.5) *Triangular pyramids having equal bases (in the sense of content) and equal height have equal volume.*

Proof. For the proof, Euclid uses the *method of exhaustion*, attributed to Eudoxus. The idea is to write both figures as an infinite union subfigures of equal volume, each one equal to more than one half of what was left over after removing the previous subfigures. In this way the subfigures "exhaust" the entire figure, and from the equality of volume of the subfigures one concludes the equality of volume of the whole figures. This last step requires an explicit use of Archimedes' axiom, because as Euclid says in the proof of (XII.5), if the two pyramids were different, let this exhaustion process be done until the remainder left over is less than the difference of the two pyramids. This is possible because each remainder is less than one half of the previous remainder, and so by repeating this process can be made smaller than any preassigned quantity.

In modern language, we would say that the volume of the whole pyramid is the *limit* of the volumes of the subfigures, and the definition of limit using ε's and δ's is nothing but a modern rewording of the argument Euclid uses here.

Now let us examine the particulars of the proof. Let $ABCD$ be one of the given triangular pyramids. Let E, F, G, H, J, K be the midpoints of the sides. Then the pyramid P is decomposed into four pieces: two smaller pyramids $P_1 = AEFG$ and $P_2 = FBHK$, which are congruent to each other and have edges equal to 1/2 the edges of P; and two triangular prisms $T_1 = FHKGJD$ and $T_2 = EFGCHJ$. Of these, T_1 is lying on its side, while T_2 is standing up, and the base of T_1 is a parallelogram $HJKD$ equal to twice the triangle CHJ (in fact it is the union of two triangles congruent to this one, if one draws the line JK).

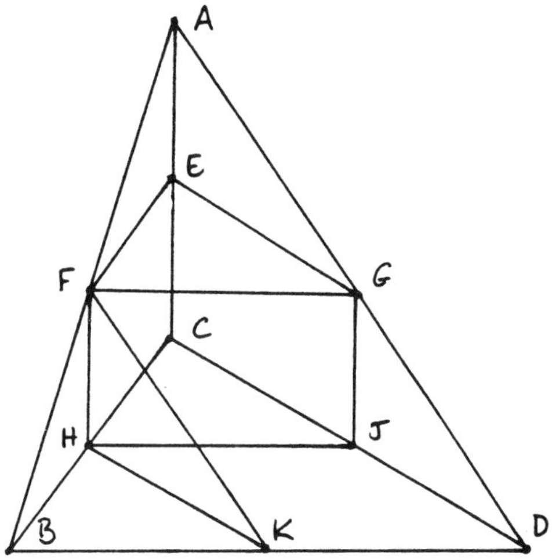

So by (XI.39), T_1 and T_2 have the same volume. Furthermore, since T_2 and P_1 have the same base EFG, and the same height, and one is a prism while the other is a pyramid, it is clear that T_2 is greater than P_1 in volume. (In fact, the pyramid $ECHJ$ is congruent to P_1 and is contained in T_2.) So we conclude that $T_1 + T_2$ is more than one half of P in volume.

Now suppose we divide the second pyramid P' similarly into two pyramids P_1', P_2' and two prisms T_1', T_2'. Since the base triangles of P and P' are equal in content, it follows that parallelepipeds on the bases of T_1 and T_1' of the same height will have equal volume (XI.39) and hence their halves T_1 and T_1' will have equal volume. So $T_1 + T_2$ and $T_1' + T_2'$ have equal volumes, which are respectively more than one half the volumes of P and P'. Moreover, the remainders $P_1 + P_2$ and $P_1' + P_2'$ are unions of pyramids of equal height and having bases of equal content.

Thus, inductively, we can repeat this process and write each of P_1, P_2, P_1', P_2' as unions of four pieces, two pyramids and two prisms, and continue in this manner.

Since the prisms constructed at each step have equal volume, and these exhaust the pyramids as explained above, in the limit, we find that P and P' have the same volume. □

Corollary 26.2 (XII.7) *A triangular pyramid has volume equal to one third of a triangular prism on the same base and of the same height.*

Proof. Let $ABCDEF$ be a triangular prism. We can regard the prism T as the union of three pyramids $P_1 = CDEF$, $P_2 = ACDE$, and $P_3 = ABCE$.

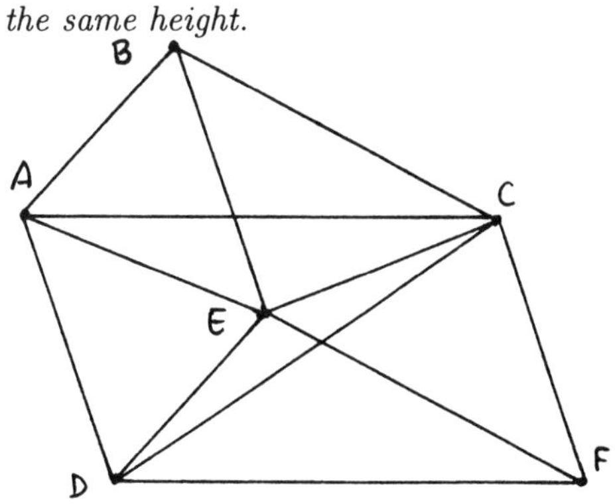

Think of P_1 and P_2 as pyramids with vertex E and bases the triangles CDF and ACD. These two triangles are halves of the parallelogram $ACDF$, hence have equal area (or content). So by (26.1) P_1 and P_2 have equal volume.

The two pyramids P_2 and P_3 can be regarded as pyramids with vertex C and bases the triangles ADE and ABE. These triangles are halves of the parallelogram $ABDE$, hence have equal content. The two pyramids have the same height, so by (26.1) again, P_2 and P_3 have the same volume.

Thus P_1, P_2, P_3 all have the same volume, so this volume is equal to one third the volume of the prism.

Euclid's use of the method of exhaustion here requires some comment. First of all, it is clear that a theory of volume based on this result will be considerably more complicated than the notion of content for plane figures. If we presuppose the existence of a volume function, and work over an Archimedean field, this is a perfectly good proof. However, if we wish to establish a purely geometric theory of volume, we will have to allow limiting processes such as this method of exhaustion in our definition, along with the earlier notions of dissection and complementation.

Secondly, we can wonder if this method is really necessary. Note that (XII.5) applies to the problem of mirror images mentioned earlier. Any figure bounded by planes can be cut up into a finite number of triangular pyramids, so it is enough to show that a triangular pyramid P has the same volume as its mirror image P'. Since P and P' have congruent bases (hence same content) and the same height, (XII.5) applies to show that P and P' have the same volume.

Gauss, in a letter to Gerling in 1844, says it is too bad that the equality of volume of two symmetrical but not congruent figures can only be proved using the method of exhaustion. Gerling, in his reply gives a direct proof that any triangular pyramid can be dissected into twelve pieces that are congruent to those in a dissection of its mirror image (Exercise 26.1). Gauss responds yes that is nice, but it is still unfortunate that the proof of (XII.5) seems to require the method of exhaustion.

These reflections, contrasted with the theorem of Bolyai and Gerwien that plane figures of equal area are equivalent by dissection (24.7), led Hilbert to include this question in his famous list of the 23 most important problems facing mathematics in the 20th century, which he presented at the International Congress of Mathematicians in 1900. The third problem was to show that the method of exhaustion really is necessary, by exhibiting two solid figures of the same volume, which cannot be subdivided in any way into a finite union of congruent pieces. This was done in the same year by Max Dehn, who showed indeed, the method of exhaustion was necessary. In particular, it is not possible to dissect a regular tetrahedron into a finite number of pieces which can be reassembled into a cube (§27).

Exercises

26.1 Gerling's proof that a triangular pyramid and its mirror image are equal by dissection.

a) Let $ABCD$ be a given triangular pyramid. Show that it can be inscribed in a sphere, i.e. there is a sphere containing the points A, B, C, D. Then by drawing planes through the center O of the sphere and the edges of the original pyramid P, show that P can be dissected into four triangular pyramids P_1, P_2, P_3, P_4 with bases ABC, etc. and such that the vertex O is equidistant from the three vertices of the base of each P_i.

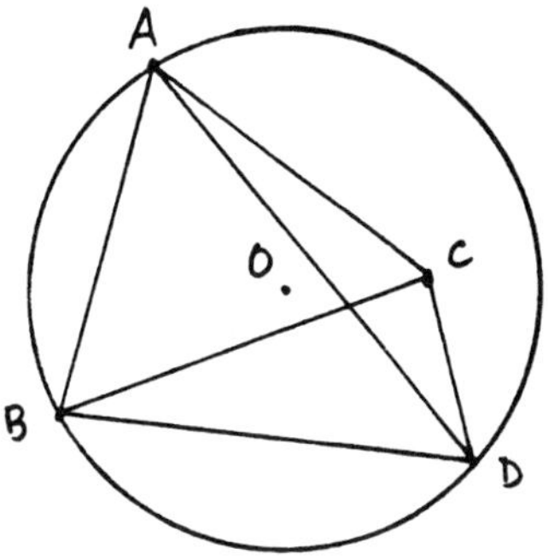

b) Now let $P_1 = OABC$ be a triangular pyramid with O equidistant from A, B, C. Drop a perpendicular OD to the plane of ABC, and show that D is equidistant from A, B, C. Then by drawing planes through OD and A, B, C respectively, show that P_1 is subdivided into three *right isosceles triangular pyramids* P_{11}, P_{12}, P_{13}. So for example $P_{11} = OABD$, and $OA \cong OB$, $DA \cong DB$, and $OD \perp DA$ and $OD \perp DB$.

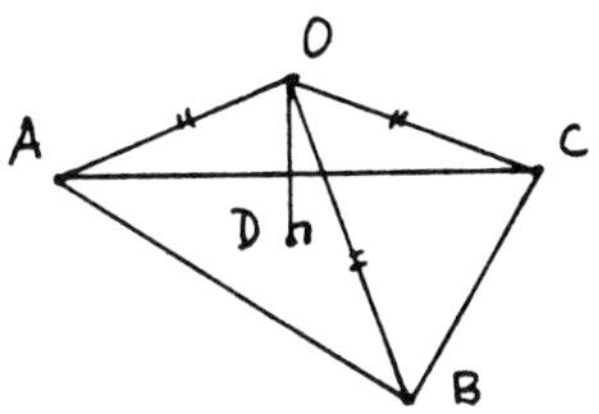

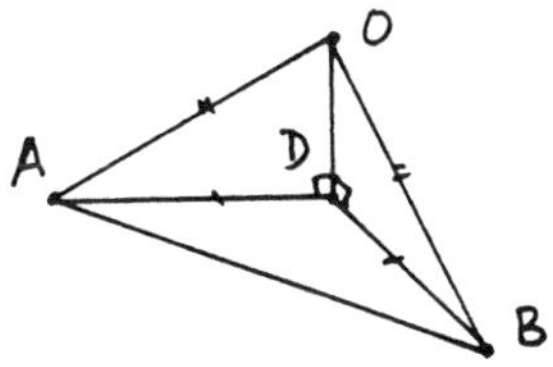

c) Show that a right isosceles triangular pyramid P_{11} as just described, is congruent to its mirror image, by a motion in 3-space.

d) If P is the original pyramid, and P' its mirror image, show that $P = \sum P_{ij}$ and $P' = \sum P'_{ij}$ are two decompositions into twelve pieces each, and $P_{ij} \cong P'_{ij}$.

26.2 Show that a regular tetrahedron of edge 1 can be dissected into four regular tetrahedra of edge 1/2 plus a regular octahedron of edge 1/2.

27 Hilbert's Third Problem

As mentioned in the last section, Hilbert's third problem was to show that the method of exhaustion really is necessary in Euclid's proof of (XII.5). More precisely the problem is to find two solid figures of equal volume, which are not equivalent by dissection, even after possibly adding on other figures which are equivalent by dissection. This problem was solved in 1900 by Max Dehn, and here we give a modern algebraic version of his solution.

This is another example of the use of methods of modern algebra to solve a purely geometric problem. To show that a certain dissection is possible, it is enough to exhibit the dissection and prove the parts are congruent. This we have done with many areas (§24), and it is a purely geometric process. However, to show that a certain dissection is not possible requires a different approach.

In contemporary mathematics, a proof that certain geometric objects are not equivalent in some way is usually accomplished by defining an invariant which is the same for equivalent figures, and then showing that the invariant of the figures in question are different. This is the philosophy behind the study of homotopy and homology groups in topology, for example. Modern algebra renders a great service in providing the tools for defining these invariants and calculating them.

In the present case, we will define a certain abelian group G, and for each polyhedral figure P we will define an element $\delta(P) \in G$, called the *Dehn invariant* of P. We will show that δ of congruent figures is the same, and that δ is additive, in the sense that $\delta(P_1 \cup P_2) = \delta(P_1) + \delta(P_2)$ for figures P_1, P_2 with non-overlapping interiors. It follows that figures which are equivalent by dissection or by complementation (the three-dimensional analogue of equal content—cf. §22) have the same invariant δ. We will compute δ of a tetrahedron, which will be non-zero, and δ of a cube, which will be zero. This will show that a tetrahedron is not equivalent to any cube by dissection or complementation.

We start with the definition of the group G.

Definition. Let G be the set of all expressions

$$(a_1, \alpha_1) + (a_2, \alpha_2) + \cdots + (a_n, \alpha_n)$$

where a_i are real numbers ($a_i \in \mathbb{R}$), and the α_i are real numbers modulo π, ($\alpha_i \in \mathbb{R}/\pi\mathbb{Z}$), modulo the equivalence relation generated by the two following types of operations:

$$\begin{aligned}(a, \alpha) + (a, \beta) &= (a, \alpha + \beta)\\ (a, \alpha) + (b, \alpha) &= (a + b, \alpha).\end{aligned}$$

We define addition by adding one expression to another to make a longer expression, and the order of the terms does not matter. Thus addition is associative and commutative.

Lemma 27.1 *G is an abelian group (called the* tensor product *$\mathbb{R} \otimes_{\mathbb{Z}} \mathbb{R}/\pi\mathbb{Z}$).*

Proof. We must show the existence of an additive identity (O) and additive inverses. First we note, for any $a \in \mathbb{R}$,

$$\begin{aligned}(a, 0) &= (a, 0) + (0, 0)\\ &= (a, 0) + (a, 0) + (-a, 0)\\ &= (a, 0) + (-a, 0)\\ &= (0, 0)\end{aligned}$$

using the rules of operation above. Similarly one can show $(0, \alpha) = (0, 0)$ for any $\alpha \in \mathbb{R}/\pi\mathbb{Z}$.

Now let $O = (0, 0)$. Then for any (a, α),

$$\begin{aligned}O + (a, \alpha) &= (0, 0) + (a, \alpha)\\ &= (a, 0) + (a, \alpha)\\ &= (a, \alpha)\end{aligned}$$

so O is an additive identity.

Given any (a, α), if we add $(-a, \alpha)$ we get $(0, \alpha) = O$, so $(-a, \alpha)$ is an additive inverse. Thus G is an abelian group.

Definition. For any polyhedral solid P in Euclidean 3-space over the real numbers, we define its *Dehn invariant* $\delta(P) \in G$ as follows. For each edge of P, let the length of the edge be a, and let α be the *dihedral angle* (measured in a plane perpendicular to the edge) in the interior of the solid between the two planes meeting along the edge.

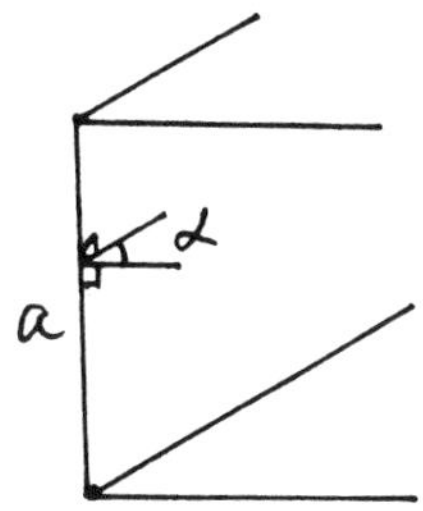

We take α in radians to be a positive number and reduce (mod π). Then we define

$$\delta(P) = \sum (a_i, \alpha_i)$$

where the sum is taken over all the edges of P.

Example 27.1.1 Let P be a cube whose edge has length a. The dihedral angle between any two faces is a right angle, so

$$\delta(P) = 12(a, \pi/2),$$

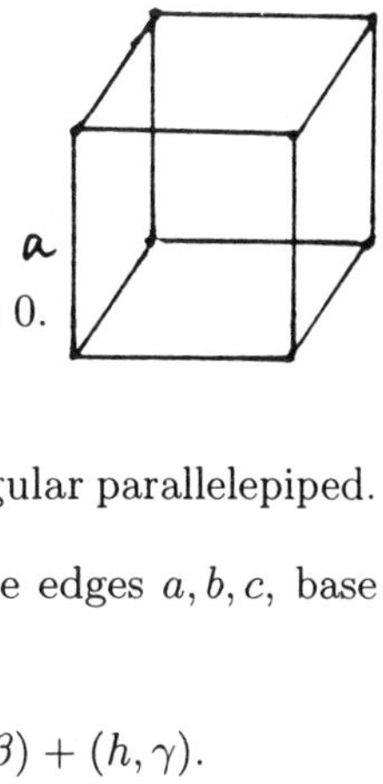

since there are 12 edges of length a. Now in the group G

$$(a, \pi/2) = (a/2, \pi/2) + (a/2, \pi/2) = (a/2, \pi) = 0.$$

Hence $\delta(P) = 0$.

In fact, the same method shows that $\delta(P) = 0$ for any rectangular parallelepiped.

Example 27.1.2 Let P be a right triangular prism, with base edges a, b, c, base angles α, β, γ, and height h. Then

$$\delta(P) = 2(a, \pi/2) + 2(b, \pi/2) + 2(c, \pi/2) + (h, \alpha) + (h, \beta) + (h, \gamma).$$

As above, $(a, \pi/2) = 0$ and similarly for b and c. Now, in G we have

$$(h, \alpha) + (h, \beta) + (h, \gamma) = (h, 2\pi) = 0.$$

So $\delta(P) = 0$.

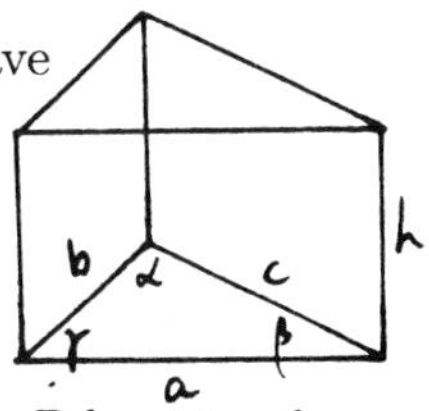

Example 27.1.3 Let P be a regular tetrahedron of edge a. Then P has six edges of length a, all having the same dihedral angle α, so

$$\delta(P) = 6(a, \alpha).$$

We can compute the angle α by drawing the altitudes of two of the faces. We get a triangle AHC, with $AC = a$, and with $AH = HC = \frac{\sqrt{3}}{2}a$, the altitude of an equilateral triangle. The perpendicular from A to the face BCD of the tetrahedron meets it a a point K equidistant from A, B, C. So K is the centroid of ABC, and $HK = 1/3\ HC$. From this we conclude that

$$\cos\alpha = \frac{1}{3}$$

and this determines the angle α.

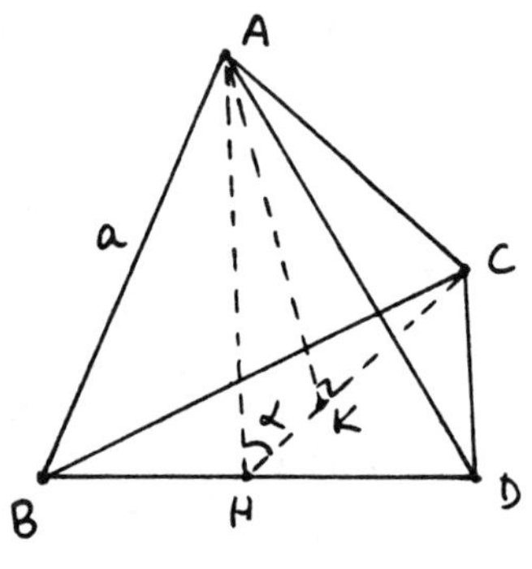

We will show later that $\delta(P) \neq 0$ in G.

Proposition 27.2 *The Dehn invariant δ has the following properties.*

a) *If P and P' are congruent, then $\delta(P) = \delta(P')$.*

b) *If P_1 and P_2 have non-overlapping interiors, then*

$$\delta(P_1 \cup P_2) = \delta(P_1) + \delta(P_2).$$

Proof. Statement a) is obvious, because congruent figures have congruent edges and congruent dihedral angles, so their lengths a_i and measure of angles α_i are equal.

For b), in comparing the union $P_1 \cup P_2$ to the two pieces P_1 and P_2, there are three ways in which the edges of $P_1 \cup P_2$ can be different than the aggregate of the edges of P_1 and P_2. In each case we will show that the contribution to δ is the same.

Case 1. An edge of P_1 and an edge of P_2 can be glued to form a single edge of $P_1 \cup P_2$. In this case the angle α of that edge in $P_1 \cup P_2$ will be the sum $\alpha_1 + \alpha_2$ of the angles in P_1 and P_2.

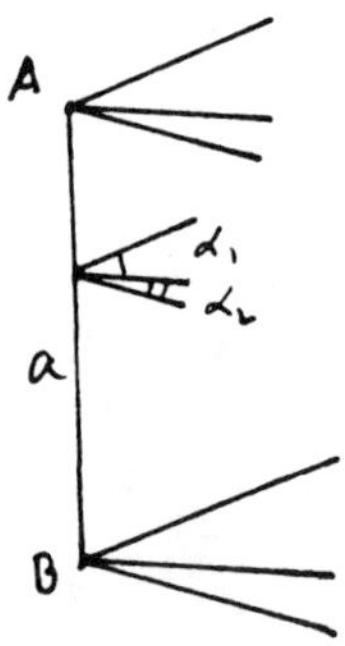

Since $(a, \alpha_1 + \alpha_2) = (a, \alpha_1) + (a, \alpha_2)$ in G, the contributions to δ are the same.

Case 2. Two edges of length a in P_1 and of length b in P_2, having the same angle α can be attached end to end to form a single edge of $P_1 \cup P_2$. Since $(a+b, \alpha) = (a, \alpha) + (b, \alpha)$ in G, the contribution to δ is the same.

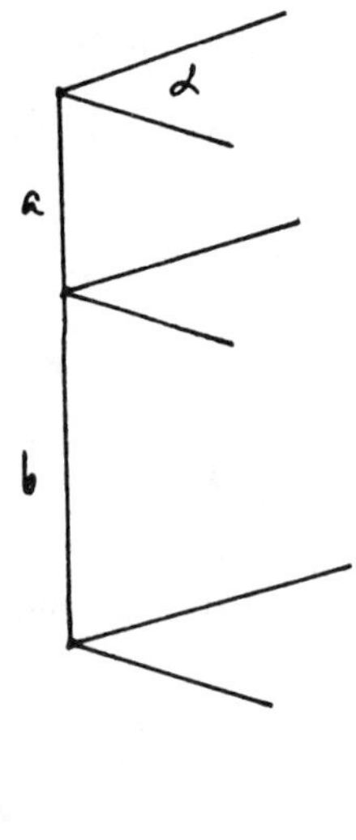

Case 3. Two edges of P_1 and P_2 can be glued so as to make a single face. In this case there is no corresponding edge of $P_1 \cup P_2$. But the dihedral angles α, β of that edge in P_1 and P_2 will add up to π. Since

$$(a, \alpha) + (a, \beta) = (a, \pi) = 0 \text{ in } G,$$

the contribution to δ is unchanged.

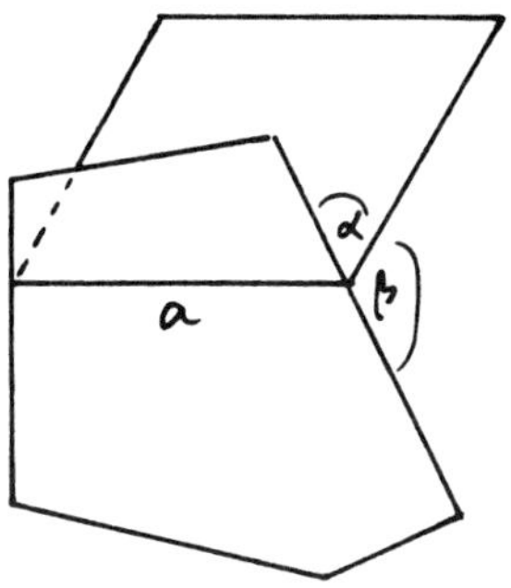

There are some further possibilities, essentially equivalent to those above, which we leave to the reader. This shows that δ is additive, as claimed.

Proposition 27.3 (Dehn's solution to Hilbert's third problem) *In Euclidean 3-space over the real numbers, a tetrahedron cannot be dissected into a cube.*

Proof. Because of the properties (27.2), any two figures equivalent by dissection must have the same Dehn invariant. We have seen that the invariant of a cube of any size is O, so it is sufficient to show that the invariant of a tetrahedron is non-zero. This will be a consequence of the following two lemmas:

Lemma 27.4 *An element of the form $(a, \alpha) \in G$ is zero if and only if $a = 0$ or α is a rational multiple of π ($\alpha \in \pi\mathbb{Q}$).*

Proof. First suppose $\alpha \in \pi\mathbb{Q}$, so $\alpha = \frac{r}{s}\pi$, with $r, s \in \mathbb{Z}$. Then in G we can write

$$(a, \alpha) = s\left(\frac{1}{s}a, \alpha\right) = \left(\frac{1}{s}a, s\alpha\right) = \left(\frac{1}{s}, r\pi\right) = 0.$$

Conversely, suppose $a \neq 0$. We will define a group homomorphism $\varphi : G \to \mathbb{R}/\pi\mathbb{Q}$ as follows. Think of $\mathbb{R}$ as a vector space over $\mathbb{Q}$. Since $a \neq 0$, $a\mathbb{Q}$ is a 1-dimensional subvector space. Choose a complementary subspace V, so that every element $b \in \mathbb{R}$ can be written uniquely as

$$b = ra + v$$

with $r \in \mathbb{Q}$ and $v \in V$. For any element $g = \sum(b_i, \beta_i) \in G$ write each $b_i = r_i a + v_i$ with $r_i \in \mathbb{Q}$, $v_i \in V$, and define

$$\varphi(g) = \sum r_i\beta_i \in \mathbb{R}/\pi\mathbb{Q}.$$

We must check that φ is well-defined. First note that if $\beta_i \in \pi\mathbb{Z}$, then $r_i\beta_i \in \pi\mathbb{Q}$, so it is well-defined on $\beta_i \pmod{\pi}$. Second we must see if φ respects the equivalence relation used to define G. If

$$(b, \beta) = (b_1, \beta) + (b_2, \beta)$$

and

$$\begin{aligned} b_1 &= r_1 a + v_1 \\ b_2 &= r_2 a + v_2 \end{aligned}$$

then

$$b = (r_1 + r_2)a + (v_1 + v_2)$$

so

$$\varphi(b, \beta) = (r_1 + r_2)\beta = \varphi(b_1, \beta) + \varphi(b_2, \beta).$$

On the other hand, if

$$(b, \beta) = (b, \beta_1) + (b, \beta_2)$$

and $b = ra + v$, then

$$\varphi(b, \beta) = r(\beta_1 + \beta_2) = \varphi(b, \beta_1) + \varphi(b, \beta_2).$$

Thus φ is well-defined.

Now observe that $a = 1 \cdot a + 0$, so $\varphi(a, \alpha) = \alpha \in \mathbb{R}/\pi\mathbb{Q}$. So if $(a, \alpha) = 0$ in G, it follows that $\varphi(a, \alpha) = 0$ in $\mathbb{R}/\pi\mathbb{Q}$, and hence $\alpha \in \pi\mathbb{Q}$, which is what we wanted to prove.

Lemma 27.5 *If α is an angle with $\cos\alpha = \frac{1}{3}$, then α is not a rational multiple of π.*

Proof. We will offer two proofs of this fact. The first is "elementary" in the sense of using nothing more than trigonometry, but does not give much insight into why the result is true. The second is more conceptual, but uses results from the Galois theory of cyclotomic extensions of $\mathbb{Q}$.

1st Proof. From a small right triangle we see that $\tan\alpha = 2\sqrt{2}$. Our proof will consist in showing that for every positive integer n, $\tan n\alpha \neq 0, \infty$. This is sufficient to show that $\alpha \notin \pi\mathbb{Q}$, because if α were a rational multiple of π, then some $n\alpha$ would be an integral multiple of π, in which case $\tan n\alpha = 0$ or ∞.

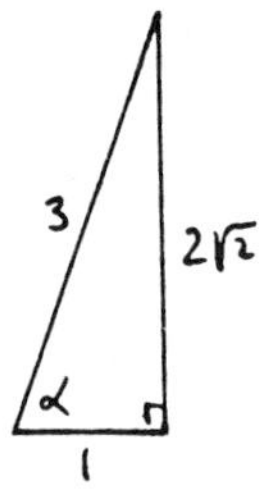

We compute $\tan n\alpha$ inductively using the angle sum formula for tangents:

$$\tan(n+1)\alpha = \frac{\tan\alpha + \tan n\alpha}{1 - \tan\alpha \tan n\alpha}.$$

In this way we find

$$\begin{aligned} \tan\alpha &= 2\sqrt{2} \\ \tan 2\alpha &= -\tfrac{4}{7}\sqrt{2} \\ \tan 3\alpha &= \tfrac{10}{23}\sqrt{2} \end{aligned}$$

and so on. More generally, if $\tan n\alpha = \frac{a}{b}\sqrt{2}$, with $a, b \in \mathbb{Z}$, then

$$\tan(n+1)\alpha = \frac{a+2b}{b-4a}\sqrt{2}.$$

We conclude, inductively, that $\tan n\alpha$ is a rational multiple of $\sqrt{2}$, for all $n \geq 1$. While the rational numbers which occur seem more and more complicated, matters are simpler if we look at the numerator and the denominator of this fraction (mod 3). To be precise, consider the transformation which takes an ordered pair (a, b) and sends it to $(a + 2b, b - 4a)$, and regard a, b as elements of $\mathbb{Z}/3\mathbb{Z}$. Starting with $a = 2, b = 1$, we find

$$(2,1) \to (1,2) \to (2,1).$$

After two steps it repeats. We conclude that for any $n \geq 1$, $\tan n\alpha = \frac{a}{b}\sqrt{2}$, and $(a, b) \equiv (2,1)$ or $(1,2)$ (mod 3). In particular, neither of a, b can be 0, so $\tan n\alpha \neq 0, \infty$, as required.

2nd Proof. If $\cos\alpha = \frac{1}{3}$, then $\sin\alpha = \frac{2}{3}\sqrt{2}$. Consider the complex number

$$\begin{aligned} z &= \cos\alpha + i\sin\alpha \\ &= \tfrac{1}{3} + \tfrac{2}{3}\sqrt{-2}. \end{aligned}$$

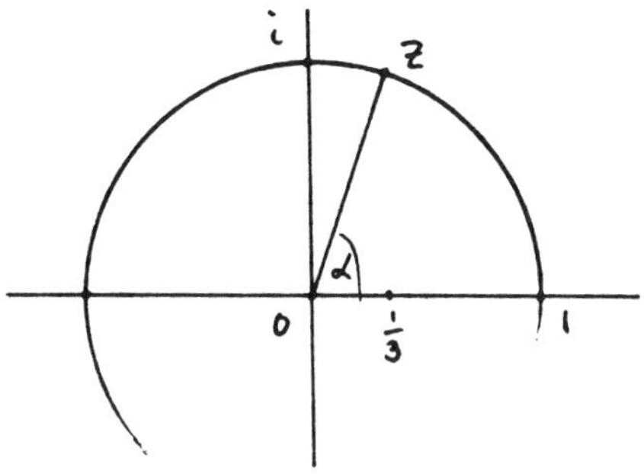

Obviously z is a root of a quadratic equation over $\mathbb{Q}$ (in fact $3z^2 + 2z + 3 = 0$), so z generates a field extension $\mathbb{Q}(z)$ of degree 2 over $\mathbb{Q}$.

Now if α is a rational multiple of π, then we can write $\alpha = \frac{p}{q} \cdot 2\pi$ with $p, q \in \mathbb{Z}$, $(p, q) = 1$. In this case z will be a primitive q^{th} root of unity. Its minimal polynomial over $\mathbb{Q}$ will be the *cyclotomic polynomial* $\Phi_q(z)$, which has degree $\varphi(q)$, where φ is the Euler φ function.

In our case z has degree 2 over $\mathbb{Q}$, so we conclude $\varphi(q) = 2$. If $q = p_1^{e_1}, \ldots, p_r^{e_r}$ is the prime factorization of q, then

$$\varphi(q) = \prod p_i^{e_i - 1}(p_i - 1).$$

So the only values of q which give $\varphi(q) = 2$ are $q = 3, 4, 6$. The corresponding field extensions are $\mathbb{Q}(\sqrt{-3})$ and $\mathbb{Q}(i)$. Neither of these is equal to $\mathbb{Q}(z) = \mathbb{Q}(\sqrt{-2})$, so we have a contradiction. This shows that α is not a rational multiple of π.

Exercises

27.1 Compute the Dehn invariant of a regular octahedron of side a. Show that this element is $\neq 0$ in G, so the octahedron cannot be dissected into a cube. Show furthermore that an octahedron cannot be dissected into a regular tetrahedron.

27.2 Show that two tetrahedrons of edge 1 cannot be dissected into a single tetrahedron of edge $\sqrt[3]{2}$. Is it possible to dissect 8 tetrahedrons of edge 1 into a single tetrahedron of side 2?

27.3 Is there any triangular pyramid with Dehn invariant 0?

27.4 Compute the Dehn invariant of a *right isosceles triangular pyramid* (Exercise 26.1) $OABC$ with $AB = AC = a$, base angle $\angle BAC = \theta$, with OA perpendicular to the plane of ABC, and dihedral angle α along BC. Use this to determine the image of the map δ, i.e. that subgroup of G consisting of elements which are the Dehn invariant of some polyhedron.

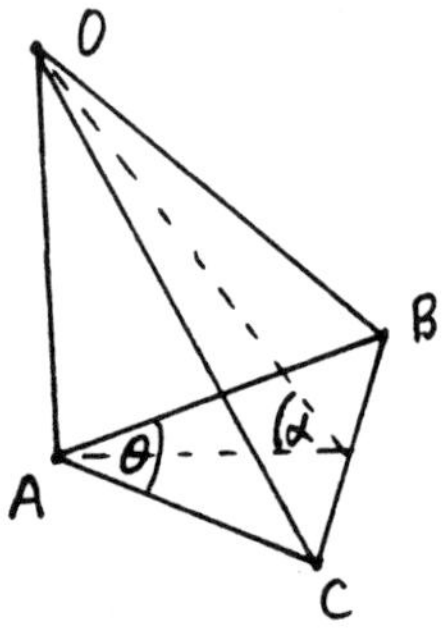

27.5 Imitate the first proof of (27.5) to show that if $\tan\alpha \in \mathbb{Q}$, and α is a rational multiple of π, then $\tan\alpha = 0, \pm1, \infty$. *Hint*: Let $\tan\alpha = \frac{r}{s}$ with $r, s \in \mathbb{Z}$, $(r, s) = 1$. If r, s are not both ±1, take an odd prime divisor p of $r^2 + s^2$, and make a calculation (mod p) similar to the one (mod 3) in (27.5).

27.6 Use the second proof of (27.5) to show that if $\cos\alpha \in \mathbb{Q}$ and if α is a rational multiple of π, then $\cos\alpha = 0, \pm\frac{1}{2}, \pm1$.

He heard the word "demonstrate" and said to himself: "What do I do when I demonstrate more than when I reason or prove?..." He looked in Noah Webster's dictionary and learned that demonstration is "proof beyond the possibility of doubt".

The definition didn't satisfy him.... He said to himself, "Lincoln, you can never make a lawyer of yourself until you understand what 'demonstrate' means".

He bought "The Elements of Euclid", a book twenty-three centuries old. It went into his carpetbag as he went out on the circuit. At night, when with other lawyers, two in a bed, eight and ten in a hotel room, he read Euclid by the light of a candle after the others had dropped off to sleep.

– from *Abraham Lincoln, the Prairie Years*
by Carl Sandburg
Harcourt, Brace, (1926), p. 303

CHAPTER 6

Construction Problems and Field Extensions

During the earlier parts of this book, we started always from Euclid's geometry, developing and expanding it using our modern mathematical awareness. Because of the construction of the field of segment arithmetic, one could even argue that the use of fields in Chapter 4 arises naturally from the geometry. In this chapter, however, we will make use of modern algebra, the theory of equations and field extensions, and in particular the Galois theory, as it developed in the late 19th and early 20th centuries.

We use this algebra as a lens to examine and interpret geometrical questions in a new language. It was this algebraic interpretation which first provided the tools for rigorously proving the unsolvability of classical construction problems. If these proofs seem easy today, just think of all the modern algebra we take for granted, which did not exist even two hundred years ago.

We begin by examining three classic construction problems, two of which we can prove to be impossible using field theory. (The third depends on the transcendence of π, whose proof would carry us too far afield.)

Next we consider the problem of constructing regular polygons, and give Gauss' proof of the constructibility of the regular 17-sided polygon.

In §30 we consider constructions made with an additional tool, the marked ruler. Using the marked ruler we show how to extract cube roots and trisect angles. Then in §31 we study the real roots of cubic and quartic equations, and show that the constructions possible with compass and marked ruler are equivalent to solving quadratic, cubic, and quartic equations.

In an appendix §32 we collect, mostly without proof, the results from field theory which we use in the earlier sections of this chapter.

28 Three Famous Problems

In this section we will discuss three classical construction problems via field theory. Recall (13.2) that a point with coordinates (α, β) in the Cartesian plane over a field F can be constructed with ruler and compass if and only if the coordinates α, β can be obtained from the initial data by a finite number of field operations and extractions of square roots of positive quantities. If we suppose that the initial data is given by rational numbers, then this is equivalent to saying that α and β

are elements of the constructible field K (16.4.1), which is just the set of those real numbers which can be obtained by a finite number of field operations and square roots from $\mathbb{Q}$. We say a real number α is *constructible* if it is an element of K. So a point (α, β) in the real Cartesian plane is constructible by ruler and compass from data given by rational numbers, if and only if α and β are constructible numbers.

For our applications, we need some criteria to decide when a number is constructible. The first is merely a restatement of what we have just said.

Proposition 28.1 *A real number $\alpha \in \mathbb{R}$ is constructible if and only if there is a tower of subfields of $\mathbb{R}$,*

$$\mathbb{Q} = F_0 \subseteq F_1 \subseteq \cdots \subseteq F_k \subseteq \mathbb{R}$$

such that each $F_i = F_{i-1}(\sqrt{a_i})$ for some $a_i \in F_{i-1}$, and $\alpha \in F_k$.

Proof. In writing α as obtained by a finite number of field operations and square roots, each time there is a square root, let F_i be the field generated by that one over the previously constructed field. Conversely, if $\alpha \in F_k$, then by construction it is written using field operations and square roots.

Corollary 28.2 *If $\alpha \in \mathbb{R}$ is constructible, then deg $\mathbb{Q}(\alpha)/\mathbb{Q}$ is a power of* 2.

Proof. Here we use the notion of the degree of a field extension (32.1). Adjoining a square root of an element which is not already a square creates a field extension of degree 2. So, assuming that the extensions in (28.1) are all non-trivial, we find deg $F_k/\mathbb{Q} = 2^k$. Since $\mathbb{Q} \subseteq \mathbb{Q}(\alpha) \subseteq F_k$, by multiplicity of degrees of field extensions (32.1) we see that deg $\mathbb{Q}(\alpha)/\mathbb{Q}$ is a power of 2.

Remark 28.2.1. This simple statement is already enough to dispose of two of the three classical construction problems. Beware, however, the converse of this statement is false (28.7.1).

Duplication of the cube

The problem is, given a cube, to construct a new cube whose volume is twice the volume of the given cube. We will interpret the phrase "given a cube" to mean that we are given its edge, as a line segment in the plane, and the problem is then to construct the edge of the bigger cube. If the first cube has edge a, then we are looking for a length b such that $b^3 = 2a^3$. This means $b = \sqrt[3]{2}a$. If b is constructible, then $\sqrt[3]{2}$ is constructible and conversely.

Theorem 28.3 *$\sqrt[3]{2}$ is not in K. Hence the duplication of the cube is not possible by ruler and compass.*

Proof. Consider $\alpha = \sqrt[3]{2}$. It is a root of the equation $x^3 - 2 = 0$. The polynomial $x^3 - 2$ is irreducible over $\mathbb{Q}$ for the following reason. Being a cubic equation, if it factors, then one factor is linear, so it has a root. Let this root be a/b with $a, b \in \mathbb{Z}$, a, b relatively prime. Then

$$\left(\frac{a}{b}\right)^3 - 2 = 0$$

so

$$a^3 = 2b^3.$$

It follows that $2 \mid a$, so a is even. Then $2^3 \mid a^3$, so $2^2 \mid b$, so b is also even. This contradicts the hypothesis that a and b were relatively prime.

Now since $x^3 - 2$ is irreducible, the field $\mathbb{Q}(\sqrt[3]{2})$ is an extension of degree 3 of $\mathbb{Q}$. Since 3 is not a power of 2, we conclude $\sqrt[3]{2} \notin K$, by (28.2).

Trisection of the angle

The problem is, given an angle in the plane, to construct by ruler and compass a new angle equal to one third of the given angle. This problem is a bit more complicated because some angles can be trisected. For example, if the given angle is a right angle, we can construct an angle equal to one third of it (30°) by first constructing an equilateral triangle (I.1) and then bisecting one of its angles. See also (Exercise 28.3).

We will show that the problem of trisecting the angle is not always possible by exhibiting one angle which exists in the Cartesian plane Π over the constructible field K, namely an angle of 60°, whose third, an angle of 20°, does not exist in Π. If an angle α exists in Π, then the trigonometric functions $\sin \alpha, \cos \alpha, \tan \alpha$ will be elements of the field K.

Theorem 28.4 *The real number* $\alpha = \cos 20°$ *is not in* K. *Hence an angle of* 60° *cannot be trisected by ruler and compass.*

Proof. Using the sum angle formulas for sin and cos, we find for any angle θ that

$$\cos 3\theta = 4\cos^3\theta - 3\cos\theta.$$

Since $\cos 60° = \frac{1}{2}$, the real number $\alpha = \cos 20°$ satisfies

$$4\alpha^3 - 3\alpha = \frac{1}{2}.$$

In other words, α is a root of the equation

$$4x^3 - 3x - \frac{1}{2} = 0.$$

We will show that this equation has no roots in K. Making the substitution $y = 2x$, it is sufficient to show that the equation

$$y^3 - 3y - 1 = 0$$

has no roots in K.

First we show it has no roots in $\mathbb{Q}$. If it did, let the root be a/b, with $a, b \in \mathbb{Z}$, relatively prime. Then

$$a^3 - 3ab^2 - b^3 = 0.$$

Consequently any prime factor p of a also divides b and conversely, any prime factor of b divides a. Since a, b are relatively prime, we conclude $a, b = \pm 1$, so $a/b = \pm 1$. But by inspection we see that neither $+1$ nor -1 is a root of the equation.

Hence $y^3 - 3y - 1$ is an irreducible polynomial over $\mathbb{Q}$, since it is of degree 3 with no roots. We conclude that $\mathbb{Q}(\alpha)$ is an extension of degree 3 of $\mathbb{Q}$. But if α

were constructible, then $\mathbb{Q}(\alpha)$ would have degree a power of 2 over $\mathbb{Q}$ (28.2). We conclude that $\alpha \notin K$, so the angle of 20° is not constructible.

Squaring of the circle

The problem is, given a circle, to construct by ruler and compass a square of area equal to the area enclosed by the circle. If the circle has radius r, its area is πr^2, so we need a square of side a with $a^2 = \pi r^2$. Thus $a = \sqrt{\pi} \cdot r$. If a were constructible from r, then $\sqrt{\pi}$, and hence π would be constructible.

Theorem 28.5 *$\pi \notin K$, so the problem of squaring the circle cannot be solved by ruler and compass.*

The fact that $\pi \notin K$ is a consequence of the fact that π is *transcendental* over $\mathbb{Q}$: there is no polynomial with rational coefficients having π as a root. But every element of K is algebraic over $\mathbb{Q}$. The proof that π is transcendental is analytic, so we will not give it here. See for example Stewart Chapter 6.

See §25 for further discussion of this problem and its history.

Up to here we have used only the necessary condition (28.2) for a number to be constructible. Using a little more field theory, we will now derive a necessary and sufficient condition for a real number to be constructible. It depends on a remarkable coincidence of three separate notions: adjoining square roots, solving quadratic equations, and making field extensions of degree 2.

Proposition 28.6 *Let $F \subseteq E$ be an extension of fields of char. $\neq 2$. Then the following conditions are equivalent:*

(i) *$E = F(\sqrt{a})$ for some $a \in F$, $\sqrt{a} \notin F$.*

(ii) *$E = F(\alpha)$, where α is a root of an irreducible quadratic polynomial over F.*

(iii) *deg $E/F = 2$.*

Proof. (i) $\Rightarrow$ (iii). If $E = F(\sqrt{a})$, then every element of E can be written uniquely in the form $c_1 + c_2\sqrt{a}$, $c_1, c_2 \in F$, so deg $E/F = 2$.

(iii) $\Rightarrow$ (ii). If deg $E/F = 2$, and if $\alpha \in E$ is any element not in F, then $1, \alpha, \alpha^2$ are linearly dependent, so α satisfies an irreducible quadratic equation with coefficients in F.

(ii) $\Rightarrow$ (i). Follows from the quadratic formula: the roots of the equation $ax^2 + bx + c = 0$ are given by $\alpha = \frac{1}{2a}(-b \pm \sqrt{b^2 - 4ac})$.

Remark 28.6.1. This is indeed an elementary result, but we thought it worth stating explicitly because of the questions it raises about generalizing to higher degrees—cf. (Exercise 28.4) and §31.

Proposition 28.7 *A real number α is constructible if and only if the Galois group of its minimal polynomial is a group of order 2^s for some s.*

Proof. Suppose α is constructible. Then there is a tower of fields

$$\mathbb{Q} = F_0 \subseteq F_1 \subseteq \cdots \subseteq F_k \subseteq \mathbb{R}$$

as in (28.1). For any isomorphism $\sigma : F_k \to F'$ to a subfield F' of $\mathbb{C}$, F' will also be the top of a similar tower

$$\mathbb{Q} = F_0 \subseteq \sigma(F_1) \subseteq \sigma(F_2) \subseteq \cdots \subseteq \sigma(F_k) = F'.$$

Hence every element of F' can be obtained by field operations and successive square roots. The subfield $\bar{F}$ of $\mathbb{C}$ generated by the fields $\sigma(F_k)$ for all possible such σ is a normal extension of $\mathbb{Q}$.

Now if α is an element of F_k with minimal polynomial $f(x)$, then all the conjugates of α lie in the various fields $\sigma(F_k)$, so the splitting field E of $f(x)$ will be a subfield of $\bar{F}$. But all the elements of $\bar{F}$ can be obtained by successive square roots, so the degree of $\bar{F}$ over $\mathbb{Q}$ is a power of 2. Hence also the degree of $E/\mathbb{Q}$, which is equal to the order of the Galois group G of $f(x)$, is a power of 2.

Conversely, let $\alpha \in \mathbb{R}$, and assume the Galois group G of its minimal polynomial has order a power of 2. An elementary result in group theory states that the center of a p-group is non-trivial (see e.g. Herstein, p. 86). Hence G has a normal subgroup G_1 of order 2. Applying the same argument to G/G_1 and continuing, we find there is a chain of subgroups

$$\{e\} = G_0 \subseteq G_1 \subseteq G_2 \subseteq \cdots \subseteq G_s = G$$

where each G_i is normal in G, and G_i/G_{i-1} has order 2.

By the fundamental theorem of Galois theory (32.4), if we denote by E the splitting field of the minimal polynomial $f(x)$ of α, there is a chain of fields

$$\mathbb{Q} = E_0 \subseteq E_1 \subseteq \cdots \subseteq E_s = E \subseteq \mathbb{C}$$

with each E_i/E_{i-1} of degree 2, and each E_i is a normal extension of $\mathbb{Q}$. Let $F_i = E_i \cap \mathbb{R}$. Then we have

$$\mathbb{Q} = F_0 \subseteq F_1 \subseteq F_2 \subseteq \cdots \subseteq F_s = E \cap \mathbb{R}.$$

Now each E_i is normal over $\mathbb{Q}$, so complex conjugation τ acts on E_i, with fixed field F_i. Hence deg $E_i/F_i = 1$ or 2 for each i, and it follows that each extension F_i/F_{i-1} has degree 2 or 1. By (28.6), $F_i = F_{i-1}(\sqrt{a_i})$ for some $a_i \in F_{i-1}$. Furthermore $\alpha \in F_s = E \cap \mathbb{R}$, so α is constructible.

Example 28.7.1 For $\alpha \in \mathbb{R}$ to be constructible, it is not sufficient that deg $\mathbb{Q}(\alpha)/\mathbb{Q}$ be a power of 2. Take for example the polynomial $f(x) = x^4 - x^3 - 5x^2 + 1$. A little curve-sketching from first-year calculus shows that it has four distinct real roots. Let α be one of them. It is shown in (32.5.1) that $f(x)$ is irreducible, and that its Galois group has order 12 or 24. Hence deg $\mathbb{Q}(\alpha)/\mathbb{Q} = 4$, but by (28.7), α is not constructible.

Exercises

28.1 A *parabola* Γ can be defined as the locus of points P equidistant from a fixed point F, the *focus*, and a fixed line l, the *directrix*.

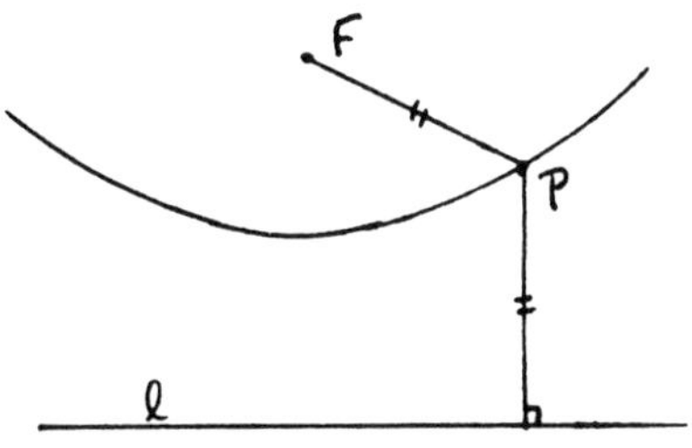

a) If the focus is $(0, 1)$ and the directrix is $y = -1$, find the equation of the parabola.

b) If $y = mx + b$ is a line with $m, b \in K$ and $b > 0$, show that the two intersection points of the line with Γ have coordinates in K.

c) Now give a ruler and compass construction for the intersection points of a line m with the parabola having focus F and directrix l. Thus, even though we cannot "draw" the parabola, yet we can find its intersection points with any line, as if it were drawn.

28.2 In the real Cartesian plane,

a) Show that the circle with center $(a, 1)$ passing through the origin, and the parabola Γ given by $y = \frac{1}{2}x^2$ intersect at the point $(2\sqrt[3]{a}, 4\sqrt[3]{a^2})$.

b) Show that if we are given a single parabola drawn in the plane, then the problem of duplication of the cube becomes possible.

c) If a parabola is given by its focus and directrix, conclude that its intersection points with a circle are not always constructible.

28.3 In the real Cartesian plane, if we are given an angle α of radian measure $2\pi a/b$ with $a, b \in \mathbb{Z}$, and b not divisible by 3, show that it is possible to construct an angle $\frac{1}{3}\alpha$ starting from the given angle.

28.4 a) Sketch the graph of $y = x^3 - 3x - 1$ in the real Cartesian plane, and use the intermediate value theorem to show that it has three real roots.

b) If α is one of the roots of $x^3 - 3x - 1$, show that the field $\mathbb{Q}(\alpha)$ is not equal to a field of the form $\mathbb{Q}(\sqrt[3]{d})$ for any $d \in \mathbb{Q}$. Thus a field extension of degree 3 need not be generated by cube roots, in contrast to the situation in degree 2 (28.6).

c) Show however that one can express roots of this equation using square roots and cube roots of complex numbers: verify that

$$\alpha = \sqrt[3]{\frac{1}{2} + \frac{1}{2}\sqrt{-3}} + \sqrt[3]{\frac{1}{2} - \frac{1}{2}\sqrt{-3}}$$

is a root of the above equation. *Hint*: Of course you could substitute directly in the equation and multiply out, but using a little geometry you should be able to show without any messy calculation, that α is a root of this equation.

28.5 *Squaring the parabola.* Let a parabola be given by its focus F and directrix l, (as in Exercise 28.1), and let another line m be given in the plane. Prove that it is possible to construct with ruler and compass a square whose area is equal to the area bounded by the line and the parabola. (You need not find the construction: just prove that it is possible.)

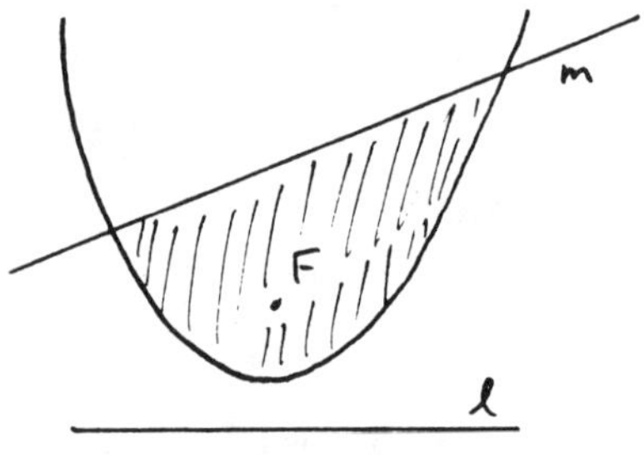

28.6 Here is a method for trisecting an angle using a marked ruler. Given the angle α with vertex O, choose any radius d and draw a circle centered at O, which cuts the sides of the angle α at A and B. Extend the line OA on the other side of O. Mark off two points on the ruler at a distance d from each other, and slide the ruler so that the two marks give points C on the line OA, D on the circle, with $CD = d$, and at the same time the ruler is aligned with B. Prove that $\angle COD = \beta$ is equal to $(1/3)\alpha$.

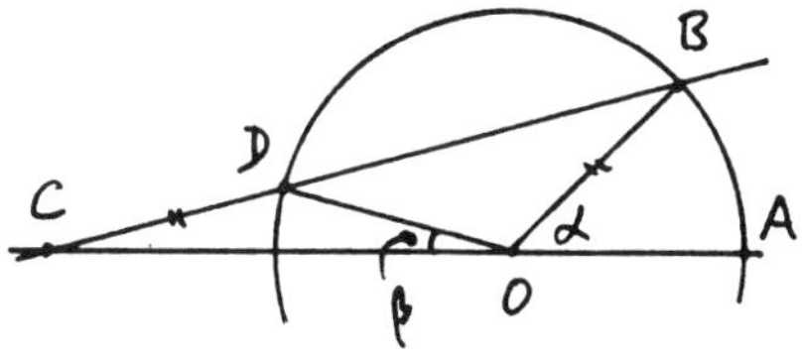

28.7 The problem of Apollonius is, given three circles $\Gamma_1, \Gamma_2, \Gamma_3$, to construct a circle Δ which is tangent to all three. Let Γ_i have center (a_i, b_i) and radius r_i. If we assume that Δ is in between the Γ_i, as in the figure, and has unknown center (x, y) and radius r, we obtain equations

$$\mathrm{dist}((x,y),(a_i,b_i)) = r + r_i$$

for $i = 1, 2, 3$. Show that these equations can be solved for x, y, r using field operations and square roots. Thus the problem is solvable by ruler and compass. (For an actual construction, see §38).

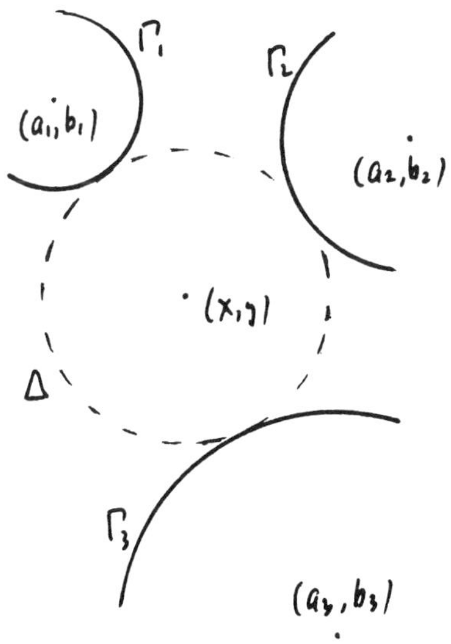

29 The Regular 17-Sided Polygon

A *regular* polygon is a polygon with all sides equal and all angles equal. Euclid knew how to construct regular polygons with n sides for $n = 3$ (I.1), $n = 4$ (I.46), $n = 5$ (IV.11), $n = 6$ (IV.15), and $n = 15$ (IV.16). By bisecting the central angles, one can construct a regular polygon of $2n$ sides, given one of n sides. Thus Euclid could construct regular polygons of n sides for $n = 2^k,\ \ 2^k{\cdot}3,\ \ 2^k{\cdot}5,\ \ 2^k{\cdot}3{\cdot}5$. For two thousand years these were the only known constructible regular polygons until Carl Friedrich Gauss in 1796, at the age of 19, made the remarkable discovery that the regular polygon of 17 sides was constructible with ruler and compass.

In this section we will explain Gauss' method using the language of modern algebra. We start with an algebraic proof of the construction of the regular pentagon, and finish with a criterion for constructibility of regular n-gons.

Our first step is to represent points of the real Cartesian plane by complex numbers. To the point (a, b) we associate the complex number $z = a + bi$. If ζ is a point on the unit circle making an angle θ with the positive real axis, then we can write

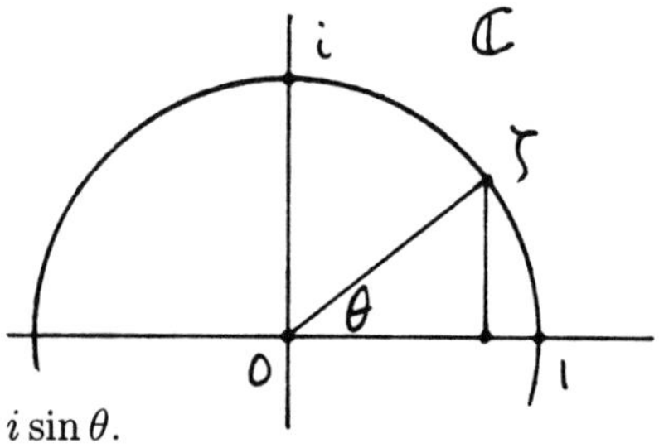

$$\zeta = \cos\theta + i\sin\theta.$$

If $\theta = 2\pi k/n$ for some integers k, n, then according to the laws of multiplication of complex numbers, $\zeta^n = 1$. In other words, ζ is a complex root of the equation $x^n - 1 = 0$. Letting $k = 0, 1, \ldots, n-1$, we obtain n distinct roots of this equation, so these are all the roots. They are called the n-th *roots of unity*.

To construct a regular polygon of n sides, it is sufficient to construct the quantity $\cos 2\pi/n$, and hence also $\sin 2\pi/n$. Since the complex number $\zeta = \cos 2\pi/n + i\sin 2\pi/n$ is then an n-th root of unity, the constructibility of the regular n-gon in the real Cartesian plane is related to the solutions of the equation $x^n - 1 = 0$ in the complex numbers.

Let us illustrate this situation by first considering the case $n = 5$. Take $\theta = 2\pi/5$, and let $\zeta = \cos 2\pi/5 + i\sin 2\pi/5$ be the complex number which corresponds to the first vertex of the pentagon after 1. Then the five vertices of the pentagon will be $1, \zeta, \zeta^2, \zeta^3, \zeta^4$. Note that $\zeta^4 = \zeta^{-1}$ is the complex conjugate of ζ. Therefore

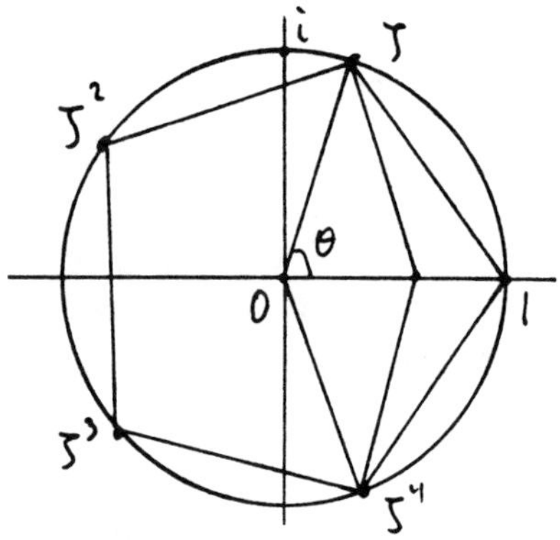

$$\zeta + \zeta^4 = 2\cos 2\pi/5.$$

On the other hand, ζ is a root of $x^5 - 1$. This polynomial factors

$$x^5 - 1 = (x-1)(x^4 + x^3 + x^2 + x + 1).$$

Since $\zeta \neq 1$, it is a root of the second factor, which is called the *cyclotomic polynomial* Φ_5 (cf. §32). Let $\alpha = \zeta + \zeta^4$. Then (remembering that $\zeta^5 = 1$)

$$\alpha^2 = \zeta^2 + 2 + \zeta^3.$$

Hence

$$\alpha^2 + \alpha = \zeta^4 + \zeta^3 + \zeta^2 + \zeta + 2 = 1,$$

using the relation $\Phi_5(\zeta) = 0$. So α is a root of the polynomial

$$x^2 + x - 1 = 0.$$

Using the quadratic formula,

$$x = \frac{1}{2}(-1 \pm \sqrt{5}).$$

Since our α is a positive real number, we conclude that $\alpha = \frac{1}{2}(\sqrt{5}-1)$. Hence

Proposition 29.1 $\cos 2\pi/5 = \frac{1}{4}(\sqrt{5}-1)$.

Remark 29.1.1. This result gives an algebraic proof of Euclid's construction of the regular pentagon (IV.11), since his construction gives $\frac{1}{2}(\sqrt{5}-1)$ for the side of the regular decagon, and that is equivalent to this by (13.3).

Now let us turn our attention to the regular 17-gon.

Theorem 29.2 (Gauss) *The regular* 17*-sided polygon is constructible with ruler and compass.*

Proof. An abstract proof of constructibility can be given using Galois theory (29.3.1). Here, however, we will give an elementary constructive proof, presumably close to Gauss' original method.

We follow the general idea of the method used above with the pentagon. Let

$$\zeta = \cos 2\pi/17 + i \sin 2\pi/17.$$

Let

$$\begin{aligned}\alpha &= \zeta + \zeta^{-1}\\ \beta &= \zeta + \zeta^4 + \zeta^{-1} + \zeta^{-4}\\ \gamma &= \zeta + \zeta^2 + \zeta^4 + \zeta^8 + \zeta^{-1} + \zeta^{-2} + \zeta^{-4} + \zeta^{-8}.\end{aligned}$$

We will show that

$$\mathbb{Q} \subseteq \mathbb{Q}(\gamma) \subseteq \mathbb{Q}(\beta) \subseteq \mathbb{Q}(\alpha)$$

and each one is quadratic over the previous one. Since $\alpha = 2\cos 2\pi/17$, this will show that the regular 17-gon is constructible.

Let us start with γ. It is natural to consider also the element γ' which is the sum of the other eight powers of ζ, namely

$$\gamma' = \zeta^3 + \zeta^5 + \zeta^6 + \zeta^7 + \zeta^{-3} + \zeta^{-5} + \zeta^{-6} + \zeta^{-7}.$$

Since ζ is a 17-th root of unity, it is a root of the cyclotomic polynomial

$$\Phi_{17} = x^{16} + x^{15} + \cdots + x + 1.$$

Thus $\gamma + \gamma' = -1$ (remember to treat exponents of ζ (mod 17) since $\zeta^{17} = 1$). We will show that $\gamma\gamma'$ is also in $\mathbb{Q}$, so that γ, which is a root of the equation

$$x^2 - (\gamma + \gamma')x + \gamma\gamma' = 0,$$

will be quadratic over $\mathbb{Q}$.

To find $\gamma\gamma'$ we must make a computation. Multiplying γ' by each term of γ in turn, we obtain the sum of ζ raised to each of the powers shown in the accompanying table. Observe that each integer $1 \leq i \leq 16$ (mod 17) occurs exactly four times in the table. Thus

$$\gamma\gamma' = 4\left(\sum_{i=1}^{16} \zeta^i\right) = -4.$$

Table of exponents of $\gamma\gamma'$

4	6	7	8	-2	-4	-5	-6
5	7	8	9	-1	-3	-4	-5
7	9	10	11	1	-1	-2	-3
11	13	14	15	5	3	2	1
2	4	5	6	-4	-6	-7	-8
1	3	4	5	-5	-7	-8	-9
-1	1	2	3	-7	-9	-10	-11
-5	-3	-2	-1	-11	-13	-14	-15

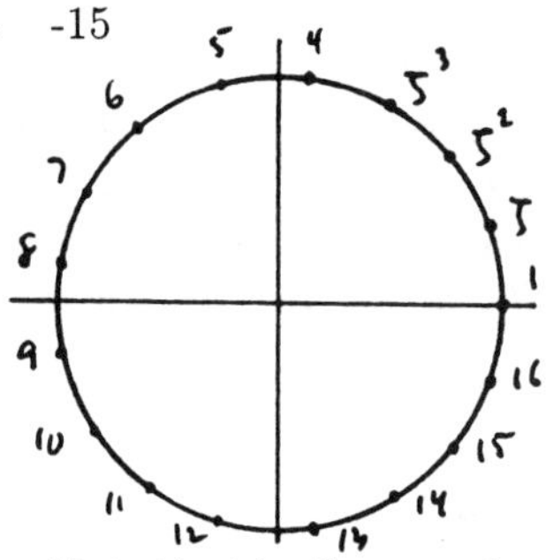

So γ is a root of the equation

$$x^2 + x - 4 = 0.$$

Using the quadratic formula,

$$x = \frac{1}{2}(-1 \pm \sqrt{17}).$$

To choose the correct sign for γ, we make an estimate. Note that in the sum for γ, the imaginary parts cancel, so γ is real. In fact

$$\gamma = 2(\cos 2\pi/17 + \cos 2 \cdot 2\pi/17 + \cos 4 \cdot 2\pi/17 + \cos 8 \cdot 2\pi/17).$$

Now looking at the approximate position of the powers $\zeta, \zeta^2, \zeta^4, \zeta^8$, it is clear that γ is positive. Therefore

$$\gamma = \frac{1}{2}(-1 + \sqrt{17}).$$

We can also confirm this choice of sign by calculating the sum of cosines above and the expression with the radical. Both come to 1.56155 to five decimals. Note, however, that the calculator check does not prove anything: it is merely a good way to detect errors in our work.

Next we consider β. If we let

$$\beta' = \zeta^2 + \zeta^8 + \zeta^{-2} + \zeta^{-8}$$

then $\beta + \beta' = \gamma$. On the other hand, $\beta\beta'$ turns out to be the sum of all the non-zero powers of ζ, so $\beta\beta' = -1$. Thus β is a root of the equation

$$x^2 - \gamma x - 1 = 0.$$

By the quadratic formula

$$x = \frac{1}{2}(\gamma \pm \sqrt{\gamma^2 + 4}).$$

Since β is positive, we choose the + sign, and find

$$\beta = \frac{1}{2}(\gamma + \sqrt{\gamma^2 + 4}).$$

Substituting for γ and simplifying gives

$$\beta = \frac{1}{4}(-1 + \sqrt{17} + \sqrt{34 - 2\sqrt{17}}).$$

Note also that by definition,

$$\beta = 2(\cos 2\pi/17 + \cos 4 \cdot 2\pi/17).$$

Both check with a calculator to give $\gamma = 2.04948$ (to 5 decimals).

Now let us look at the field extensions created so far. Clearly $\mathbb{Q}(\gamma) = \mathbb{Q}(\sqrt{17})$. We claim that $\mathbb{Q}(\beta) = \mathbb{Q}(\sqrt{34 - 2\sqrt{17}})$, which clearly contains $\mathbb{Q}(\gamma)$. It is clear that

$$\mathbb{Q}(\beta) = \mathbb{Q}(\sqrt{17} + \sqrt{34 - 2\sqrt{17}}).$$

If we let

$$x = \sqrt{17} + \sqrt{34 - 2\sqrt{17}},$$

then squaring twice to get rid of radicals gives the equation

$$x^4 - 6 \cdot 17x^2 + 8 \cdot 17x + 13 \cdot 17 = 0.$$

By Eisenstein's criterion (32.8) this is irreducible, so the degree of the field extension $\mathbb{Q}(\beta)/\mathbb{Q}$ is four. Since β is clearly contained in the field $\mathbb{Q}(\sqrt{34 - 2\sqrt{17}})$, which has degree at most four, we conclude that $\mathbb{Q}(\beta) = \mathbb{Q}(\sqrt{34 - 2\sqrt{17}})$, as claimed.

Next let us consider α. Let

$$\alpha' = \zeta^4 + \zeta^{-4}.$$

Then

$$\alpha + \alpha' = \beta$$

and

$$\alpha\alpha' = \zeta^3 + \zeta^5 + \zeta^{-3} + \zeta^{-5}.$$

Let us denote $\alpha\alpha'$ by β'', and let

$$\beta''' = \zeta^6 + \zeta^7 + \zeta^{-6} + \zeta^{-7}.$$

Then

$$\beta'' + \beta''' = \gamma'$$

and

$$\beta''\beta''' = \sum_{i=1}^{16} \zeta^i = -1.$$

Thus β'' is a root of the equation

$$x^2 - \gamma' x - 1 = 0.$$

A calculation similar to the one used for β then gives

$$\beta'' = \frac{1}{4}(-1 - \sqrt{17} + \sqrt{34 + 2\sqrt{17}}).$$

Note that

$$\sqrt{34 - 2\sqrt{17}} \cdot \sqrt{34 + 2\sqrt{17}} = \sqrt{34^2 - 4 \cdot 17} = 8\sqrt{17}.$$

Therefore $\beta'' \in \mathbb{Q}(\sqrt{34 - 2\sqrt{17}}) = \mathbb{Q}(\beta)$.

Now α satisfies the equation

$$x^2 - \beta x + \beta'' = 0,$$

so α is quadratic over $\mathbb{Q}(\beta)$. Using the quadratic formula, substituting, and choosing the sign correctly, we find

$$\begin{aligned} \alpha &= 2\cos 2\pi/17 \\ &= \tfrac{1}{8}(-1 + \sqrt{17} + \sqrt{34 - 2\sqrt{17}} + 2\sqrt{17 + 3\sqrt{17} + \sqrt{170 - 26\sqrt{17}} - 4\sqrt{34 + 2\sqrt{17}}}). \end{aligned}$$

This checks by calculator, giving $\alpha = 1.86494$. In particular, this shows that α is a constructible number, and hence the regular 17-gon is constructible by ruler and compass.

Although we have just completed the proof of the theorem, we have not yet shown that $\mathbb{Q}(\beta) \subseteq \mathbb{Q}(\alpha)$ as claimed earlier. To do this, write $\alpha = \zeta + \zeta^{-1}$ and then compute

$$\begin{aligned} \alpha^2 &= \zeta^2 + 2 + \zeta^{-1} \\ \alpha^4 &= \zeta^4 + 4\zeta^2 + 6 + 4\zeta^{-2} + \zeta^{-4}. \end{aligned}$$

Therefore β, which was defined to be $\zeta + \zeta^4 + \zeta^{-1} + \zeta^{-4}$, can be obtained as

$$\beta = \alpha^4 - 4\alpha^2 + \alpha + 2.$$

Therefore $\beta \in \mathbb{Q}(\alpha)$, as required.

Corollary 29.3 *The side of the regular* 17*-gon inscribed in the unit circle is*

$$\frac{1}{4}\sqrt{34 - \sqrt{17} - \sqrt{34 - 2\sqrt{17}} - 2\sqrt{17 + 3\sqrt{17} + \sqrt{170 - 26\sqrt{17}} - 4\sqrt{34 + 2\sqrt{17}}}}.$$

Proof. Use (13.3) and the expression for α above.

Remark 29.3.1. If we only wish to show that the regular 17-gon is constructible, without an explicit formula for $\cos 2\pi/17$, we can proceed as follows. The question is whether $\alpha = 2\cos 2\pi/17$ is a constructible real number. Now α is contained in the field $\mathbb{Q}(\zeta)$ of 17-th roots of unity. By (32.7) this is a normal extension of $\mathbb{Q}$ of degree 16 and with Galois group $\mathbb{Z}_{17}^*$, which is cyclic of degree 16. Therefore the Galois group of the splitting field of the extension $\mathbb{Q}(\alpha)$ will be a quotient of this one, and its order will be a power of 2. By (28.7) therefore, α is constructible.

The proof of (29.2) given above actually illustrates the proof of (28.7) and at the same time gives a nice example of the fundamental theorem of Galois theory (32.4) by showing the correspondence between the subgroups of the Galois group and the subfields of the field. For inside the group $\mathbb{Z}_{17}^*$ we have a chain of subgroups

$$\{1,-1\} \subseteq \{1,-1,4,-4\} \subseteq \{1,2,4,8,-1,-2,-4,-8\}.$$

The fixed fields of these subgroups are precisely $\mathbb{Q}(\alpha)$, $\mathbb{Q}(\beta)$ and $\mathbb{Q}(\gamma)$. And of course it was these subgroups of $\mathbb{Z}_{17}^*$ which provided the motivation for choosing α, β, γ the way we did.

Remark 29.3.2. The explicit expressions for α, β, γ found in the proof of (29.2) can without too much difficulty be turned into an actual construction of the 17-gon. Here is a particularly simple one:

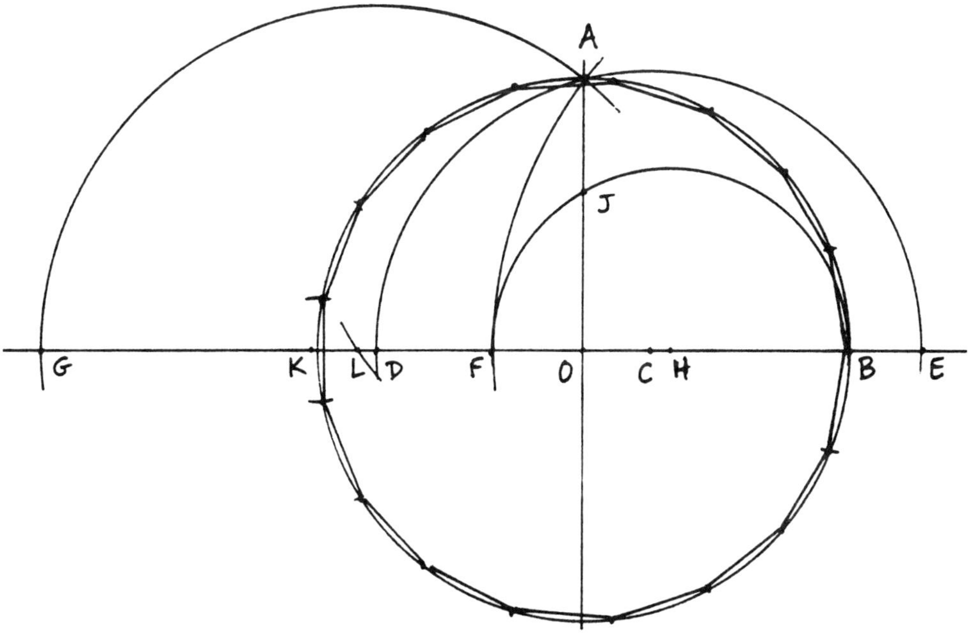

Let the circle with center O be given. Let OA and OB be two orthogonal radii. Take $OC = \frac{1}{4}OB$. Let the circle CA meeet OB in D and E. Let the circle EA give F, and let the circle DA give G. Take H the midpoint of BF, and let the circle HB meet OA in J. Take K to be the midpoint of OG. Let the circle with center J and radius OK meet OB in L. Then KL is the side of the inscribed 34-gon. See (Exercise 29.4).

Using the ideas of this section it is not too hard now to obtain a complete determination of the possible constructible regular n-gons.

Theorem 29.4 *The regular polygon of n sides is constructible by ruler and compass if and only if n is a number of the form*

$$n = 2^r p_1 \dots p_s, \quad r, s \geq 0,$$

where the p_i are distinct odd primes, each of which is a prime of the form

$$p = 2^{2^k} + 1.$$

Proof. For any integer n, let

$$\zeta = \cos 2\pi/n + i \sin 2\pi/n$$

and let

$$\alpha = \zeta + \zeta^{-1}.$$

Then

$$\alpha = 2\cos 2\pi/n$$

and ζ is a root of the equation

$$x^2 - \alpha x + 1 = 0$$

with coefficients in $\mathbb{Q}(\alpha)$. Therefore deg $\mathbb{Q}(\zeta)/\mathbb{Q}(\alpha) = 2$. Since $\mathbb{Q}(\zeta)$ is the field of n-th roots of unity, it is a normal field extension of $\mathbb{Q}$ with Galois group $\mathbb{Z}_n^*$, which is an abelian group of order $\varphi(n)$, the Euler φ-function (32.7). Since deg $\mathbb{Q}(\zeta)/\mathbb{Q}(\alpha) = 2$, the orders of the Galois groups of $\mathbb{Q}(\alpha)$ and $\mathbb{Q}(\zeta)$ differ at most by 2. We conclude from (28.7) that the regular n-gon is constructible if and only if the order of the Galois group of $\mathbb{Q}(\zeta)$, which is $\varphi(n)$, is a power of 2.

Now let us write

$$n = 2^k p_1^{e_1} \dots p_s^{e_s}$$

with distinct odd primes p_i. Then

$$\varphi(n) = 2^{k-1} \prod p_i^{e_i - 1}(p_i - 1).$$

In order for this to be a power of 2, we must have $e_i = 1$ for each odd prime occurring, and $p_i - 1$ must be a power of 2, so

$$p_i = 2^{t_i} + 1.$$

One sees easily that in order for p_i to be prime, it is necessary that t_i be a power of 2. Thus p_i is of the form

$$p = 2^{2^k} + 1$$

as required. This argument is reversible, and so the theorem is proved.

Remark 29.4.1. Prime numbers of the form $p = 2^{2^k} + 1$ were studied by Fermat, who hoped he had discovered a formula for generating prime numbers. If we denote by F_k the k-th *Fermat number* $2^{2^k} + 1$, then $F_0 = 3$, $F_1 = 5$, $F_2 = 17$, $F_3 = 257$, and $F_4 = 65537$ are all primes. For $k \geq 5$ there are no further Fermat numbers which are known to be prime. As of this date, F_k is known to be composite for $5 \leq k \leq 23$, so F_{24} is the first unknown case.

Exercises

29.1 To illustrate that the square root of a complex number with constructible coordinates also has constructible coordinates, write $\sqrt{5+2\sqrt{-5}}$ in the form $a+bi$ where a and b are constructible real numbers in standard form (cf. Exercise 13.4).

29.2 Let $\zeta = \cos 2\pi/7 + i \sin 2\pi/7$, and let $\alpha = \zeta + \zeta^{-1}$.

a) Find the minimal polynomial for α over $\mathbb{Q}$ (it is a cubic).

b) Show that $\mathbb{Q}(\zeta)$ contains a unique subfield E of degree 2 over $\mathbb{Q}$. Find an integer d for which $E = \mathbb{Q}(\sqrt{d})$.

29.3 An investigation of the Fermat number F_{24}. Using a *hand calculator* only, find

a) How many digits does F_{24} have in its decimal expansion?

b) What are the first six digits of F_{24}?

c) What are the last six digits of F_{24}?

In each case, explain your method, and include a discussion of your calculator's round-off error and how reliable you believe your answer to be.

29.4 a) Carry out the construction of the 17-gon described in (29.3.2) (Par = 20 steps to get KL; 25 more to mark the vertices and draw the edges of the 17-gon.)

b) Prove that the construction works by showing (in the notation of the proof of (29.2)) that $OD = 2\gamma$, $OG = \beta$, $OF = \beta''$, $KL = \alpha'$. Then show that $\alpha' = \zeta^4 + \zeta^{-4}$ is equal to the side of the inscribed regular 34-gon.

30 Constructions with Compass and Marked Ruler

Up to now we have studied the classical ruler and compass constructions of Euclid's *Elements*. We have seen that there are some problems which cannot be solved by ruler and compass constructions, such as the duplication of the cube (28.3), the trisection of the angle (28.4), or the construction of a regular 7-sided polygon (29.4). From classical times people have attempted to effect such constructions using additional tools, such as a given parabola in the plane (Exercise 28.2), or the marked ruler (Exercise 28.6).

In this section we will consider the use of the marked ruler in a limited sense, and we will see that with its help we can trisect angles and extract cube roots. As an application we give a construction of the regular heptagon (7-sided polygon). We will also show that analytically, a single application of the marked ruler corresponds to finding a root of a certain quartic equation. In the next section we will discuss the associated field theory and show that the geometrical use of the compass and marked ruler is equivalent to the algebraic solution of cubic and quartic equations.

First let us make clear what the marked ruler can do. You can make two marks on the ruler corresponding to a given distance, and then you can slide the ruler along so that the marks lie on two given lines, while at the same time the ruler passes through a given point.

In other words, given two lines l, m, given a distance d, and given a point O, you can draw a line OAB with $A \in l$, $B \in m$, and such that $AB = d$. This will count as one marked ruler step. We do not allow any other use, such as the one in (Exercise 28.6). Meanwhile the ruler can still be used as an ordinary ruler in the old sense.

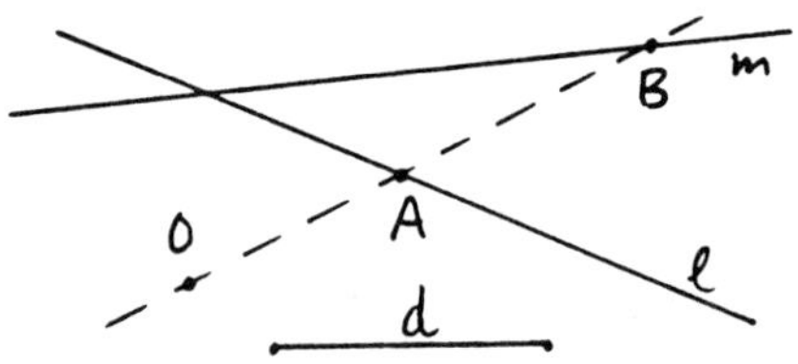

Proposition 30.1 *Using compass and marked ruler, it is possible to trisect any angle.*

Proof. Let AOB be the given angle. Drop a perpendicular AC from A to OB. Draw a line l through A parallel to OB. Now use the marked ruler to draw a line ODE such that $D \in AC$, $E \in l$, and $DE = 2AO$. This line will be the trisector of the original angle.

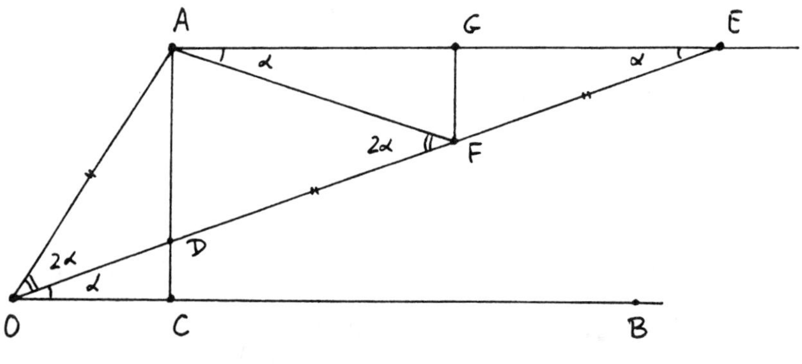

To see this, let F be the midpoint of DE, and let G be the midpoint of AE. Then FG is perpendicular to AE, so by (SAS) the triangles EFG and AFG are congruent. Now the new angle $EOB = \alpha$ is equal to $\angle AEO$ by parallel lines and to $\angle EAF$ by congruent triangles. So $\angle AFO = 2\alpha$ since it is an exterior angle to the triangle AEF.

Now DE was constructed equal to $2AO$, so $AO = EF = AF$. Hence the triangle AOF is isosceles, and so $\angle AOD = 2\alpha$. Thus the original angle $AOB = 3\alpha$, and α is one-third of it, as required.

Proposition 30.2 *Given segments of lengths 1 and a, it is possible with compass and marked ruler to construct a segment of length $\sqrt[3]{a}$.*

Proof. Let AB be the given segment of length a. Using the segment of length 1, choose $b = 2^{3k-1}$ for suitable k, such that $b > a$. Make an isosceles triangle ABC with $CA = CB = b$, and extend CA to D with $AD = b$. Draw the line DB. Now use the marked ruler to draw CEF with $E \in DB$ and $F \in AB$ and $EF = b$. Let $BF = y$. Then $\sqrt[3]{a} = y/2^{2k}$.

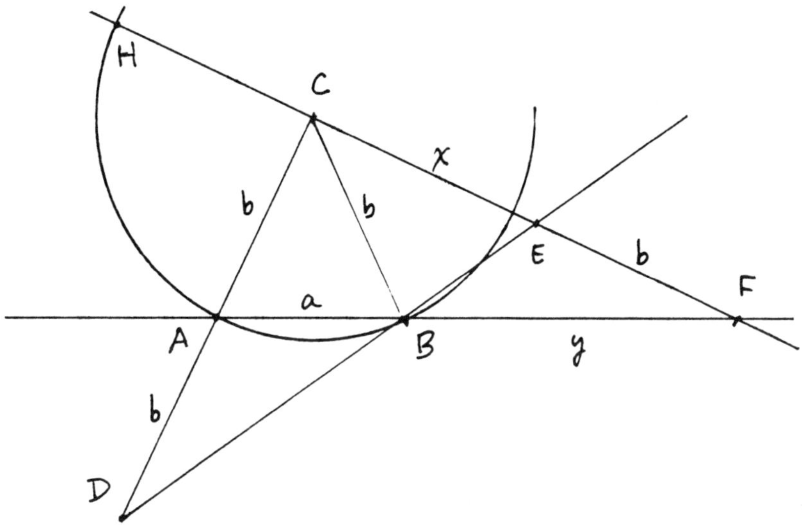

To see this, we first apply Menelaus' theorem (20.8) to the triangle ACF and the transversal line DBE. Letting $CE = x$, starting with vertex A, and going clockwise, it says

$$\frac{b}{2b} \cdot \frac{x}{b} \cdot \frac{y}{a} = 1.$$

This gives us

$$xy = 2ab.$$

Then we apply (III.36) to the circle with center C and radius b, and the point F outside, and the two chords FBA and FGH. Note that $CG = b$, so by subtraction, $FG = x$. Thus we obtain

$$y(y + a) = x(x + 2b).$$

Eliminating x from our two equations we find

$$y^3 = 4ab^2.$$

Substituting $b = 2^{3k-1}$ we find

$$y^3 = 2^{6k} \cdot a.$$

Therefore $\sqrt[3]{a} = y/2^{2k}$ can be obtained from y by bisecting $2k$ times.

Proposition 30.3 *With compass and marked ruler it is possible to construct a regular heptagon.*

Proof. We follow the method of §29. Let

$$\zeta = \cos 2\pi/7 + i \sin 2\pi/7$$

be a seventh root of unity. Let

$$\alpha = \zeta + \zeta^{-1}.$$

Then

$$\begin{aligned} \alpha^2 &= \zeta^2 + 2 + \zeta^{-1} \\ \alpha^3 &= \zeta^3 + 3\zeta + 3\zeta^{-1} + \zeta^{-3}. \end{aligned}$$

Using the fact that ζ is a root of the cyclotomic polynomial

$$\Phi_7 = x^6 + x^5 + x^4 + x^3 + x^2 + x + 1,$$

we find that

$$\alpha^3 + \alpha^2 - 2\alpha - 1 = 0.$$

So $\alpha = 2\cos 2\pi/7$ is a root of the equation

$$x^3 + x^2 - 2x - 1 = 0. \tag{1}$$

Our next step is to bring this equation into a form where it can be solved by trisecting a certain angle. Substituting

$$x = y - \frac{1}{3}$$

we get

$$y^3 - \frac{7}{3}y - \frac{7}{27} = 0. \tag{2}$$

Next, substituting $y = \frac{1}{3}\sqrt{7}z$ we get

$$z^3 - 3z - \frac{1}{7}\sqrt{7} = 0. \tag{3}$$

Using the trigonometric identity

$$\cos 3\theta = 4\cos^3\theta - 3\cos\theta$$

as in the proof of (28.4) we see that for any θ, $2\cos\theta$ is a root of the equation

$$z^3 - 3z - 2\cos 3\theta = 0.$$

In other words, changing notation, if we let θ be an angle with

$$2\cos\theta = \frac{1}{7}\sqrt{7}, \tag{4}$$

then $2\cos\frac{1}{3}\theta$ will be a root of the equation (3) above.

Working backwards, we find

$$y = \frac{2}{3}\sqrt{7}\cos\frac{1}{3}\theta$$

is a root of (2), and

$$\alpha = \frac{2}{3}\sqrt{7}\cos\frac{1}{3}\theta - \frac{1}{3}$$

is a root of (1). Therefore

$$\cos 2\pi/7 = \frac{1}{3}\sqrt{7}\cos\frac{1}{3}\theta - \frac{1}{6}. \tag{5}$$

So, to construct a regular heptagon, we first construct an angle θ with $\cos\theta = \frac{1}{14}\sqrt{7}$, to satisfy (4). Then we use (30.1) to trisect the angle θ and thus obtain $\cos\frac{1}{3}\theta$. From there we can construct $\cos 2\pi/7$ and hence the heptagon.

Remark 30.3.1. Here is an actual construction of the regular heptagon, based on the proof of (30.3).

1. Circle OA. (Call it radius 1.)
2. Line OA.
3. Circle AO, get B, C.
4. Circle CA, get D.
5. Line BC, get E. ($BC \perp OA$, and $OE = \frac{1}{2}$.)
6. Line CD. (CD is parallel to OA.)
7. Circle BE, get F. (Then $CF = \frac{3}{2}\sqrt{3}$.)
8. Circle center C, radius OE, get G.
9. Line GF, get H. (Then $GF = \sqrt{7}$, $GH = \frac{1}{3}\sqrt{7}$, and $\angle FGC = \theta$.)
10. Circle, center C, radius EH, get K. ($GK = \frac{1}{6}$.)
11. Line HK ($HK \perp CD$.)
12. Use the marked ruler to draw GLM with $LM = FH$. (Since $FH = 2GH$, the angle MGC is $\frac{1}{3}\theta$.)
13. Circle GH, get N. (The projection of GN on $OA = \frac{1}{3}\sqrt{7}\cos\frac{1}{3}\theta$.)
14. Circle center N, radius GL, get P. (The projection of GP on OA will then be $= \cos 2\pi/7$.)

15–17. Line through P, parallel to BC, get S, T. (Then A, S, T are vertices of the heptagon.)

18–21. With radius AS, mark off U, V, W, X.

22–28. Draw the sides of the heptagon $ASUVWXC$.

Note: It is an accident that AT appears to pass through N.

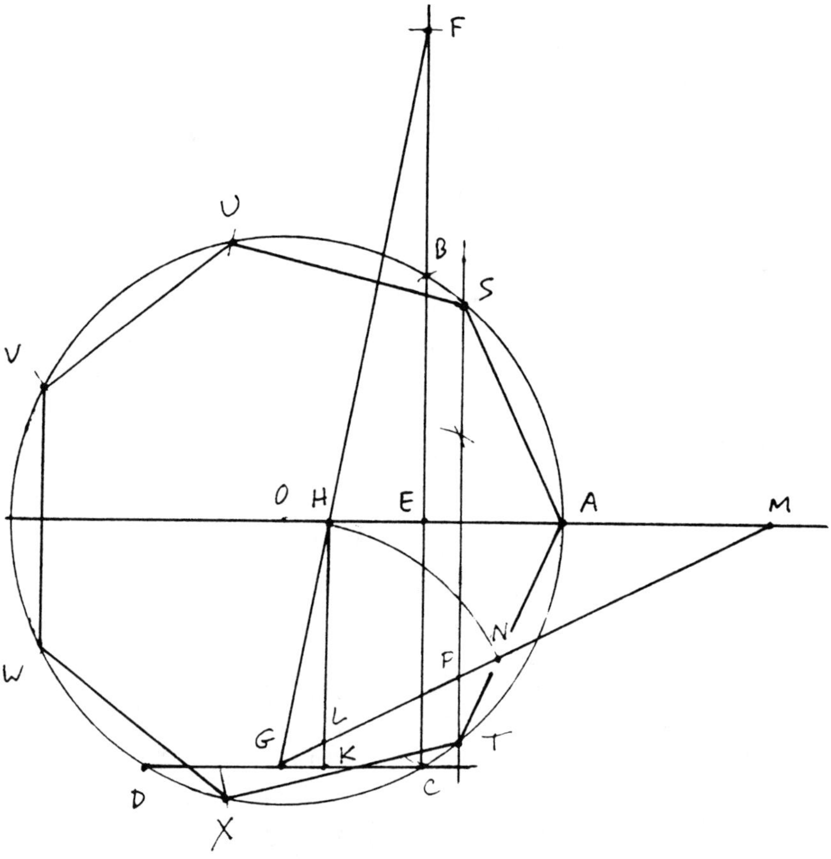

Remark 30.3.2. Since AB is a side of the regular hexagon in the above construction, BS will be a side of a regular 42-gon, twice it will give the regular 21-gon, and three times it will give the regular 14-gon.

Proposition 30.4 *Let lines l, m, a point O, and a segment of length d be given in the Cartesian plane over a field F. Suppose OAB is a line with $A \in l$, $B \in m$, and $AB = d$. Then the coordinates of A and B lie in a field $F(\alpha)$, where α is a root of a certain quartic polynomial with coefficients in F.*

Proof. By a linear change of coordinates, we may assume that $O = (0, 0)$ is the origin, and that l is the line $y = b$, for some b. We consider the locus of all points P such that OP cuts l in a point Q and $PQ = d$. This locus is a curve, called the *conchoid of Nicomedes*. To find the line OAB of the proposition is equivalent to finding the intersection of the second line m with the conchoid. This we will now do analytically.

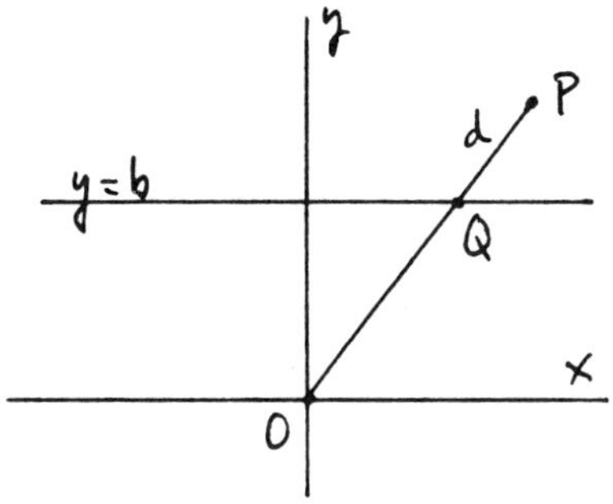

First we need to find the equation of the conchoid.

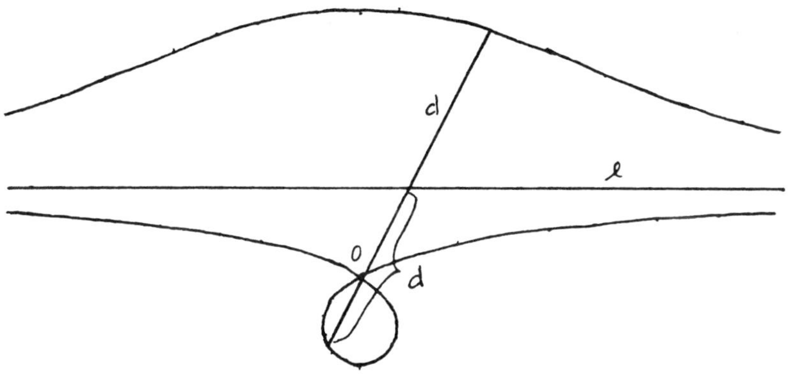

Take an arbitrary line $y = ax$ through the origin. This line meets l at the point $Q = (b/a, b)$. Let P have variable coordinates $P = (x, y)$. Then the condition $PQ = d$ gives

$$(x - b/a)^2 + (y - b)^2 = d^2.$$

But P also lies on the line $y = ax$. We use this equation to eliminate the variable a from the equation. Substituting $a = y/x$ and simplifying, we obtain the equation of the conchoid,

$$(x^2 + y^2)(y - b)^2 = d^2y^2.$$

To find the intersection B of the conchoid with the line M, we substitute the linear equation of m in the equation of the conchoid. This gives a quartic equation in x. If α is a root of that equation, the coordinates of B are then $x = \alpha$ and y is linear expression in α. From there we get the line OB and the coordinates of A.

Remark 30.4.1. It is clear from this proposition that the use of the marked ruler is equivalent to being given a single conchoid, drawn somewhere in the plane, and being allowed to intersect it with lines. Indeed, by rigid motions and similarities, any application of the marked ruler can be reduced to finding an intersection point of the conchoid with a certain line.

Exercises

30.1 Construct a regular nonagon (= 9-sided polygon) using compass and marked ruler. (par = 21 steps, including one marked ruler step.)

30.2 Show that a regular 13-gon can be constructed with compass and marked ruler by the following steps.

a) Find a cubic and a quadratic equation whose successive solution will give $\cos 2\pi/13$.

b) Show that the cubic equation in a) can be solved by trisecting a certain angle θ, and find $\cos\theta$.

30.3 Let AB be a given segment. Make ABM a right angle, and $ABN = 120°$. With the marked ruler, draw ACD cutting off a segment $CD = AB$ between the lines BM and BN. Show that $AC = \sqrt[3]{2} \cdot AB$.

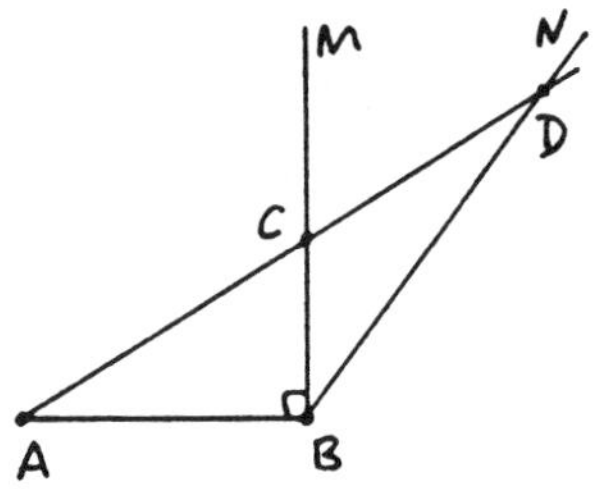

31 Cubic and Quartic Equations

Our purpose in this section is to investigate the roots of cubic and quartic equations, and to show that the use of the compass and marked ruler is equivalent to the successive solution of such equations.

In the quadratic case, there is a happy coincidence between solving quadratic equations, adjoining square roots of field elements, and considering degree 2 field extensions (28.6). For cubic and quartic equations, the situation is more complicated. A real root of a cubic equation may not be expressible in terms of cube roots (Exercise 28.4). A root of a quartic equation may give a degree 4 field extension which has no intermediate degree 2 subfield, and so α may not be expressible by square and fourth roots (32.5.1).

If we ignore questions of reality, it can be shown that arbitrary cubic and quartic equations can be solved by successive square roots and cube roots *of complex numbers.* This gives rise to Cardan's formulas for cubic and quartic equations which you can find in any algebra book.

Since this is a course in geometry, we are interested in what happens in the real Cartesian plane, and so we will be concerned with real roots of these equations. We will show that real roots of cubic and quartic equations can be expressed by three types of field extensions of a subfield $F \subseteq \mathbb{R}$:

1) $F(\sqrt{a})$ where $a \in F$, $a > 0$.
2) $F(\sqrt[3]{a})$ where $a \in F$.
3) $F(\cos \frac{1}{3}\theta)$ where $\cos\theta \in F$.

Roughly speaking, we will say these equations can be solved by taking square roots, cube roots, and trisecting angles. In this way we will see that solving cubic and quartic equations is equivalent to the use of the compass and marked ruler.

It is a remarkable tribute to the wisdom of the Greek geometers that two of the problems they highlighted, the duplication of the cube, and the trisection of the angle, provide precisely the tools needed for the solution of cubic and quartic equations.

For the reader who is already curious about quintic polynomials, see (Exercise 31.5).

We begin our work with some equations we do know how to solve. Throughout these discussions we fix a subfield F of the real numbers. We consider polynomials with coefficients in F, and will be concerned with the field extensions necessary to find roots. Of course the equation $x^3 - a = 0$ can be solved by $x = \sqrt[3]{a}$. Another type of equation we can solve by trisecting an angle.

Proposition 31.1 *Suppose given an equation*

$$x^3 - 3x - b = 0$$

with $b \in F$ and $|b| < 2$. Let θ be an angle with $\cos\theta = \frac{1}{2}b$. Then $\alpha = 2\cos\frac{1}{3}\theta$ is a root of the equation.

Proof. We have already encountered this equation twice before (28.4) and (30.3). It is a consequence of the trigonometric identity

$$\cos 3\psi = 4\cos^3\psi - 3\cos\psi$$

where we put $\theta = 3\psi$, $x = 2\cos\psi$, and $b = 2\cos 3\psi$. The restriction $|b| < 2$ is necessary to find a θ with $\cos\theta = \frac{1}{2}b$.

Proposition 31.2 *A real root of the equation*

$$x^3 - 3x - b = 0$$

with $b \in F$ and $|b| \geq 2$ can be found by a square root and a cube root.

Proof. The solution of this equation comes from the following observation. If a is any quantity, then

$$(a + a^{-1})^3 = a^3 + 3a + 3a^{-1} + a^{-3}.$$

Thus $a + a^{-1}$ is a root of the equation

$$x^3 - 3x - b = 0 \tag{1}$$

where $b = a^3 + a^{-3}$.

So now, working backward, given the equation (1) let us first solve

$$b = y + y^{-1}$$

which can be rewritten

$$y^2 - by + 1 = 0.$$

By the quadratic formula,

$$c = \frac{1}{2}(b + \sqrt{b^2 - 4})$$

is a root. Here we use the hypothesis $|b| \geq 2$ to ensure that c is real.

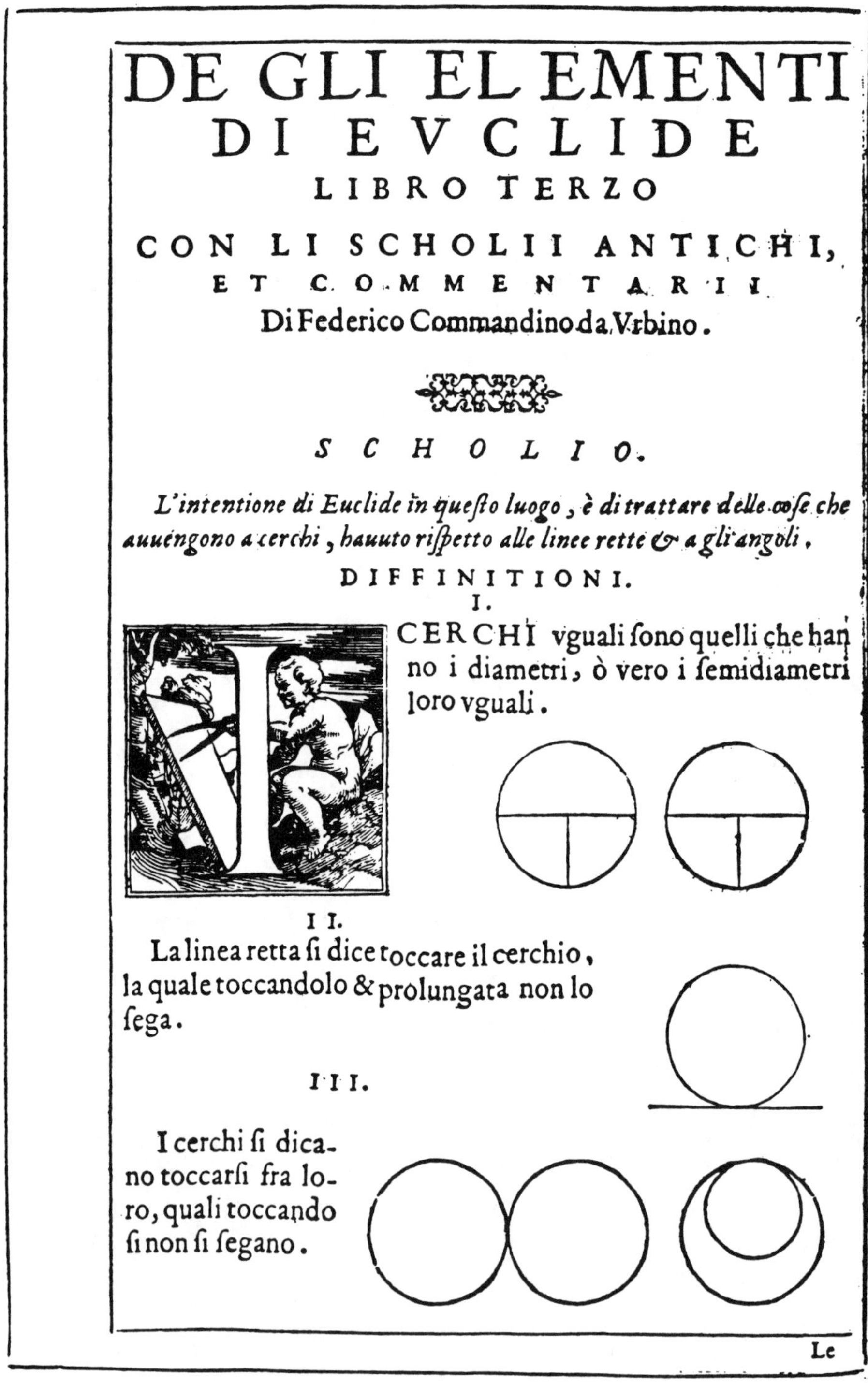

DE GLI ELEMENTI
DI EVCLIDE
LIBRO TERZO
CON LI SCHOLII ANTICHI,
ET COMMENTARII
Di Federico Commandino da Vrbino.

SCHOLIO.

L'intentione di Euclide in queſto luogo, è di trattare delle coſe che auuengono a cerchi, hauuto riſpetto alle linee rette & a gli angoli.

DIFFINITIONI.

I.

CERCHI vguali ſono quelli che hanno i diametri, ò vero i ſemidiametri loro vguali.

II.

La linea retta ſi dice toccare il cerchio, la quale toccandolo & prolungata non lo ſega.

III.

I cerchi ſi dicano toccarſi fra loro, quali toccando ſi non ſi ſegano.

Le

Plate IX. First page of Book III from Commandino (1575). The initial capitals in these notes are taken from Commandino's book.

Then $x = \sqrt[3]{c} + (\sqrt[3]{c})^{-1}$ is a root of the original equation. So we needed the square root of $b^2 - 4$ and the cube root of c. Note that if c' was the other root of the quadratic equation, then $cc' = 1$, so $c' = c^{-1}$ and we get the same root x of the original equation.

Proposition 31.3 *A real root of the equation*

$$x^3 + 3x - b = 0$$

for any $b \in F$ can be found by a square root and a cube root.

Proof. This is similar to the previous case. From the identity

$$(a - a^{-1})^3 = a^3 - 3a + 3a^{-1} - a^{-3}$$

we see that $a - a^{-1}$ is a root of the equation

$$x^3 + 3x - b = 0$$

where $b = a^3 - a^{-3}$.

In this case, given b we first solve

$$b = y - y^{-1}$$

or

$$y^2 - bx - 1 = 0.$$

This gives

$$c = \frac{1}{2}(b + \sqrt{b^2 + 4})$$

which is always real.

Then $x = \sqrt[3]{c} - (\sqrt[3]{c})^{-1}$ is a root of the original equation.

Theorem 31.4 *A real root of any cubic equation with coefficients in F can be found by successive (real) square roots and either a cube root or a trisection of an angle.*

Proof. Let the equation be

$$x^3 + a_1x^2 + a_2x + a_3 = 0$$

with $a_1, a_2, a_3 \in F$. By the substitution $x = y - \frac{1}{3}a$, we can eliminate the x^2 term and get a new equation of the form

$$y^3 + b_1y + b_2 = 0$$

with $b_1, b_2 \in F$.

Now, if $b_1 = 0$, we need only take a cube root of b_2. If $b_1 > 0$ we make the substitution $y = \frac{1}{3}\sqrt{3b_1}z$ and get a new equation

$$z^3 + 3z - b = 0$$

where $b \in F(\sqrt{3b_1})$. This equation can be solved by a square root and a cube root, by (31.3).

If $b_1 < 0$ we make the substitution $y = \frac{1}{3}\sqrt{-3b_1}z$ and obtain a new equation

$$z^3 - 3z - b = 0$$

with $b \in F(\sqrt{-3b_1})$. If $|b| < 2$ a root can be found by trisecting a certain angle (31.1), and if $|b| \geq 2$, it can be found by a square root and a cube root (31.2).

Proposition 31.5 *If α is a real root of a quartic polynomial with coefficients in F, then α can be found by solving a certain cubic equation with coefficients in F, followed by successive quadratic equations.*

Proof. Let the quartic polynomial be $f(x)$, and let α be a real root. We may assume that $f(x)$ is irreducible, for otherwise the result is trivial. Then the field $F(\alpha)$ has degree 4 over F. Let G be the Galois group of its splitting field. If the order of G is a power of 2, we can obtain α by successive square roots (28.7). Otherwise, since G is a subgroup of S_4, it will have order 12 or 24. Let E be the splitting field of $F(\alpha)$ over F, and let $E_0 = E \cap \mathbb{R}$. By the fundamental theorem of Galois theory (32.4), E_0 corresponds to a subgroup $G_0 \subseteq G$ which will have order 1 or 2, depending whether $E = E_0$ or not. Let H be a Sylow 2-group of G, containing G_0, and let F' be the fixed field of H. Then H has index 3 in G, and H contains G_0, so F' has degree 3 over F, and $F' \subseteq E_0$. Therefore F' can be written as $F(\beta)$, where β is a real root of some cubic polynomial with coefficients in F.

The Galois group of α over F' will be H, whose order is a power of 2, so α can be obtained by successive quadratic extensions from F' (28.7).

Theorem 31.6 *Let F be a subfield of $\mathbb{R}$ and let $\alpha \in \mathbb{R}$. The following conditions are equivalent:*

(i) *There exists a tower of subfields*

$$F = F_0 \subseteq F_1 \subseteq \cdots \subseteq F_k \subseteq \mathbb{R}$$

with $\alpha \in F_k$, and for each i, F_i is obtained from F_{i-1} by adjoining an element $\beta_i = \beta$, where either

1) $\beta = \sqrt{a}$, *with $a \in F_{i-1}$, $a > 0$, or*

2) $\beta = \sqrt[3]{a}$, *with $a \in F_{i-1}$, or*

3) $\beta = \cos \frac{1}{3}\theta$, *with $\cos\theta \in F_{i-1}$.*

(ii) *There exists a tower of subfields*

$$F = F_0 \subseteq F_1 \subseteq \cdots \subseteq F_k \subseteq \mathbb{R}$$

with $\alpha \in F_k$, and each F_i is obtained from F_{i-1} by adjoining a root of a quadratic, a cubic, or a quartic polynomial.

(iii) *The quantity α can be constructed by compass and marked ruler from data with coordinates in F.*

Proof. (i) $\Rightarrow$ (iii). The three types of extensions are constructible with compass and marked ruler, the first by ordinary ruler and compass construction, the second by (30.2), and the third by (30.1). (Note that $\cos\theta \in F$ does not necessarily imply that the angle θ can be realized by lines in the Cartesian plane over F. We may have to make a quadratic extension to obtain $\sin\theta$ first, before we have an actual angle to trisect.

(iii) $\Rightarrow$ (ii). Regular ruler and compass constructions correspond to quadratic equations, and each use of the marked ruler can be accomplished by solving a quartic equation (30.4).

(ii) $\Rightarrow$ (i). A quartic polynomial can be reduced to cubic and quadratic equations by (31.5), and cubic and quadratic polynomials can be solved by the three types of extensions listed, by (31.4).

Corollary 31.7 *If the quantity $\alpha \in \mathbb{R}$ is constructible by compass and marked ruler from data in the field F, then deg $F(\alpha)/F$ is $2^r \cdot 3^s$ for some $r, s \geq 0$.*

Proof. Since $F(\alpha) \subseteq F_k$ of (i) in the theorem, and each of the indicated extensions is of degree 2 or 3, it follows that deg F_k/F, and hence also deg $F(\alpha)/F$, is of the form $2^r 3^s$ for some $r, s \geq 0$.

Remark 31.7.1. In fact, considering all of the conjugates of α over F and taking the subfield of $\mathbb{C}$ generated by them, we obtain the stronger result that the order of the Galois group of α must be of the form $2^r 3^s$ also. This shows that the condition of the corollary is not sufficient for α to be constructible by compass and marked ruler, because for example α could be a root of a 6-th degree equation with Galois group S_6, whose order is divisible by 5.

Exercises

31.1 Show that it is not possible to construct a regular 11-gon with compass and marked ruler.

31.2 Show that it is possible to construct a regular 19-gon with compass and marked ruler.

31.3 Show that it is not possible to extract 5-th roots with compass and marked ruler.

31.4 Show that it is not possible to quintisect (divide in 5) a general angle with compass and marked ruler.

31.5 Suppose, in addition to the compass and marked ruler, we were given tools to extract 5-th roots and quintisect angles. Show that even with these new tools, it is still not possible to solve a general fifth degree equation. *Hint*: Show that the Galois group of any extension obtained with these new tools is still solvable, and then take a quintic equation with Galois group S_5, such as (32.4.4).

31.6 Verify that the following construction using a parabola will trisect an angle. Suppose given in the xy-plane the parabola Γ defined by $y = x^2$.

Suppose given also an angle AOB, with A and lying on the unit circle and $B = (1, 0)$. Draw a vertical line through A, to meet the line $y = 2$ at C. Draw a circle with center C, passing through the origin O, and let it meet the parabola Γ at D. Drop a perpendicular DE from D to the x-axis. Draw a circle with center E and radius 1 to meet the unit circle at F. Then $\angle FOB$ will be one third of the angle $\angle AOB$.

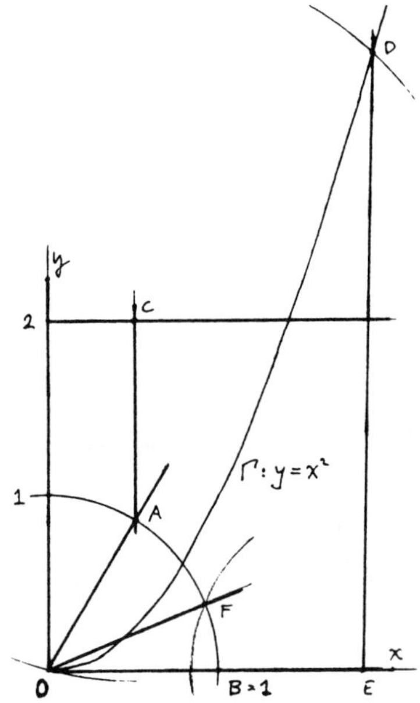

31.7 Show that the constructions one can accomplish given the fixed parabola $y = x^2$ in the xy-plane, and being allowed to intersect it with lines and circles, are precisely equivalent of those one can achieve with compass and marked ruler.

31.8 The problem of Alhazen. Given a circle Γ, and given two points A, B, find a point $C \in \Gamma$ so that the lines AC and BC make equal angles with the circle. Show that this problem leads to a quartic equation, and so can be solved with marked ruler and compass.

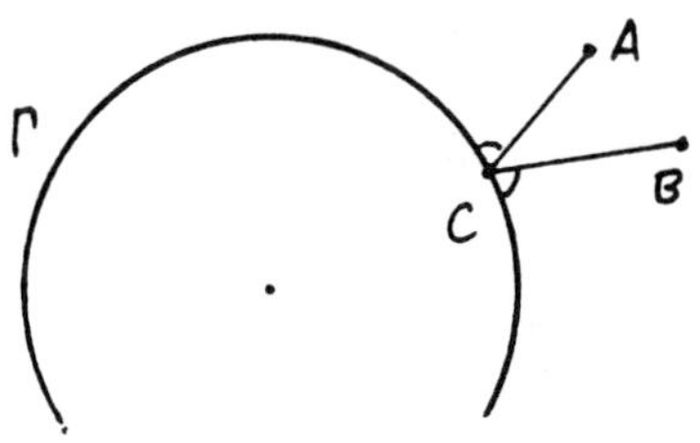

31.9 In the proofs of Propositions 5–9 of his book on Spirals, Archimedes makes use of the following construction: Given a circle Γ, given a chord l, given a point P on the circle, and given a segment d, to draw a line through P, such that the segment AB cut off by the line and the circle is equal to the given segment d. This is not a legitimate use of the marked ruler in the limited sense we have considered. Show however

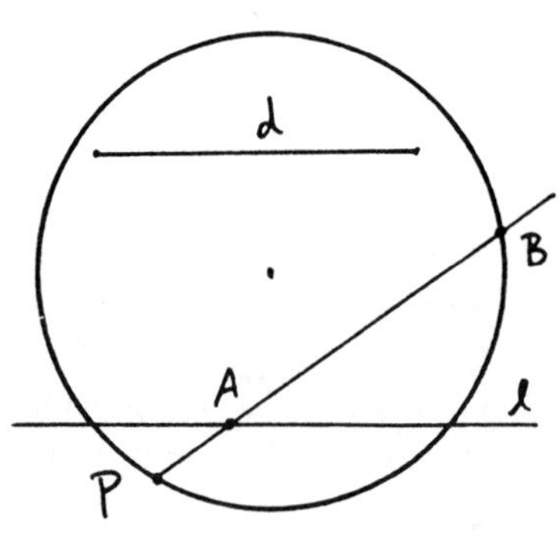

a) This construction can be made with compass and marked ruler in our sense.

b) If l is perpendicular to the diameter of the circle passing through P, then the construction is even possible with ruler and compass.

31.10 In (31.4), show that those cubic equations which can be solved with square roots and an angle trisection are the equations with three real roots; while those equations needing square roots plus a cube root are the equations with one real and two complex roots.

31.11 If $f(x)$ is an irreducible cubic equation over the field $F \subseteq \mathbb{R}$ with one real and two complex roots, then its Galois group is S_3.

31.12 We have seen (28.4) that $\alpha = 2\cos 20^\circ$ is a root of the equation $x^3 - 3x - 1 = 0$. Show that the other two roots of this equation are contained in the field $\mathbb{Q}(\alpha)$ and so its Galois group is $\mathbb{Z}_3$.

31.13 We have seen (30.3) that $\alpha = 2\cos 2\pi/7$ is a root of the equation $x^3 + x^2 - 2x - 1 = 0$. Show that the other two roots of this equation are also in the field $\mathbb{Q}(\alpha)$, so that it is a normal extension with Galois group $\mathbb{Z}_3$. Note that in this case however, it is only after adjoining $\sqrt{7}$ that we can express α using a trisection of an angle.

31.14 *The discriminant.* Let $f(x)$ be an irreducible cubic polynomial with coefficients in the field $F \subseteq \mathbb{R}$, and let its roots be $\alpha_1, \alpha_2, \alpha_3$ in its splitting field. We define the *discriminant* of $f(x)$ to be

$$\Delta = (\alpha_1 - \alpha_2)^2(\alpha_1 - \alpha_3)^2(\alpha_2 - \alpha_3)^2.$$

a) Show that $\Delta \in F$.

b) Show that $\Delta > 0$ if and only if $f(x)$ has 3 real roots, while $\Delta < 0$ if and only if $f(x)$ has one real and two complex roots.

c) Show that the Galois group of $f(x)$ is $\mathbb{Z}_3$ if and only if $\sqrt{\Delta} \in F$; otherwise it is S_3.

31.15 If $f(x) = x^3 - 3x - b$, show that its discriminant is $\Delta = 27(4 - b^2)$, as follows: Let α be one root. Then show that the remaining two roots of $f(x)$ are roots of the quadratic equation

$$x^2 + \alpha x + (\alpha^2 - 3) = 0.$$

Solve this using the quadratic formula. Then put the three roots into the definition of Δ (Exercise 31.14) and simplify.

31.16 Consider the equation $x^3 - 3x - \frac{1}{2} = 0$ over $\mathbb{Q}$. If we take θ such that $\cos\theta = \frac{1}{4}$, then $\alpha = 2\cos\frac{1}{3}\theta$ is a root, by (31.1). Show that in this case, the other two roots of the equation are not contained in the field $\mathbb{Q}(\alpha)$, so the Galois group will be S_3. Show that the splitting field of $\mathbb{Q}(\alpha)$ is $\mathbb{Q}(\alpha, \sqrt{5})$.

31.17 In this exercise we investigate when two different polynomials can give rise to the same field extension. For simplicity let us consider an irreducible polynomial $f(x) = x^3 - 3x - b$ over $\mathbb{Q}$. Let α be a root of this polynomial, and denote by $\Delta(\alpha)$ its discriminant.

a) If β is any other element of the field $\mathbb{Q}(\alpha)$, show that $\Delta(\beta)$, the discriminant of its minimal polynomial, satisfies

$$\Delta(\beta) = \Delta(\alpha) \cdot a^6(f(c))^2$$

for some $a, c \in \mathbb{Q}$. In particular, $\Delta(\beta)/\Delta(\alpha)$ is a square in $\mathbb{Q}$.

b) If $f(x) = x^3 - 3x - 1$ with root α, and $g(x) = x^3 - 3x - \frac{1}{2}$ with root β, show that $\Delta(\beta)/\Delta(\alpha)$ is not a square, and hence $\mathbb{Q}(\alpha) \neq \mathbb{Q}(\beta)$.

c) Now let $f(x) = x^3 - 3x - 1$ have root α and $g(x) = x^3 - 3x - 13/7$ have root β. In this case show that $\Delta(\beta)/\Delta(\alpha)$ is a square, but nevertheless, $\mathbb{Q}(\alpha) \neq \mathbb{Q}(\beta)$. *Hint*: Use the criterion of a) to reduce to a certain cubic Diophantine equation over $\mathbb{Z}$, and show that it has no solutions.

32 Appendix: Finite Field Extensions

Our purpose in this section is to review the basic facts about finite field extensions which we use in studying geometric constructions. For the most part we will not give proofs, which can be found in any standard algebra textbook. If you are learning this material for the first time, it may make sense to simply accept these results and let their use in geometry provide some motivation before studying their abstract algebraic proofs.

Field Extensions

When one field is contained in another, $F \subseteq E$, we can think of F as a subfield of E, or we can think of E as an extension field of F, depending on the point of view. For example, when we see $\mathbb{Q} \subseteq \mathbb{Q}(\sqrt{2})$, the field of rational numbers $\mathbb{Q}$ is more familiar, so we will think of $\mathbb{Q}(\sqrt{2})$ as a field extension. Or, when a field F is given, and we are looking for an extension field E with some particular property, we think of E as an extension field.

If $F \subseteq E$ are two fields, and if $\alpha_1, \dots, \alpha_n \in E$, then we denote by $F(\alpha_1, \dots, \alpha_n)$ the smallest subfield of E which contains F and the elements $\alpha_1, \dots, \alpha_n$. We call it the subfield of E *generated* by the elements $\alpha_1, \dots, \alpha_n$ over F.

When one field is contained in another, $F \subseteq E$, we can regard the larger field as a vector space over the smaller field. The dimension of this vector space is called the *degree* of the field extension, written $\deg E/F$.

Example 32.0.1 Every element of $\mathbb{Q}(\sqrt{2})$ can be written uniquely in the form $a + b\sqrt{2}$ with $a, b \in \mathbb{Q}$. Thus $\mathbb{Q}(\sqrt{2})$ is a vector space of dimension 2 over $\mathbb{Q}$, so $\deg \mathbb{Q}(\sqrt{2})/\mathbb{Q} = 2$.

A field extension may have infinite degree. For example if $\bar{\mathbb{Q}}$ is the set of all *algebraic numbers* (those complex numbers which are roots of polynomial equations with coefficients in $\mathbb{Q}$), then $\deg \bar{\mathbb{Q}}/\mathbb{Q}$ is (countably) infinite. If $\mathbb{R}$ is the real numbers, $\deg \mathbb{R}/\mathbb{Q}$ is (uncountably) infinite.

Proposition 32.1 *If $F \subseteq E \subseteq G$ are three fields, each contained in the next, then*

$$\deg G/F = (\deg G/E) \cdot (\deg E/F).$$

Proof. If $\alpha_1, \dots, \alpha_n$ is a vector space basis for E over F, and if $\beta_1, \dots, \beta_m$ is a vector space basis for G over E, then $\{\alpha_i\beta_j \mid i = 1, \dots, n;\ j = 1, \dots, m\}$ is a basis for G over F.

Example 32.1.1 Consider $\mathbb{Q} \subseteq \mathbb{Q}(\sqrt{2}) \subseteq \mathbb{Q}(\sqrt{2}, \sqrt{3})$. Adjoining a square root of an element which was not already a square in a field makes a field extension of degree 2. We know $\sqrt{2} \notin \mathbb{Q}$. We can also show $\sqrt{3} \notin \mathbb{Q}(\sqrt{2})$ as follows. If $\sqrt{3} = a + b\sqrt{2}$, then

$$a^2 + 2ab\sqrt{2} + 2b^2 = 3.$$

This equation takes place in $\mathbb{Q}(\sqrt{2})$ where each element is *uniquely* written as $c + d\sqrt{2}$. Hence

$$\begin{aligned} a^2 + 2b^2 &= 3 \\ 2ab &= 0. \end{aligned}$$

Thus $a = 0$ or $b = 0$, so either $a^2 = 3$ or $2b^2 = 3$, both of which are impossible in $\mathbb{Q}$. Hence $\sqrt{3} \notin \mathbb{Q}(\sqrt{2})$, so $\mathbb{Q}(\sqrt{2}, \sqrt{3})$ is an extension of degree 4 of $\mathbb{Q}$. For a basis, we can take the elements $1, \sqrt{2}, \sqrt{3}, \sqrt{6}$. In other words, every element of this last field can be written uniquely as $a + b\sqrt{2} + c\sqrt{3} + d\sqrt{6}$, with $a, b, c, d \in \mathbb{Q}$.

Roots of Polynomials

Let F be a field, and let

$$f(x) = a_0x^n + a_1x^{n-1} + \cdots + a_n$$

be a polynomial with coefficients $a_i \in F$. A *root* of f means an element α in some extension field E of F for which $f(\alpha) = 0$.

Proposition 32.2 *For any field F and any nonconstant polynomial $f(x)$ there always exists an extension field E which contains a root α of $f(x)$. If $f(x)$ is irreducible of degree n, then the degree of the field extension $F(\alpha)/F$ is n. If $f(x)$ is irreducible, and if $\alpha_1 \in E_1$ and $\alpha_2 \in E_2$ are roots of $f(x)$ in two extension fields of F (which may be the same), then there is an isomorphism $\varphi : F(\alpha_1) \to F(\alpha_2)$ of the subfields of E_1, E_2 generated by α_1, α_2, respectively, which sends α_1 to α_2 and leaves every element of F fixed.*

This proposition is usually proved using an abstract field extension E constructed by taking the polynomial ring $F[x]$ and dividing out by the ideal I generated by $f(x)$. Since $f(x)$ is irreducible, I will be a maximal ideal, and so the quotient ring $E = F[x]/I$ is a field. The image $\alpha \in E$ of $x \in F[x]$ then becomes a root of $f(x)$.

In the case which will concern us most, polynomials with rational coefficients, there is another more concrete way of finding roots. Since $\mathbb{Q} \subseteq \mathbb{C}$, the field of complex numbers, and since every polynomial with complex coefficients has a root in $\mathbb{C}$ (this is the so-called "fundamental theorem of algebra", which is not really a theorem of algebra at all, because its proof requires complex analysis), it follows that every polynomial $f(x) \in \mathbb{Q}[x]$ has a root in $\mathbb{C}$ (and sometimes even a root in $\mathbb{R}$). So if we write $\sqrt{2}$, for example, it can be assumed that we are referring to the real number $\sqrt{2} = 1.414\ldots$, and not to some abstract element in some abstract field extension of $\mathbb{Q}$ in which the polynomial $x^2 - 2$ has a root.

Example 32.2.1 The polynomial $x^2 - 2$ is irreducible over $\mathbb{Q}$, since $\sqrt{2}$ is irrational, as was shown by Euclid (X.117). It has a root $\sqrt{2} \in \mathbb{R}$, so we can denote by $\mathbb{Q}(\sqrt{2})$ the subfield of $\mathbb{R}$ consisting of all real numbers of the form $a + b\sqrt{2}$, with

$a, b \in \mathbb{Q}$. This is a field extension of degree 2 of $\mathbb{Q}$, with $1, \sqrt{2}$ as a basis. Now the polynomial $x^2 - 2$ also has another root $-\sqrt{2}$. The field extension generated by this root, $\mathbb{Q}(-\sqrt{2})$ is equal to $\mathbb{Q}(\sqrt{2})$. According to (21.2) there is an isomorphism (in this case an *automorphism*) $\varphi : \mathbb{Q}(\sqrt{2}) \to \mathbb{Q}(\sqrt{2})$ which leaves all rational numbers fixed and sends $\sqrt{2}$ to $-\sqrt{2}$. Indeed, you can verify that the map φ defined by $\varphi(a + b\sqrt{2}) = a - b\sqrt{2}$ is a field automorphism of $\mathbb{Q}(\sqrt{2})$, namely, it satisfies $\varphi(\alpha + \beta) = \varphi(\alpha) + \varphi(\beta)$ and $\varphi(\alpha\beta) = \varphi(\alpha) \cdot \varphi(\beta)$ for all $\alpha, \beta \in \mathbb{Q}(\sqrt{2})$.

Example 32.2.2 For a slightly more complicated example, consider the polynomial $x^3 - 2$ over $\mathbb{Q}$. It has a root $\sqrt[3]{2} = 1.2599\ldots$ in $\mathbb{R}$, so we can consider the field $\mathbb{Q}(\sqrt[3]{2})$. This is an extension field of degree 3 of $\mathbb{Q}$, with basis $1, \sqrt[3]{2}$, and $\sqrt[3]{4}$. But the equation $x^3 - 2$ also has a complex root $\omega\sqrt[3]{2}$, where $\omega = \frac{1}{2}(-1 + \sqrt{-3})$ is a cube root of unity. This gives a different extension field $\mathbb{Q}(\omega\sqrt[3]{2})$ of $\mathbb{Q}$ in which the polynomial $x^3 - 2$ has a root. This is also an extension of degree 3 of $\mathbb{Q}$, with basis $1, \omega\sqrt[3]{2}, \omega^2\sqrt[3]{4}$. According to the proposition, these two field extensions are isomorphic *as field extensions* of $\mathbb{Q}$, even though one is contained in $\mathbb{R}$ and the other is not. The isomorphism $\varphi : \mathbb{Q}(\sqrt[3]{2}) \to \mathbb{Q}(\omega\sqrt[3]{2})$ sends $1 \to 1$, $\sqrt[3]{2} \to \omega\sqrt[3]{2}$, and $\sqrt[3]{4} \to \omega^2\sqrt[3]{4}$.

Example 32.2.3 In some other cases we cannot write roots explicitly, but we can show they exist. Try $x^5 - 5x - 1$ for example. Substituting $x = y + 1$ gives

$$y^5 + 5y^4 + 10y^3 + 10y^2 - 5$$

which is irreducible by Eisenstein's criterion (32.8), so $x^5 - 5x - 1$ is irreducible. A little elementary calculus shows the graph has a relative maximum at $(-1, 3)$ and a relative minimum at $(1, -5)$ and no other relative extreme. Therefore, it has exactly 3 real roots and 2 complex roots, though we cannot write them explicitly. Nevertheless, the roots exist. Say $\alpha_1, \alpha_2, \alpha_3$ are the real roots and α_4, α_5 are the complex roots. In this case the field extensions $\mathbb{Q}(\alpha_1)$, $\mathbb{Q}(\alpha_2)$, $\mathbb{Q}(\alpha_3)$, $\mathbb{Q}(\alpha_4)$, $\mathbb{Q}(\alpha_5)$ are actually five *distinct* subfields of $\mathbb{C}$, three of which are contained in $\mathbb{R}$, and which are all isomorphic as field extensions of $\mathbb{Q}$ (Exercise 32.7).

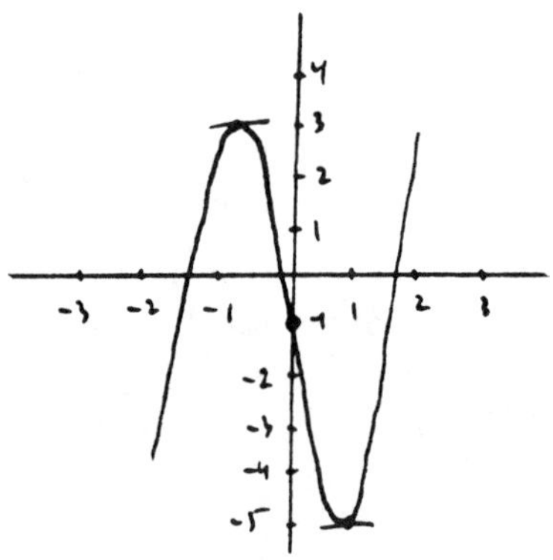

If instead of starting with the field F and the polynomial $f(x)$, we start with a field extension $F \subseteq E$ and an element $\alpha \in E$, and if E is a *finite* field extension, meaning deg E/F is finite, then α will be *algebraic* over F, that is, there exists a polynomial of which α is a root. Indeed, if deg $E/F = n$, the $n+1$ elements $1, \alpha, \alpha^2, \ldots, \alpha^n$ of E must be linearly dependent over F. In other words, there are elements $a_0, \ldots, a_n \in F$, not all 0, such that

$$a_0\alpha^n + a_1\alpha^{n-1} + \cdots + a_n = 0.$$

Thus α is a root of the polynomial

$$f(x) = a_0x^n + a_1x^{n-1} + \cdots + a_n.$$

Once we know there is some polynomial of which α is a root, we can find one of minimal degree, and this is called the *minimal polynomial* of α (usually normalized to have leading coefficient 1). If $g(x)$ is the minimal polynomial of α, then $g(x)$ is necessarily irreducible, so by the previous proposition, $\deg F(\alpha)/F = \deg g(x)$.

Example 32.2.4 Take $\alpha = \frac{1}{2}\sqrt{10-2\sqrt{5}} \in \mathbb{R}$ for example, which is the side of the regular pentagon. Let us find its minimal polynomial. Set

$$x = \frac{1}{2}\sqrt{10-2\sqrt{5}}.$$

Squaring

$$4x^2 = 10 - 2\sqrt{5}.$$

Regrouping

$$2x^2 - 5 = -\sqrt{5}.$$

Squaring again

$$4x^4 - 20x^2 + 25 = 5.$$

So

$$x^4 - 5x^2 + 5 = 0.$$

This procedure produces for us a polynomial of which α is a root. In fact, running the same calculation backward shows that the four roots of this polynomial are $\pm\frac{1}{2}\sqrt{10 \pm 2\sqrt{5}}$. To see if this polynomial is the minimal polynomial of α, we must decide if it is *irreducible*. We first note that none of the four roots is rational, so that the polynomial has no linear factor. If it had quadratic factors, say

$$x^4 - 5x^2 + 5 = (x^2 + ax + b)(x^2 + cx + d)$$

then from the absence of cubic and linear terms we find $c = -a$ and $d = b$. Then

$$(x^2 + ax + b)(x^2 - ax + b) = x^4 + (2b - a^2)x^2 + b^2,$$

so we must have

$$\begin{aligned} 2b - a^2 &= -5 \\ b^2 &= 5 \end{aligned}$$

which is not possible for $a, b \in \mathbb{Q}$.

Therefore $x^4 - 5x^2 + 5$ is irreducible, and it is the minimal polynomial of α.

We could also have used Eisenstein's criterion (32.8) with $p = 5$.

Splitting Fields

Given a field F and a polynomial $f(x) \in F[x]$, one can always find a root α in some extension field E as we have seen above. Then in the polynomial ring $E[x]$ one can factor out a linear factor, so $f(x) = (x - \alpha)g(x)$, where $g(x)$ is a new polynomial, of degree one less than $f(x)$, and having coefficients in the new field E.

Repeating this process, one can find a root of $g(x)$ in some extension field of E, and split off another linear factor. In this way, one eventually finds an extension field E' of F in which the polynomial $f(x)$ splits into linear factors $f(x) = c\prod(x - \alpha_i)$, where $\alpha_1, \ldots, \alpha_n$ are $n =$ deg $f(x)$ roots of $f(x)$ in E', and $c \in F$ is a constant. Note that the α_i need not all be distinct in general, although if f is irreducible and the characteristic of the field F is zero, they will be.

A *splitting field* for the polynomial $f(x)$ over the field F is a field extension E of F, such that over E one can factor $f(x) = c\prod(x - \alpha_i)$ as above, with $c \in F$, $\alpha_i \in E$, and furthermore, $E = F(\alpha_1, \ldots, \alpha_n)$, in other words, the field E is minimal with this property.

Proposition 32.3 *Let a field F and a polynomial $f(x) \in F[x]$ be given. Then there exists a splitting field for $f(x)$ over F. Any two splitting fields are isomorphic as field extensions of F. If $f(x)$ is irreducible, if E is a splitting field for $f(x)$ over F, and if $\alpha_1, \alpha_2 \in E$ are any two roots of $f(x)$ in E, then there exists an automorphism $\varphi : E \to E$, leaving elements of F fixed and sending α_1 to α_2.*

Example 32.3.1 If $f(x) = x^2 - 2$, the two roots are $\pm\sqrt{2}$, and $\mathbb{Q}(\sqrt{2})$ is a splitting field.

Example 32.3.2 If $f(x) = x^3 - 2$, the roots are $\sqrt[3]{2}$, $\omega\sqrt[3]{2}$, $\omega^2\sqrt[3]{2}$, so the splitting field is $\mathbb{Q}(\sqrt[3]{2}, \omega\sqrt[3]{2}, \omega^2\sqrt[3]{2})$. It is easy to see that this field is the same as the field $\mathbb{Q}(\sqrt[3]{2}, \sqrt{-3})$ which has degree 6 over $\mathbb{Q}$. In general, as we see from the method of constructing a splitting field, the splitting field of a polynomial of degree n will have degree $\leq n!$. In this case we have equality: $6 = 3!$.

Example 32.3.3 Let $f(x) = x^4 - 5x^2 + 5$. We saw earlier that the roots of this polynomial are $\pm\frac{1}{2}\sqrt{10 \pm 2\sqrt{5}}$, so the splitting field will be $\mathbb{Q}(\sqrt{10 + 2\sqrt{5}}, \sqrt{10 - 2\sqrt{5}})$. Since

$$\sqrt{10 + 2\sqrt{5}} \cdot \sqrt{10 - 2\sqrt{5}} = \sqrt{80} = 4\sqrt{5},$$

the splitting field is actually equal to $\mathbb{Q}(\sqrt{10 + 2\sqrt{5}})$, because this field already contains $\sqrt{5}$. We saw earlier that this polynomial is irreducible, so the splitting field has degree 4. This is an example where the degree of the splitting field is considerably less than $4! = 24$.

Example 32.3.4 For $x^5 - 5x - 1$ which we looked at earlier, the splitting field has degree $120 = 5!$ over $\mathbb{Q}$ (32.4.4).

Normal Extensions and Galois Groups

For this part, we will restrict our attention to fields of *characteristic* 0 for simplicity, since in the applications to geometry we will be dealing mainly with extension fields of $\mathbb{Q}$.

A finite field extension E/F is called *normal* if it is equal to the splitting field of some polynomial $f(x) \in F[x]$. In this case we denote by G the group of all automorphisms of E which leave elements of F fixed. G is called the *Galois group* of the field extension E/F. Or, if E is the splitting field of a polynomial $f(x) \in F[x]$, G is also called the Galois group of the polynomial $f(x)$. A normal extension E/F is also sometimes called a Galois extension.

Theorem 32.4 (Fundamental theorem of Galois theory) *Let E/F be a normal field extension with Galois group G. Then*

a) *The order of G is equal to the degree of the extension E/F.*

b) *The only elements of E fixed under all elements of G are the elements of F.*

c) *There is a 1–1 inclusion-reversing correspondence between subgroups $H \subseteq G$ and intermediary fields $F \subseteq K \subseteq E$ given as follows: to a subgroup $H \subseteq G$ we associate the field E^H of elements of E left fixed by all elements of H. Conversely, to an intermediary field K we associate the subgroup $H \subseteq G$ of those elements of G which leave all elements of K fixed.*

d) *Under the correspondence just described, the subgroup $H \subseteq G$ will be a normal subgroup of G if and only if the associated field K is a normal extension of F, and in that case the quotient group G/H is isomorphic to the Galois group of K/F.*

Note that if $f(x)$ is an irreducible polynomial of degree n over a field F of characteristic 0, then its roots $\alpha_1, \ldots, \alpha_n$ are all distinct. Let $E = F(\alpha_1, \ldots, \alpha_n)$ be a splitting field, and G its Galois group. For any root, say α_1, the equation $f(\alpha_1) = 0$ must be preserved by any element of G. The coefficients of f are elements of the base field F, so they are fixed by elements of G. Thus the image of α_1 by an element of G must be another root of $f(x)$, namely one of $\alpha_1, \ldots, \alpha_n$. In this way we see that an element of G *permutes* the set $\alpha_1, \ldots, \alpha_n$ of the n roots of $f(x)$. Since E is generated by these roots, the action of an element of G is completely determined by its action on the α_i. In this way we see that G can be regarded as a *subgroup* of the symmetric group S_n of all permutations of the set $\{\alpha_1, \ldots, \alpha_n\}$.

Example 32.4.1 Consider the polynomial $x^2 - 2$ over $\mathbb{Q}$. Its splitting field is $\mathbb{Q}(\sqrt{2})$. This is a normal extension of $\mathbb{Q}$, and the Galois group consists of the identity and the automorphism φ which takes $\sqrt{2}$ to $-\sqrt{2}$. In this case G is equal to the symmetric group S_2, which is isomorphic to $\mathbb{Z}_2 = \mathbb{Z}/2\mathbb{Z}$.

Example 32.4.2 Consider $x^3 - 2$ over $\mathbb{Q}$. First look at the field extension $\mathbb{Q}\sqrt[3]{2}$. Any automorphism of this field would have to take $\sqrt[3]{2}$ to another root of $x^3 - 2$. But the other two roots are complex numbers, not contained in the field $\mathbb{Q}(\sqrt[3]{2})$. Thus the only automorphism of this field is the identity. It follows from (32.4) that $\mathbb{Q}(\sqrt[3]{2})$ is *not* a normal extension of $\mathbb{Q}$. Of course we know it is not the splitting field of the polynomial $x^3 - 2$, but this shows, assuming the result (32.4a), that $\mathbb{Q}(\sqrt[3]{2})$ cannot be the splitting field of any polynomial over $\mathbb{Q}$.

The splitting field of $x^3 - 2$ is $\mathbb{Q}(\sqrt[3]{2}, \sqrt{-3})$, which has degree 6 over $\mathbb{Q}$. Hence the Galois group will be equal to S_3. There are three intermediary fields of degree 3 over $\mathbb{Q}$, namely $\mathbb{Q}(\sqrt[3]{2})$, $\mathbb{Q}(\omega\sqrt[3]{2})$ and $\mathbb{Q}(\omega^2\sqrt[3]{2})$. These correspond to the three subgroups $\{e, (12)\}$, $\{e, (13)\}$, $\{e, (23)\}$ of order 2 of S_3. The field $\mathbb{Q}(\sqrt{-3})$, which is a normal extension of $\mathbb{Q}$, corresponds to the normal subgroup $A_3 = \{e, (123), (132)\}$ of S_3.

Example 32.4.3 Consider again the polynomial $x^4 - 5x^2 + 5$ over $\mathbb{Q}$. We saw it is irreducible with roots

$$\begin{aligned}\alpha_1 &= \tfrac{1}{2}\sqrt{10+2\sqrt{5}}\\ \alpha_2 &= \tfrac{1}{2}\sqrt{10-2\sqrt{5}}\\ \alpha_3 &= -\tfrac{1}{2}\sqrt{10+2\sqrt{5}}\\ \alpha_4 &= -\tfrac{1}{2}\sqrt{10-2\sqrt{5}}.\end{aligned}$$

We saw that the splitting field is $\mathbb{Q}(\sqrt{10+2\sqrt{5}})$, which has degree 4 over $\mathbb{Q}$. Thus the Galois group G will be a subgroup of order 4 of the symmetric group S_4. Which subgroup of order 4 is it?

To investigate this question, we must describe explicitly the automorphisms of the field $E = \mathbb{Q}(\sqrt{10+2\sqrt{5}})$. According to (32.3), there exists an element $\sigma \in G$ such that $\sigma(\alpha_1) = \alpha_2$. Then $\sigma(\alpha_1^2) = \alpha_2^2$, and since elements of $\mathbb{Q}$ are fixed, it follows that $\sigma(\sqrt{5}) = -\sqrt{5}$. Now we make use of the identity $\alpha_1\alpha_2 = \sqrt{5}$. Applying σ we find $\sigma(\alpha_1)\cdot\sigma(\alpha_2) = \sigma(\sqrt{5}) = -\sqrt{5}$. But $\sigma(\alpha_1) = \alpha_2$ by choice of σ, so we find $\alpha_2\cdot\sigma(\alpha_2) = -\sqrt{5}$. From the equation $\alpha_1\alpha_2 = \sqrt{5}$ it follows that $\sigma(\alpha_2) = -\alpha_1 = \alpha_3$. Finally, $\sigma(\alpha_3) = \sigma(-\alpha_1) = -\alpha_2 = \alpha_4$. Thus $\sigma = (1234)$. So G is the subgroup of S_4 generated by (1234), which is a *cyclic* group of order 4.

Perhaps from this example it is already clear that the determination of the splitting field of a polynomial and its Galois group is not a straightforward matter. It depends in each case on using some special information about that particular example, which may not be an easy task.

Example 32.4.4 Coming back to the example $x^5 - 5x - 1$ we studied earlier, let us see if we can determine the degree of its splitting field and its Galois group. The polynomial is irreducible, so if we adjoin one root α, then $\mathbb{Q}(\alpha)$ will be an extension of degree 5 of $\mathbb{Q}$. Let E be the splitting field. Then $\mathbb{Q} \subseteq \mathbb{Q}(\alpha) \subseteq E$. Therefore deg $E/\mathbb{Q}$ is a multiple of 5. So the Galois group G will be a subgroup of S_5 whose order is a multiple of 5. Hence (from abstract group theory), G contains an element of order 5. In S_5, the only elements of order 5 are the 5-cycles. Thus G contains a 5-cycle. On the other hand, we saw earlier that $f(x)$ has 3 real roots, and hence two conjugate complex roots. Complex conjugation in $\mathbb{C}$ then induces an automorphism of E which fixes the three real roots, and permutes the two complex roots. This is an element of G whose image in S_5 is therefore a transposition (a 2-cycle). Now one can show (again using abstract group theory) that a subgroup of S_5 which contains a 5-cycle and a transposition must be the full symmetric group. Hence $G = S_5$, and the degree of $E/\mathbb{Q}$ is $5! = 120$.

Note the indirect nature of the reasoning, and how we were able to prove that $G = S_5$ without having any explicit representation of the roots of $f(x)$.

Reduction mod p

A useful technique for obtaining information about the Galois group of a polynomial with integer coefficients is the following.

Proposition 32.5 *Let $f(x)$ be an irreducible monic polynomial with integer coefficients. Fix a prime p, and assume that the polynomial $\bar{f}(x)$ with coefficients reduced $(\text{mod } p)$ has distinct roots in a splitting field $\bar{E}$ for $\bar{f}(x)$ over the prime field $\mathbb{F}_p$. Then there is a 1–1 correspondence between the roots of $f(x)$ in its splitting field*

E over $\mathbb{Q}$ and the roots of $\bar{f}(x)$ in $\bar{E}$, such that the Galois group of $\bar{f}$ over $\mathbb{F}_p$, considered as a group of permutations of the roots, corresponds to a subgroup of the Galois group of f over $\mathbb{Q}$.

Example 32.5.1 Consider the polynomial $f(x) = x^4 - x^3 - 5x^2 + 1$. It has no roots in $\mathbb{Q}$, because a root would have to be an integer dividing 1, and neither $+1$ nor -1 is a root. Reducing (mod 2) we obtain $\bar{f}(x) = x^4 + x^3 + x^2 + 1 = (x+1)(x^3+x+1)$. Note that $x^3 + x + 1$ is irreducible because it has no roots. It follows that $f(x)$ cannot be a product of two quadratic polynomials, because then $\bar{f}(x)$ would be also. Thus $f(x)$ is irreducible, and so 4 divides the order of the Galois group.

Now let us apply the proposition. The polynomial $\bar{f}(x)$ has distinct roots (since the derivative of $x^3 + x + 1$ is non-zero), and its Galois group is cyclic of order 3. We conclude from the proposition that the Galois group of $f(x)$ contains a 3-cycle. Hence its order is divisible by 3, and we see that the Galois group of $f(x)$ must have order 12 or 24.

Roots of Unity

We will examine in some more detail the Galois groups of the polynomials $x^n - 1$ over $\mathbb{Q}$, since these are closely related with the question of constructing regular polygons of n sides.

If $\alpha = 2\pi/n$ is an angle equal to one n-th of a complete rotation, then the complex number

$$\zeta = e^{2\pi i/n} = \cos\alpha + i\sin\alpha$$

is an n-th root of unity. In fact, $1, \zeta, \zeta^2, \ldots, \zeta^{n-1}$ are the n roots of the equation $x^n - 1$ in the field $\mathbb{C}$ of complex numbers. Among these we will designate as *primitive* n-th roots of unity those which are not a d-th root of unity for any $d \mid n$.

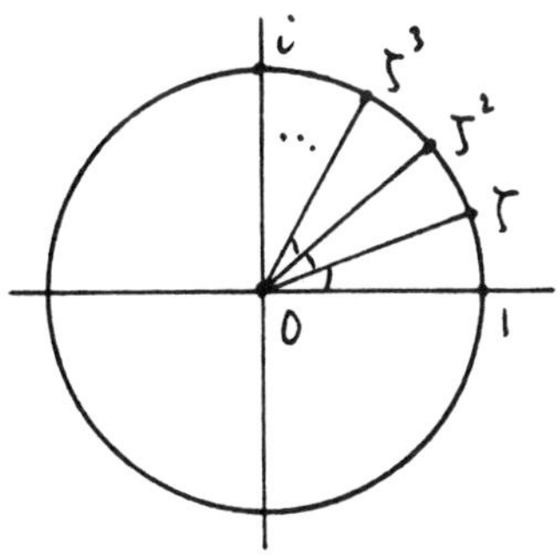

In other words, the primitive n-th roots of unity are those powers ζ^r of ζ for which $(r, n) = 1$. The product

$$\Phi_n(x) = \prod_{(r,n)=1} (x - \zeta^r)$$

is called the n-th *cyclotomic polynomial*. From this description, it is not obvious where its coefficients lie, but from the expression

$$\Phi_n(x) = (x^n - 1)/(\prod_{\substack{d|n \\ d\neq n}} \Phi_d(x))$$

it follows inductively that Φ_n has coefficients in $\mathbb{Q}$, and in fact (using Gauss' lemma), coefficients in $\mathbb{Z}$.

The degree of the polynomial $\Phi_n(x)$ is the Euler φ-function: $\varphi(n) = \#\{1 \leq r < n \mid (r,n) = 1\}$.

Proposition 32.6 *For any n, the cyclotomic polynomial $\Phi_n(x)$ is irreducible over $\mathbb{Q}$.*

Example 32.6.1 The first few cyclotomic polynomials are

$$\begin{aligned}
\Phi_1 &= x - 1 \\
\Phi_2 &= x + 1 \\
\Phi_3 &= x^2 + x + 1 \\
\Phi_4 &= x^2 + 1 \\
\Phi_5 &= x^4 + x^3 + x^2 + x + 1 \\
\Phi_6 &= x^2 - x + 1 \\
\Phi_7 &= x^6 + x^5 + x^4 + x^3 + x^2 + x + 1 \\
\Phi_8 &= x^4 + 1.
\end{aligned}$$

(However, do not conclude from this list that the coefficients are always ± 1: there may be other integer coefficients.)

Let $\zeta = e^{2\pi i/n}$ again. Since Φ_n is irreducible, it will be the minimal polynomial of ζ, and we find degree of $\mathbb{Q}(\zeta)/\mathbb{Q} = \varphi(n)$, the Euler φ-function. Furthermore, since the other roots of Φ_n are all powers of ζ, the splitting field of Φ_n is this same field $\mathbb{Q}(\zeta)$. From (32.3) we find that for each $1 \leq r < n$ with $(r,n) = 1$, there is an element φ_r of the Galois group G for which $\varphi_r(\zeta) = \zeta^r$. If we compose two of these, $\varphi_s\varphi_r(\zeta) = \varphi_s(\zeta^r) = \zeta^{rs}$. Reducing $rs \pmod{n}$ to t with $1 \leq t < n$, it follows that $\varphi_s\varphi_r = \varphi_t$. In this way we see that the Galois group G is isomorphic to the group of integers, $1 \leq r < n$, relatively prime to n, under multiplication (mod n). This group is usually denoted $\mathbb{Z}_n^*$. The field $\mathbb{Q}(\zeta)$ is called the n-th *cyclotomic extension* of $\mathbb{Q}$, or the field of n-th roots of unity. Summarizing we have

Proposition 32.7 *The cyclotomic field of n-th roots of unity $\mathbb{Q}(\zeta)$ is generated by $\zeta = e^{2\pi i/n}$. It has degree $\varphi(n)$ over $\mathbb{Q}$, and its Galois group is isomorphic to $\mathbb{Z}_n^*$. In particular, if n is a prime number p, then $\mathbb{Q}(\zeta)$ has degree $p-1$, and its Galois group is $\mathbb{Z}_p^*$, which is a cyclic group of order $p-1$.*

Example 32.7.1 $n = 3$. $\Phi_3 = x^2 + x + 1$. Its roots are $\omega = \frac{1}{2}(-1 + \sqrt{-3})$ and $\omega^2 = \frac{1}{2}(-1 - \sqrt{-3})$. The cyclotomic field $\mathbb{Q}(\omega) = \mathbb{Q}(\sqrt{-3})$, with Galois group isomorphic to $\mathbb{Z}_2$.

Example 32.7.2 $n = 4$. $\Phi_4 = x^2 + 1$. Roots are $\pm i$. The field is $\mathbb{Q}(i)$ with Galois group $\mathbb{Z}_2$.

Example 32.7.3 $n = 5$. $\Phi_5 = x^4 + x^3 + x^2 + x + 1$. The cyclotomic field is $\mathbb{Q}(\zeta)$ where

$$\begin{aligned}
\zeta = e^{2\pi i/5} &= \cos 2\pi/5 + i \sin 2\pi/5 \\
&= \tfrac{1}{4}(\sqrt{5} - 1) + i\tfrac{1}{4}\sqrt{10 + 2\sqrt{5}}.
\end{aligned}$$

The Galois group is isomorphic to $\mathbb{Z}_4$.

Example 32.7.4 $n = 8$. $\Phi_8 = x^4 + 1$. Its roots are $\zeta, \zeta^3, \zeta^5, \zeta^7$ where

$$\zeta = e^{2\pi i/8} = \frac{\sqrt{2}}{2} + i\frac{\sqrt{2}}{2}.$$

The Galois group is $\mathbb{Z}_8^*$ which is $\{1, 2, 5, 7\}$ under multiplication (mod 8). This group is isomorphic to a Klein Viergruppe $\mathbb{Z}_2 \times \mathbb{Z}_2$.

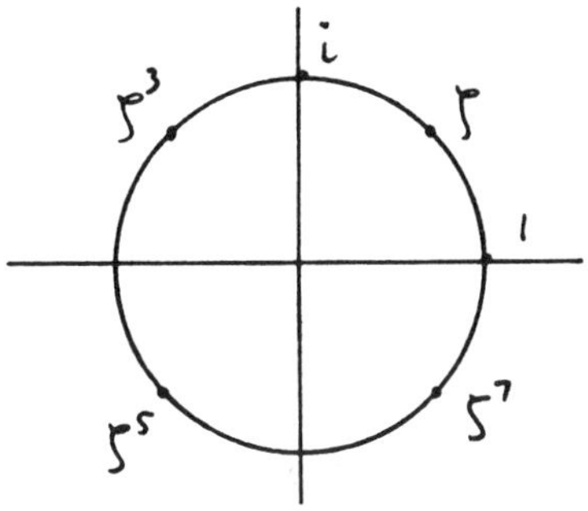

For reference we include the following useful criterion for a polynomial to be irreducible.

Proposition 32.8 (Eisenstein) *Let*

$$f(x) = x^n + a_1x^{n-1} + \cdots + a_n$$

be a monic polynomial with integer coefficients $a_i \in \mathbb{Z}$. Suppose for some prime number p, that p divides all the a_i, and p^2 does not divide a_n. Then $f(x)$ is irreducible over $\mathbb{Q}$.

Exercises

32.1 For each of the following real numbers, find the minimal polynomial over $\mathbb{Q}$ (be sure to prove it is irreducible), find the degree of the splitting field of the polynomial, and determine the Galois group (up to isomorphism). *Hint*: Expect to find groups of orders 2, 4 and 8.

a) $\sqrt{2+\sqrt{2}}$

b) $\sqrt{3+\sqrt{2}}$

c) $\sqrt{3+2\sqrt{2}}$.

In the following exercises, we will investigate the *Galois group of a construction problem*, defined as follows. For a construction problem in the real Euclidean plane $\mathbb{R}^2$, starting with data defined over $\mathbb{Q}$, the construction will create various points. Let F be the field $F = \mathbb{Q}(\beta_1, \ldots, \beta_m)$ obtained by adjoining the coordinates of all the points constructed. Let E be the smallest normal field extension containing F, which can be obtained as follows. Let $f_i(x)$ be the minimal polynomial of β_i, $i = 1, \ldots, m$, let $g(x) = \prod_{i=1}^m f_i(x)$, and take E to be the splitting field of $g(x)$. The Galois group G of $E/\mathbb{Q}$ we will then call the Galois group of the construction problem.

32.2 Given the unit segment A, B, where $A = (0,0)$, $B = (1,0)$, construct an equilateral triangle with side AB. Determine the associated field F, the normal field E, and the Galois group G of the construction problem. *Answer*: $F = E = \mathbb{Q}(\sqrt{3})$. $G \cong \mathbb{Z}_2$.

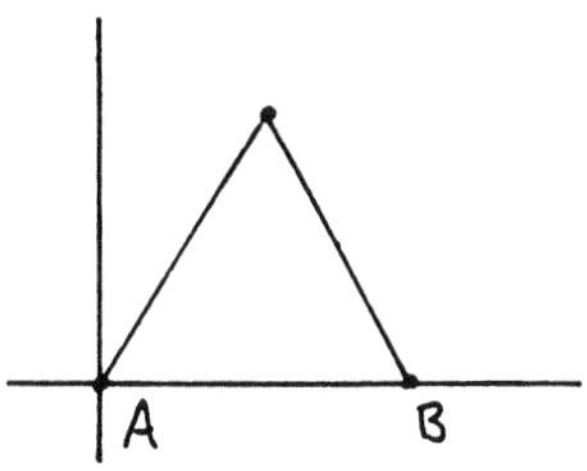

32.3 Construct a square with one corner at $(0,0)$, with sides along the x- and y-axes, and with area equal to the area of the triangle in # 2. Find F, E, G as above.

32.4 Construct a regular pentagon inscribed in the unit circle, having one vertex at $(1,0)$. Find F, E, G as above.

32.5 Construct an equilateral triangle with one side on the x-axis, and with area equal to twice the area of the triangle in # 2 above. Find F, E, G as above.

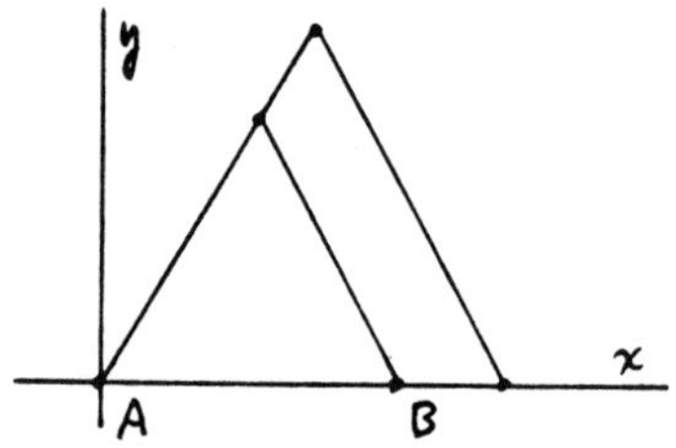

32.6 a) Compute Φ_9, the ninth cyclotomic polynomial, and find the Galois group of the cyclotomic field $\mathbb{Q}(\zeta)$ where $\zeta = e^{2\pi i/9}$.

b) Let $\alpha = \frac{1}{2}\cos 40°$. Find the minimal polynomial of α. Show that $\mathbb{Q}(\alpha)$ is a normal field extension of $\mathbb{Q}$ of degree 3 and that $\mathbb{Q}(\alpha) \subseteq \mathbb{Q}(\zeta)$.

c) Show that $\zeta + \zeta^4 + \zeta^7 = \sqrt{-3}$, so that $\sqrt{-3} \in \mathbb{Q}(\zeta)$.

d) Finally, show that $\mathbb{Q}(\zeta) = \mathbb{Q}(\alpha, \sqrt{-3})$.

32.7 Using facts about the Galois group from (32.4.4), show that the five fields $\mathbb{Q}(\alpha_i)$ mentioned in (32.2.3) are all distinct.

32.8 Let $\mathbb{Q}(\zeta)$ be the cyclotomic field of 5-th roots of unity.

a) Show that $\sqrt{5} \in \mathbb{Q}(\zeta)$.

b) Show that $\mathbb{Q}(\zeta) = \mathbb{Q}(\delta)$, where $\delta = \frac{i}{2}\sqrt{10 + 2\sqrt{5}}$.

c) What is the minimal polynomial of δ?

32.9 For $p = 5, 7, 11, 13, 17$, show explicitly that $\mathbb{Z}_p^*$ is cyclic by finding a suitable element and proving that it generates the group.

As lines, so Loves oblique may well
  Themselves in every Angle greet:
But ours so truly Parallel,
  Though infinite can never meet.

– from *The Definition of Love*
by Andrew Marvell (1621–1678)

CHAPTER 7

Non-Euclidean Geometry

Certainly one of the greatest mathematical discoveries of the 19th century was that of non-Euclidean geometry: seen but not revealed by Gauss, and developed in all its glory by Bolyai and Lobachevsky. The purpose of this chapter is to give an account of this theory, but we do not always follow the historical development. Rather with hindsight we use those methods which seem to shed the most light on the subject. For example, continuity arguments have been replaced by a more axiomatic treatment.

There are actually three different approaches presented here. One begins with the theorem of Saccheri–Legendre, using Archimedes' axiom, in §34. The second is the analytic model of a non-Euclidean geometry given in §39. Third is Hilbert's axiomatic approach based on the axiom of limiting parallels (L) in §40.

We start with a historical introduction to the problem of the parallels, and the various futile attempts to prove Euclid's fifth postulate from the other axioms. Then we begin to explore this strange new world where the sum of the angles of a triangle can be less than two right angles. The defect of this angle sum provides a measure of area which we exploit in §36.

To explain the Poincaré model of a non-Euclidean geometry, we need the Euclidean technique of circular inversion. This is developed in §37. It is a technique with many applications in Euclidean geometry. In particular we make a digression in §38 to show how it can provide a solution to the classical problem of Apollonius, to construct a circle tangent to three given circles.

In §40 we present a development of non-Euclidean geometry based on the axiom of existence of limiting parallel rays, proposed by Hilbert. This allows us to avoid the appeal to continuity invoked by the founders of the subject, and free ourselves from dependence on the real numbers. Then we give Hilbert's brilliant construction of an abstract field from the set of common ends of limiting parallel rays. This allows us to characterize hyperbolic planes by their associated fields without using the techniques of projective geometry.

We follow the principle, established earlier in this book, of systematically avoiding the use of real numbers. There is a slight cost, in that some familiar results will look different here, but I believe this approach is justified by keeping the intrinsic geometry in the foreground. For example, instead of taking logarithms to define a distance function, we use a multiplicative distance function μ. Then Bolyai's famous formula for the angle of parallelism α of a line segment PQ takes the form $\tan \alpha/2 = \mu(PQ)^{-1}$ (39.12) and (41.9). The arbitrary constant k which appears in some books, coming from the choice of a base for the logarithms in the distance function, is absent: in our approach, any two hyperbolic planes over the same field are isomorphic. Also, the hyperbolic trigonometric functions sinh, cosh, tanh do not appear in our formulae of hyperbolic trigonometry (42.2) and (42.3). As a result of this approach, the solution of any problem we consider can be found constructively, by ruler and compass, or, equivalently, by solving linear and quadratic equations in the coefficient field.

33 History of the Parallel Postulate

To set the background for the discovery of non-Euclidean geometry, a kind of geometry where there may be many lines through a point, parallel to a given line, let us trace the history of attitudes toward the parallel postulate.

We have seen already that Euclid's fifth postulate, which we refer to as the parallel postulate, was of a much more sophisticated nature than the other postulates and axioms. Euclid seems to have recognized this himself, since he postponed using it as long as possible, and was careful to develop the standard congruence theorems for triangles without the parallel postulate.

Euclid was criticized for making this a postulate and not a theorem. Proclus (410–485), who represented the school of Plato in fifth-century Athens, has left an extensive commentary on the first book of Euclid's *Elements*. His opinion on the fifth postulate is unambiguous:

"This ought to be struck from the postulates altogether. For it is a theorem—one that invites many questions, which Ptolemy proposed to resolve in one of his books—and requires for its demonstration a number of definitions as well as theorems" [Proclus (1970), p. 150].

In his commentary on (I.29), Proclus gives Ptolemy's proof and points out its flaws, and then proceeds to give his own proof of the fifth postulate. First, he says, we must accept an axiom which was used earlier by Aristotle:

Aristotle's Axiom: "If from a single point two straight lines making an angle are produced indefinitely, the interval between them... will exceed any finite magnitude" [Proclus (1970), p. 291]. In other words, given any angle BAC, and given a segment DE, there exists a point F on the ray AB so that the perpendicular FG from F to the line AC will be greater than DE.

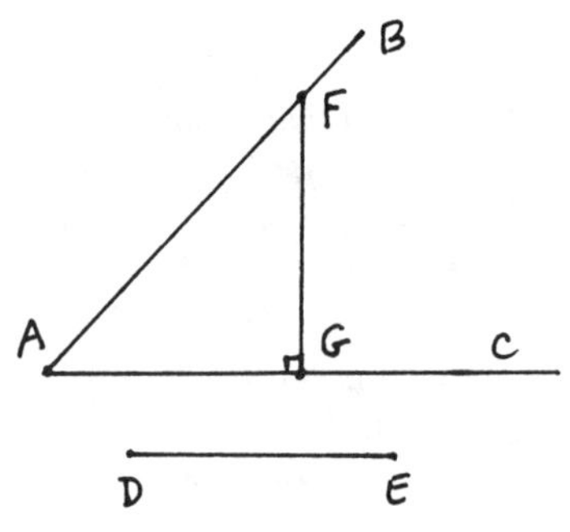

Then Proclus proposes to prove the

Lemma of Proclus *If a straight line cuts one of two parallel lines, it cuts the other also.*

His proof goes like this. If AB and CD are two parallel lines, and if EF cuts AB, with F on the side toward CD, then we apply Aristotle's axiom to the angle BEF. As we extend the ray EF indefinitely, its interval from the line AB will exceed the distance between the parallel lines, and so it must cut the line CD.

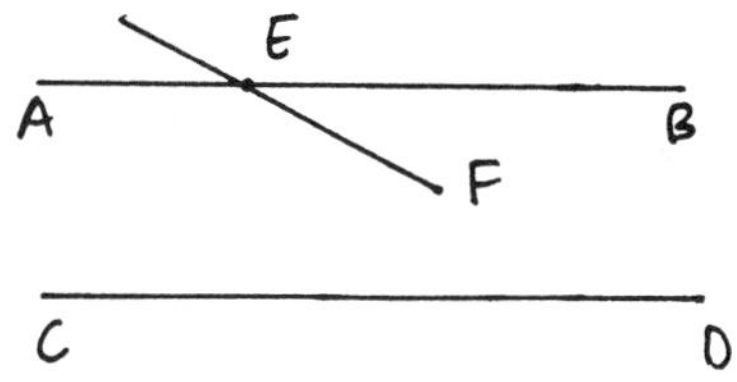

From this lemma (which is essentially the same as what we now call Playfair's axiom), Proclus easily proves the parallel postulate.

Proclus' reasoning was apparently accepted for some time, since it is reproduced without critical comment by F. Commandino in his edition of Euclid (1575).

We can observe two things about the argument of Proclus. First of all, he assumes another axiom (the axiom of Aristotle) in the course of his proof. This is not uncommon in various attempted proofs of the parallel postulate. Often one ends up assuming (consciously or unconsciously) something else which turns out to be equivalent to the parallel postulate. In this particular case, it is not so bad: we will see that Aristotle's axiom is a consequence of Archimedes axiom, and does not imply the parallel postulate by itself (35.5).

The more serious flaw in Proclus' argument is that he speaks of "the distance between the parallel lines" as if all the points of one line were at the same distance from the other line. Since the definition of parallel lines is lines in the same plane which do not meet, however far extended, it does not follow from the definition that they are at a constant distance from each other. In fact, this assumption of constant distance is enough to prove the parallel postulate (in the presence of Aristotle's axiom) as Proclus shows. Thus, in view of (I.34) it is equivalent to the parallel postulate.

This confusion of the definition of parallel lines as lines which do not meet with the common-sense notion of parallel lines as equidistant from each other (like railroad tracks) has persisted. For example, in the edition of Euclid's first six books by J. Peletier (1557), definition 35 says "Parallels, or equidistant straight lines, are those which being in the same plane, and extended arbitrarily in either direction, do not meet". However, Peletier follows Euclid's proofs in Book I, and does not make use of the equidistant property.

A more striking example is the very popular edition of the *Elements of Geometry* by the Jesuit Andrea Tacquet (1612–1660), first published in 1654 and reprinted many times over the next hundred and fifty years. Tacquet's book is not a strict translation of Euclid, but an arrangement, to make the study of geometry easier for beginners. Though he preserves the numbering of Euclid's propositions, he takes great liberties with their proofs. For example, he says there is no point in proving (I.16), because it is a special case of (I.32)! He apparently does not care about the fact that Euclid's proof of (I.16) is independent of the parallel postulate, while (I.32) depends on it.

Tacquet says since there are various species of lines (such as the hyperbola and a straight line) which approach each other indefinitely but never meet, so Euclid's definition of parallel lines does not satisfactorily reflect the nature of parallels.

He takes as his definition that two lines are parallel if the points of one are all equidistant from the other, as measured by perpendiculars from points on the first line to the second line.

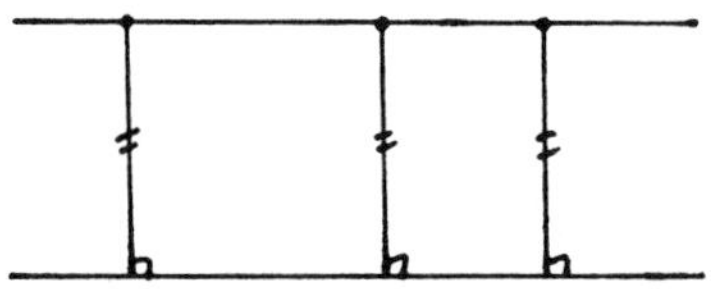

There is no harm, of course, in using any definition you like of parallel lines, though this one will clearly place a great burden on any proof of existence of parallel lines. Tacquet misses this subtlety however, because in the next sentence he says you can generate parallel lines as the locus of points at a fixed distance from a given line as the perpendicular moves along. Here he is implicitly using another axiom, which was in fact stated explicitly and used earlier by Christoph Clavius (1537–1612) as a substitute for Euclid's parallel postulate:

Clavius' Axiom: The set of points equidistant from a given straight line, on one side of it, form a straight line.

This axiom, as one can easily show, is equivalent to the parallel postulate which Tacquet was trying to avoid.

The French mathematician Alexis Claude Clairaut (1713–1765) wrote an *Eléments de Géométrie* (first published in 1741) in which he tried to make geometry more accessible for students. He complained about the usual method of teaching the elements, in which "one always starts with a great number of definitions, postulates, axioms, and first principles, which appear to offer nothing but dryness to the reader". He thought that Euclid's careful reasoning was merely to satisfy a fussy audience: "That Euclid went to the trouble to prove that two circles which cut each other do not have the same center; that a triangle contained inside another triangle has the sum of its sides less than that of the triangle in which it is enclosed—one should not be surprised. For this geometer had to convince the obstinate sophists who glorified in finding fault with the most evident truths: so it was necessary that geometry, like logic, make use of proper reasoning, to close the mouths of its critics."

Clairaut's purpose is to introduce the concepts of geometry simply and naturally in the context of practical questions such as measurement of terrain. So he talks of straight lines to measure the distance between points, and how to construct perpendicular lines. Then he says, what is more easy than to use this method to construct a rectangle? One has only to take a segment AB, and at its endpoints raise perpendiculars AC and BD of equal length, and then join CD. From here he develops the theory of parallels. The hidden assumption is that his construction makes a rectangle. So we will call this

Clairaut's Axiom: Given a segment AB, let AC and BD be equal segments, perpendicular to AB. Then the angles at C and D are right angles, i.e. $ABCD$ is a rectangle.

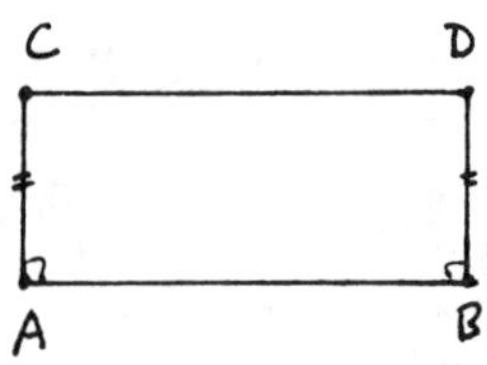

Robert Simson, M.D. (1687–1768), professor of mathematics in the University of Glasgow, made an important edition of Euclid's elements, in Latin and in English, first published in 1756, which went through some thirty successive editions. Simson

railed against the errors introduced by earlier editors, and wished to "restore the principal Books of the Elements to their original Accuracy... This I have endeavored to do by taking away the inaccurate and false Reasonings which unskilful Editors have put into the place of some of the genuine Demonstrations of Euclid, who has ever been justly celebrated as the most accurate of Geometers, and by restoring to him those Things which Theon or others have suppressed, and which have these many ages been buried in Oblivion" [Simson, Preface]. Simson's restorations were not so much based on textual studies as on his faith that anything mathematically true and accurate must have been Euclid's, while anything incorrect or not rigorous must have been inserted by "some unskilful editor". About the parallel postulate, he says "it seems not to be properly placed among the Axioms, as, indeed, it is not self-evident; but it may be demonstrated thus". Simson then introduces an axiom,

Simson's Axiom: "A straight line cannot first come nearer to another straight line, and then go further from it, before it cuts it; and, in like manner, a straight line cannot go further from another straight line, and then come nearer to it; nor can a straight line keep the same distance from another straight line, and then come nearer to it, or go further from it" [Simson, p. 295].

From this axiom, Simson proves (correctly) five propositions, implicitly making use of Archimedes axiom, of which the last is Euclid's parallel postulate.

So here we have a clear case of an author substituting another axiom, which seems more natural to him, and then using it to prove the parallel postulate.

John Playfair (1748–1819), professor of natural philosophy, formerly of mathematics, in the University of Edinburgh, published a new edition of the first six books of Euclid's *Elements* which first appeared in 1795. He says Dr. Simson has done a fine job of restoring Euclid's *Elements*, and that his purpose in presenting a new edition is to give them the form which may "render them most useful". He says "A new axiom is also introduced in the room of the 12th [which we call the fifth postulate], for the purpose of demonstrating more easily some of the properties of parallel lines" [Playfair, Preface]. This is

Playfair's Axiom: Two straight lines which intersect one another, cannot be both parallel to the same straight line.

In his notes to (I.29), Playfair has an interesting discussion of the problem of parallels. He agrees with Proclus that Euclid's postulate should be proved, and not taken as an axiom. He then reviews the three methods by which geometers "have attempted to remove this blemish from the *Elements*...

1) by a new definition of parallel lines

2) by introducing a new Axiom concerning parallel lines, more obvious than Euclid's

3) by reasoning merely from the definition of parallels, and the properties of lines already demonstrated, without the assumption of any new Axiom" [Playfair, p. 300].

In the next section we will see another approach, still having the same goal of proving the parallel postulate, which is to suppose that the parallel postulate were false, and then to seek a contradiction. It was this approach which eventually led to the discovery of non-Euclidean geometry.

Exercises

Throughout these exercises, we assume the axioms of a Hilbert plane.

33.1 Show that the lemma of Proclus is equivalent to Playfair's axiom (P).

33.2 Consider the following special case of Euclid's parallel postulate, which we will call the **Right Triangle Axiom**: Given a right angle ABD and an acute angle $\alpha = CAB$ on the same side of the line AB, the ray AC when extended will meet the ray BD extended.

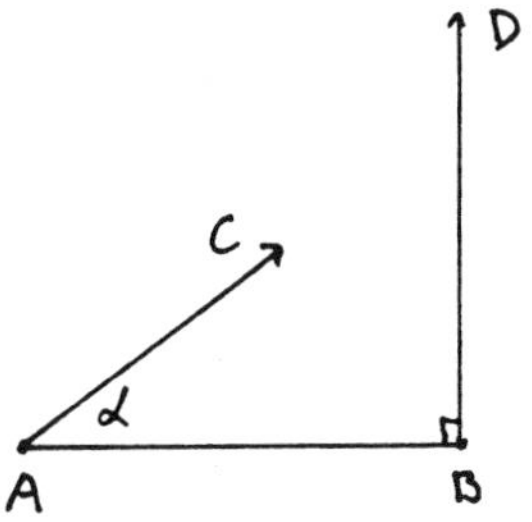

Show that the right triangle axiom is equivalent to (P).

33.3 Show directly that the right triangle axiom implies the special case of Euclid's parallel postulate which says, given acute angles $\alpha = CAB$ and $\beta = ABD$ on the same side of the line AB, the rays AC and BD will meet.

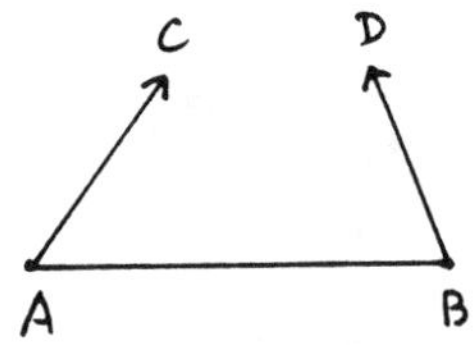

33.4 Discuss the following "proof" of the right triangle axiom due to Franceschini (1756–1840): Given A, B, C, D as in (Exercise 33.2), drop a perpendicular CE from C to the line AB. Since α is an acute angle, E will lie between A and B. Now take a point F further out on the ray AC. Drop a perpendicular FG from F to AB. Then G is between E and B. As the point F moves out the ray AC without bound, so the point G must move along the ray AE without bound, and thus it must eventually reach B. Then F will be the intersection of AC and BD.

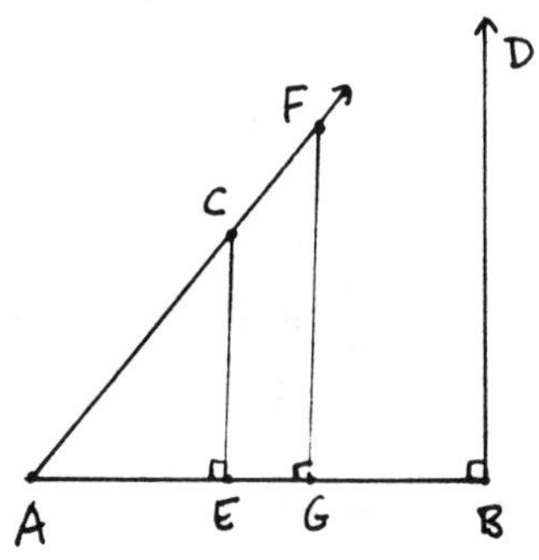

33.5 John Wallis (1616–1703) gave a proof of the parallel postulate based on the principle that to every figure there is always a similar figure of arbitrary size. To be precise, we state

Wallis' Axiom: Given a triangle ABC and given a line segment DE, there exists a *similar* triangle $A'B'C'$ (that is, a triangle with the same angles as the triangle ABC) having side $A'B' \geq DE$.

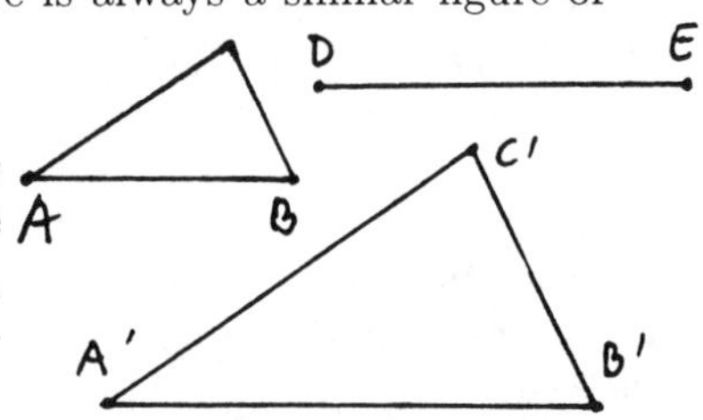

a) Show that Wallis' axiom implies (P).

b) In the non-Archimedean geometry of (18.4.3) show that there are similar triangles of different sizes, but that Wallis' axiom fails. (We will see later that in an Archimedean non-Euclidean geometry, the only similar triangles are congruent triangles (Exercise 34.1).)

33.6 In a Hilbert plane, show that opposite sides of a *rectangle* (i.e. a figure with four right angles) are equal. *Hint*: Bisect one side AB at E, erect a perpendicular to AB at E and use the accompanying diagram. Your goal: to show $AB \cong CD$.

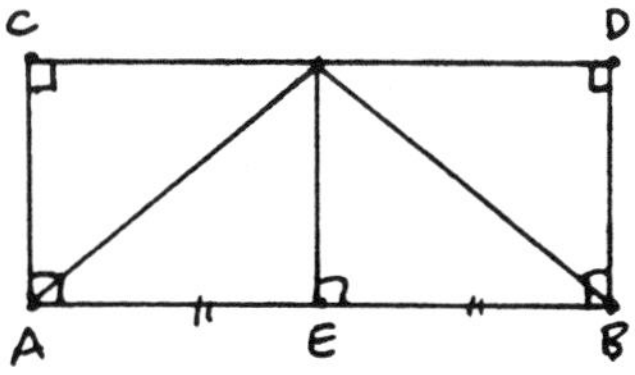

33.7 In this exercise we explore the consequences of Clavius' axiom.

a) Let l be a line, and let m be a set of equidistant points, which by Clavius' axiom is a line. Thus for points A, B, C in m, the perpendiculars AA', BB', CC' to l are all equal. Show that the angles at A, B, C are also right angles.

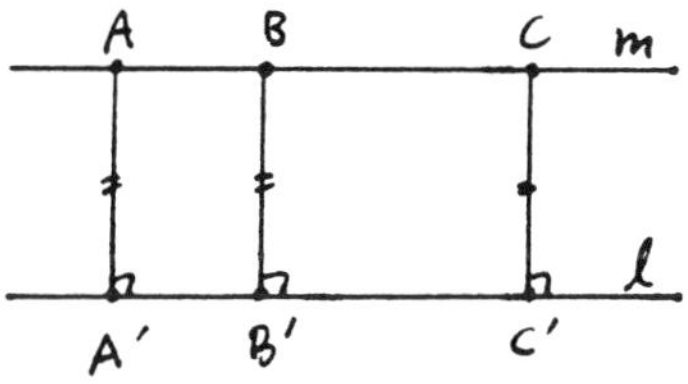

b) Let ABC be a right triangle. Extend AB to D so that $AB \cong BD$, and drop the perpendicular DE to AC. Assuming Clavius' axiom, show that $DE \cong 2BC$.

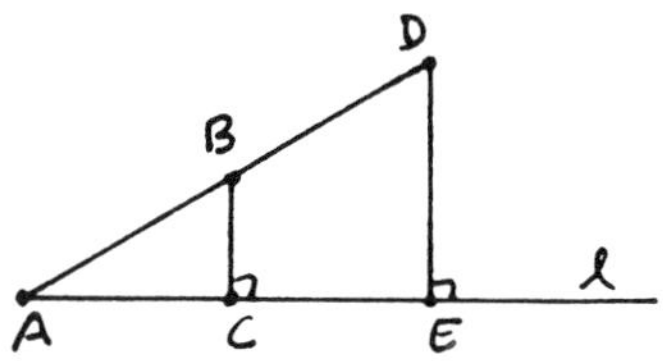

c) Show that Clavius' axiom, together with Archimedes axiom (A), implies (P).

d) Show that Clavius' axiom holds in the non-Archimedean plane of (18.4.3) even though (P) does not.

33.8 a) Show that Aristotle's axiom holds in the Cartesian plane over a field F, even if F is not Archimedean.

b) Show that Aristotle's axiom fails in the plane of (18.4.3).

33.9 Show that Clairaut's axiom is equivalent to Clavius' axiom.

33.10 Show that Simson's axiom is equivalent to Clavius' axiom.

33.11 Farkas Bolyai, the father of Janos, proposed the following

Bolyai's Axiom: For any three non-collinear points A, B, C there exists a circle containing them.

a) Use the following construction to show that Bolyai's axiom implies Euclid's parallel postulate. Given two lines l, m and a transversal AB, assume that the angles α, β on one side add up to less than two right angles. Let C be the midpoint of AB. From C drop perpendiculars to l and m, and extend each an equal distance on the far side to obtain D and E. Show that C, D, E are not collinear, and then use Bolyai's axiom to prove that l and m must meet.

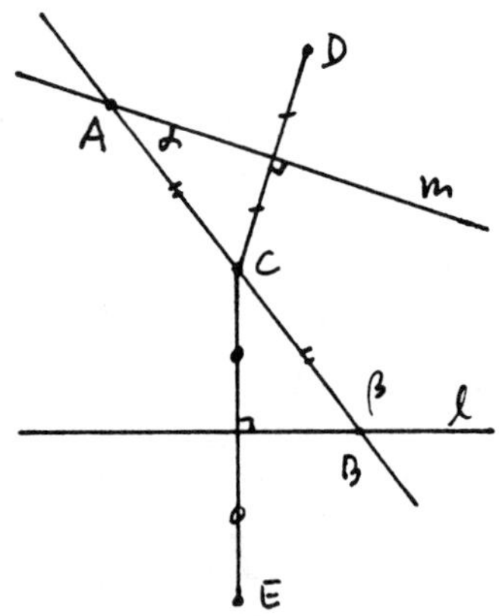

b) Show that Bolyai's axiom holds in any Hilbert plane with (P).

33.12 Dr. Anton Bischof in his thesis (1840) proposed to free the theory of parallels from its dependence on Euclid's parallel postulate by giving a different definition of parallel lines. Discuss his theory, which goes like this: lines are parallel if they have the same direction.

The direction of a line can be measured by the angle it makes with another line. So we define "parallelism is the equality of direction of similar lines against every other straight line". In other words, two lines are parallel if they make equal angles with every other line which meets them both.

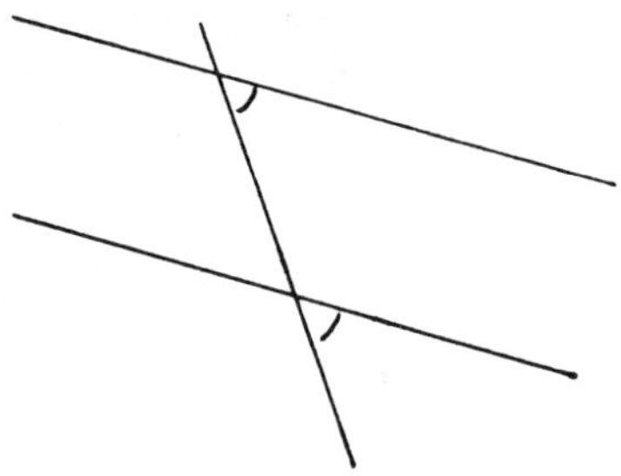

Then it is clear that parallel lines cannot meet, because a transversal line through the point of intersection would make the same angle with both of them, so they would be equal. By the same reasoning it is clear that there can be only one parallel to a given line through a given point. If two lines make the same angle with a line which cuts them, they will be parallel. "Similarly one obtains all the other corollaries which one finds in all the textbooks."

33.13 Discuss the following "proof" that the sum of the angles of a triangle is equal to two right angles, independent of the theory of parallels, due to Thibaut (1775–1832):

Let ABC be the given triangle. Take a segment AD on the line AC, pointing away from C. Rotate it to the position AE on the line AB. Then slide it along the line AB into the position BF. Rotate to BG, slide to CH, rotate to CI, and slide back to AD.

In this process, the segment AD has made one complete rotation, which is 4 right angles. But the amount it has rotated is equal to the sum of the exterior angles DAE, FBG, and HCI. Replacing these by their supplementary angles, we find the sum of the three interior angles of the triangle is equal to two right angles.

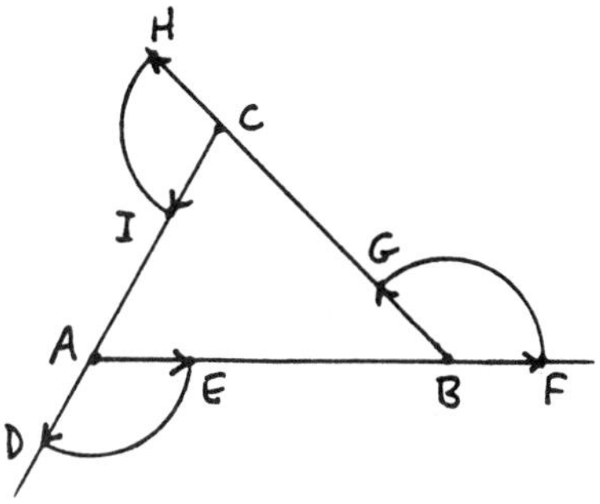

34 Neutral Geometry

Sir Henry Savile, in his public lectures on Euclid's *Elements* in Oxford in 1621 said, "In this most beautiful body of Geometry there are two moles, two blemishes, and no more, as far as I know, for whose removal and washing away, both older and more recent authors have shown much diligence". He was referring to the theory of parallels and the theory of proportion. Euclid's theory of proportion has been thoroughly vindicated, and receives its modern expression in the segment arithmetic which we have explained in Chapter 4.

The work on the theory of parallels, however, did not lead to the expected result. Instead of confirming Euclid's as the one true geometry, these researches showed that Euclid's was only one of many possible geometries. The others are what we now call non-Euclidean geometries. The story of this discovery is one of the most fascinating chapters in the history of mathematics, and has been amply told elsewhere. Here we will confine ourselves to the briefest outline.

We can distinguish four periods. The first, which we have elaborated in the previous section, might be called "dissatisfaction with Euclid". While fully accepting Euclid's *Elements* as the true geometry, critics said only, his treatment of this topic could have been better. So they tried to better Euclid, either by proving the parallel postulate, or by replacing it with some other more natural assumption.

The second period, exemplified by the work of Saccheri, Legendre, and Lambert, was based on the attitude, let us suppose the parallel postulate is false and see what conclusions we can draw. In this way they developed a collection of results which would be true if the parallel postulate were false, still expecting ultimately to find a contradiction and thus vindicate Euclid. So strong was the power of tradition that even after meticulously proving a whole series of propositions in this new geometry, each of these authors fell into error and deluded himself into thinking he had found a contradiction.

What a small step of the imagination, with what great consequences, was the transition to the third period! All it required was to think, yes it is possible to have a geometry in which the parallel postulate is false, and these are its first theorems. This step was taken independently by Carl Friedrich Gauss (1777–1855) in Germany, János Bolyai (1802–1860) in Hungary, and Nicolai Ivanovich Lobachevsky (1793–1856) in Russia. Although Gauss was the first to realize the existence of this new geometry, he published nothing of his researches, leaving Bolyai and Lobachevsky to believe each was the inventor of this new geometry. Bolyai exclaimed, in a letter to his father, "Out of nothing I have created a strange new universe".

The fourth period contains the confirmation of these new geometries by providing models for the axiom systems to show their consistency. This occurred only later, with the work of Beltrami, Klein, and Poincaré.

In this section we will describe some work of the second period. Then in later sections we will give a model of the non-Euclidean geometry due to Poincaré, and a fuller axiomatic development of the theory, containing the results of Bolyai and Lobachevsky in a logical framework provided by Hilbert.

A geometry satisfying Hilbert's axioms of incidence, betweenness, and congruence, in which we neither affirm nor deny the parallel axiom (P), will be called a *neutral geometry.* This is the same as a Hilbert plane, but the terminology emphasizes that we do not assume (P). Recall from §10 that the results of Euclid, Book I, up through (I.28), with the possible exception of (I.1) and (I.22) also hold in neutral geometry. A Hilbert plane in which (P) does not hold will be called a *non-Euclidean geometry.* We have already seen one example of a non-Euclidean geometry (18.4.3), but that one is non-Archimedean, and in this section we will be concerned with neutral geometry which also satisfies Archimedes' axiom.

The results of this section are mainly due to Girolamo Saccheri (1667–1733) and Adrien Marie Legendre (1752–1833). Saccheri's book, *Euclides ab omni naevo vindicatus* was published in 1733. The title "Euclid freed of every blemish" recalls the quote from Savile above. The first 32 propositions are a marvel of mathematical exposition. Unfortunately after that his previously impeccable rigor lapses, and he says he has proved the parallel postulate, because if it were false, there would be two lines having a common perpendicular at infinity, which is "contrary to the nature of a straight line".

Saccheri's work was perhaps before its time, because it did not receive the recognition it deserved, and lay hidden in obscurity until the end of the 19th century. Essentially equivalent results were discovered independently half a century later by Legendre, whose book, *Eléments de Géométrie* was first published in 1794. It was followed by many new editions, reprints, and translations, which had a wide influence on the teaching of geometry and revitalized interest in the question of parallels.

Our first result is the remarkable theorem that in neutral geometry, the sum of the angles of a triangle is always less than or equal to two right angles. Saccheri's proof of this result uses a continuity argument, so we prefer the method of Legendre, using a repeated application of the construction Euclid used in (I.16) for the proof of the exterior angle theorem. In either case, the proof makes essential use of Archimedes' axiom. To begin with, we show that the analogue of Archimedes' axiom holds for angles.

Lemma 34.1 *In a Hilbert plane with* (A), *let* α, β *be given angles. Then there exists an integer* $n > 0$ *so that* $n\alpha > \beta$.

Proof. First we make a reduction. Given the angle β at O, measure off equal segments OA and OB on the two arms, and draw AB. The line OC joining O to the midpoint of AB will bisect the angle β and will make a right angle at C.

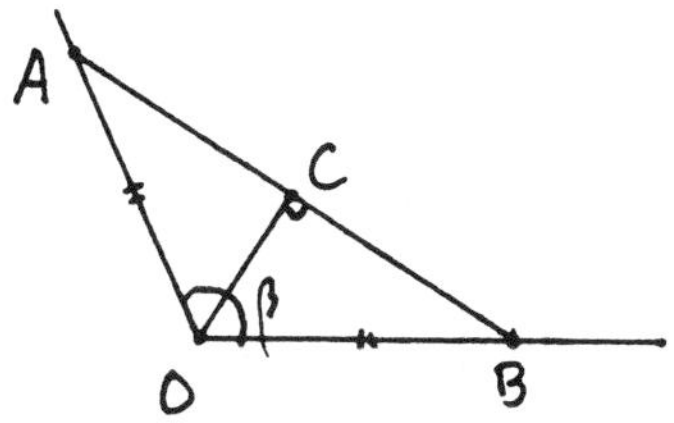

Since it is just as good to prove the lemma for $\frac{1}{2}\beta$, we reduce to studying the case of an angle contained in a right triangle.

So now let OAB be a right triangle with the angle β at O and a right angle at A. Lay off the angle α inside the triangle, and let that angle cut off a segment AA_1 on the line AB. Again lay off the angle α at O to cut a segment A_1A_2 on the line AB. Continuing in this manner, we obtain a sequence of points $A_1, A_2, A_3, \ldots$ on the ray AB, with each successive segment A_iA_{i+1} subtending an angle α to O. We must show that A_n lies beyond B for some n.

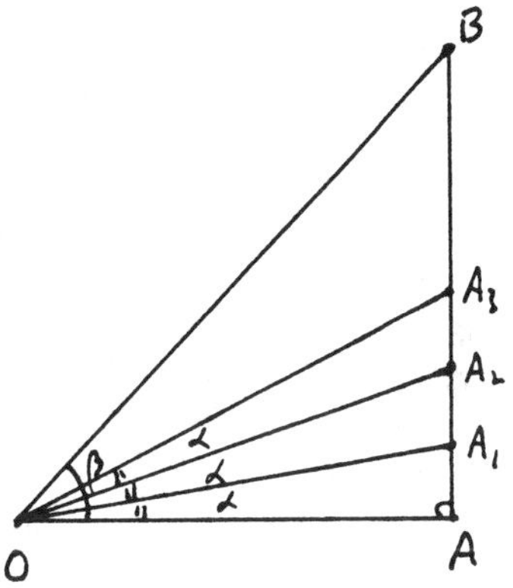

I claim that $AA_1 < A_1A_2 < A_2A_3 < \ldots$. Consider any three consecutive A_i and call them C, D, E. Let DF be drawn such that $\angle CDO = \angle FDO$. Then the triangles ΔCDO and ΔFDO are congruent by (ASA), and hence $CD = DF$. Now the angle $\delta = \angle EFD$ is an exterior angle to the triangle ΔFDO, and so $\delta > \gamma = \angle FDO = \angle CDO$, by (I.16). On the other hand, $\gamma = \angle CDO$ is an exterior angle to the triangle ΔDEO, and so $\gamma > \varepsilon = \angle DEO$, by (I.16). We conclude that $\delta > \varepsilon$. Then by (I.19), the larger angle subtends the larger side, so $DE > DF = CD$, as required.

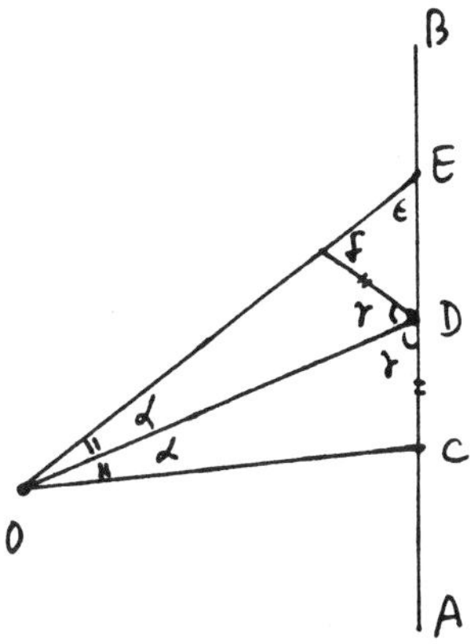

By the way, in order for the drawing to be accurate, i.e. for F to lie in the segment OE, we were tacitly assuming that γ is acute. This is true for the first triangle OAA_1, and follows inductively for the rest from (I.16).

Now, the axiom of Archimedes (A) implies that there is some integer n such that $n \cdot AA_1 > AB$. Since the successive segments $A_1A_2, A_2A_3, \ldots$ are each bigger than the one before, it follows for a stronger reason that $AA_n > AB$. In other words, the point A_n falls beyond B on the ray AB, and so the angle $n\alpha$ is bigger than β.

Theorem 34.2 (Saccheri–Legendre) *In a Hilbert plane with Archimedes axiom* (A), *the sum of the angles of a triangle is less than or equal to two right angles.*

Proof. Suppose to the contrary there was a triangle ΔABC whose angle sum is greater than two right angles, say two right angles plus ε where ε is some non-zero angle. Then we will get a contradiction by replacing the triangle ΔABC by another triangle ΔDEF, which has the same angle sum as ΔABC, and furthermore has one angle α very small, less than ε. Then the remaining two angles would be more than two right angles, which contradicts (I.17).

As a first step we show, given any triangle ΔABC, and having chosen one of its angles, say $\angle A$, that there is another triangle ΔAEC, having the same angle sum as ΔABC, and one of whose angles is $\leq \frac{1}{2}$ (angle A).

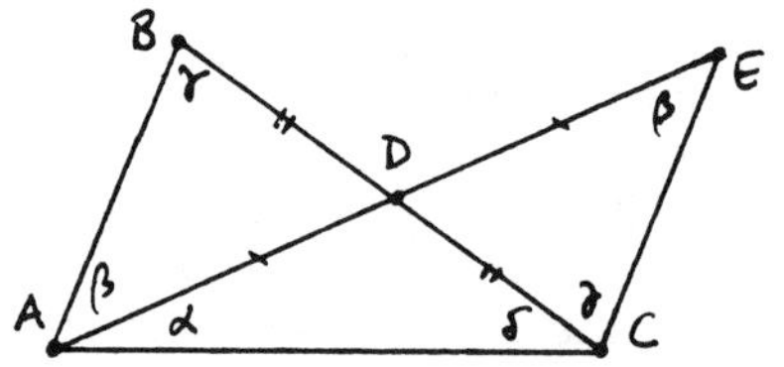

We use the construction of (I.16). Let D bisect BC, draw AD and extend that line to a point E so that $AD \cong DE$, and draw EC. Now the vertical angles at D are equal (I.15) and $BD \cong DC$ and $AD \cong DE$ by construction, so by (SAS) the triangles ΔABD and ΔAEC are congruent. In particular, angle $\beta = \angle BAD$ is equal to $\angle DEC$. Also $\gamma = \angle ABD = \angle DCE$.

Now ΔABC and ΔAEC both have the same angle sum, which is equal to $\alpha + \beta + \gamma + \delta$, where $\alpha = \angle DAC$ and $\delta = \angle DCA$. On the other hand, the angle at A is $\alpha + \beta$, so of the two angles α, β in the new triangle, one of them satisfies α (or β) $\leq \frac{1}{2}$(angle A).

Now let us go back to our original triangle ΔABC having angle sum equal to two right angles plus ε. At least one angle of this triangle is acute, so by repeating the above process we can obtain a sequence of triangles

$$T_0 = \Delta ABC,\ T_1,\ T_2,\ T_3, \ldots$$

each of which has angle sum the same as ΔABC, and where T_n has one angle $\leq \left(\frac{1}{2^n}\right)$(right angle). Now using Archimedes principle for angles (34.1) we see that for some n, T_n will have one angle $< \varepsilon$, which gives the desired contradiction to (I.17).

Remark 34.2.1. In the above construction, two angles of the triangle T_{n+1} are less than the chosen angle in triangle T_n. Thus T_{n+1} has two angles $\leq \left(\frac{1}{2^n}\right)$(right angle).

Corollary 34.3 *In any triangle, the exterior angle is greater than or equal to the sum of the opposite interior angles.*

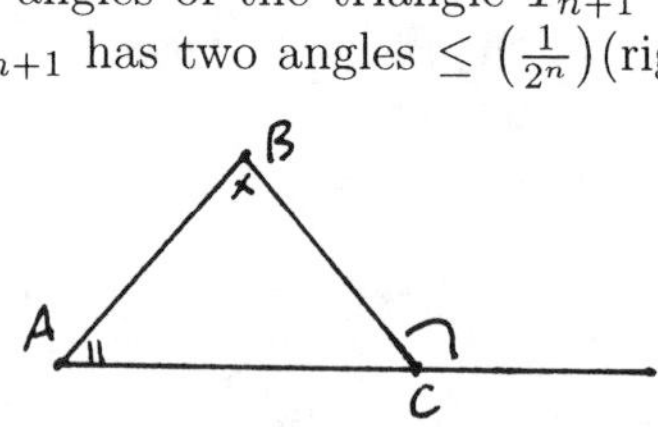

Proof. Indeed, the exterior angle plus the third interior angle is equal to two right angles. Since the sum of all the angles is ≤ 2 right angles, the sum of the two opposite angles is $\leq$ exterior angle.

Definition. Now that we have seen that the sum of the angles in any triangle is less than or equal to two right angles (2RA for short), we define the *defect* δ of a triangle to be the *angle* 2RA $-$ (sum of angles of the triangle) if that is a positive angle, or $\delta = 0$ if the sum of the angles is equal to 2RA. Thus the defect of a triangle measures how far it is from being a *Euclidean triangle*, meaning one whose sum is equal to 2RA (I.32). If $\delta \neq 0$, we will say it is a *non-Euclidean triangle.*

Lemma 34.4 *If a triangle ABC is cut into two triangles by a single transversal BD, the defect of the big triangle is equal to the sum of the defects of the two small triangles:*

$$\delta(ABC) = \delta(ABD) + \delta(BCD).$$

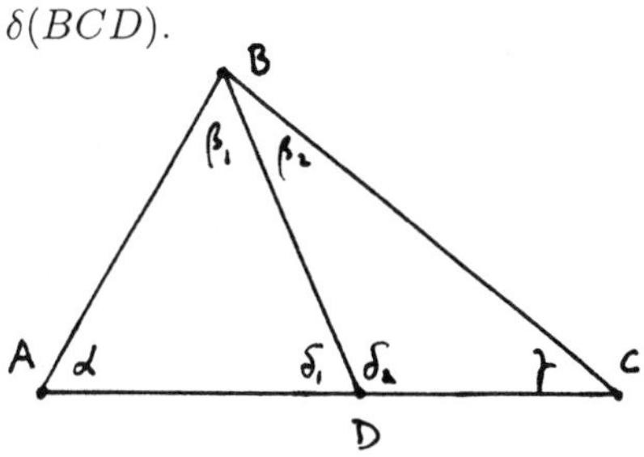

Proof. Label the angles as shown in the diagram. Then

$$\begin{aligned}\delta(ABD) &= 2\text{RA} - \alpha - \beta_1 - \delta_1\\ \delta(BCD) &= 2\text{RA} - \beta_2 - \delta_2 - \gamma.\end{aligned}$$

Since $\delta_1 + \delta_2 = 2\text{RA}$, by adding we find

$$\delta(ABD) + \delta(BCD) = 2\text{RA} - \alpha - \beta_1 - \beta_2 - \gamma = \delta(ABC)$$

as required.

Our next goal is to show that the angle sum of a triangle is either always 2RA (the Euclidean case) or never equal to 2RA (the non-Euclidean case). To prove this, we will make use of a figure extensively studied by Saccheri, which goes back to Clavius, in his commentary on Euclid's (I.29), where he proposes the axiom which we discussed earlier (§33). Since it was Clavius' edition of Euclid which was recommended to Saccheri by the Jesuit mathematician Tommaso Ceva, we may assume that Saccheri was inspired by Clavius to study this figure further.

Lemma 34.5 *In a Hilbert plane, suppose that two equal perpendiculars AC, BD stand at the ends of an interval AB, and we join CD. (This is called a* Saccheri quadrilateral.*) Then the angles at C and D are equal, and furthermore, the line joining the midpoints of AB and CD is perpendicular to both.*

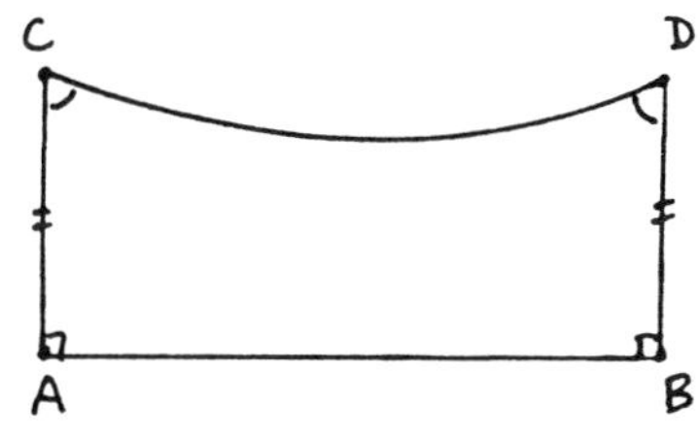

Proof. Given $ABCD$ as above, let E be the midpoint of AB and let EF be the perpendicular to AB at E. By (SAS) the triangles AEF and BEF are congruent. Hence the angles $\angle FAE$ and $\angle FBE$ are equal, and $AF = FB$.

By subtraction from the right angles at A and B we find the angles $\angle CAF$ and $\angle DBF$ are equal. So by (SAS) again, the triangles CAF and DBF are congruent. This shows that the angles at C and D are equal, and that F is the midpoint of CD.

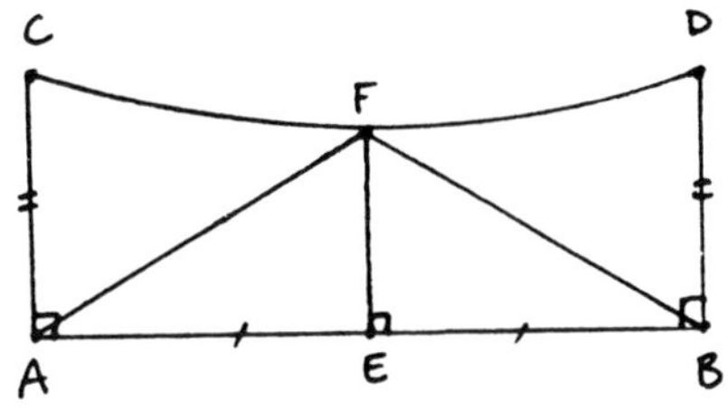

The two pairs of congruent triangles also imply that the angles $\angle CFE$ and $\angle DFE$ are equal. So by definition, both of these angles are right angles.

Remark 34.5.1. From the equality of the angles at C and D, Saccheri distinguished three cases, which he called the hypothesis of the acute angle, the hypothesis of the right angle, and the hypothesis of the obtuse angle, according to whether C and D were acute, right, or obtuse. He showed that if any one of these holds for one such quadrilateral, it holds for all. He was also able to eliminate the case of the obtuse angle (cf. 34.8.1).

Proposition 34.6 *Given a triangle ABC in a Hilbert plane, there is a Saccheri quadrilateral whose defect is equal to the defect of the triangle. (Here we define the* defect *of a quadrilateral to be* 4RA *minus the sum of the four angles.)*

Proof. Let ABC be the given triangle. Let D and E be the midpoints of AB and AC, and draw the line DE, which we call the *midline* of the triangle. Drop perpendiculars BF, AG, CH to DE.

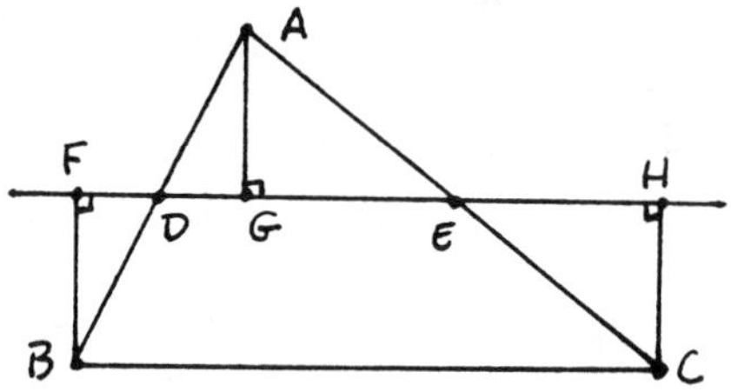

Now $AD = DB$, and the vertical angles at D are equal, so by (AAS) the triangles ADG and BDF are congruent. Similarly $AE = EC$ and the vertical angles at E are congruent, so the triangles AEG and CEH are congruent. From congruent triangles we find $BF = AG = CH$. The quadrilateral $FHBC$ has right angles at F and H, so it is a Saccheri quadrilateral (upside down). The angles of the quadrilateral at B and C are composed of the angles of the triangle at B and C, plus angles which are congruent to the two parts of the angle at A, divided by the line AG. Hence the angles at B and C of the quadrilateral equal the angle sum of the triangle. It follows that the triangle and the quadrilateral have equal defect.

If G happens to fall outside the interval FH, the same argument works, except that we use differences instead of sums of angles.

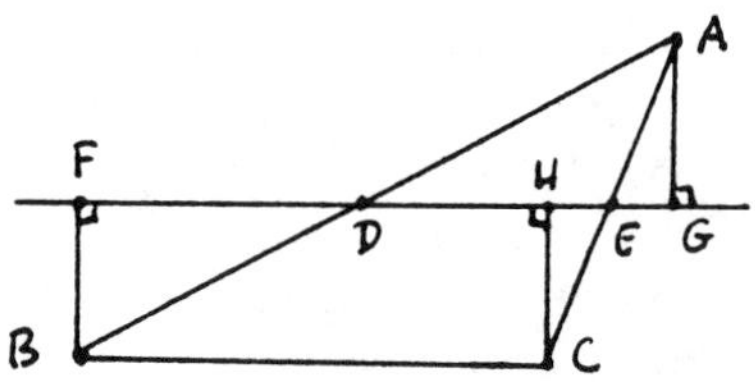

Proposition 34.7 *In a Hilbert plane, the following conditions are equivalent:*

(i) *There is at least one Euclidean triangle.*

(ii) *There is at least one* rectangle *(meaning a quadrilateral with four right angles—cf. Clairaut's axiom).*

(iii) *Every Saccheri quadrilateral is a rectangle.*

(iv) *Every triangle is Euclidean.*

Proof. (i) $\Rightarrow$ (ii). Suppose there is a Euclidean triangle ABC. According to (34.6) there is a Saccheri quadrilateral $FHBC$ with the same defect, namely zero. Since the angles at B and C are equal (34.5), they are both right angles, so it is a rectangle.

(ii) $\Rightarrow$ (iii). Suppose $ABCD$ is a rectangle. Take any point E on AB, and drop a perpendicular EF to CD. Reflecting in the line AC we obtain another point G on AB, and the reflection of EF becomes a perpendicular GH to CD, and $EF = GH$. Reflecting EF in the line BD, we obtain another point I on AB, together with a perpendicular IK to CD, and $IK = EF$.

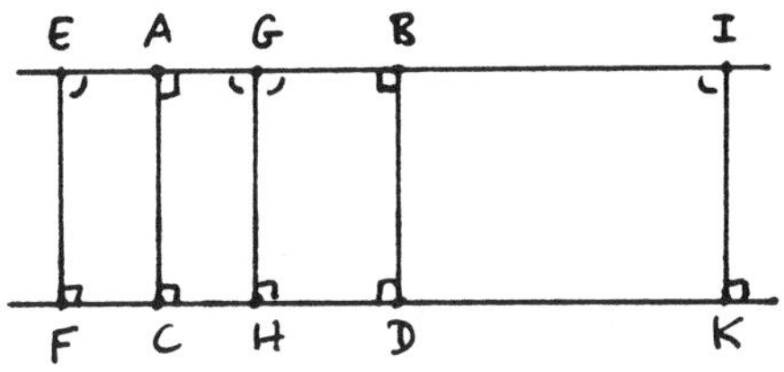

Thus $FHEG$, and $FKEI$, and $HKGI$ are all Saccheri quadrilaterals. By (34.5) their upper angles are equal. In particular, supposing G is between E and I, the two angles at G are equal, so they are right angles. Thus the angle at the original point E is a right angle.

Thus for any point E on AB, we have shown that $AECF$ is a rectangle. So there exist rectangles whose one side is the original AC, and whose other side AE is arbitrary.

Repeating this construction after interchanging tops and sides, we find that there exist rectangles with any two given sides.

Now we can show that any Saccheri quadrilateral is a rectangle. Indeed, there exists a rectangle with the same base and sides, so its top side is equal to the top side of the quadrilateral, so it is a rectangle.

(iii) $\Rightarrow$ (iv). Just use the argument of (i) $\Rightarrow$ (ii) in reverse.

(iv) $\Rightarrow$ (i) is trivial.

Thus we see that either every triangle is Euclidean, or every triangle is non-Euclidean. You should then not be too surprised to find that this property distinguishes Euclidean from non-Euclidean geometry:

Proposition 34.8 *In a Hilbert plane with Archimedes axiom* (A), *the equivalent conditions of* (34.7) *are also equivalent to the parallel axiom* (P).

Proof. Euclid's proof of (I.32) shows that (P) implies all triangles are Euclidean.

For the reverse implication, we will prove the contrapositive assertion that ($\sim$P) implies ($\sim$iv). In other words, if there is a line l and a point P not on l through which there are two or more lines parallel to l, then there exists a non-Euclidean triangle.

Given such a line l and point P, drop a perpendicular from P to L at A, and let m be the line through P perpendicular to that perpendicular. Then m is parallel to l (I.27). Now let m' be another line through P parallel to l. Then on one side of AP, the ray m' will be inside the right angle formed by AP and m. Let it form an angle β with m at P.

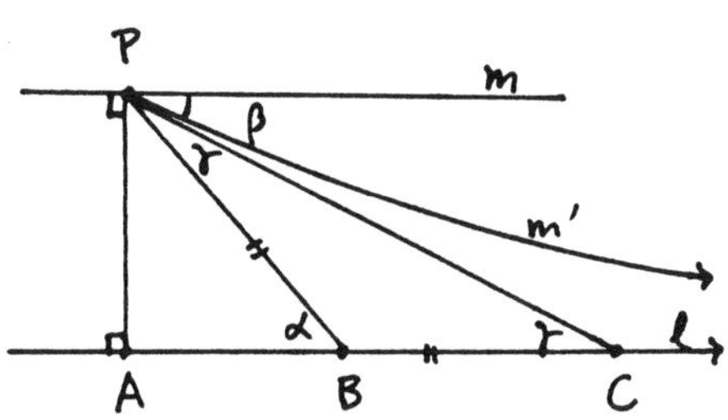

Now take any point B on l on the same side of AP as m'. Let the angle $\angle PBA$ be α. Choose C or l so that $BC \cong PB$, and draw PC. Then we have an isosceles triangle BPC, so the angles γ shown are equal. But α is an exterior angle to the triangle BPC, so $\alpha \geq 2\gamma$ by (34.3). Repeating this process n times, we can find a point F further out on the ray l so that the angle PFA is $\leq \left(\frac{1}{2^n}\right)\alpha$. By (34.1) we can take n large enough so that $\angle PFA < \beta$.

Consider the triangle PFA. It has one right angle, one angle $< \beta$, and one angle, $\angle APF$, which is less than RA $-\ \beta$, because the ray m', being parallel to l, cannot lie inside the triangle PFA. Thus the angle sum of the triangle PFA is strictly less than 2RA. So (∼P) ⇒ (∼iii), which concludes the proof.

Remark 34.8.1. Summing up, we see that in a Hilbert plane with (A), either (P) holds (the Euclidean case) or every triangle has positive defect. In this latter case it follows that every Saccheri quadrilateral has positive defect, and so the top angles are acute (which Saccheri called the hypothesis of the acute angle). Indeed, by (34.6) we have only to show that every Saccheri quadrilateral arises from some triangle. Using the same notation as in (34.6), suppose given the Saccheri quadrilateral $FHBC$. Let D be any point on FH, extend BD to A so that $BD = DA$, and joint AC. Let AC meet FH at E. Now the triangles BDF and ADG are congruent by (AAS). So $AG = BF$. But $BF = CH$ since it is a Saccheri quadrilateral. Therefore triangles AEG and CEH are congruent by (AAS). It follows that E is the midpoint of AC. Thus $FHBC$ is the Saccheri quadrilateral associated to ABC.

Now we can explain how Legendre thought to prove the parallel postulate. He suggested another postulate, which we call

Legendre's Axiom: Given an angle α, and given a point P in the interior of the angle α, there exists a line l through P, which meets both sides of the angle.

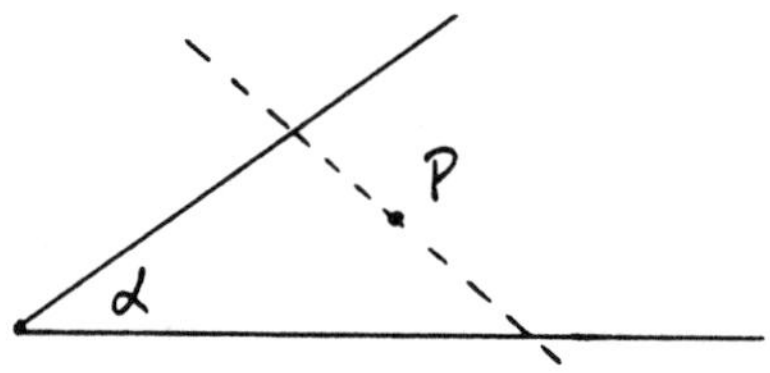

He shows that this implies the parallel postulate (see (34.7) below). While this axiom seems reasonable enough in itself, he shows further that if it fails, then for any angle α, however small, there would have to be a line l entirely contained inside the angle (Exercise 34.7).

Then he says "or, il répugne à la nature de la ligne droite"—it is repugnant to the nature of a straight line—that such a line should be entirely contained within an angle!

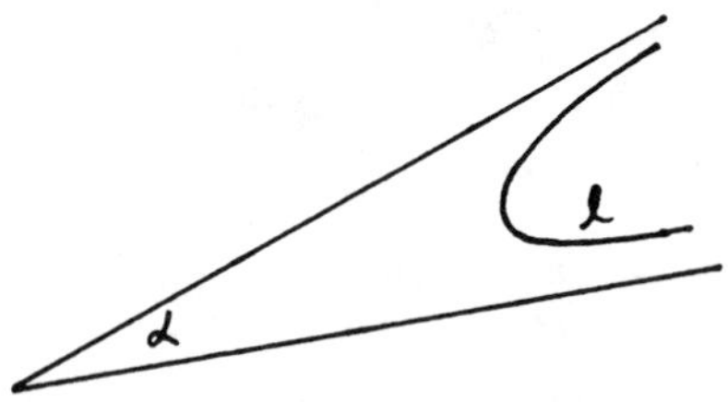

Proposition 34.9 *In a Hilbert plane with* (A), *if Legendre's axiom holds for a single angle* α, *then also* (P) *holds.*

Proof. Since Legendre's axiom for an angle α implies the same axiom for any smaller angle, we may assume it holds for a given acute angle α. Let the vertex of the angle be A, take any two points B, C on the sides of the angle, and draw BC.

NOTE II. 279

pouvait être nié. (1) C'est cette considération qui nous a fait revenir dans la 9ème édition, à la simple marche d'Euclide, en renvoyant aux notes pour la démonstration rigoureuse.

En examinant les choses avec plus d'attention nous sommes resté convaincu que pour démontrer complètement notre *postulatum* il fallait déduire de la définition de la ligne droite une propriété caractéristique de cette ligne qui exclût toute ressemblance avec la forme d'une hyperbole comprise entre ses deux asymptotes. Voici quel est à cet égard le résultat de nos recherches.

Soit BAC *un angle donné, et* M *un point donné au dedans de cet angle; divisez l'angle* BAC *en deux également par la droite* AD *et du point* M *menez* MP *perpendiculaire sur* AD*: je dis que la droite* MP *prolongée dans un sens et dans l'autre, rencontrera nécessairement les deux côtés de l'angle* BAC. fig. 274.

Car si elle rencontre un des côtés de cet angle, elle rencontrera l'autre, tout étant égal des deux côtés à partir du point P; si elle ne rencontrait pas un côté, elle ne rencontrerait pas l'autre par la même raison; ainsi, dans ce dernier cas elle devrait être renfermée tout entière dans l'espace compris entre les côtés de l'angle BAC; or, il répugne à la nature de la ligne droite qu'une telle ligne, indéfiniment prolongée, puisse être renfermée dans un angle.

En effet, toute ligne droite AB tracée sur un plan, et indéfiniment prolongée dans les deux sens, divise ce plan en deux parties qui étant superposées coïncident dans toute leur étendue et sont parfaitement égales. La partie AMB du plan total, située d'un côté de AB, est égale en tout à la partie AN[illegible] de l'autre côté; car si l'on prend un point fig. 275.

(1) On voit dans un article du *Philosophical magazine* de mars 1822, qu'un savant géomètre a essayé de perfectionner cette démonstration et de la rendre indépendante de tout *postulatum*; mais la construction employée pour démontrer la seconde partie consiste à mener d'un point donné différentes droites à tous les sommets d'une ligne qu'on doit considérer comme polygonale, pour raisonner dans l'hypothèse de celui qui nie la proposition : or la convexité de cette ligne, si elle avait lieu, ne permettrait pas de continuer indéfiniment la construction de l'auteur, comme il le faudrait pour l'exactitude de sa démonstration.

Plate X. Legendre's "proof" that through any point in the interior of an angle, there exists a straight line meeting both sides of the angle. From Legendre (1823).

Now repeat the angle ACB at B, lay off $BD = AC$, and join CD. Then by (SAS) the new triangle BCD is congruent to the triangle ABC. Applying Legendre's axiom for the angle α to the point D, let l be a line through D meeting the sides of the angle α at E and F. One sees easily that E and F must lie beyond B and C, so that the new triangle AEF contains both the triangles ABC and BCD.

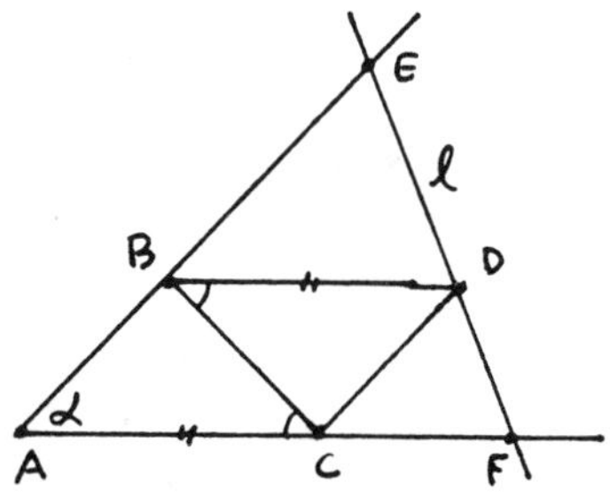

Suppose that (P) does not hold. Then by (34.5) and (34.6) the triangle ABC has a positive defect $\delta > 0$. By the additivity of defect (34.4), the new triangle AEF will have defect $> 2\delta$. Repeating this process and using (34.1) we would eventually have a triangle with defect greater than two right angles, which is absurd. Therefore (P) must hold.

Remark 34.9.1. We will give in the exercises some examples to show the necessity of Archimedes' axiom in the results of this section: see Exercises 34.10, 34.11, 34.12.

Exercises

Unless specified otherwise, the following problems take place in a non-Euclidean (i.e. (P) is false) Hilbert plane satisfying Archimedes' axiom (A).

34.1 The AAA congruence theorem: Show that if two triangles have their three angles respectively equal, the two triangles are congruent.

34.2 Show that the angle at C of a triangle in a semicircle is less than or equal to a right angle.

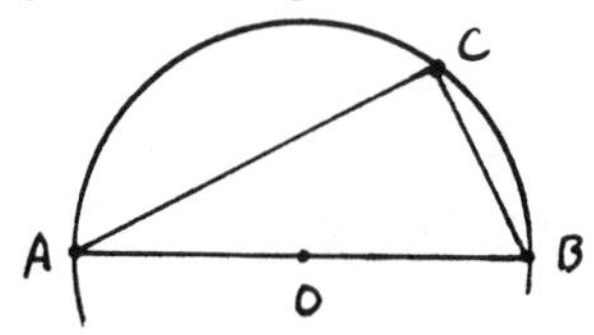

34.3 Show that for every line l, and for every point P not on l, there is more than one parallel to l through P.

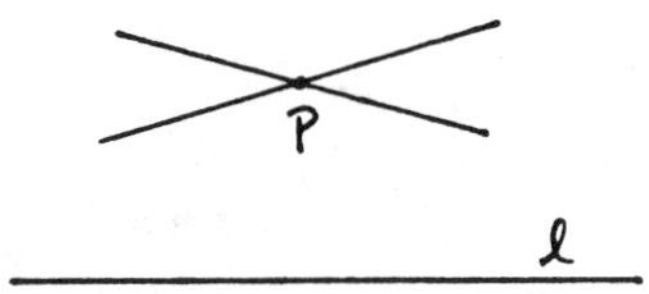

34.4 Prove: given any angle $\varepsilon > 0$, there exists a triangle with defect $\delta < \varepsilon$.

34.5 Discuss the following "proof", due to Legendre, that the angle sum of a triangle is two right angles: we have seen that for any triangle ABC, there is a triangle T_n with the same angle sum as ABC, and where T_n has two angles $\leq (\frac{1}{2^n})$(right angle) (34.2.1). In the limit, the two small angles will become zero, so the triangle becomes a straight line, and the third angle will be 2RA. Thus the angle sum of the original triangle must be 2RA.

34.6 Show that for any angle α, however small, there exists a line l entirely contained in the inside of the angle. *Hint*: For any point P inside the angle α, consider the line through P, perpendicular to the angle bisector of α.

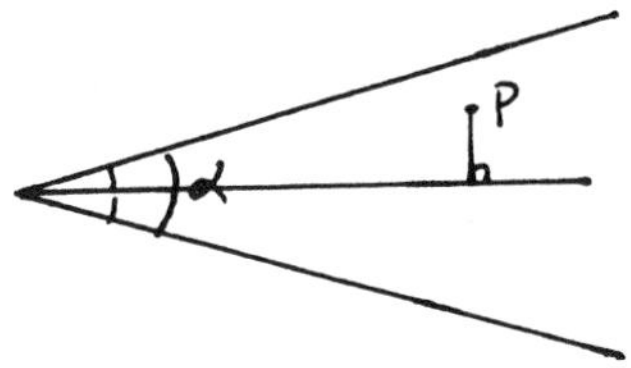

34.7 Given an angle α with vertex A, and given a ray l inside α, show that there exists a point P on the ray l, such that for any two points B, C on the two arms of α, the line BC meets the ray l inside the interval AP.

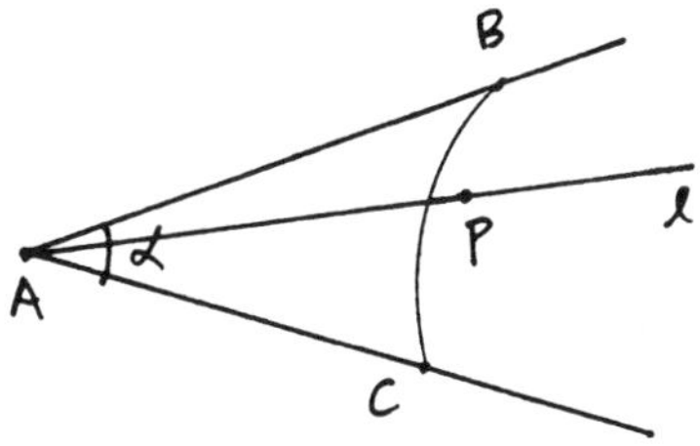

34.8 Given an angle $\varepsilon > 0$, show that there exists a triangle with angles α, β, γ, all three smaller than ε. *Hint*: Use (Exercise 34.6).

34.9 Show that (34.1) is false in the Cartesian plane over a non-Archimedean field.

34.10 Use (18.4.3) to show that the hypothesis (A) is necessary in the proof of (34.6).

34.11 *Spherical geometry.* Let F be an ordered field satisfying (**) of (16.2). In the Cartesian 3-space over F consider the sphere of radius r given by the equation $x^2+y^2+z^2=r^2$. We can make a geometry, called *spherical geometry*, as follows. Our *s-points* are the points of F^3 on the surface of the sphere. Our *s-lines* are *great circles* on the sphere, that is to say, the intersections of the sphere with planes of F^3 passing through the origin $O=(0,0,0)$. On any piece of an s-line which is less than half of a great circle, we can define betweenness by projecting the points from O into any plane. We say two segments of s-lines are congruent if the chords joining their endpoints, as line segments of F^3 inside the sphere, are congruent. We say angles are congruent if the projected angles on the tangent planes to the sphere at their vertices are congruent.

a) Which of Hilbert's axioms hold in this geometry? You will see right away that (I1) fails and betweenness does not make very good sense, so it is not a Hilbert plane. Show however that the congruence axioms (C1)–(C6) and (ERM) do hold.

b) Now suppose that we take F to be a non-Archimedean field, such as the one in (18.4). Let t be an infinite element in F, take the sphere of radius t, and take our geometry to consist of only those points on the surface of the sphere which are at finite distance from a fixed point A on the sphere. Show that this geometry satisfies all of Hilbert's axioms, so that it is a Hilbert plane. Show also that the sum of the angles of any triangle in this geometry is *greater* than two right angles. This shows the necessity of the hypothesis (A) in the theorem of Saccheri–Legendre (34.2).

35 Behavior of Parallels

Now that we have the fundamental results about the angle sum in non-Euclidean triangles, we will discuss as applications two other topics, the behavior of parallel lines, and (in the next section) the theory of area in non-Euclidean geometries.

In Euclidean geometry we know that parallel lines are equidistant. In non-Euclidean geometry we will see that parallel lines either converge or diverge from each other. We will use the theory of Saccheri quadrilaterals, already mentioned in (34.5).

Lemma 35.1 *In a Hilbert plane with* (A), *let $ABCD$ be a Saccheri quadrilateral* (34.5). *Then $CD \geq AB$, with equality only if* (P) *holds.*

Proof. First suppose $AB = CD$. Then, in the diagram of (34.5), also $EB = FD$. Applying (34.5) to the quadrilateral $FEDB$ with base FE, it would follow that the angle at D is equal to the angle at B, hence is a right angle. Then $ABCD$ is a rectangle, which implies (P) (34.8).

So now we have only to show that $CD < AB$ is impossible. In that case also $FD < EB$. Extend FD to G so that $FG = EB$. Then we apply (34.5) to the quadrilateral $FEGB$ with base EF. It follows that the angles at G and B are equal.

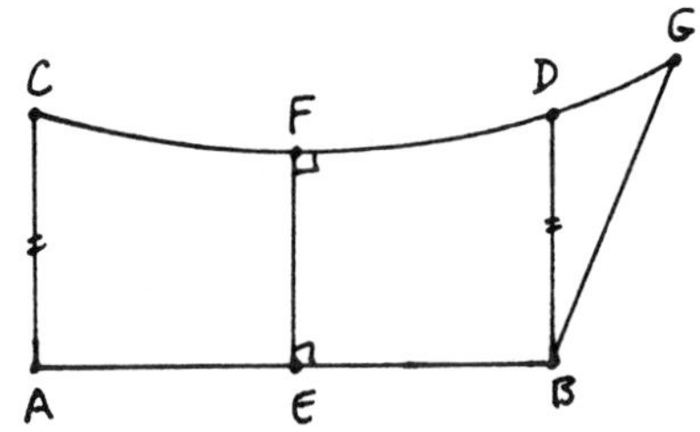

But the angle at B is greater than a right angle, while the angle at G is less than the exterior angle $\angle FDB$ by (I.16), and this angle is acute (34.8.1). From this contradiction we conclude that $CD \geq AB$.

Corollary 35.2 *In a quadrilateral $ABCD$ with right angles at A and B, and unequal sides AC, BD, the fourth side CD is greater than AB.*

Proof. Suppose that $BD > AC$, and choose E on the side BD so that $AC = BE$. Then $CE \geq AB$ by (35.1), and the angle CEB is acute (34.8.1). Hence the angle CED is obtuse, so $CD > CE$ by (I.19).

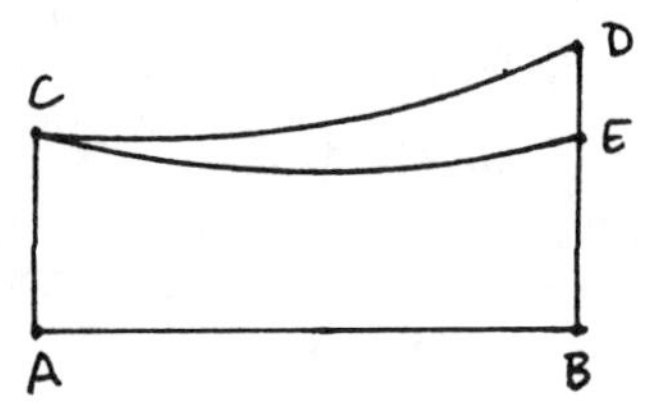

Lemma 35.3 *In any Hilbert plane, let $ABCD$ be a quadrilateral with right angles at A and B, and unequal sides AC, BD. Then the angle at C is greater than the angle at D if and only if $AC < BD$.*

Proof. Suppose $AC < BD$, and use the same diagram as for (35.2). Then using (34.5) and (I.16), $\angle DCA > \angle ECA = \angle CEB > \angle CDB$. Similarly if $AC > BD$ we find the angle at C is less than the angle at D. So the inequality of angles implies also the inequality of sides.

Proposition 35.4 *In a Hilbert plane with* (A), *Aristotle's axiom holds, namely, given any angle, the perpendicular from a point on one arm of the angle to the other arm can be made to exceed any given segment.*

Proof. Using (A), it will be sufficient to show that if BC is one perpendicular from one arm of the angle to the other, then there exists another such perpendicular with $DE \geq 2BC$. To do this, we proceed as follows. First mark off $BD = AB$, and drop the perpendicular DE. Then make a perpendicular to BC at B and let it meet DE at F. Mark off $BG = BF$ and join AG.

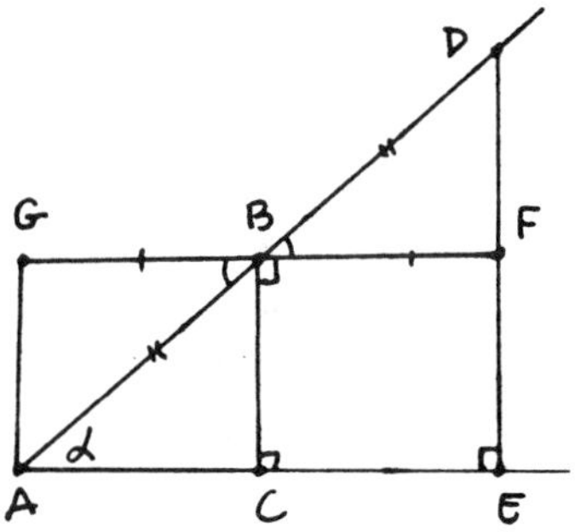

Now $EF \geq BC$ by (35.1) and (35.2). On the other hand, the vertical angles at B are equal, so the triangles BDF and BAG are congruent by (SAS), so $DF = AG$. But $AG \geq BC$ by another application of (35.1) and (35.2), so $DE \geq 2BC$ as required.

Proposition 35.5 *Let P be a point not on a line l, let PQ be the perpendicular to l, and let m, n be any two rays emanating from P, on the same side of PQ, and both parallel to l. Suppose n lies between l and m. Then the perpendicular AC from a point A on m to l can be made to exceed any given segment. (We will say in this case that m and l* diverge.*)*

Proof. From a point A on m, let AB be the perpendicular to n, let AC be the perpendicular to l, and let AC meet n at D. Then D is between A and C, by hypothesis. Now $AC > AD > AB$, because the hypotenuse of a right triangle is bigger than its legs (I.19). On the other hand, we can choose A so that AB exceeds any segment by (35.4). Thus AC grows without bound.

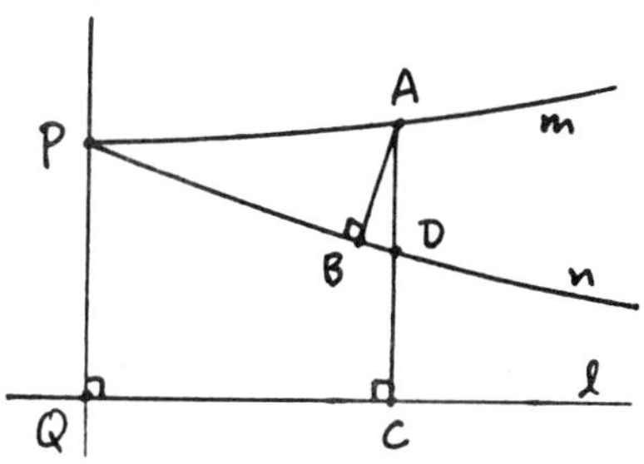

This proposition suggests that among all possible parallels to l through P, we should look for one which is "closest" to l. So we make the following

Definition. Let P be a point not on a line l, with orthogonal PQ to l. A ray m emanating from P is a *limiting parallel* to l if it is parallel (i.e. does not meet l) and furthermore, any other ray m', inside the angle defined by PQ and m, does meet l.

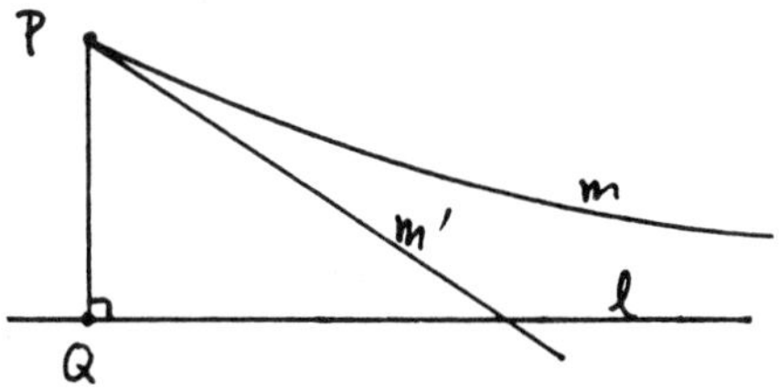

Remark 35.5.1. Note that this definition says nothing about the existence of limiting parallels. Both Bolyai and Lobachevsky used a continuity argument to

claim they exist. Since this is foreign to our approach, we will actually take the existence of limiting parallels as an axiom, following Hilbert, in our treatment of hyperbolic geometry in §40. For the present, we will merely discuss the properties of limiting parallels, if they exist.

Note also that there is no uniformity of terminology in the literature. Some authors use the terms "parallel" and "non-intersecting" for what we call "limiting parallel" and "parallel". So I repeat, in this book, parallel lines are simply two lines which do not meet.

Remark 35.5.2. Bolyai also gave a ruler and compass construction for the limiting parallel, but unfortunately I do not know any proof that his construction works, without first assuming the existence of the limiting parallels—cf. (41.10).

Proposition 35.6 *Let m be a limiting parallel ray to a line l through a point P, in a non-Euclidean Hilbert plane with* (A). *Then the perpendicular distance from a point A on m to l is strictly decreasing as A moves along m away from P.*

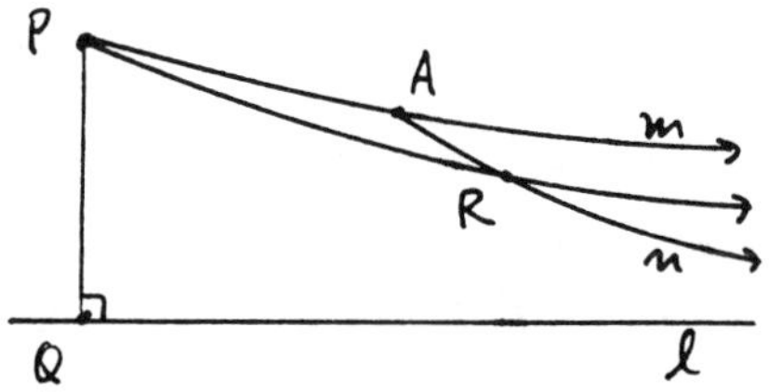

Proof. Let m be a limiting parallel to l through P. First we show that for any other point A on the ray m, that ray m is also a limiting parallel to l through A. If not, there would be another ray n through A parallel to l, between m and l. Let R be a point on that ray.

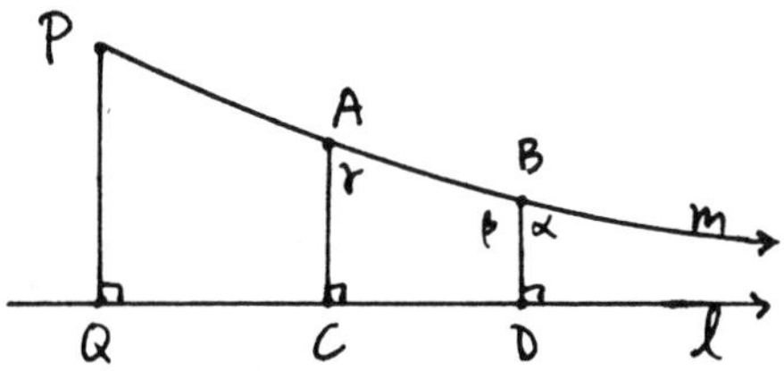

Then the ray PR would lie above n after it passes R, and so would be a parallel to l through P, below m. This contradicts the hypothesis that m was a limiting parallel. Now, to prove the proposition, let A, B be any two points on the ray m, with A between P and B.

Let AC and BD be the perpendiculars to l. We must show that $AC > BD$.

Since m is also a limiting parallel for B to l, by what we have just shown, it follows that the angle α formed by BD and m is acute. Otherwise one could make a perpendicular to BD at B, which is parallel to l by (I.27) and it would lie beneath m. So the supplementary angle β will be obtuse. The same argument shows that the angle γ formed by AC and m is acute. So $\beta > \gamma$, and we conclude from (35.3) that $AC > BD$.

Remark 35.6.1. We would like to say in this proposition that the distance AC can be made to be smaller than any preassigned segment, but we have not been able to find a proof of this without assuming the existence of limiting parallels—cf. (Exercise 40.8).

Corollary 35.7 *In a non-Euclidean Hilbert plane with* (A), *if m is a ray through a point P, parallel to a line l, then the perpendicular distance from points on m to l either (eventually) grows without bound, or is strictly decreasing. There is at most one ray of the latter sort on each side of PQ, which will be the limiting parallel to l through P, if it exists.*

Exercises

The following exercises, unless indicated to the contrary, take place in a non-Euclidean Hilbert plane with (A).

35.1 If m is a ray emanating from a point P, and if A is a point on m for which m is a limiting parallel to a line l, then m is also a limiting parallel from P to l.

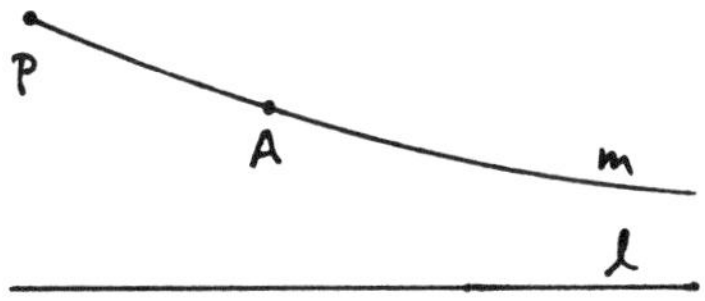

35.2 If m is a limiting parallel from a point P to a line l, then l is also a limiting parallel from any point on it to m. Because of this and the previous exercise, it makes sense to say simply l and m are limiting parallel rays, without specifying a point or an order of the two lines.

35.3 If l, m, n are three rays, with l limiting parallel to m (we will write $l|||m$) and $m|||n$, then also $l|||n$.

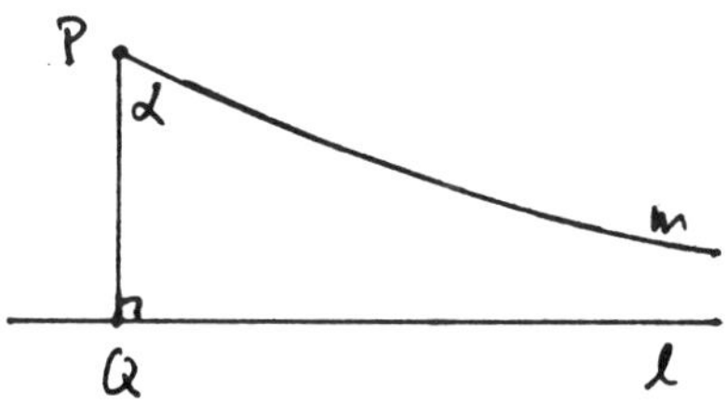

35.4 If m is a limiting parallel through P to a line l, and if PQ is the perpendicular to l, we call the angle α formed by PQ and m the *angle of parallelism* for the segment PQ. Show that the angle of parallelism depends only on PQ up to congruence, namely if P', Q', m', l', α' is another such situation, and $PQ \cong P'Q'$, then $\alpha \cong \alpha'$.

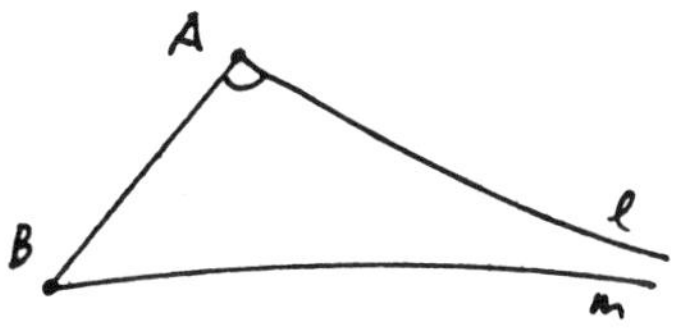

35.5 The ASL congruence theorem: Suppose given two *limit triangles*, that is segments with limiting parallel rays at each end, say $ABlm$ and $A'B'l'm'$. Assume the angles at A and A' are equal, and assume the sides AB and $A'B'$ are equal. Then prove that the triangles are congruent, namely that the angles at B and B' are also equal.

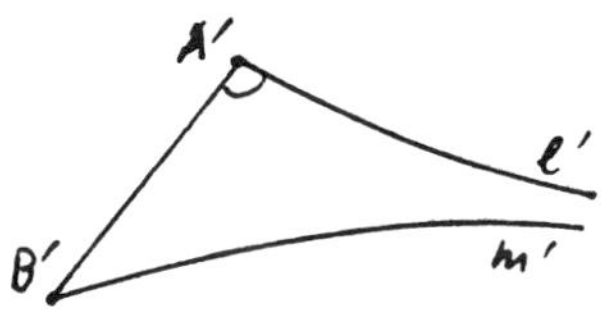

35.6 The exterior angle theorem for limit triangles: Given a limit triangle $ABlm$, show that the exterior angle β at B is greater than the interior angle α at A.

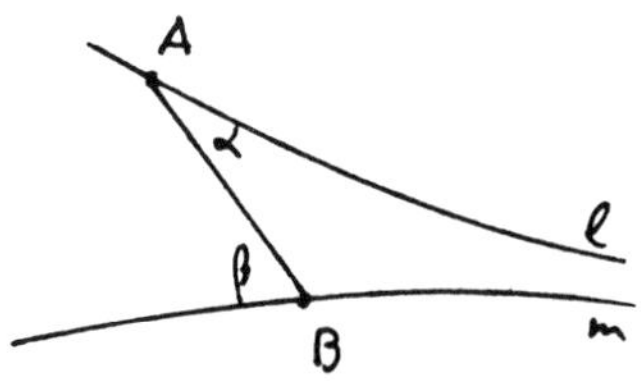

35.7 The AAL congruence theorem: Suppose given two limit triangles $ABlm$ and $A'B'l'm'$. Assume the angles at A and B are congruent to the angles at A' and B', respectively. Then show that the triangles are congruent, namely $AB \cong A'B'$.

35.8 Show that the sum of the two interior angles in a limit triangle is less than two right angles.

35.9 Show that limiting parallel lines do not exist in the non-Archimedean example of (18.4.3).

35.10 If l, m are two parallel rays which are not limiting parallels, show that there is a point P on l such that the feet of all perpendiculars from points of m to l lie on the same side of P.

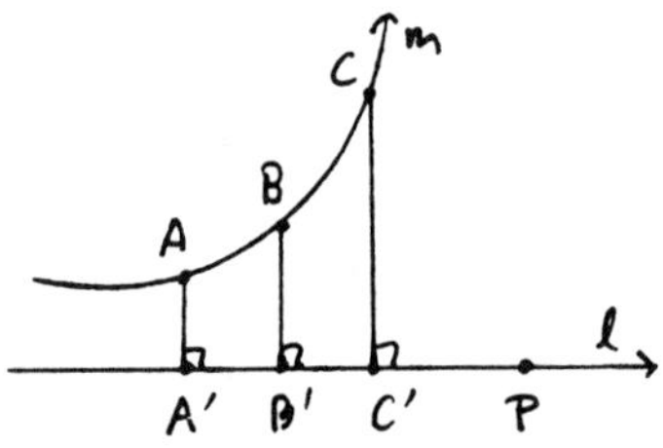

36 Non-Euclidean Area

The discussion of area arose quite early in the development of non-Euclidean geometry. Gauss had already thought about area as early as 1799 (see Exercise 36.1), and Bolyai's fundamental treatise of 1832 has a number of results on area. The defect of a triangle, as we will see, provides a natural measure of area, which is *absolute* in that it does not depend on any arbitrary choice of unit interval, as is the case for the Euclidean area function.

To make our treatment rigorous, we will follow Hilbert's method, which we used already for Euclidean area in §22, 23. So recall that in any Hilbert plane, two figures are called *equidecomposable* if they can be written as the unions of congruent triangles. Two figures have *equal content* if one can add to them equidecomposable figures so that the whole becomes equidecomposable. We showed in §22 that both of these notions are equivalence relations.

In §23 we defined the notion of a measure of area function. In most texts this is taken to have values in the real numbers, but true to our principle of not imposing the real numbers on geometry, we prefer to have it take values in a group which arises naturally. For the non-Euclidean case, this is an ordered abelian group whose elements are constructed out of finite sums of angles.

To be precise, we proceed as follows. Recall first that in our development of a Hilbert plane, an *angle* is simply two rays emanating from a point, which do not lie on the same line. Thus there is no zero angle, and every angle is less than two right angles. We have defined addition of angles only when the sum is less than 2RA (cf. §9). We define a set

$$A = \{0\} \cup \{\text{angles less than RA}\},$$

and we take

$$G = \mathbb{Z} \times A$$

to be the direct product set. Define addition on G by

$$(n_1, \alpha_1) + (n_2, \alpha_2) = \begin{cases} (n_1 + n_2, \alpha_1 + \alpha_2) & \text{if } \alpha_1 + \alpha_2 < \text{RA} \\ (n_1 + n_2 + 1, \alpha_1 + \alpha_2 - \text{RA}) & \text{if } \alpha_1 + \alpha_2 \geq \text{RA}. \end{cases}$$

In this definition, if either α_1 or α_2 is the symbol 0, we interpret it by setting $0 + \alpha_2 = \alpha_2$, and $0 + 0 = 0$, etc. Define the lexicographic order on the set G by setting

$$(n_1, \alpha_1) < (n_2, \alpha_2)$$

if either $n_1 < n_2$, or $n_1 = n_2$ and $\alpha_1 < \alpha_2$.

Proposition 36.1 *In any Hilbert plane, the set G defined above, with the operation + and the relation <, is an ordered abelian group.*

Proof. This is pretty much obvious, using the known properties of angles from §9. The element $(0,0)$ acts as zero element for the group. The inverse of an element (n, α) is $(-n, 0)$ if $\alpha = 0$, otherwise it is $(-n-1, \text{ RA } - \alpha)$. The trichotomy for the order relation follows from the corresponding fact for angles (9.5).

Remark 36.1.1. There is a natural homomorphism of the group G to the group R of rotations around a fixed point O of the plane, defined as follows.

Fix a ray OA emanating from O, and let the ray OC make a right angle with OA. For any angle $\alpha <$ RA, choose a ray OB inside the right angle to represent α. Now send $(1,0) \in G$ to the rotation which sends A to C; send $(0, \alpha) \in G$ to the rotation which sends A to B, and extend by linearity (cf. (Exercise 17.4) for facts about rotations).

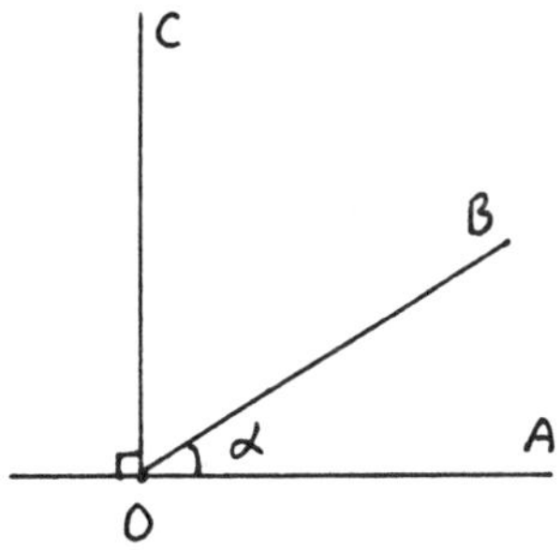

This homomorphism is surjective, with kernel the subgroup $\mathbb{Z}$ generated by $(4,0)$. Thus the elements of G correspond to "rotations with winding number". We will call G the *unwound circle group* of the given Hilbert plane (cf. Exercise 17.6).

Example 36.1.2. In the real Cartesian plane, the group G is isomorphic to $(\mathbb{R}, +)$, by sending $(1,0)$ to $\pi/2$ and $(0, \alpha)$ to the radian measure of α.

Now that we have a suitable group for it to take values in, we can show the existence of a measure of area function in non-Euclidean geometry.

Theorem 36.2 *In a non-Euclidean Hilbert plane satisfying Archimedes' axiom* (A), *there is a measure of area function α with values in the unwound circle group G of the plane. It is uniquely determined by the additional condition that for any triangle T, its value is equal to the defect δ of the triangle. (Here for any angle α, we identify α with the element* $(0, \alpha) \in G$ *if* $\alpha <$ RA, *or* $(1, \alpha -$ RA$)$ *if* $\alpha \geq$ RA.*)*

Proof. For any figure P, we write P as a union of triangles T_i, and define $\alpha(P) = \Sigma\delta(T_i)$, where δ denotes the defect of a triangle (§34) and the sum is taken in the group G. Because of the Saccheri–Legendre theorem and its consequences (§34), this will be a positive element of the group G. It gives the same value for congruent triangles, and it is additive for non-overlapping figures, so the only problem is to show that it is well-defined. Any two triangulations of a figure can be refined by a third (23.5), so we have to show it is additive for an arbitrary subdivision of a triangle into smaller triangles. The case of a triangle cut in two by a single transversal is given in (34.4). This corresponds to Step 1 in the proof of (23.4). The remaining steps of the proof of (23.4) are valid in our case also, and this completes the proof.

Remark 36.2.1. It now follows from (23.1) that equidecomposable figures and figures of equal content have the same area. The property (Z) of §22, de Zolt's axiom, also holds.

Remark 36.2.2. We will see later (Exercise 42.10) that for any angle δ, there exists a triangle with area $= \delta$, so that the image of the measure of area function α is just the set of positive elements of G.

To illustrate this theory of area, we will give the neutral geometry analogue of Euclid's (I.37), that "triangles on the same base and in the same parallels are equal".

Proposition 36.3 *In a Hilbert plane, any triangle has equal content to a suitable Saccheri quadrilateral. If furthermore we assume* (A), *the two are equidecomposable.*

Proof. We use the method of (34.6). The construction given there shows that the triangle ABC is equidecomposable with the Saccheri quadrilateral $FHBC$ if G lies between F and H. If G lies outside the interval FH, the proof given shows they have equal content.

Now let us assume (A), and let ABC be any given triangle, with associated quadrilateral $FHBC$. Join CD and extend to A' so that $CD = DA'$. Join $A'B$. Then by (SAS) the triangles ACD and $BA'D$ are congruent. Leaving BCD fixed, this gives a dissection of ABC into $A'BC$. Furthermore, they have the same midline, and the same Saccheri quadrilateral, and $D'D = DE = \frac{1}{2}FH$ (Exercise 36.6).

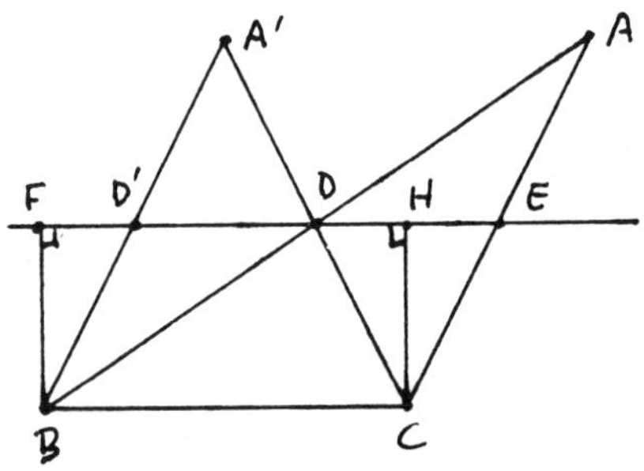

Repeating this process a finite number of times, and using (A), we transform the original triangle ABC by a finite succession of dissections into a new triangle A^*BC, such that the foot G^* of the perpendicular from A^* to FH lies between F and H. Then the method of (34.6) gives a dissection to the Saccheri quadrilateral.

Corollary 36.4 *Triangles on the same base and with the same midline have equal content. Furthermore, assuming* (A), *they are equidecomposable.*

Proof. Let ABC and $A'BC$ be two triangles with the same base BC and the same midline l. Drop perpendiculars BF, CH to l. Then by the proposition, both triangles have content equal to the Saccheri quadrilateral $FHBC$, so they have equal content to each other (22.3). In the Archimedean case, they are equidecomposable.

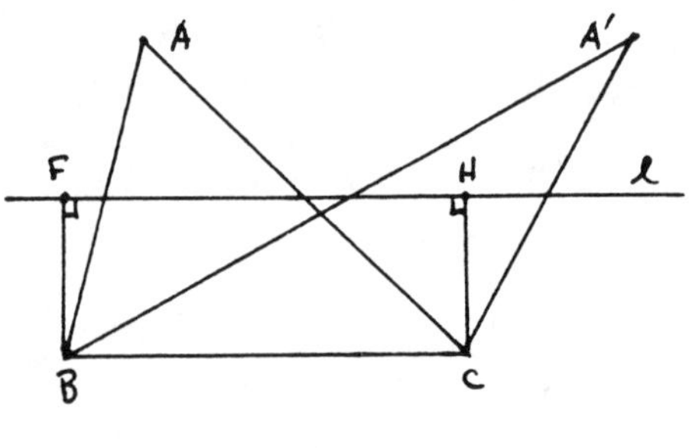

Remark 36.4.1. Note how the Euclidean hypothesis "lying in the same parallels" of (I.37) has been replaced by "having the same midline". Of course in the Euclidean case this is equivalent, because A and A' will lie on the same line parallel to BC. But in non-Euclidean geometry, the locus of points A for which the triangle ABC has midline l is not a straight line: it is the set of points having the same distance from l as B and C, but on the opposite side of the line l. Given a line l and a segment d, we call the set of points at distance d from l an *equidistant curve* (or *hypercycle*). It has two branches, one on either side of l. So we could rephrase this result as "Triangles on the same base BC, and having all three vertices A, B, C on the same equidistant curve, B, C on one branch, A on the other, have equal content".

Using ideas from the above proofs, we can now establish the non-Euclidean analogue of the theorem of Bolyai–Gerwien (24.7). We start with the case of triangles.

Proposition 36.5 *In a non-Euclidean Hilbert plane satisfying* (A) *and* (E), *any two triangles of equal area are equidecomposable.*

Proof. We use the area function of (36.2), so that equal area means equal defect.

First let us consider the case where one side of the first triangle is equal to one side of the second triangle. We label the equal sides BC, so we can call the triangles ABC and $A'BC$. By (36.3) we can find Saccheri quadrilaterals $FHBC$ and $F'H'BC$ which are equidecomposable with the given triangles. Furthermore, the defects of these quadrilaterals are equal to the defects of the triangles (34.6), and so are equal to each other. Since the top angles of a Saccheri quadrilateral are equal to each other (34.5), it follows that the angles at B and C of the two quadrilaterals are the same, so the sides of the quadrilaterals lie on the same lines.

If FG is not equal to $F'G'$, then $FGF'H'$ is a rectangle, which is impossible in the non-Euclidean case (34.8). So $FG = F'G'$, the two Saccheri quadrilaterals are equal, and the two triangles are equidecomposable.

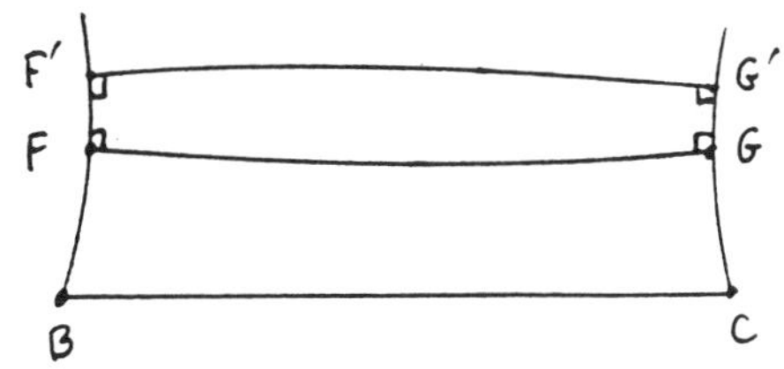

Now consider the case of two arbitrary triangles ABC and $A'B'C'$. Suppose $AB < A'B'$. Then $\frac{1}{2}A'B' > BD$, so using the axiom (E) we can find a point D^* on the midline DE, such that $BD^* = \frac{1}{2}A'B'$. Extend BD^* to A^* so that $BD^* = D^*A^*$. Join A^*C.

Then ABC and A^*BC are triangles with the same base BC and the same midline. Therefore they are equidecomposable (36.4), and also they have the same defect. On the other hand, A^*BC and $A'B'C'$ have one side equal $A^*B = A'B'$, and they have the same defect, so by the previous case, they are equidecomposable.

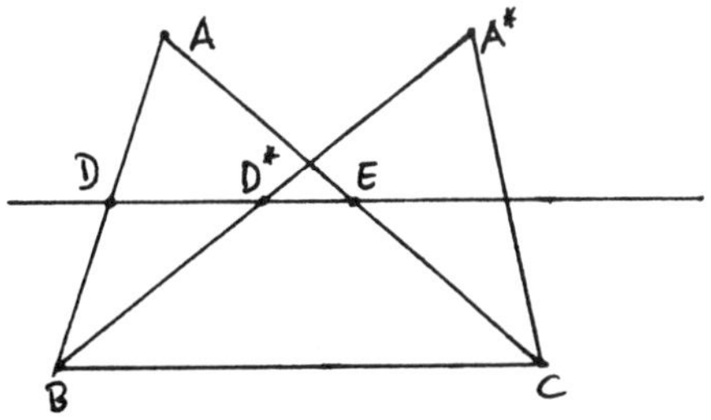

Theorem 36.6 *In a non-Euclidean Hilbert plane with* (A) *and* (E), *any two rectilineal figures with the same area are equidecomposable.*

Proof. Suppose rectilineal figures P and P' have the same area (36.2). Let us write P as a union of triangles $T_1, \ldots, T_n$, and P' as the union of triangles $T_1', \ldots, T_m'$. We will work by induction on the total number of triangles $n+m$. If that number is 2, then P and P' are triangles, and we need only apply (36.5).

If there are two triangles among the T_i which both have area $<$ RA, then we use the lemma below to replace them by a single triangle, thus reducing the sum $n+m$, and we win by induction. The same holds if there are two triangles among the T_i' with area $<$ RA.

So we may assume there is at most one triangle among the T_i and at most one among the T_i' with area $<$ RA. If there are no triangles with area $\geq$ RA we are then done by (36.5). So suppose $\alpha(T_1) \geq$ RA. Then there must be also a T_1' with $\alpha(T_1') \geq$ RA. If $\alpha(T_1) = \alpha(T_1')$ we use (36.5) and treat the difference figures $P - T_1$ and $P' - T_1'$ by induction. So suppose $\alpha(T_1) < \alpha(T_1')$.

Using the technique of the proof of (36.5), we may assume that T_1 and T_1' have a common side BC. We will show that each triangle is equidecomposable with an isosceles triangle having the same base.

Given a triangle ABC, consider the associated Saccheri quadrilateral $FHBC$, and let KL be the line joining the midpoints of the top and bottom (34.5). Choose D' on KF so that $KD' = \frac{1}{2}DE$. Then extend BD' to A' so that $BD' = D'A'$, and join $A'C$. As we have seen above, ABC and $A'BC$ are equidecomposable. But furthermore, by construction, $A'BC$ is isosceles (Exercise 36.6).

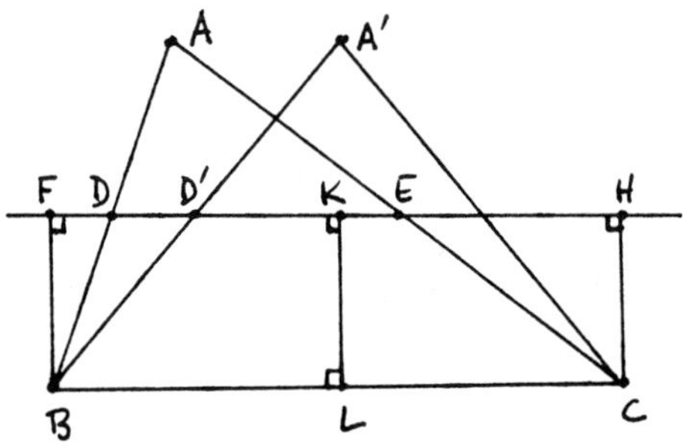

Applying this construction to both T_1 and T_1', we may assume that they are both isosceles triangles on the same base BC. Then we can place T_1 inside T_1'. Let $T_1 = ABC$ and $T_1' = A'BC$. Then we subdivide T_1' into three triangles ABC, $AA'B$, and $AA'C$. We can remove $T_1 = ABC$, obtaining new figures $P - T_1$ and $P' - ABC$, and a new decomposition with one less triangle on the first side, but one more triangle on the second side.

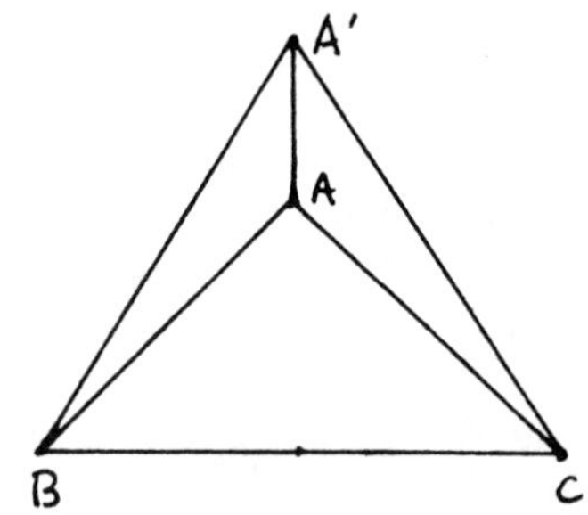

The sum $n' + m'$ is equal to $n+m$, so we cannot use induction yet, but we have gained something. Because of our hypothesis RA $\leq \alpha(T_1) < \alpha(T_1')$ and the fact in any case $\alpha(T_1') < 2$ RA, we see that $\alpha(AA'B) + \alpha(AA'C) <$ RA. So then we apply the lemma, and we are done by induction.

Lemma 36.7 *If T_1 and T_2 are triangles, both having area < RA, then the union $T_1 \cup T_2$ is equidecomposable with a single triangle T.*

Proof. Our strategy is to replace T_1 and T_2 by equidecomposable right triangles having a common side, which we can then glue together along their common side into a single bigger triangle.

The first step is to show that any triangle with area < RA is equidecomposable with a triangle having an obtuse angle. We use the method of proof of the Saccheri–Legendre theorem (34.2). Each step is a dissection of one triangle into the next, so we can replace the given triangle T by another, two of whose angles are arbitrarily small. Since the sum of three angles is > RA, the remaining third angle then will be obtuse. Note the use of (A) in this step.

Next, if ABC is a triangle with an obtuse angle at C, then using the technique of proof of (36.5), we can find an equidecomposable triangle $A'BC$, where $A'C$ is any segment we like bigger than AC, and the new triangle will still have an obtuse angle at C.

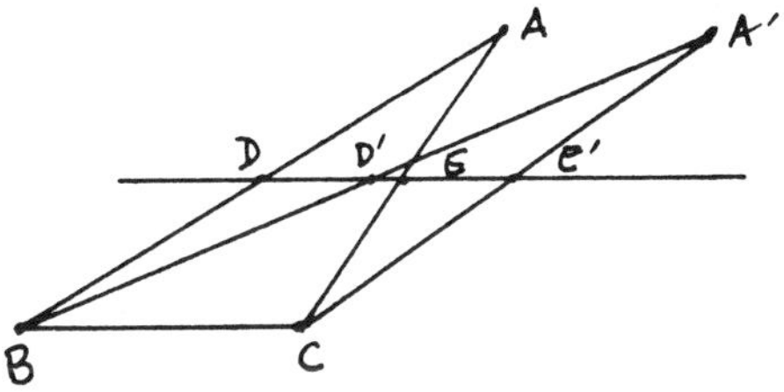

Thus we may assume that our two triangles T_1 and T_2, changing notation, have the same base BC, and both have obtuse angles at C.

Now let ABC be a triangle with an obtuse angle at C. If we draw a line at right angles to BC at C, this line will fall inside the angle HCE, because the angle of the Saccheri quadrilateral $FHBC$ at C is acute (34.8.1), and the angle BCA is obtuse. Then by the crossbar theorem (7.3) it must meet the line FH at a point E'. Extend CE' to A' so that $CE' = E'A$. Then $A'BC$ is a right triangle with side BC and equidecomposable to ABC.

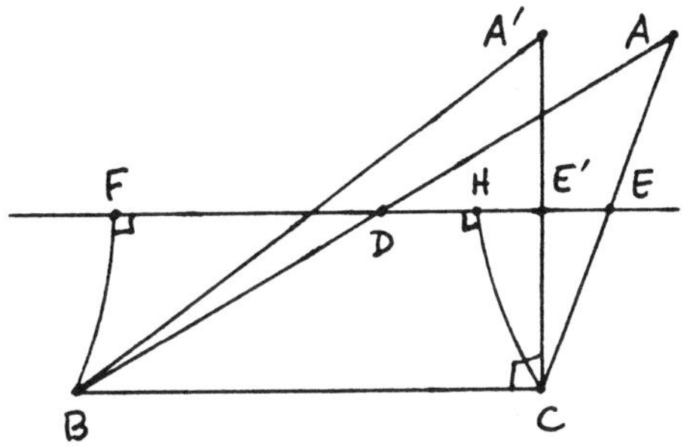

Do this to both T_1 and T_2. Thus they are replaced by equivalent right triangles with a common side, and these we can glue together into a bigger single triangle.

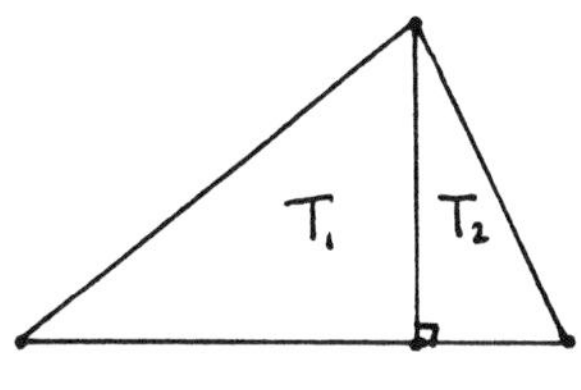

Remark 36.7.1. One could simplify the above proof considerably if one could show the following: Given a triangle $T = ABC$ and another triangle T' with $\alpha(T') < \alpha(T)$, then there exists a point $D \in BC$ such that $\alpha(ABD) = \alpha(T')$.

This statement follows easily from continuity and so can be used if you admit Dedekind's axiom (D). It can also be proved using hyperbolic trigonometry (Exercise 42.12), but I do not know any direct proof under the hypotheses of (36.6).

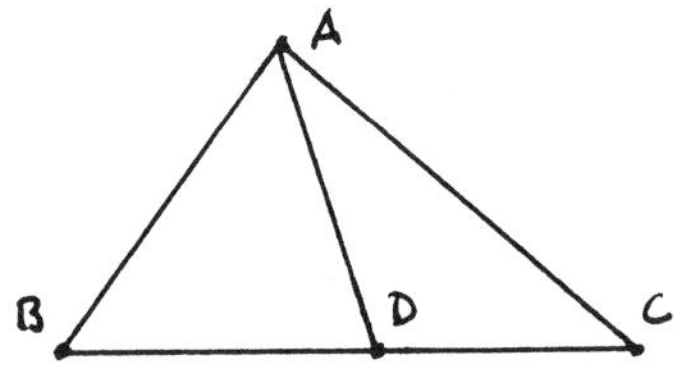

Remark 36.7.2. It follows from (36.6) that if two figures in a non-Euclidean Hilbert plane with (A) and (E) have equal content, they are equidecomposable—cf. (22.1.4). Indeed, they will have the same area (23.1), using the area function of (36.2), and then the theorem applies.

Exercises

The following exercises, unless otherwise indicated, take place in a non-Euclidean Hilbert plane with (A).

36.1 Gauss, in a letter to W. (= Farkas) Bolyai in 1799 said "if one could prove that there is a triangle whose area is bigger than any given figure, then I could prove the whole geometry". Let us call this **Gauss' axiom**, that for any rectilineal figure, there exists a triangle whose content is bigger than that of the figure. Show that Gauss' axiom, assuming (A), implies (P), by showing that

a) the area of any triangle is < 2RA, and

b) there exist (even convex) rectilineal figures with arbitrarily large area.

36.2 Let ABC be an isosceles triangle, let D, E be the midpoints of the sides, and let AG be the altitude which bisects DE and BC.

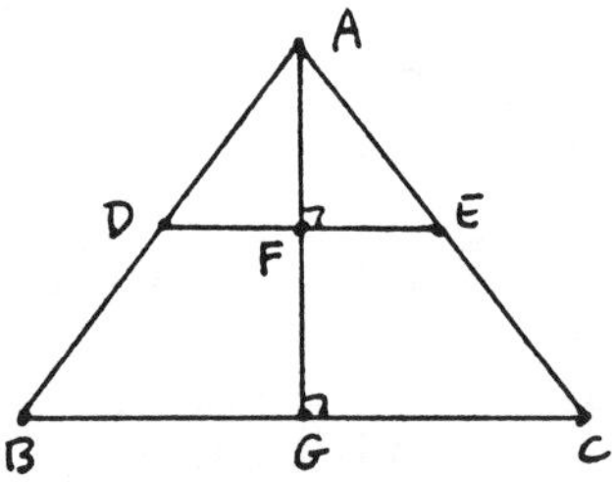

a) Show that $AG < 2AF$.

b) Show that $BC > 2DE$.

c) In the Euclidean case, the area of ABC would be four times the area of ADE. In the non-Euclidean case, show that there is no such multiple estimate for comparing the two areas, in the following sense: For any given constant $k > 1$, there exists an isosceles triangle ABC as above such that the area of ABC is $< k \cdot$ (area of ADE). (In order for multiplication by k to make sense in the unwound circle group G, take k to be a *dyadic rational* number, i.e. of the form $a/2^b$ for a, b integers.)

36.3 Let P be a polygon inscribed in a circle of radius r. Show that there is a polygon P' inscribed in a circle of radius $2r$, having twice as many sides as P, and whose area is at least twice the area of P. This shows that the area of a circle (if properly defined!) can become arbitrarily large.

36.4 If a Saccheri quadrilateral $ABCD$ is cut by a diagonal AD, show that the upper triangle ACD has area greater than the lower triangle ABD. *Hint:* First show that $\beta' < \beta$.

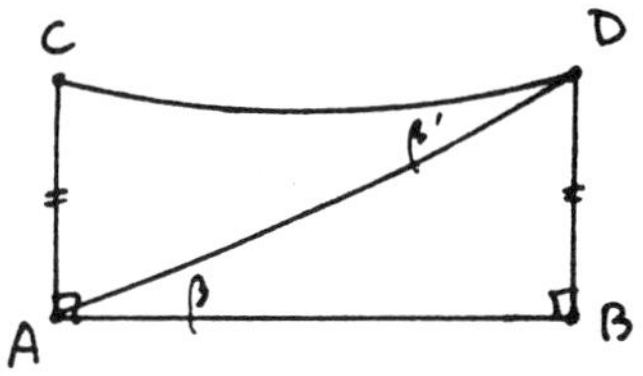

36.5 Let a triangle ABC be cut by its median AD. Assuming that the triangle is not isosceles, show that the half (in this drawing ABD) containing the acute angle α at D has greater area than the other half.

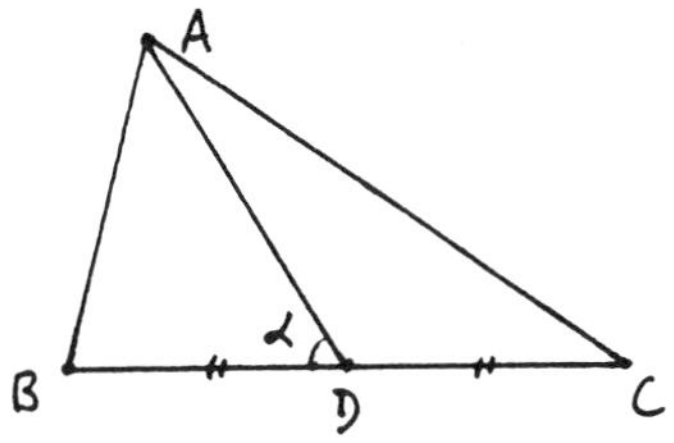

36.6 a) In the situation of (34.6), show that $DE = \frac{1}{2}FH$.

b) In the proof of (36.6), verify that the triangle claimed to be isosceles is indeed so.

36.7 We have seen that in the real Cartesian plane, the unwound circle group G is isomorphic to $(\mathbb{R}, +)$ (36.1.2). This group is also isomorphic to the multiplicative group $(\mathbb{R}_{>0}, \cdot)$ via the exponential map.

In the case of an abstract ordered field F, we have three ordered abelian groups naturally associated with F, the additive group $(F, +)$, the multiplicative group $(F_{>0}, \cdot)$, and the unwound circle group G of the Cartesian plane over F. We can ask which of these may be isomorphic, as ordered abelian groups.

a) Let F be the constructible field (16.4). Show that $(F, +)$ is not isomorphic to either $(F_{>0}, \cdot)$ or G as an ordered abelian group. *Hint:* Think of dividing by 3 in each group.

b) Again with F the constructible field, show that if $(F_{>0}, \cdot)$ is isomorphic to G, then there can be only finitely many Fermat primes (29.4.1). (I have no reason to believe that these two groups should be isomorphic, so don't get your hopes up that this might be a way to prove the finiteness of the set of Fermat primes!)

37 Circular Inversion

In this section we will study circular inversion, which is a kind of transformation of the plane which leaves the points of a given circle fixed, and sends points inside the circle to points outside and vice versa. While this study belongs in the context of Euclidean geometry, it is a technique not used by Euclid. Perhaps this is because the idea of a *transformation* of the plane, moving points to other points, was foreign to the way of thinking of the Greeks. Euclid does use the "method of superposition" to compare triangles in his proof of (SAS), but there is no evidence that he thought of a rigid motion moving the whole plane onto itself. Given that perspective it seems even less likely that the Greeks would have seen any value in a transformation of the plane which does not even preserve distances, in fact which does not even preserve proportion in figures. It seems that the notion of transformation in geometry is a relatively recent notion, which has come to serve very important roles, as we have seen in the usefulness of the existence of rigid motion (ERM).

The theory of circular inversion can be developed purely geometrically using the results of Euclid's Book III, but it will be more efficient to use the theory of proportion (similar triangles) in our proofs. So in this section we work in the Cartesian plane over an ordered field F satisfying (**) (16.2). Thus we have Hilbert's axioms, including (P) and (E), and we can use the theory of similar triangles (§20).

Definition. Let Γ be a fixed circle in the plane (over the field F as above...), with center O and radius r. For any point $A \neq O$, draw the ray OA, and let A' be the unique point on the ray OA such that $OA \cdot OA' = r^2$. (This equation means products of lengths in the field F.) Then we say that A' is obtained from A by *circular inversion* with respect to the circle Γ.

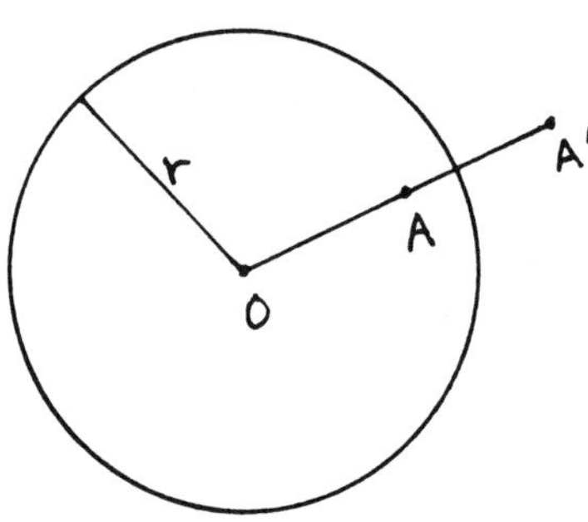

Briefly, we will say that A' is the *inverse* of A (with respect to the circle Γ). Since the condition $OA \cdot OA' = r^2$ is symmetric in A and A', this notion is reciprocal: A' is the inverse of A if and only if A is the inverse of A'. We think of circular inversion in Γ as a transformation $\rho = \rho_\Gamma$ defined for all points $A \neq O$, which thus transforms the plane $\Pi - \{O\}$ into itself by sending each point to its inverse. From the definition it is clear that any point *on* Γ is sent to itself. Points inside Γ are sent to points outside Γ, and vice versa. As a point A approaches the center of the circle O, its inverse gets farther and farther away, so in the limit, the point O would have to go to infinity. Since we don't have infinity in our geometry, we simply say ρ is *undefined* at O. (Or, if you like, you can imagine completing our plane by a single point called infinity, and then O and ∞ are interchanged—cf. (Exercise 37.1) on stereographic projection for another interpretation of this idea.)

Proposition 37.1 *Let A be a point inside the circle Γ. Draw the ray OA. Let PQ be the chord of the circle perpendicular to OA at A. Then the tangents to Γ at P and Q will meet the ray OA at the point A' which is the inverse of A with respect to Γ.*

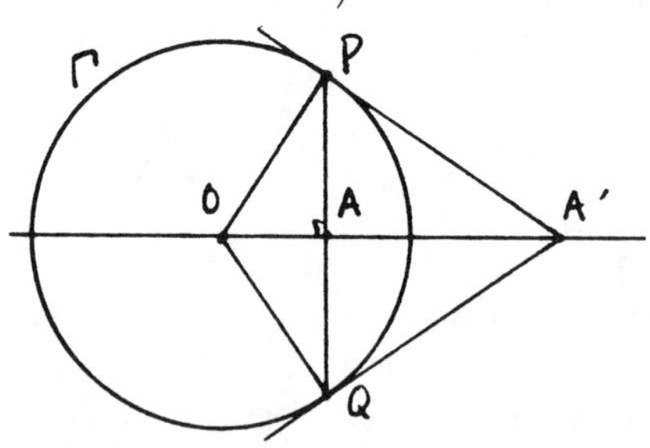

Proof. First note that the two tangents will both meet OA in the same point A', by symmetry. Now the right triangles ΔOAP and $\Delta OPA'$ have the angle at O in common, so they are similar. Hence corresponding sides are proportional, in particular

$$\frac{OA}{OP} = \frac{OP}{OA'}.$$

Cross multiplying

$$OA \cdot OA' = OP^2 = r^2.$$

Hence A and A' are circular inverses in Γ.

Remark 37.1.1. This proposition gives us a method of constructing circular inverses by ruler and compass: If A is given inside Γ, draw OA, construct the perpendicular to OA at A, let it meet Γ at P, draw the radius OP, draw the perpendicular to OP at P, which will be the tangent line, and let this line meet OA at A' (9 steps). Conversely, if the point A' is given outside the circle, draw the two tangent lines from A' to Γ, join their points of tangency P, Q, and let the line PQ meet OA' at A (6 steps).

Next we will investigate what circular inversion does to lines and circles in the plane.

Proposition 37.2 *A line through O is transformed into itself by circular inversion. A line not passing through O will be transformed into a circle passing through O, and conversely.*

Proof. A line through O is transformed into itself by definition of circular inversion.

Now let l be a line not through O. Let OA be the perpendicular from O to l. Let A' be the inverse of A, and let γ be the circle with diameter OA'. I claim that the inverses of points on l all lie on γ and vice versa. So let B be any point on l. Draw OB and let it meet γ at B'. Then $OB'A'$ is a right triangle (III.31). It has the angle at O in common with the right triangle OAB, so we have similar triangles. Therefore the sides are proportional:

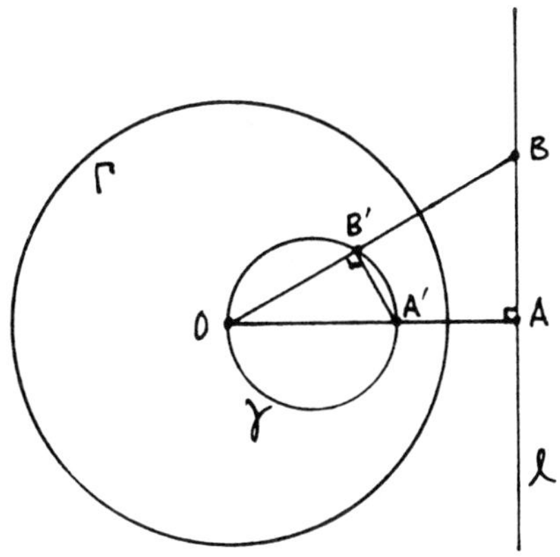

$$\frac{OB'}{OA'} = \frac{OA}{OB}.$$

Cross multiplying

$$OB \cdot OB' = OA \cdot OA'.$$

But A' was chosen to be the inverse of A, so $OA \cdot OA' = r^2$. Hence also $OB \cdot OB' = r^2$, so B and B' are inverse to each other. This shows that the circular inversion transforms the points of the line l to the points of the circle γ (except O) and vice versa.

Definition. When two circles meet (or when a circle meets a line) by the *angle* between them we mean the angle between their tangent lines at that point (resp. the angle between the tangent line and the other line).

Note that when two circles meet in two points, the angle between them is the same at both points, because the two circles are symmetrical about the line joining their two centers.

Proposition 37.3 *If a circle γ is perpendicular to Γ (at its intersection points), then γ is transformed into itself by circular inversion in Γ. Conversely, if a circle γ contains a single pair A, A' of inverse points, then γ is perpendicular to Γ and is sent into itself.*

Proof. First suppose that γ is perpendicular to Γ, and let γ meet Γ at P and Q. Then the radius OP is tangent to γ, because radius and tangent of any circle are perpendicular (III.18). Let A be another point of γ and let OA meet γ again at A'. Applying (III.36) (cf. 20.7) to γ we find $OP^2 = OA \cdot OA'$. (Actually Euclid meant the square on OP has content equal to the rectangle formed by OA and OA', but since we are working over the field F, we interpret this statement as lengths and products (20.7).) Since $OP = r$, this shows that A and A' are inverses. This holds for any A on γ, so γ is sent into itself.

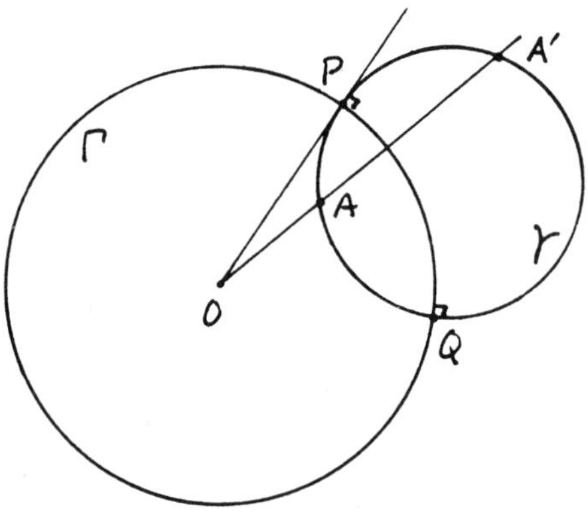

Now (using the same picture) suppose conversely that γ is any circle passing through some pair of inverse points A, A'. Let γ meet Γ at P, and draw OP. Since OP is a radius and A, A' are inverse points, we have $OA \cdot OA' = OP^2$. But now by (III.37), the converse of (III.36) it follows that OP is tangent to γ, which means that γ and Γ are perpendicular at P (and hence also at their other point of intersection Q).

Proposition 37.4 *If γ is a circle, not passing through the center O of Γ, then the transform of γ by circular inversion is another circle γ'.*

Proof. This result is not so easy to prove directly (you can try if you like) so we will resort to a trick. Suppose given a circle γ not passing through O, and let us assume O is outside γ, for the moment.

Draw OP tangent to γ, and let Γ' be a new circle with center O, passing through P. Then by construction γ is sent into itself by $\rho_{\Gamma'}$. Thus $\rho_\Gamma(\gamma) = \rho_\Gamma \cdot \rho_{\Gamma'}(\gamma)$ and we are led to consider the new transformation of the plane $\theta = \rho_\Gamma \cdot \rho_{\Gamma'}$. Let $r =$ radius of Γ, and $r' =$ radius of Γ'. For any point A, let $A' = \rho_{\Gamma'}(A)$, $A'' = \rho_\Gamma(A')$. Then $\theta(A) = A''$. By definition of inversion, $OA \cdot OA' = r'^2$ and $OA' \cdot OA'' = r^2$. Dividing, we find

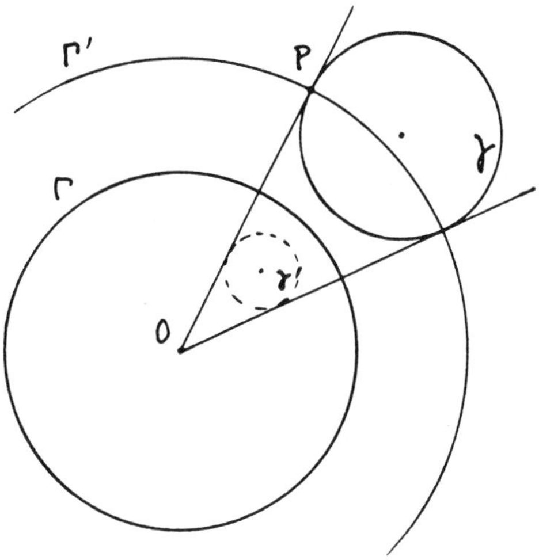

$$\frac{OA''}{OA} = \frac{r^2}{r'^2}$$

so

$$OA'' = k \cdot OA \text{ where } k = r^2/r'^2.$$

So this is a transformation which leaves O fixed, and stretches points toward (or away from) O in a fixed ratio k. This is called a *dilation* with center O and ratio k. In rectangular coordinates with center at the origin it would be expressed by

$$\begin{cases} x' = kx \\ y' = ky \end{cases}$$

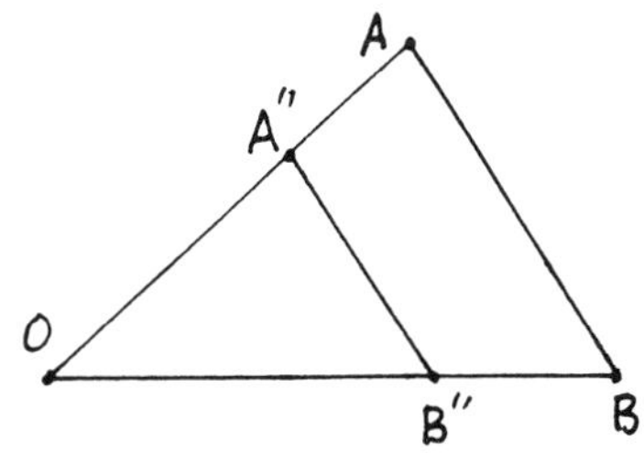

It follows, either from thinking of the distance formula in terms of coordinates, or by using the (SAS) criterion for similar triangles (20.4) that all distances are changed by the same ratio k.

In particular, a dilation sends any circle (and its center) into another circle and its center. It follows that $\rho_\Gamma(\gamma) = \rho_\Gamma \cdot \rho_{\Gamma'}(\gamma) = \theta(\gamma)$ is a circle. (Warning: Even though $\rho_\Gamma(\gamma)$ is a circle, in general ρ_Γ does not send the center of γ to the center of γ'.)

In this proof we were assuming O outside γ. If O is inside γ, we leave you to construct an analogous proof in (Exercise 37.4).

Now that we have seen that circular inversion preserves lines and circles (every line or circle is transformed into another line or circle), sometimes turning a line into a circle, or a circle into a line, the next step is to show that circular inversion is *conformal*, i.e. preserves angles.

Proposition 37.5 *Circular inversion is* conformal*: whenever two curves meet (here "curve" means line or circle), their transforms under circular inversion meet again at the same angle.*

Proof. First suppose that $P \notin \Gamma$, and that two curves (not shown) meet at P with tangent lines l, m. Let P' be the inverse of P. Then we can find a circle γ, through P, P' and with tangent m at P; and we can find a circle γ_2 through P, P' with tangent line l at P. Now by (37.3) γ_1 and γ_2 are transformed into themselves. Therefore the original curves are transformed into curves at P' tangent to γ_1 and γ_2, so they make the same angle as at P, because when two circles γ_1, γ_2 intersect they have the same angle at both intersections.

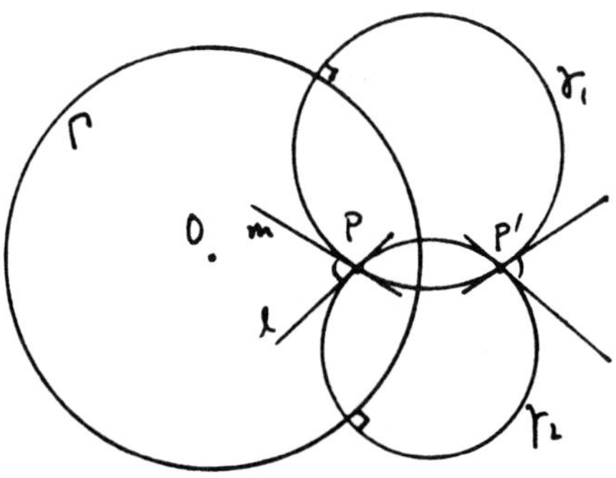

If $P \in \Gamma$, we leave the special case to you (Exercise 37.5).

For our last general result about the properties of circular inversion, we look at what happens to distances. Of course distances are not preserved, because a very small distance near O will be transformed into a very large distance far away. Even ratios of distances are not preserved, as you can see by simple examples. However, a remarkable fact is that if we take four points, a certain ratio of ratios of distances, called their cross-ratio, is preserved.

Definition. Let A, B, P, Q be four distinct points in the plane. Their *cross-ratio* is defined to be the ratio of ratios

$$(AB, PQ) = \frac{AP}{AQ} \div \frac{BP}{BQ},$$

which can also be written

$$\frac{AP}{AQ} \cdot \frac{BQ}{BP}.$$

Proposition 37.6 *Let A, B, P, Q be four distinct points in the plane, $\neq O$. Then the circular inversion in Γ preserves the cross-ratio: if their inverses are A', B', P', Q', then*

$$(AB, PQ) = (A'B', P'Q').$$

Proof. Given two points A, P and their inverses A', P', we know by definition that

$$OA \cdot OA' = r^2 = OP \cdot OP'.$$

Thus

$$\frac{OA}{OP} = \frac{OP'}{OA'}.$$

Case 1. Suppose O, A, P are not collinear. Since the triangles ΔOAP and $\Delta OP'A'$ have the angle at O in common, they are similar (20.4), and we conclude

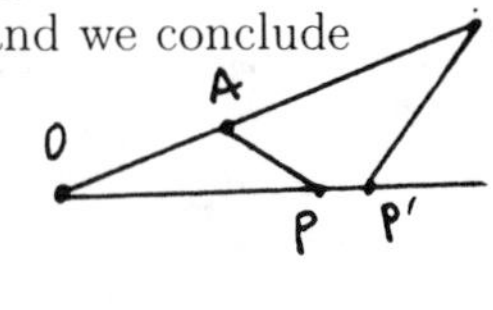

(1) $$\frac{AP}{A'P'} = \frac{OA}{OP'}.$$

Case 2. If O, A, P are collinear , then $AP = OP - OA$ and $A'P' = OA' - OP'$. Using the fact that in a field F, $\frac{a}{b} = \frac{c}{d} \Rightarrow \frac{a}{b} = \frac{c-a}{d-b}$ we conclude same result (1).

Now if Q is another point we find similarly

(2) $$\frac{AQ}{A'Q'} = \frac{OA}{OQ'}.$$

Dividing we get

(3) $$\frac{AP}{A'P'} \div \frac{AQ}{A'Q'} = \frac{OQ'}{OP'}.$$

Now let B be another point. Working with P and Q as before we find similarly

(4) $$\frac{BP}{B'P'} \div \frac{BQ}{B'Q'} = \frac{OQ'}{OP'}.$$

So the expressions (3) and (4) are equal. Moving the primed letters to one side and the unprimed letters to the other side shows the cross ratios (AB, PQ) and $(A'B', P'Q')$ are equal.

Remark 37.6.1. At this point I can just hear someone asking "What is the geometrical significance of the cross-ratio?". Although I first encountered cross-ratios as a senior in high school, and have dealt with them many times since then, I must say frankly that I cannot visualize a cross-ratio geometrically. If you like, it is magic. Here is this algebraic quantity whose significance it is impossible to understand, and yet it turns out to do something very useful. It works. You might say it was a triumph of algebra to invent this quantity which turns out to be so valuable, and could not be imagined geometrically. Or, if you are a geometer at heart, you may say it is an invention of the devil and hate it all your life.

Let me say a few words in defense of the poor cross-ratio.

In the present context of transformations of the Euclidean plane, there are the rigid motions which preserve distance. Then there are the dilations and stretch-rotations of §17, which do not preserve distance, but do preserve ratios of distances. Then there is the circular inversion which does not preserve distances or even ratios of distances. Since it does preserve the cross-ratio, that particular ratio of ratios, it is the best we can do. It is something to hang on to, a pillar of support, when all the distances and their ratios are changing all around us. In §39 we will use this cross-ratio to define the notion of distance in the Poincaré model of non-Euclidean geometry: it plays an essential role there.

In projective geometry the cross-ratio is also important. There is the notion of a *projectivity* from one line to another, defined as a composition of a finite number of projections from one line to another in the plane. A projectivity preserves neither distances nor ratios of distances, but it does preserve the cross-ratio. In fact, a transformation of one line to another in the projective plane is a projectivity *if and only if* it preserves the cross-ratio of every set of four distinct points on the line.

If you have studied complex variables, these projectivities correspond to the fractional linear transformations of the Riemann sphere $\mathbb{C} \cup \{\infty\}$ into itself, given by

$$z' = \frac{az+b}{cz+d} \qquad ad - bc \neq 0.$$

If you are given four points A, B, C, D, then there is always a fractional linear transformation sending A, B, C to $0, 1, \infty$. In that case the image of D, say λ, is essentially the cross-ratio of the original four points.

Finally, if you have four points on a line, and you take signed distances (+ or − depending on a chosen preferred direction), then the cross-ratio is equal to -1 if and only if the four points form a set of *four harmonic points*. This notion of harmonic points is also important in projective geometry.

According to Klein (1927), the notion of cross-ratio already occurs in the work of Pappus (300 AD). It came into prominence again in the early 19th century with the projective geometry of Poncelet and Monge.

Exercises

Unless otherwise noted, the following exercises take place in the Cartesian plane over an ordered field F satisfying (**).

37.1 *Stereographic projection.* In three-dimensional space, imagine our plane Π and a circle Γ of radius r and center O. Now take a sphere of radius $\frac{1}{2}r$, and set it on the plane Π so that its south pole is at O. The *stereographic projection* associates to each point B of the sphere, $B \neq N =$ north pole, that point of Π obtained by drawing the line NB and intersecting with Π. (In the limit, N would go to infinity, so you can think of the sphere as a completion of the plane by adding the point N.) Under this projection, the equator of the sphere is mapped to the circle Γ.

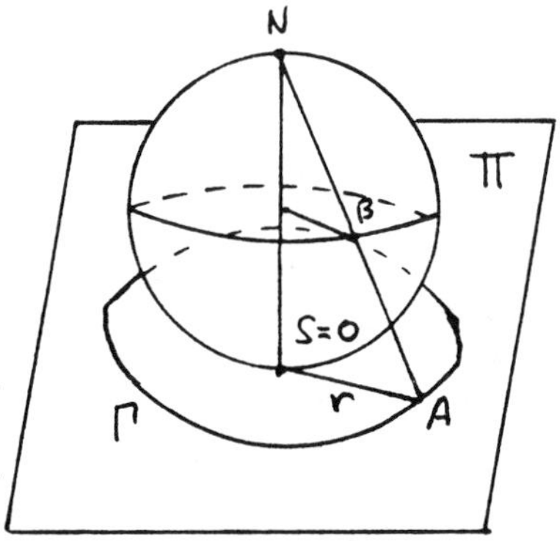

Show that circular inversion in the circle Γ corresponds to the operation of *reflection in the equator* of the sphere, which interchanges the northern and southern hemispheres. In other words, if B is a point on the sphere, and B' its reflection in the equator (same longitude, but latitude has changed from N to S or from S to N), then the projected points A, A' are inverses under inversion in Γ.

37.2 Prove that the following construction with compass alone gives the inverse of A in Γ (provided $OA \geq \frac{1}{2}r$): Draw a circle with center A through O, to meet Γ at P and Q. Then draw circles with centers P and Q through O to meet at A' (3 steps).

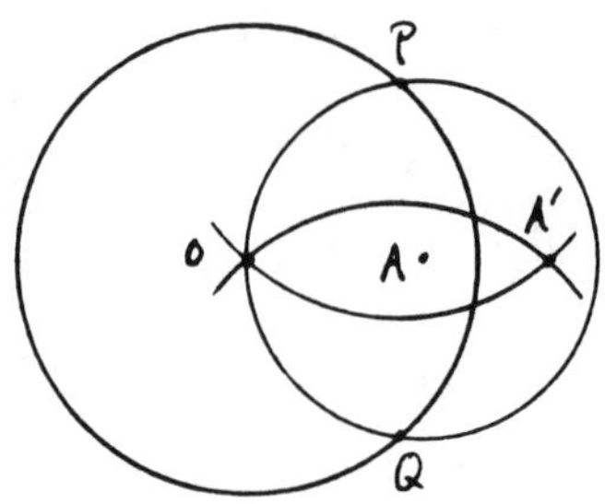

37.3 Let l be a line which meets the circle Γ in two points A, B. Let γ be the (unique) circle through O, A, B. Prove that γ is the transform of l under inversion in Γ.

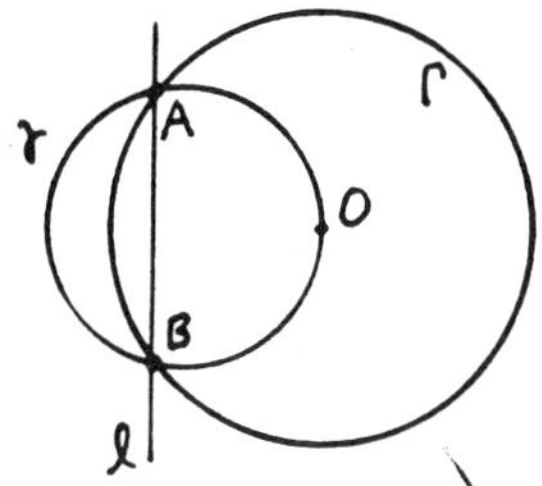

37.4 Prove the other case of (37.4), namely, if γ is a circle containing O, then $\rho_\Gamma(\gamma)$ is a circle. For any points A, B on γ, let A', B' be their inverses in Γ, and let A'', B'' be the points where the lines OA, OB meet γ again. By (III.35) (cf. (20.6))

$$OA \cdot OA'' = OB \cdot OB'' = c$$

is a constant independent of the points A, B, depending only on O, γ.

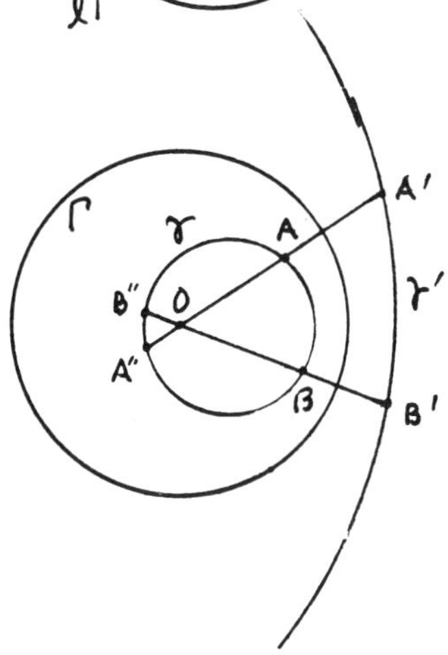

Let us use signed lengths from O, so that $c < 0$. Since $OA \cdot OA' = r^2$, then show that $OA' = k \cdot OA''$ for a certain constant $k < 0$. Thus γ' is obtained from the circle γ by dilation with a negative constant k. Conclude that $\rho_\Gamma(\gamma) = \gamma'$ is a circle.

37.5 If two curves meet at a point $P \in \Gamma$, show that their two transforms by circular inversion in Γ meet at the same angle at P.

37.6 If we identify the real Euclidean plane $\mathbb{R}^2$ with the complex numbers $\mathbb{C}$, show that the transformation $z' = 1/\bar{z}$ (where $z = a + bi$, $\bar{z} = a - bi$) is just inversion in the unit circle $|z| = 1$.

37.7 If A is a point inside the circle Γ, improve the ruler and compass construction of the inverse of A given in (37.1.1) by constructing the circle through O, P, Q instead of constructing the tangent line at P. (Par = 6 steps).

37.8 Given the circle Γ and its center O, and given a line l, give a ruler and compass construction of the circle $\rho_\Gamma(l)$. (Par = 7 steps).

37.9 Given Γ, given its center O, and given a circle γ passing through O (but not given the center of γ), construct the line $\rho_\Gamma(\gamma)$. (Par = 15 steps).

37.10 Given Γ and its center O, and given a circle γ not through O, construct the circle $\rho_\Gamma(\gamma)$. (Par = 15 steps).

37.11 Verify the following ruler only construction of the inverse of a point A (7 steps): Draw OA, get R, S. Draw any line l through A, meeting Γ in P, Q. Draw RP and SQ to meet at T. Draw RQ and PS to meet at U. Draw TU to meet OA at A'.

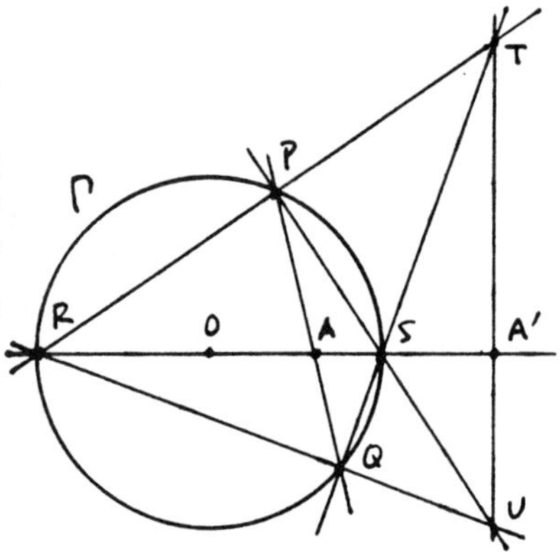

37.12 Verify the following 5-step construction for the inverse of a point A with respect to the circle Γ. Take a circle of any radius with center A, to meet Γ at P and Q. Let AP and AQ meet Γ in further points R, S. Join PS and RQ. Their intersection is A'. (This works equally well if A is inside Γ.)

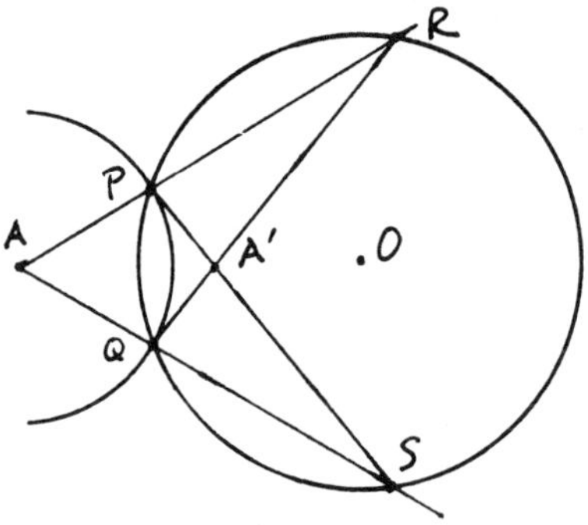

37.13 If F is an ordered field satisfying $(*)$ but not $(**)$ of §16, we can still define inversion in a circle Γ by the same method as above. If γ is another circle, with $\rho_\Gamma(\gamma) = \gamma$, show that γ does meet Γ in two points, even though we do not have the axiom (E).

Poor John has lost his ruler. Can you help him do his construction problems (following) using his compass alone?

37.14 Given two points A, B construct a third collinear point C with $AB = BC$. (Par $= 4$ steps).

37.15 Given two points A, B, construct the midpoint C of the segment AB. (Par $= 7$ steps). *Hint*: Use (Exercise 37.2).

37.16 Given two points A, B, construct C, D so that $ABCD$ will be a square. (Par $= 8$ steps).

37.17 Given points A, B, C, O, with O, A, B not collinear, construct the intersection points of the line AB with the circle OC (assuming they meet). (Par $= 4$ steps).

37.18 Given non-collinear points A, B, C, construct the foot of the perpendicular from C to the line AB. (Par $= 9$ steps).

37.19 Given non-collinear points O, A, B, show that it is possible to construct the intersection of the circle OA with the line OB using compass alone. *Hint*: Perform a circular inversion which leaves the circle OA fixed and transforms the line OB into a circle. (About 25 steps).

37.20 Given points A, B, C, D show that it is possible to construct the intersection point of the lines AB and CD using compass alone. *Hint*: Use a circular inversion to transform the two lines into circles. (About 44 steps).

37.21 Using the experience gained in the previous exercises, prove the theorem of Mascheroni: Any point which can be constructed from given data by ruler and compass construction can also be constructed using compass alone.

37.22 Make a new proof of (Exercise 1.15) by doing a circular inversion with center P and suitable radius, and solving the transformed problem.

38 Digression: Circles Determined by Three Conditions

In order to specify a circle in the plane, you give its center, which is a point, and its radius, which is a line segment, or distance. A point moves in a 2-dimensional plane, while the length of a line segment is a 1-dimensional quantity, so we can say roughly that the set of all circles forms a three-dimensional family, or that a circle depends on three parameters. If we are working in the Euclidean plane over a field F, then a general circle has an equation

$$(x-a)^2 + (y-b)^2 = r^2$$

where a, b, r are elements of the field F. In this case the circle is determined by the three quantities $a, b, r \in F$, so again it makes sense to say that a circle depends on three parameters.

From this informal discussion it is reasonable that we can impose three conditions and then expect to find a unique circle satisfying those three conditions. Of course there will be situations where this does not work. For example, you cannot find a circle passing through three collinear points (unless you allow the line containing the points to count as a limiting case of a circle). Another example is, you cannot find a circle tangent to two given parallel lines l, m and passing through a point P not between the two lines.

In general, however, we can expect to find a circle satisfying three conditions. Among conditions we can impose are

P) to require that the circle pass through a given point A

L) to require that the circle be tangent to a given line l

C) to require that the circle be tangent to a given circle γ.

Furthermore, it is natural to expect that the required circle be constructible by ruler and compass. In this way, taking all possible combinations of three conditions of types P), L), and C), we find the following ten construction problems.

IACOBI PELETARII
CENOMANI IN EVCLIDIS
ELEMENTA GEOMETRICA
DEMONSTRATIONVM
LIBER SECVNDVS.

Parallelogrammum Rectangulum.

OMne Parallelogrammum Rectangulum, ſub duabus rectum angulum conſtituentibus rectis lineis dicitur contineri.

Parallelogrammum Rectangulum fit ex ductu lineæ rectæ in lineam rectam: idq; inſtar Numerorum: facilitatis, vt diximus, gratia. Vbi etiam meminiſſe oportet, Ducere maiorem lineam in minorem, idem eſſe ac ſi minorem in maiorem ducamus: ſicut 3 in 4, & 4 in 3, idem efficiunt.

Gnomon.

In Parallelogrammis, alterutrum Parallelogrammorum quæ circa Dimetientem conſiſtunt cum duobus Supplementis, Gnomon vocatur.

Hoc iam antè in quadrageſimatertia Primi explicauimus. Si enim in Figura adſcripta, Parallelogrammũ H F cum duobus Supplementis C K & K B ſumamus, efficietur Gnomon E D G: ſeu, vt alij deſignant, C D G K: quod vnum & idem eſt. Quòd ſi Parallelogrammum E G cum iiſdem Supplementis ſumpſerimus, fiet Gnomon F A C K. Deeſt itaque Gnomoni alterutrum interiorum Parallelogrammorum quæ Dimetiens media ſecat: quæq; Complementa in quadrageſimatertia Primi nominauimus.

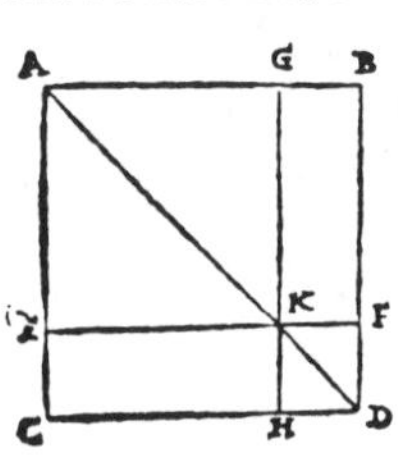

Plate XI. Beginning of Book II in Peletier's edition of Euclid (1557).

PPP. To construct a circle through three given points A, B, C. This is the *circumscribed* circle to the triangle ABC.

LLL. To construct a circle tangent to three given lines l, m, n. This is the *inscribed* circle of the triangle formed by l, m, n.

The remaining eight we designate as PPL, PLL, PPC, PLC, PCC, LLC, LCC, and CCC, according to the conditions imposed. The last one, to find a circle tangent to three given circles, is classically known as the problem of Apollonius. Apollonius of Perga (c. 262–c. 200 BC) is best known for his book on conics, but he also wrote a book *On Tangencies*, which is now lost, in which he discussed this problem of the three circles. We know of Apollonius' book from the commentary of Pappus, which has survived. Based on this, Vieta made a restitution of the lost book of Apollonius, and later Camerer (1795) edited both Pappus' commentary and Vieta's restitution with added work of his own. One of Pappus' constructions was given in (5.11).

In this section we give another method for solving these problems, using circular inversion.

Advice to the reader: The best way to study this material is to stop reading now, and attempt to solve each of these problems yourself. Then come back for hints if you need some help. If you read on further, you will find my solutions, but I expect you will find other perhaps even better solutions on your own.

As hints let me list the general techniques we will use. First of all, we use the basic mini-constructions of 3 or 4 steps each which we have often used before. With these techniques you should be able to solve problems PPP, PPL, PLL, and LLL.

For the problems involving circles, there is sometimes an easy method to reduce one problem to another. For example, in LLC, given two lines l, m, and a circle γ with center O, we are trying to find the small dotted circle tangent to l, m, and γ. Let l' and m' be lines parallel to l and m, and at a distance from l and m equal to the radius of γ. Then the circle through O (the center of γ) and tangent to l' and m' will have the same center as the circle we are looking for (verify!). Thus we can reduce LLC to PLL.

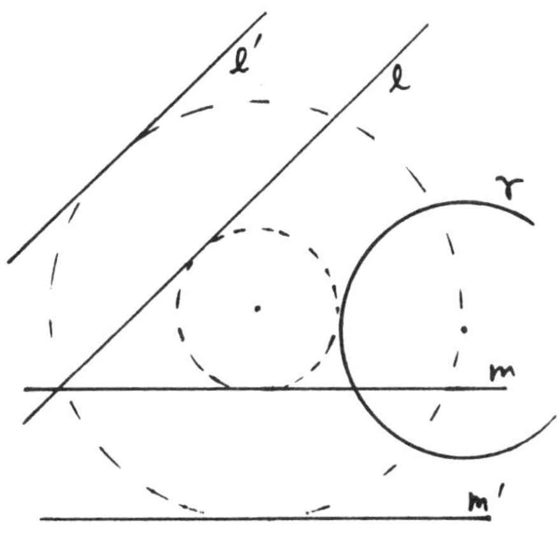

In the most difficult problems we can make use of circular inversion. For example, in CCC, the problem of Apollonius, by adding or subtracting the same quantity to the radius of each circle, we reduce to a problem where two circles meet. Let O be an intersection point of two circles, and let Γ be a circle with center O. Perform a circular inversion in the circle Γ. Then the two circles through O are transformed into straight lines, while the third circle becomes another circle. By LLC we can find a circle tangent to the two lines and circle obtained by circular inversion. Applying a circular inversion Γ to this newly constructed circle, we obtain a solution to the original problem.

Of course these constructions may get rather long, and in carrying them out on a piece of paper with pencil or pen, the accumulated error may be greater than what one could achieve by guessing. But that is not the point. The point is to have a correct theoretical procedure which gives a mathematically exact answer when carried out in ideal conditions.

Construction 38.1 (PPP). To construct a circle through three given points A, B, C. The center of this circle must lie on the perpendicular bisectors of AB and BC. So construct the perpendicular bisectors (3 steps each) to get the center O, and then draw the circle (7 steps). However, we can save one step by making one of our circles do double duty:

1. Circle center B, radius BC.
2. Circle center A, same radius. Get E, D.
3. Draw line ED.
4. Circle center C, same radius. Get F, G.
5. Line FG. Get O.
6. Circle center O, radius OA, passes through B, C.

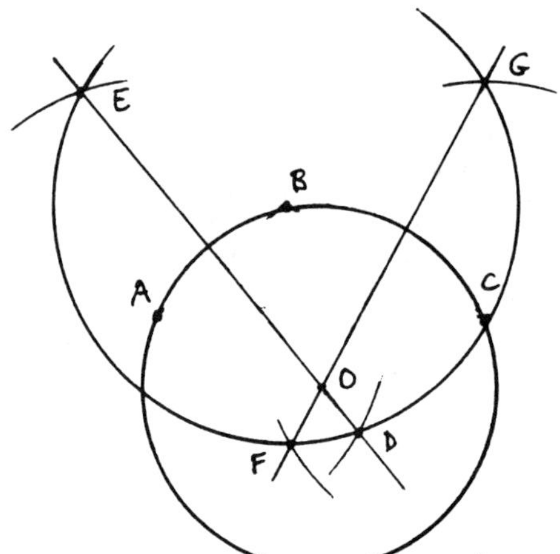

Construction 38.2 (LLL). To find a circle tangent to three given lines l, m, n. Let ABC be the triangle formed by the three lines. The center of the inscribed circle must lie on the angle bisectors of the three angles. So we bisect the angles at A and B to get the center of the circle O. Then we drop a perpendicular from O to one side to find the radius, then draw the circle. Adding miniconstructions comes to 13 steps. But with care, we reduce the construction to 10 steps:

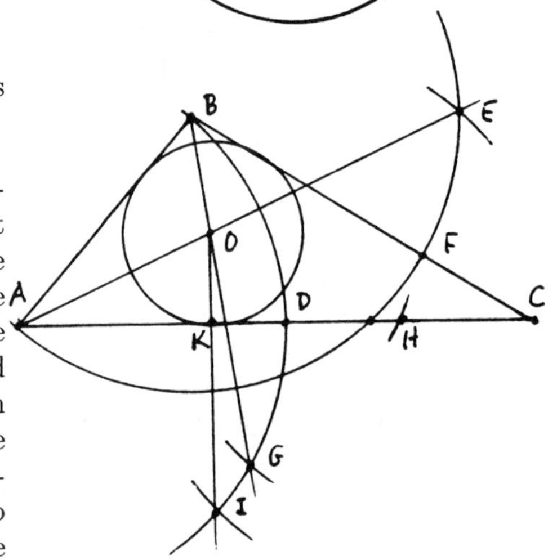

Bisect angle A, using circles centered at A, B, D, all of radius AB (4 steps).
Bisect angle at B, using F, G (2 more steps).
Perpendicular from O to AC, using H, I, K (3 more steps).
Circle center O, radius OK (1 step).

Construction 38.3 (PPL). To construct a circle through two given points A, B, tangent to a given line l.

First draw the line AB, to meet l at P (1 step). If we imagine the circle already drawn, being tangent to l at Q, then PQ is a tangent to the circle and PAB cuts the circle, so by (III.36) $PA \cdot PB = PQ^2$. So our method will be to use this fact to find Q, and then draw the circle. According to (III.36) again, if we take any circle through AB, the tangent from P to that circle will have the same length as PQ. So let us bisect AB, using CD (3 steps) and get its midpoint E. Draw the circle with center E passing through A and B (1 step). To find the tangent from P to this circle, bisect PE, using F, G (3 steps) and get H. Draw the circle with center H through P, E (1 step) and let it meet the circle through AB at K. Then PK will be tangent to the circle. Find Q and l with $PQ = PK$ (1 step). Now to get the circle through A, B, Q, bisect AQ (2 more steps), get O, and draw the circle with center O through A, B, Q (1 step). Total: 13 steps.

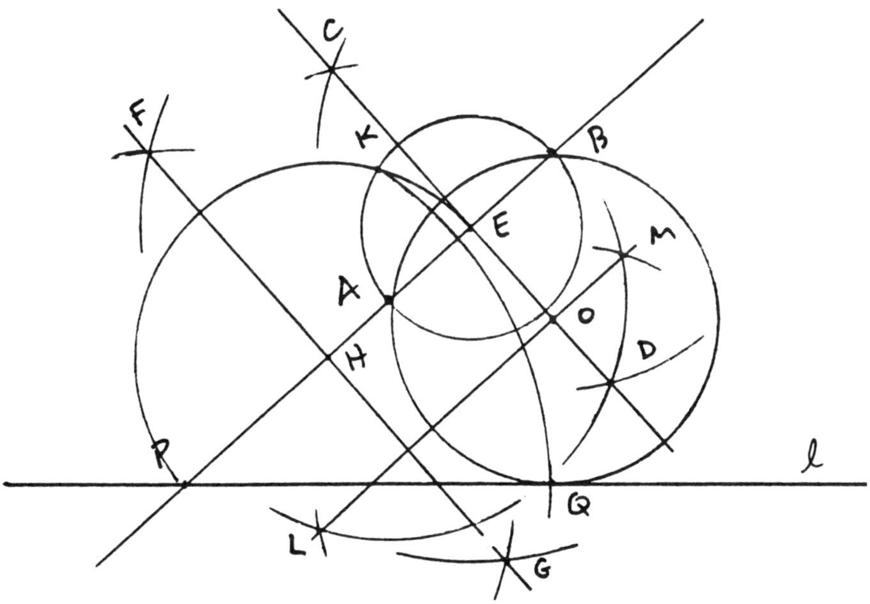

Construction 38.4 (PLL). To construct a circle through a given point A and tangent to two given lines l, m.

One method is to reduce to PPL as follows: Imagine a reflection in the angle bisector of the angle at O. This will give a new point A', symmetrically located, and any circle through A, tangent to l and m must also pass through A'. Thus we need only construct the circle through A, A', tangent to l, by PPL. We can construct A' in 2 steps: Circle center O through A, get B, C. Circle through C, radius AB. Get A'. Then do PPL. Total: 15 steps.

Another method is to use the fact that circles tangent to two lines l, m are all related by dilations from center O where l and m meet. So this time our strategy is to construct any other circle tangent to l and m, and then scale it down or up so as to pass through A. First bisect the angle at O, using B, C, D and circles all of the same radius (4 steps). Get E, F where the circle with center B through O meets l, m. Draw EF (1 step) which will be perpendicular to l since it is contained in a semicircle (III.31). Get G, and draw a circle with center G and radius GF (1 step). This is our comparison circle γ'. Draw OA to meet γ' at A' (1 step). The desired circle γ will have center K on OD, and with KA parallel to GA'. So construct AH parallel to GA' (3 steps), and get K. Draw the circle γ with center K and radius KA (1 step). Total: 11 steps.

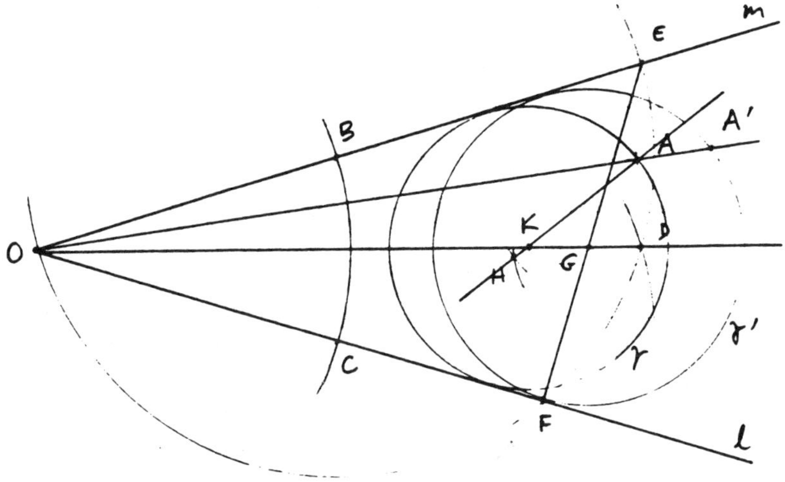

Construction 38.5 (PPC). As an application of the use of circular inversion, we will construct a circle passing through two given points A, B, and tangent to a given circle γ. Since circular inversion sends circles to circles (or lines) and preserves incidence and tangencies, any circular inversion will transform one of our ten problems into another of our ten problems, and solving one is equivalent to solving the other, since the constructed circle can also be transformed by the same circular inversion. The point is to make a judicious choice of which circular inversion to perform so that our job becomes as simple as possible.

In our case, let A, B be the given points, and let γ be the given circle. Draw the perpendicular bisector of AB (using C, D, 3 steps) and let it meet γ at O. (If it does not meet γ, then construction will be longer.) Take the circle Γ with center O, passing through A, B (1 step) as our circle to invert in. Let Γ meet γ at P, R. Draw the line l through P, R (1 step). Since γ passes through O, l will be the inverse of the circle γ, while A, B are fixed, since they lie on Γ. So we have the new problem to find the circle δ through A, B, tangent to l. Suppose this is solved by PPL (13 steps, not shown), and let Q be the point of tangency of δ with l. We find the inverse Q' of Q by drawing the line OQ and intersecting with γ (1 step). So the desired circle is the circle through A, B, and Q'. This is PPP. We need only construct the perpendicular bisector of AQ' (using E, F, 3 steps) to find G, and then draw the circle with center G through A, B, Q' (1 step). Total: 23 steps. Actually, in applying PPL, we already have the perpendicular bisector of AB, so that saves 3 steps. Also, we do not need the actual circle γ, only its point of tangency Q with l. That saves 3 more steps, so the count becomes 17 steps.

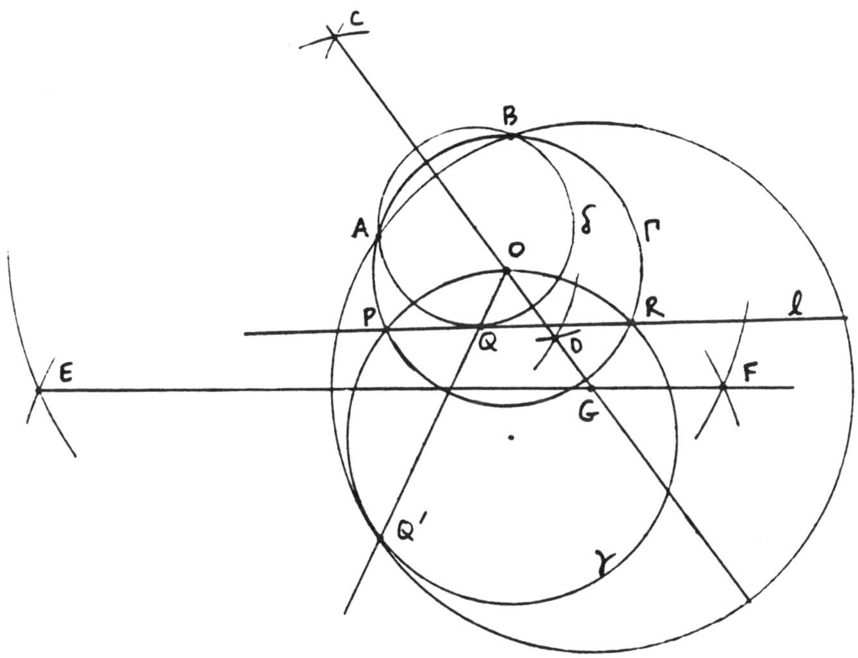

To complete this section, we will do a special case of LLC, and then show how to use that to solve CCC, the problem of Apollonius. The remaining problems we will leave as exercises for the reader.

Construction 38.6. Given two parallel lines l, m, and given a circle γ, with center O, to find a circle tangent to l, m, and γ.

The idea is to construct the line parallel to l and m which is halfway between l and m. The center of the desired circle will be a point on this midline whose distance from the center O of γ is equal to (its radius) $\pm$ (radius γ) depending on whether we want a circle containing γ or not.

Pick any point A on l, and draw circle with large enough radius to intersect m at B, C (1 step). Bisect BC (3 steps) using A, D, get E. Circle with center E, radius AB to get F, G (1 step). Draw FG (1 step) and get H. This is the midline. Make $HI =$ radius of γ (1 step). Center O, radius EI, find K on the midline (1 step). Center K, radius EH, draw the required circle σ_1 (1 step). Total: 9 steps. To get another solution σ_2, center O, radius AI to get L. Center L, radius EH is

σ_2. Of course there are two more solutions with centers K', L' on the other side of O.

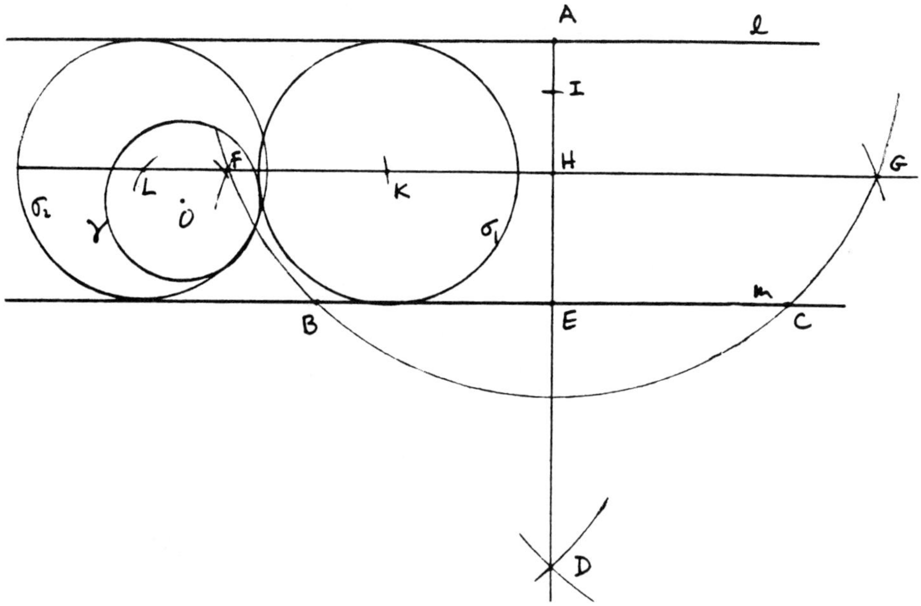

Construction 38.7 (The problem of Apollonius). Given three circles γ_1, γ_2, and γ_3, with their centers O_1, O_2, O_3, to find a circle tangent to all three given circles.

Note first that if we were not given the centers of the circles, we could find them with little constructions of 5 steps each, so to make life a little easier for ourselves, we suppose O_1, O_2, O_3 also given.

The idea is to expand the three circles by adding a fixed length to each of their radii. This will not change the center of the circle in the middle tangent to all 3. (If we wanted a circle containing one of the γ_i, then we should subtract from its radius.) Doing this expansion carefully, we can arrange that two of the new circles are tangent to each other.

This first part of the construction goes as follows: Draw O_1O_2 (1 step) get A, B. Bisect AB, using C, D (3 steps) get E. Draw O_3E, get F' (1 step). Make $FF' = AE$ (1 step). Now draw the new circles γ_1': center O_1, through E; γ_2' center O_2, through E, and γ_3', center O_3, through F (3 steps). Then, by this operation, we have reduced to a special case of CCC when two of the circles are tangent (9 steps).

For the next stage of the construction, we will perform a circular inversion in a circle Γ with center E. This will transform γ_1' and γ_2' into two parallel straight lines, and γ_3' will become another circle. It seems worthwhile to choose Γ orthogonal to γ_3', because then γ_3' will be sent into itself, saving us the trouble of finding its image under the circular inversion.

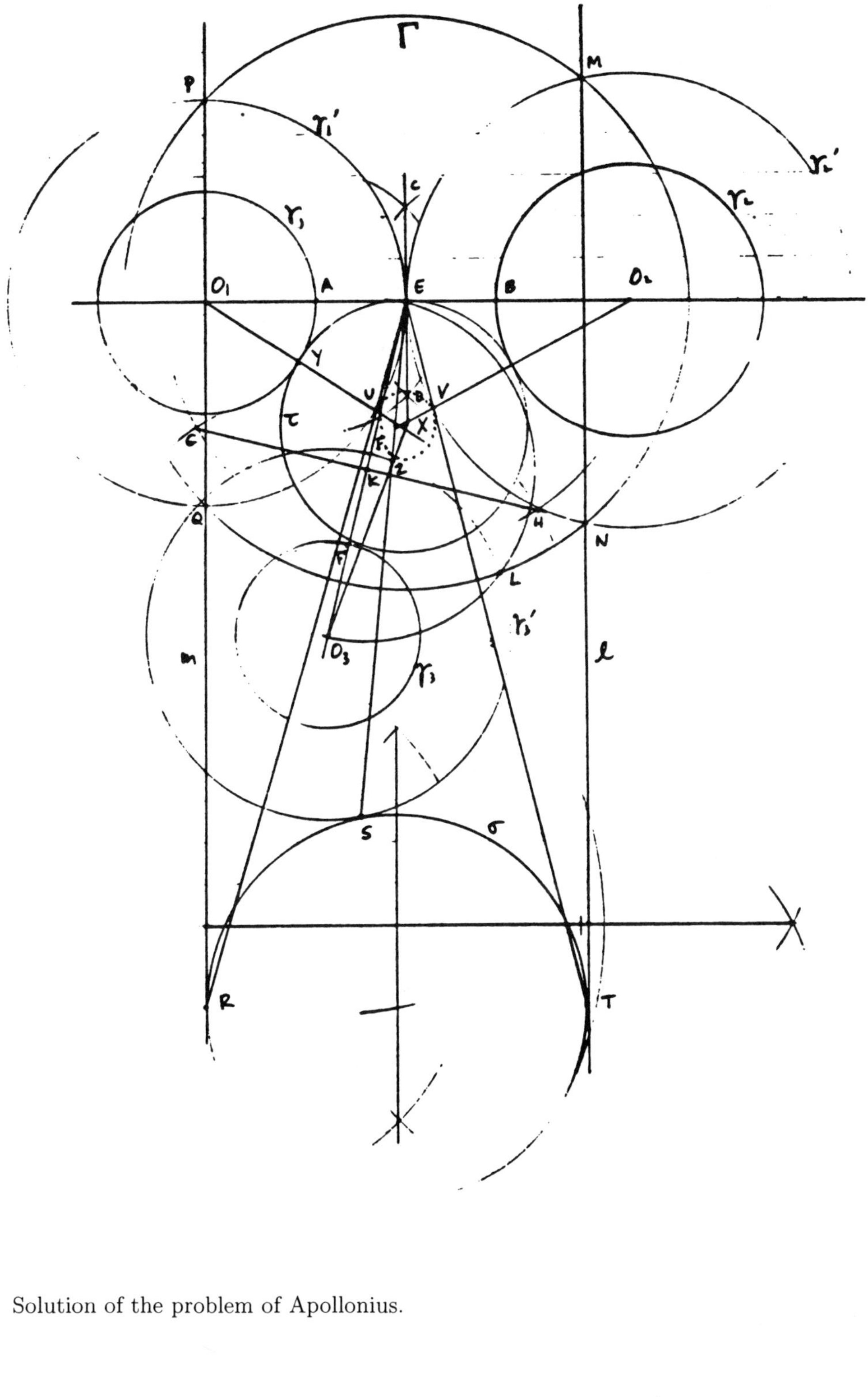

Solution of the problem of Apollonius.

So we proceed as follows: Bisect O_3E, using G, H (3 steps) and get K. Draw the circle with center K, radius KE (1 step) and let it intersect γ_3' at L. Draw the circle Γ with center E through L (1 step). Then Γ will be orthogonal to γ_3', and so will be left fixed by circular inversion in Γ. Let Γ meet γ_2' in M, N, and draw the line $l = MN$ (1 step). Let Γ meet γ_1' in P, Q, and draw the line $m = PQ$ (1 step). Thus we have transformed $\gamma_1', \gamma_2', \gamma_3'$ into l, m, and γ_3' and we have the new problem of finding a circle tangent to these three. Furthermore, since γ_1' and γ_2' were tangent to O, their transforms l, m do not meet. In other words, l and m are parallel, so we have a case of (38.6) treated above. This portion of our construction was (7 steps).

Now perform (38.6) to find a circle σ tangent to l, m, γ_3', and let the points of tangency be R, S, T (9 steps). Actually, since we already have a line O, O_2 perpendicular to l and m, we can get the midline in 3 steps instead of 6, thus saving 3 steps. So this part of the construction counts (6 steps).

The last stage of the construction is to transport back σ by the circular inversion in Γ to get a circle tangent to $\gamma_1', \gamma_2', \gamma_3'$. Then for the same center we can draw the desired circle τ tangent to $\gamma_1, \gamma_2, \gamma_3$.

It is actually sufficient to pull back two of the points of tangency. Draw ER and let it intersect γ_1' at U (1 step). Then U is the inverse of R in Γ. Draw ET and let it meet γ_2' at V (1 step). Now U, V are two of the points of tangency of a circle (dotted) tangent to $\gamma_1', \gamma_2', \gamma_3'$. To find its center, draw O_1U and O_2V and let them meet at X (2 steps). Now the circle τ with center X and radius XY is the desired circle (1 step). This last part of the construction is (5 steps). (In the drawing I also found the inverse Z of S and draw O_3Z to check for accuracy, but this is not really part of the theoretical construction.) Total: 27 steps.

Exercises

Carry out the following ruler and compass constructions.

38.1 PLC. Treat as a special case of LCC.

38.2 LLC. Follow hint given earlier in text.

38.3 PCC. Treat as a special case of CCC.

38.4 LCC. Use a technique similar to the one we used for CCC to reduce to (38.6).

38.5 PPC. Do the general case, where the perpendicular bisector of AB does not meet γ.

38.6 Describe how you would construct all eight solutions to the problem of Apollonius.

39 The Poincaré Model

In this section we will show the *existence* of a non-Euclidean geometry, and hence the *consistency* of the axioms of non-Euclidean geometry, by exhibiting a model for a non-Euclidean geometry. Ironically, our model of a non-Euclidean geometry will be constructed within the logical framework of Euclidean geometry. So what we must do is to give an *interpretation* of the undefined notions of geometry in the model: point, line, betweenness, and congruence for line segments and angles, and then we must prove that the axioms all hold in this interpretation.

Our starting point will be the Cartesian plane Π over an ordered field F, satisfying (∗∗) of (16.2). In this plane we fix a circle Γ with center O.

The *points* of our model (which we will call P-*points* if necessary to avoid confusion) will be the set of points of Π *inside* Γ, not counting the points on Γ. A P-*line* will be the set of all P-points lying on a circle γ which is orthogonal to Γ, or which lie on a line through O. (To keep our language straight, the words point, line, circle will refer to the Euclidean notions in Π, and we will prefix a P to any word to mean the corresponding concept in the model we are building.)

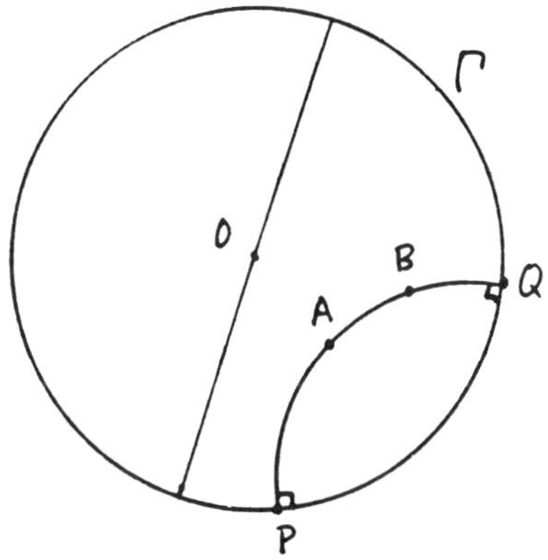

Having thus defined the P-points and P-lines of our model, we can verify the incidence axioms (I1), (I2), (I3).

Proposition 39.1 *The* P-*model satisfies* (I1), (I2), *and* (I3).

Proof. For (I1), suppose given two P-points, A, B. If the line AB passes through O, then it is a P-line containing them and is the only such. If A, B, and O are not collinear let A' be the inverse of A under inversion in the circle Γ (cf. §37). Then there is a unique circle γ passing through A, A', and B. By (37.3), γ is orthogonal to Γ, so that portion of γ which is inside Γ becomes a P-line containing A and B. It is unique, because again by (37.3), any circle γ orthogonal to Γ which contains A also contains A'.

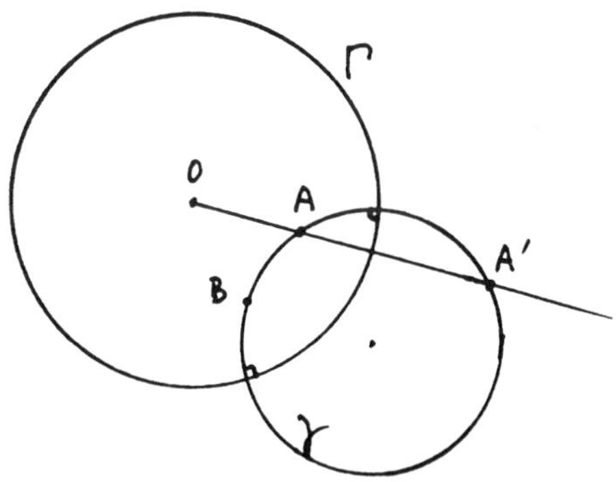

The other two axioms (I2), existence of at least two points on a line, and (I3) existence of three non-collinear prints, are obvious.

We see immediately that this geometry will be non-Euclidean because the parallel axiom (P) does not hold.

Proposition 39.2 *The parallel axiom* (P) *does not hold in the* P*-model: there is a* P*-line* γ *and a* P*-point* A *such that there is more than one* P*-line through* A *which is* P*-parallel to* γ*. (Of course* P*-parallel means the two* P*-lines do not intersect.)*

Proof. Take a P-line γ in one part of our P-plane and take a point A far away. Let A' be the inverse of A. Then (by (37.3)), any circle through A and A' will be orthogonal to Γ, so gives a P-line passing through A. There are lots of these which do not meet γ, and these are all P-lines through A which are P-parallel to γ.

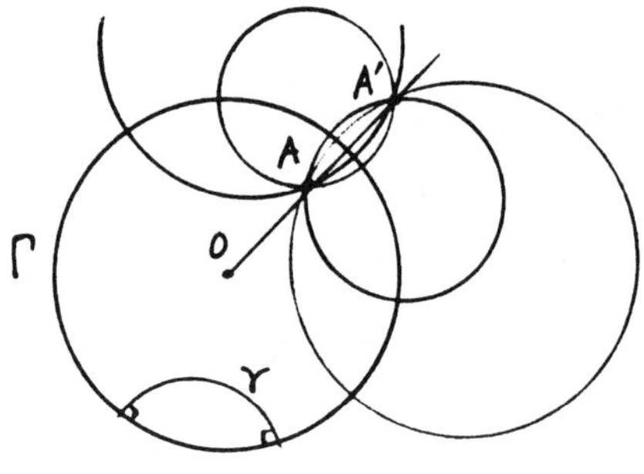

Definition. If A, B, C are P-points on a P-line γ, we define the P-*betweenness* relation $A * B * C$ as follows. Let O' be the center of γ (which is always outside Γ), draw the line PQ, and project the points A, B, C to points $A', B', C' \in PQ$ from the point O'. Then we will say $A * B * C$ (P-betweenness) if and only if $A' * B' * C'$ on the line PQ (usual betweenness). If A, B, C lie on a P-line which is an ordinary line through O, we take the usual notion of betweenness.

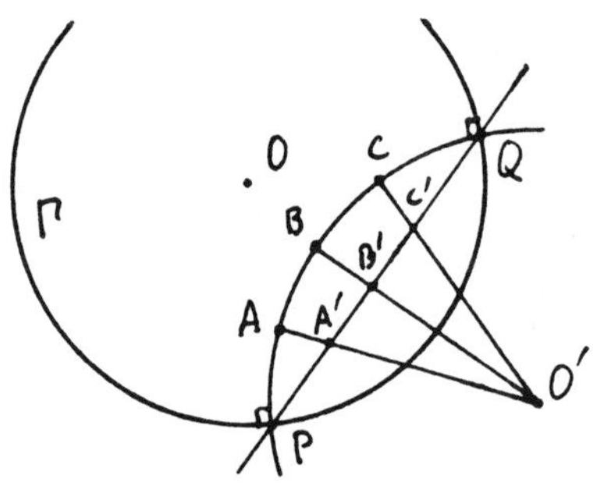

Proposition 39.3 *The notion of* P*-betweenness for* P*-points satisfies axioms* (B1)–(B4).

Proof. Axioms (B1), (B2), and (B3) follow immediately from the corresponding statements for ordinary betweenness on the line PQ. For (B4), taking into account the circle-circle intersection property (E) in Π, and noting that two circles orthogonal to Γ can meet at most once inside Γ, we see that to say P-points A, B are on the same P-side of a P-line γ is equivalent to saying that A, B as ordinary points are either both inside γ or both outside γ. Thus we can define the inside of a P-triangle, and (B4) is clear.

Definition. We define *congruence* in our P-model as follows. Two P-angles are P-*congruent* if the Euclidean angles they define are congruent in the usual sense. For line segments, we proceed as follows. Given two P-points, let the P-line joining them be the circle γ orthogonal to Γ. Let γ meet Γ in two points P, Q, and label them so that P is the one closer to A. For another pair of points A', B' lying on a P-line γ', label P', Q' similarly. Then we say the P-segment AB is P-*congruent* to the P-segment $A'B'$ if the cross-ratio (AB, PQ) is equal to the cross-ratio $(A'B', P'Q')$ (cf. §37 for cross-ratios).

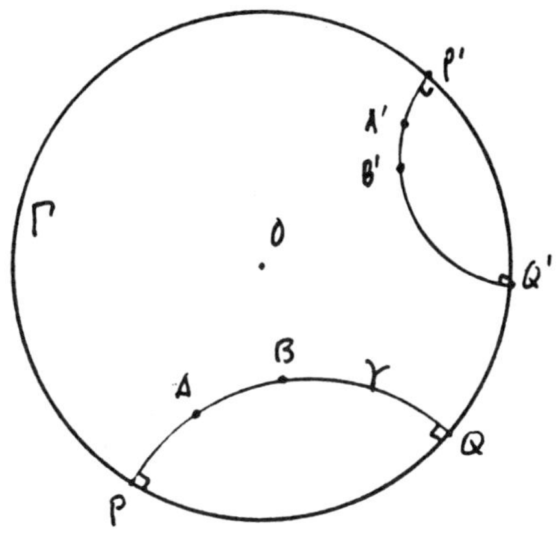

Now the real work begins, to verify the congruence axioms. We start with the easy ones.

Proposition 39.4 P-*congruence satisfies axioms* (C2)–(C5).

Proof. (C2) is obvious from the definition, since congruence of segments is defined by equivalence of associated quantities in the field.

(C3) requires a calculation. From the definition of cross-ratio it follows that $(AB, PQ) \cdot (BC, PQ) = (AC, PQ)$ (verify!). So when two segments are added together, the associated cross-ratios multiply. From this (C3) follows immediately.

To prove (C4), laying off angles, first suppose given a point A inside Γ, and a tangent direction m at A. Let A' be the inverse of A. Then there exists a unique circle γ passing through A and A' and tangent to the line m. By (37.3) γ is orthogonal to Γ. This shows that there exists a P-line at A with any given tangent direction.

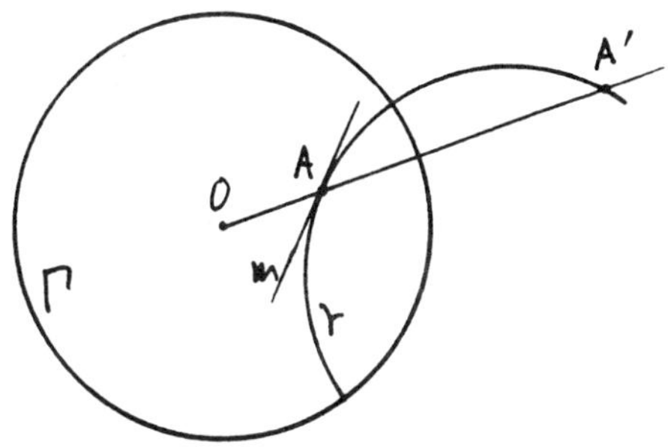

Now if an angle α is given and a P-line δ given at A, by (C4) in Π there is a unique line m forming the angle α with δ at A (and on a given side of δ). Then the P-line with tangent m gives the required P-angle at A, and is unique.

(C5) follows from the same statement for Euclidean angles, because congruence of angles is the same.

Before proceeding to a discussion of the remaining axioms (C1), (C6), (E), (A), and (D), we will establish the existence of rigid motion (ERM) in this model. Recall from §17 that a *rigid motion* is a transformation of the geometry which preserves the undefined notions of point, line, betweenness, and congruence. In our case, a P-*rigid motion* will be a transformation of the set of points inside Γ, which is 1–1 and onto, sends P-lines to P-lines, and preserves P-betweenness and P-congruence of angles and segments.

Proposition 39.5 (Existence of Rigid Motions (ERM) for the Poincaré model). *There are enough* P*-rigid motions of the Poincaré model so that*

1) *For any two* P*-points* A, A', *there is a* P*-rigid motion sending* A *to* A'.

2) *Given* P*-points* A, B, B' *for which* AB *is* P*-congruent to* AB', *there is a* P*-rigid motion leaving* A *fixed and sending* B *to* B'.

3) *For any* P*-line* γ *there is a* P*-rigid motion leaving all the points of* γ *fixed and interchanging the two sides of* γ.

Proof. We start with the last property. Given a P-line γ, let ρ_γ be the circular inversion in γ. Since Γ is orthogonal to γ, ρ_γ sends Γ to itself (37.3). Also the inside of Γ is sent to the inside of Γ, so that the P-plane is mapped to itself, in a way which is clearly 1–1 and onto. Since circular inversion sends circles into circles (37.4) and is conformal (37.5), a circle orthogonal to Γ will be sent to another circle orthogonal to Γ, in other words, ρ_γ sends P-lines into P-lines. (Note this works also for the limiting case of a line through O, which is also orthogonal to Γ.)

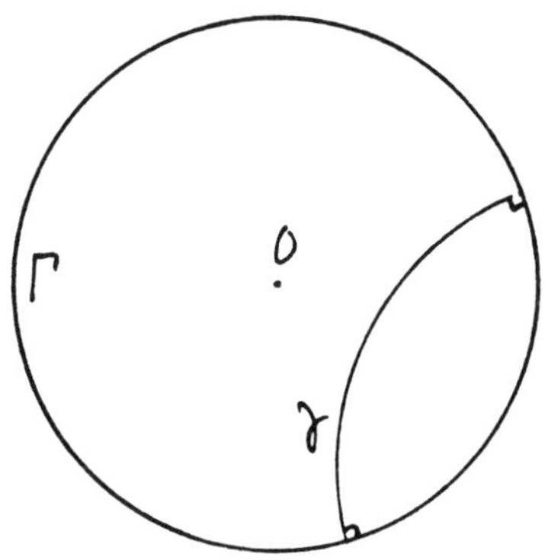

Circular inversion clearly preserves betweenness (Exercise 39.1). It preserves P-congruence of angles because this is the same as usual congruence of angles, and inversion is conformal (37.5). Also ρ_γ preserves P-congruence of P-segments, because this is defined by the cross-ratio, which is invariant under circular inversion (37.6). Finally, note that ρ_γ interchanges that part of the P-plane which is inside γ with that part which is outside γ, so ρ_γ is a P-rigid motion as required for the third statement of (ERM). Since it leaves the points of γ fixed and interchanges the sides of γ, it is the P-*reflection* in γ (§17).

Next we will show that for any $A \neq O$, there is a circle γ orthogonal to Γ (a P-line) such that the P-reflection in γ interchanges O and A. Let A' be the inverse of A, let γ be the circle with center A' which is orthogonal to Γ. Then the construction (37.1) for the circle γ, using the same diagram (!) shows that inversion in γ sends A to O. Thus the P-reflection in γ interchanges A and O.

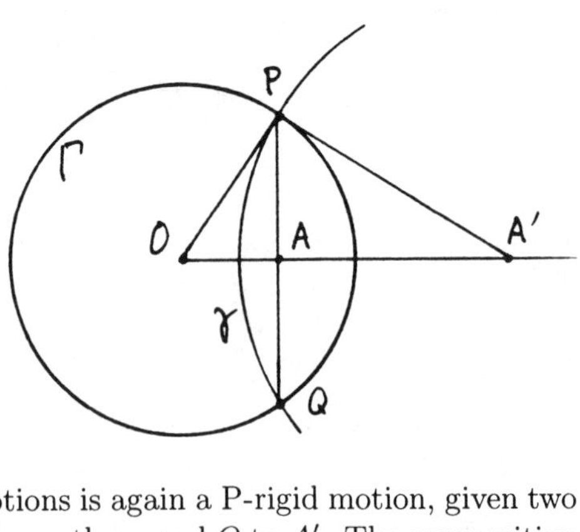

Now, since a composition of P-rigid motions is again a P-rigid motion, given two points A, A', we can first send A to O as above, then send O to A'. The composition of these two reflections will be a P-rigid motion sending A to A', which proves 1).

Now suppose given A, B, B' for which AB is P-congruent to AB'. Let ρ be a P-rigid motion taking A to O, and let $\rho(B) = C$, $\rho(B') = C'$. Then OC is P-congruent to OC'. If we can solve problem 2) for O, C, C', in other words, if there is a P-rigid motion θ leaving O fixed and sending C to C', then $\rho^{-1} \circ \theta \circ \rho$ will solve the problem 2) for A, B, B'. So we reduce to solving the problem for O, C, C'.

Lemma 39.6 *If C, C' are two points inside Γ, not equal to the center of Γ, O, then the* P*-segment OC (which is equal to the Euclidean segment OC, since the* P*-line joining O and C is just the usual line OC) is* P*-congruent to the* P*-segment OC' if and only if OC is congruent to OC' in the ambient Euclidean plane* Π.

Proof. Let P and Q be the endpoints of the diameter of Γ passing through O and C. Then the P-congruence of OC is determined by the cross-ratio

$$(OC, PQ) = \frac{OP}{OQ} \div \frac{CP}{CQ}.$$

Let r = radius of Γ and let x = distance from O to C. Then the cross-ratio is

$$\frac{r}{r} \div \frac{r+x}{r-x} = \frac{r-x}{r+x}.$$

If C' is another point, and if the distance from O to C' is y, then we find similarly

$$(O, C', P'Q') = \frac{r-y}{r+y}.$$

Thus, to say that OC is P-congruent to OC' is to say

$$\frac{r-x}{r+x} = \frac{r-y}{r+y}.$$

Cross-multiplying

$$r^2 - rx + ry - xy = r^2 + rx - ry - xy.$$

So

$$2rx = 2ry.$$

Since our field has characteristic 0, this is equivalent to $x = y$, i.e. OC is congruent to OC' in the usual sense.

Proof of (39.5) **continued.** Now let l be the angle bisector of angle COC'. Then l is a line through O, which is also a P-line. The ordinary reflection in l is clearly a P-rigid motion which leaves O fixed and sends C to C', since the hypothesis that OC is P-congruent to OC' implies, by the lemma, that $OC \cong OC'$ in the usual sense.

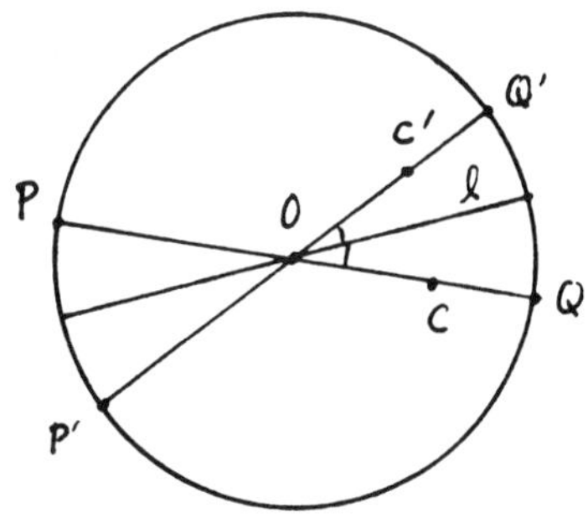

This completes the proof of (ERM) for the Poincaré model.

Proposition 39.7 *Axioms* (C1) *and* (C6) *hold in the Poincaré model.*

Proof. Suppose it is required to find a point B' on a P-ray emanating from a point A', such that $A'B'$ is P-congruent to a given P-segment AB. By (ERM) there is a P-rigid motion taking A' to O. Then the given ray will become a radius of the circle Γ. There is also a rigid motion taking A to O, so the P-segment AB will become a segment originating from O. Since P-congruence from O is equivalent to ordinary congruence (39.6) the result follows from (C1) in the ambient Euclidean plane Π.

To show that (C6) holds, since we have already established (C1)–(C5), we simply apply (17.1) which shows that under those circumstances, (ERM) implies (C6).

In order to discuss (E) in the Poincaré model, we first need to identify what is a P-circle. By definition, of course, it is the set of all P-points B' such that the P-segment $A'B'$, for a certain fixed point A', is P-congruent to a given P-segment AB. Since the definition of P-congruence of segments is not very intuitive, it is not easy to see immediately what kind of curves these are.

Proposition 39.8 *Every* P*-circle is an ordinary circle which is entirely contained in the inside of* Γ, *and conversely every circle entirely inside* Γ *is a* P*-circle. (*Warning*: The* P*-center of a* P*-circle is usually not equal to its ordinary center.)*

Proof. Given a P-circle ζ with P-center A' and P-radius AB, consider a rigid motion θ which takes A' to O. This will transform ζ into a P-circle with P-center O. Since P-congruence and ordinary congruence are the same for segments beginning at O (39.6), this image $\theta(\zeta)$ is an ordinary circle with center O. Then θ^{-1} will carry this ordinary circle back to the given P-circle ζ. Now observe that in the proof of (ERM), all the rigid motions we needed were made out of compositions of P-reflections (which are circular inversions in suitable circles) or reflections in a line through O. Since all of these transformations send circles into circles (37.4) it follows that ζ is a circle. Since the transformed circle was a circle around O, entirely contained inside Γ, the image is also entirely contained inside Γ.

Conversely, given an ordinary circle ζ completely contained inside Γ, with (ordinary) center O', draw OO'. Let it meet ζ at A, B. P-bisect the segment AB at A', and choose a P-reflection ρ_γ which sends A' to O. Then $\rho_\gamma(\zeta)$ will be a circle, the images of A and B will be equidistant from O, and this circle will be symmetric about the line $l = OO'$, which is sent into itself by ρ_γ. Hence $\rho_\gamma(\zeta)$ is a circle with center O, which is also a P-circle. Applying ρ_γ^{-1}, it follows that the original circle ζ is a P-circle with P-center A'.

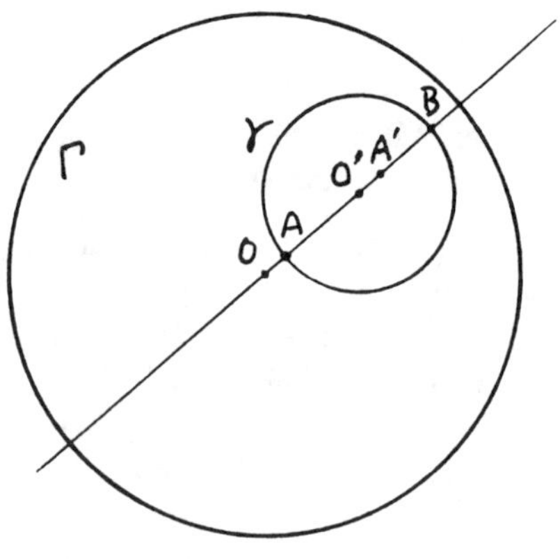

Proposition 39.9 *The circle-circle intersection property* (E) *holds in the Poincaré model.*

Proof. Since P-lines and P-circles are all either usual circles, or lines through O, and since betweenness is the same in the P-model as in the ambient Euclidean space, (E) in the P-model follows directly from (E) in the Euclidean plane. Since P-circles are usual circles entirely contained inside Γ, there is no problem about any of the intersections falling outside Γ.

For the next proposition, it will be convenient to introduce the notion of a distance function. In ordinary Euclidean geometry the distance function assigns to each interval a positive number, and adding segments corresponds to adding numbers. More generally we make the following definition.

Definition. A *distance function* on a Hilbert plane is a function d which to each segment assigns an element of an ordered abelian group G, such that
1) $d(AB) > 0$ for any segment AB
2) $d(AB) = d(A'B')$ if and only if $AB \cong A'B'$
3) if $A * B * C$ then $d(AC) = d(AB) + d(BC)$.

If the group happens to be written multiplicatively, we will call it a *multiplicative* distance function. The usual distance function on the Cartesian plane over a field F (§16) is an additive distance function with values in the additive group of the field $(F, +)$.

Lemma 39.10 *In the Poincaré model over a field F, the function $\mu(AB) = (AB, PQ)^{-1}$ is a multiplicative distance function with values in the multiplicative group of the field $(F_{>0}, \cdot)$.*

Proof. Because of our convention that P is the endpoint closer to A, the cross-ratio (AB, PQ) is in the interval $(0, 1)$ in F. Therefore $\mu(AB) > 1$. We have already used it to define congruence, and we have seen that it is multiplicative (39.4). Hence μ is a multiplicative distance function.

Proposition 39.11 *Archimedes' axiom* (A) *will hold in the* P*-model if we assume Archimedes' axiom* (A′) *for the field F. Similarly, Dedekind's axiom* (D) *will hold if we assume* (D′) *in the field.*

Proof. Using the multiplicative distance function μ of (39.10), Archimedes' axiom in the P-plane is equivalent to the following statement in F: given $c, d \in F$, $c, d > 1$, $\exists n > 0$ such that $c^n > d$.

We will show that this property is a consequence of the Archimedes' axiom (A′) for F. Write $c = 1 + x$, so $x \in F$, $x > 0$. Then

$$c^n = (1+x)^n = 1 + nx + \text{ positive terms } \geq 1 + nx.$$

Now (A′) says that for some n, $nx > d$. Hence also $c^n > d$, as required.

For Dedekind's axiom, (D′) in F implies (D) in Π and this clearly implies (D) in the P-plane because of the way we defined betweenness by projecting onto a line segment.

One sees easily that limiting parallels (cf. §35) exist in the Poincaré model. Using a little Euclidean geometry in the ambient Cartesian plane, we can derive a marvelous relationship between the length of a segment and its angle of parallelism.

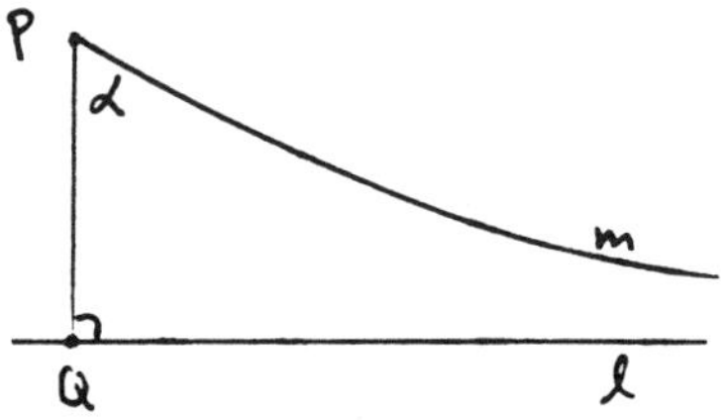

Proposition 39.12 *Suppose given in the Poincaré model a point P, a line l, the perpendicular PQ to l, and a limiting parallel line m, making an angle α with PQ.*

Then

$$\tan \alpha/2 = \mu(PQ)^{-1},$$

where the tangent is understood to be of the corresponding Euclidean angle, and μ is the multiplicative distance function above. The equality takes place in the field F.

Proof. We may assume that the Poincaré model is made with a circle Γ of radius 1 (cf. Exercise 39.20). We can move P, Q, l, m so that Q becomes the center of Γ, the line l becomes a radius QA, and P lies on an orthogonal radius QB. The limiting parallel through P to l will be part of a circle Δ, orthogonal to Γ at A. Its center therefore is at a point $C = (1, c)$ on the line $x = 1$. Let P be the point $(0, y)$. Then $CP = CA$, so

$$c^2 = (c - y)^2 + 1.$$

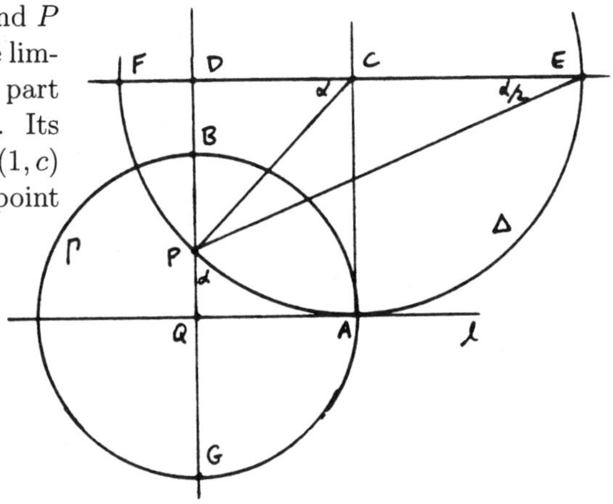

Therefore

$$c = \frac{1 + y^2}{2y}. \tag{1}$$

Draw a diameter EF of Δ parallel to the x-axis. Then the angle α between our limiting parallel and PQ, the *angle of parallelism*, is equal to the angle PCF. If we draw EP, then the angle $PEF = \alpha/2$ (III.20). Now

$$\begin{aligned} \tan \alpha/2 &= DP/DE \\ &= \frac{c - y}{c + 1}. \end{aligned}$$

Substituting from (1) we find

$$\tan \alpha/2 = \frac{1 - y}{1 + y}. \tag{2}$$

On the other hand, the multiplicative distance function is

$$\begin{aligned} \mu(PQ) &= (PQ, BG)^{-1} \\ &= \left(\tfrac{PB}{PG} \div \tfrac{QB}{QG}\right)^{-1} \\ &= \left(\tfrac{1-y}{1+y} \div \tfrac{1}{1}\right)^{-1} \\ &= \frac{1 + y}{1 - y}. \end{aligned} \tag{3}$$

From (2) and (3) we conclude that

$$\tan\alpha/2 = \mu(PQ)^{-1}$$

as required.

Remark 39.12.1. From this it follows that given any angle α less than a right angle, there exists a segment PQ with angle of parallelism equal to α. Indeed, $\tan\alpha/2$ will be an element of the field F, and then we can find a $y \in F$ satisfying (2) above. In particular if we take $\alpha = \frac{1}{2}$ RA, there will be a corresponding segment PQ uniquely determined up to congruence. In this sense there is an absolute standard of length in the Poincaré model, whereas in Euclidean geometry the choice of unit length is arbitrary.

Exercises

All exercises take place in the Poincaré model over a field F satisfying (**), unless otherwise noted. Proofs should be based on the Euclidean geometry of the Cartesian plane over F. In particular, do not use any of the results of §34 or §35 which depend on Archimedes' axiom.

39.1 Verify that circular inversion preserves betweenness in the Poincaré model (cf. proof of (39.5)).

39.2 Show that the angle sum of any triangle in the Poincaré model is $<$ 2RA.

39.3 For any angle α, show the existence of a line entirely contained inside the angle α (cf. Exercise 34.6).

39.4 Show that for any angle $\alpha < 60°$, there exists an equilateral triangle with all of its angles equal to α.

39.5 If an equilateral triangle has sides equal to AB and angles equal to α, show that

$$\frac{2a}{1+a^2} = \frac{2t^2}{1-t^2}$$

where $a = \mu(AB)$ is its multiplicative length, and where $t = \tan\alpha/2$.

39.6 Given any three angles α, β, γ with $\alpha + \beta + \gamma <$ 2RA, show there exists a triangle with angles α, β, γ in the Poincaré model. *Hint*: First show in the Cartesian plane that you can find an angle α meeting a circle at angles β and γ. Then shrink or expand this figure so that it becomes a triangle in the Poincaré model.

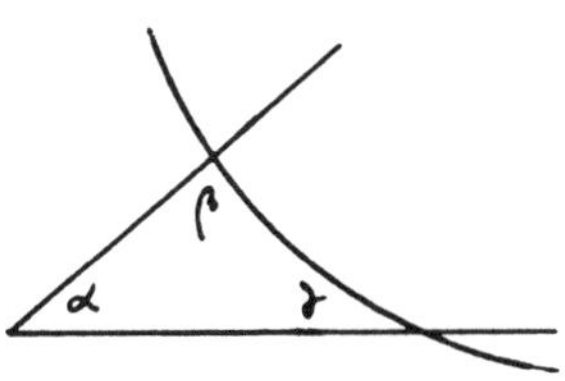

39.7 Show that for any point P and any line l there are two limiting parallels through P to l, one in each direction. Show that the property of being limiting parallel is an equivalence relation on rays.

39.8 If two lines are parallel, but not limiting parallel, then they have a unique common orthogonal line.

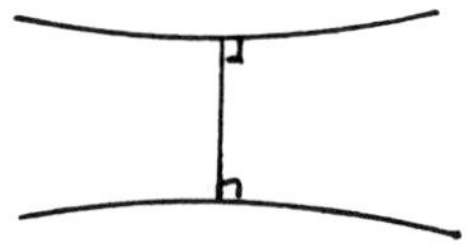

39.9 For any angle α, there is an *enclosing line*, which is a line limiting parallel to both arms of α.

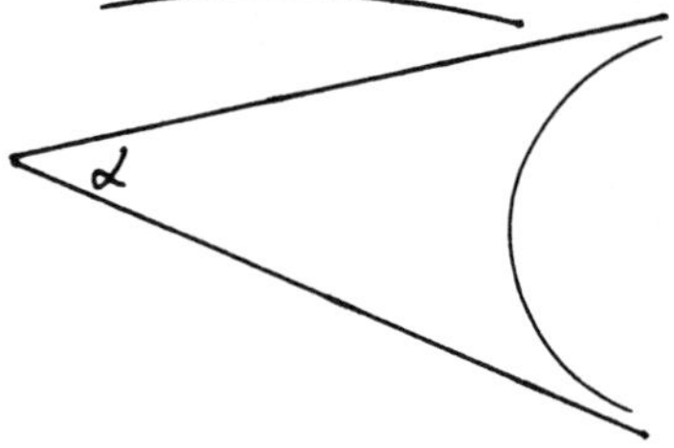

39.10 Give an alternate proof of (C1) in the Poincaré model without using rigid motions, as follows. Given a point A, a P-line γ, and given a quantity $b \in F$, $0 < b < 1$, we need to find a point $B \in \gamma$ so that

$$(AB, PQ) = b.$$

Do this by showing that in Euclidean geometry, the locus of points B such that BP/BQ is a given ratio $k \in F$ is a *circle*. Then use (E), in the Cartesian plane, to show this circle intersects γ and thus find the required B.

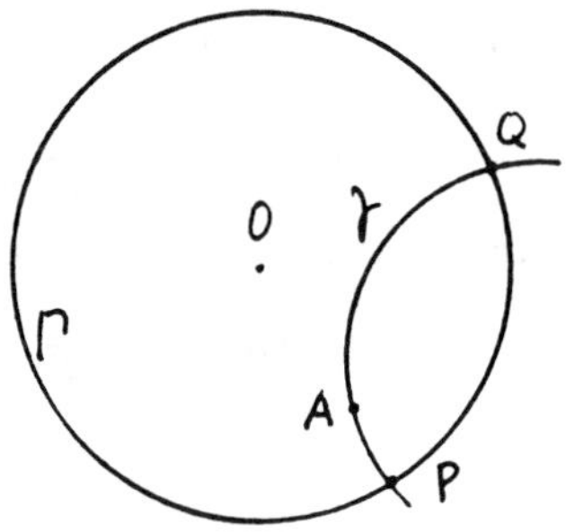

39.11 Given the circle Γ, its center O, and another circle ζ entirely contained inside Γ, give a ruler and compass construction (in the ambient Euclidean plane) of the P-center of ζ regarded as a P-circle (cf. (39.8)).

39.12 (Euclidean geometry). Find all possible ways of filling the entire Euclidean plane with triangles satisfying the following conditions

a) The triangles are all congruent to each other. There is no overlap, and they fill the entire plane.

b) At each vertex of the triangulation, all the angles are the same (though they may be different from the angles at a different vertex).

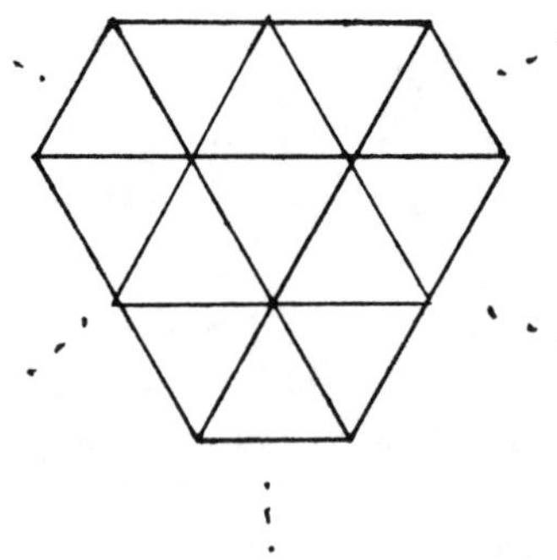

We consider two "ways" of filling the plane "the same" if one can be moved to the other by a dilation followed by a rigid motion.

One such triangulation is shown, where the angles at each vertex are all 60°. This is the only possibility if all angles are equal. Expect to find three more ways, allowing angles at different vertices to be different, and *prove* that you have found all possibilities.

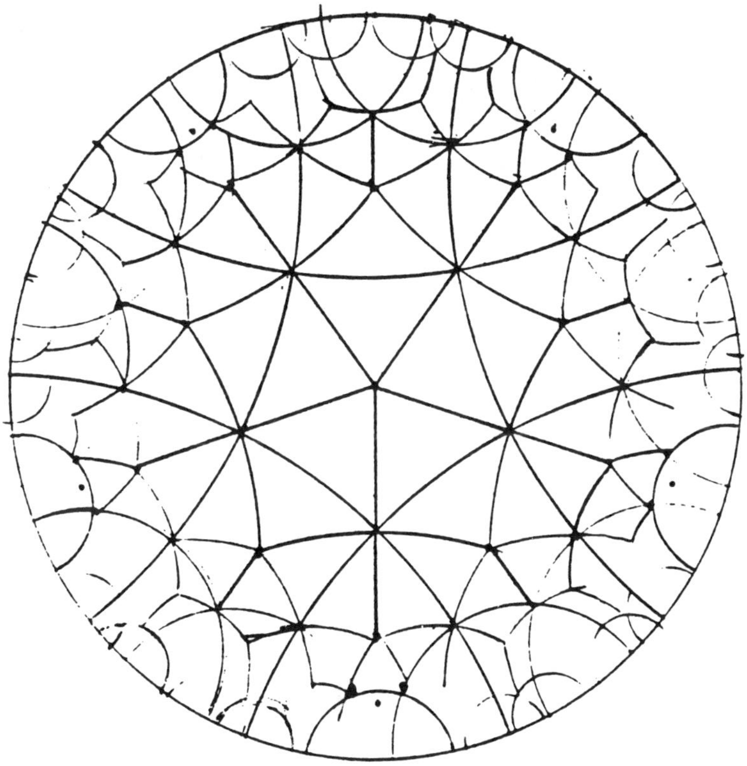

Congruent, isosceles, $72° - 45° - 45°$ triangles, filling up the Poincaré model of the non-Euclidean plane (cf. Exercise 39.13).

39.13 In the Poincaré model of non-Euclidean geometry, show, in contrast to the Euclidean situation described in Exercise 39.12 above, that there are infinitely many different ways to cover the P-plane by congruent P-triangles satisfying properties a) and b).

In particular, prove that the plane can be covered by equilateral triangles with all angles equal to 45° and with eight meeting at each vertex. If AB is a side of one of these triangles, find $\mu(AB)$.

Draw a big circle Γ on a piece of paper, and then draw accurately enough of these P-triangles inside Γ to show how they cover the whole P-plane. (This drawing can be accomplished entirely by ruler and compass, but don't bother listing the steps, except to show how you got the first triangle.)

39.14 In the Poincaré model made inside a circle Γ in the Cartesian plane over F, we have seen that any Euclidean circle γ entirely contained inside Γ is a P-circle (39.8).

a) If γ is a Euclidean circle inside Γ and tangent to Γ, show that there is a pencil of limiting parallel lines (a *pencil* means the set of all lines which are mutually limiting parallels at one end), such that the curve γ is orthogonal to all the lines of the pencil. Such a curve is called a *horocycle* in the Poincaré model.

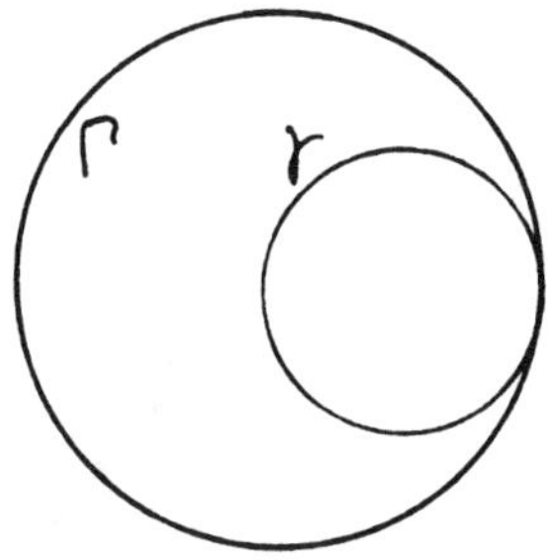

b) If γ is a Euclidean circle or a Euclidean line which cuts Γ at points P, Q, let l be the P-line having the endpoints P, Q. Show that the points of γ inside Γ form a curve of points equidistant from the P-line l. Such a curve is called an *equidistant curve* or *hypercycle*.

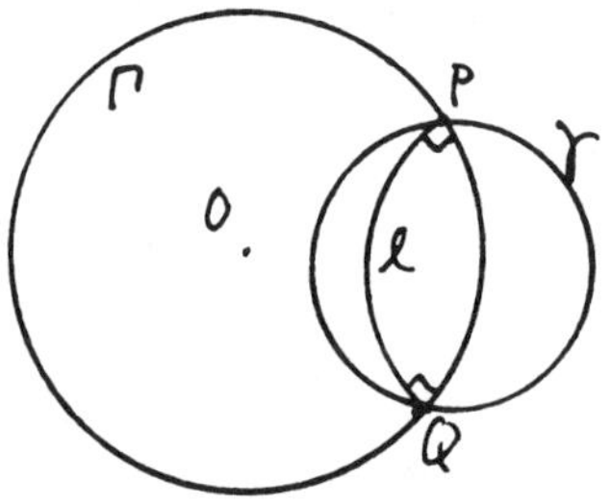

c) Given any three distant points A, B, C in the Poincaré model, show that they are contained in a unique P-line or P-circle or horocycle or hypercycle. (Contrast to Euclidean geometry where only the first two possibilities occur.)

39.15 Show in the Poincaré model that it is in general not possible to trisect an angle (i.e. if α is an angle, the angle $(1/3)\alpha$ may not exist) (cf. §28).

39.16 Show in the Poincaré model, in contrast to the Euclidean case (Exercise 2.14), that it is in general not possible to trisect a line segment (i.e. the 3-division points may not exist).

39.17 In the Poincaré model, show that if two altitudes of a triangle meet in a point, then the third altitude also passes through that point. Here is a method. Let the triangle be ABC, and suppose that the altitudes from A to B meet. By a rigid motion of the Poincaré plane we move that meeting point to the center O of the defining circle Γ. Then those altitudes become Euclidean lines through O. We must show that the line OC is orthogonal to the side AB.

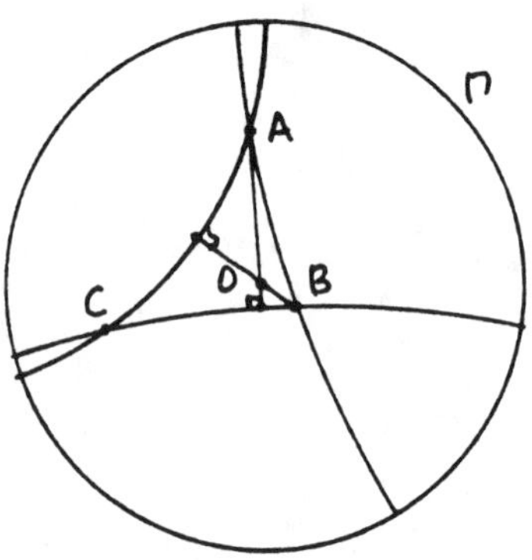

The P-lines AB, AC, BC are Euclidean circles orthogonal to Γ. Let D, E, F be the centers of these circles. Show that the altitudes of the P-triangle ABC are at the same time altitudes of the Euclidean triangle DEF. Then use the Euclidean theorem that the altitudes of a triangle meet (5.6) to finish the proof.

Note: This is a curious method, whereby the Euclidean result is used to show (via Euclidean geometry) that the same result holds in the non-Euclidean Poincaré model. Since we now know that this result holds in both Euclidean and non-Euclidean geometry, it would be nice to have a single proof in neutral geometry which applies to both cases (see Exercise 40.17).

39.18 Show that the result of (Exercise 1.15) is also valid in the Poincaré model, by moving the figure so that P becomes the center of Γ and using the Euclidean result already proved. Can you find a proof in neutral geometry which will cover both cases at once?

39.19 There is another model of a non-Euclidean geometry due to Felix Klein (1927) constructed as follows. In the Cartesian plane over a field F, fix a circle Δ. Then the K-points are the points inside Δ, and the K-lines are chords of Euclidean lines contained inside Δ. In this model the incidence axioms (I1)–(I3) and the betweenness axioms (B1)–(B4) are immediate, taking betweenness to be the same as in the Cartesian plane.

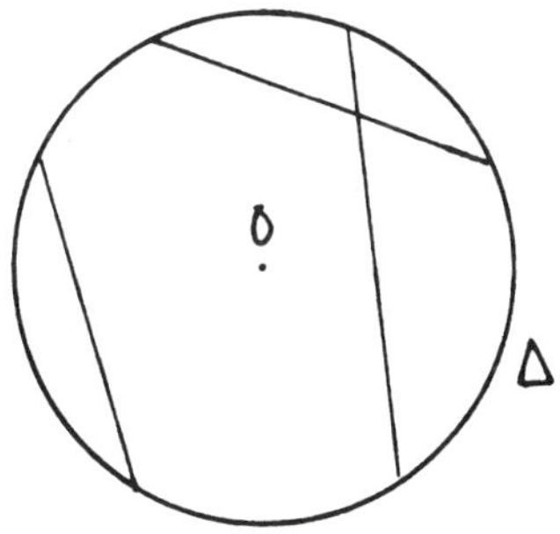

However, the model is not conformal (i.e. angles are not the same as Euclidean angles), so the definition and properties of congruence for line segments and for angles is more complicated. Rather than doing this directly, we will show in this exercise how to obtain the Klein model from the Poincaré model.

Let Δ be a circle of radius 1 centered at the origin, and in the Cartesian 3-space, place a sphere of radius 1 on the plane, with its south pole at the origin (cf. Exercise 37.1). Let Γ be the circle of radius 2 centered at the origin. For each K-point inside Δ, project it straight up to obtain a point of the southern hemisphere of the sphere, and then use the stereographic projection (Exercise 37.1) from the north pole to obtain a P-point inside Γ.

Show that this transformation gives a 1–1 correspondence between the points of the K-plane inside Δ with the points of the P-plane inside Γ, which sends K-lines to P-lines and vice versa. Then we can transport the notions of congruence for P-segments and P-angles to the K-plane, so that the K-plane becomes a model of a non-Euclidean Hilbert plane, isomorphic to the Poincaré model.

39.20 If ABC is a triangle having a circumscribed circle, prove that the medians of ABC meet in a point, as follows. Use the Klein model (Exercise 39.19) and place the center of the circumscribed circle at the center O of the circle Δ. Then the perpendicular bisectors of the sides of ABC become diameters of the circle Δ. Conclude that the K-midpoints of the sides of the triangle are equal to the Euclidean midpoints, and then use the Euclidean theorem about medians in the ambient plane.

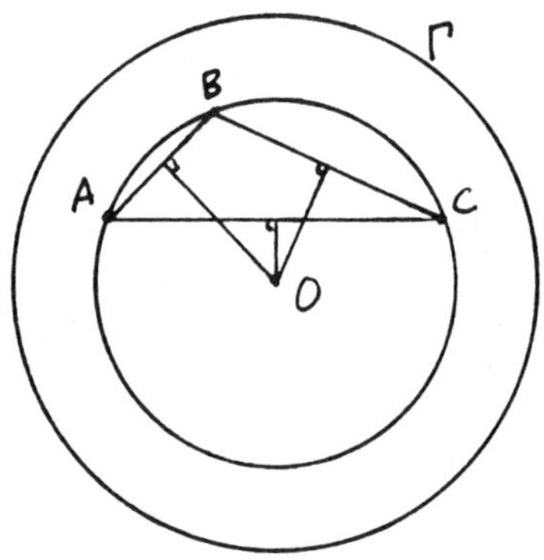

39.21 In the Cartesian plane over the field F, let Γ be a circle of radius r centered at the origin, and let Γ' be a concentric circle of radius r'. Consider the map φ from the set of points inside Γ to the set of points inside Γ' given by

$$\begin{cases} x' = kx \\ y' = ky \end{cases}$$

where $k = r'/r$. Show that φ gives an isomorphism of the Poincaré model made with Γ and the Poincaré model made with Γ', which preserves the multiplicative distance function of (39.10). Conclude that if Γ and Γ' are *any* two circles in the Cartesian plane over F, the associated Poincaré models are isomorphic.

39.22 Let F be a non-Archimedean field satisfying $(**)$ such as the one described in (18.4). Let Π be the Poincaré model over F, as described in this section, and let Π^0 be the subset of points which are at finite multiplicative distance μ from some fixed point O. Show that Π^0 is a non-Euclidean Hilbert plane with properties

a) The angle sum of any triangle is $<$ 2RA.

b) Limiting parallel lines do not exist.

Compare (Exercise 18.3) and (Exercise 18.4).

40 Hyperbolic Geometry

In the earlier sections of this chapter we have seen something of the development of neutral geometry, and the study of the angle sum of a triangle using Archimedes' axiom. We have also seen the Poincaré model of a non-Euclidean geometry over a field. For the full development of the geometry of Bolyai and Lobachevsky, we need the limiting parallels. The existence of these limiting parallels, which we have seen in the Poincaré model, does not follow in the axiomatic treatment from what we have done so far. Therefore, following Hilbert, we will take the existence of the limiting parallels as an axiom. This axiom is quite strong. It will allow us to develop the non-Euclidean geometry independently of Archimedes' axiom. It also allows the construction of an ordered field out of the geometry (in the next section), and a proof that the abstract geometry is isomorphic to the Poincaré model over this field. Using this field as coordinates we can develop non-Euclidean analytic geometry and trigonometry.

So at this point we will start a new axiomatic development of non-Euclidean geometry, almost completely independent of the material in the earlier sections of this chapter. In particular, we will not use the circle-circle intersection axiom (E) nor Archimedes' axiom (A). We base our work on Hilbert's axioms of incidence, betweenness, and congruence, namely the axioms of a Hilbert plane, plus one further axiom which we now describe.

We recall from §35 the definition of a limiting parallel ray. If P is a point not on a line l, and if PQ is the perpendicular from P to l, a *limiting parallel ray* m to l through P is a ray m emanating from P, parallel to l (i.e. not meeting l), and such that any other ray from P inside the angle formed by PQ and m must meet the line l.

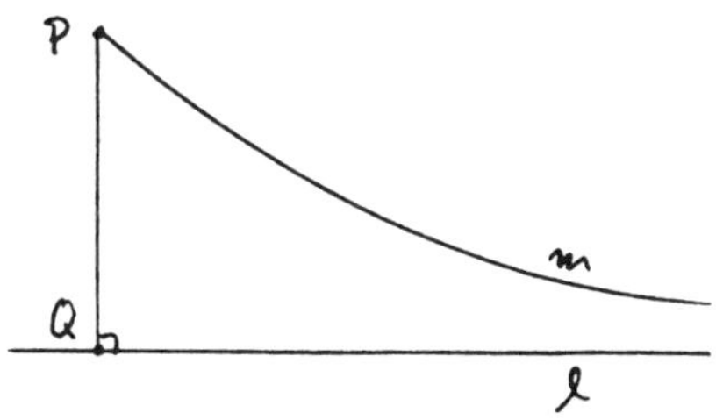

From the definition it is clear that there can be at most one such ray through P on each side of the line PQ. We can now state the *axiom of limiting parallels.*

L. For each line l and for each point P not on l, there are two limiting parallel rays through P to l, one on each side of the perpendicular PQ, and these two rays do not lie on the same line.

Note that (L) immediately implies the geometry is non-Euclidean, because in the Euclidean case there is a unique parallel line through P, and both its rays (which are limiting parallels in that case!) lie on the same line.

A Hilbert plane satisfying (L) we will call a *hyperbolic plane* or a *hyperbolic geometry.* The results of this section all take place in a hyperbolic plane.

Proposition 40.1 *In a hyperbolic plane, let P be a point, and let m be one of its limiting parallel rays to a line l. Let m make an angle α with the perpendicular PQ to l. Let P', l', m', Q', α' be another such situation. Suppose $PQ \cong P'Q'$. Then $\alpha \cong \alpha'$.*

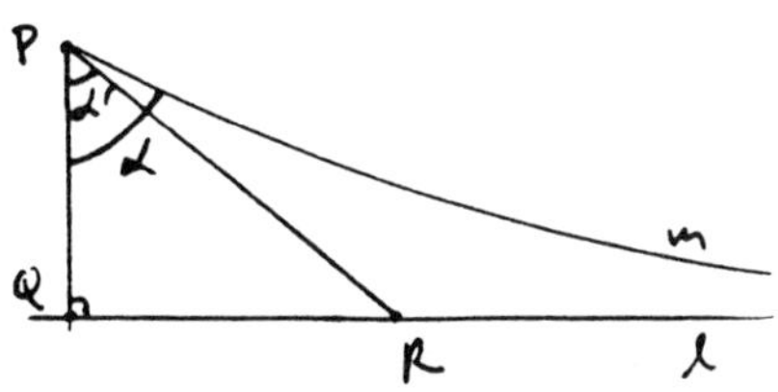

Proof. If not, we may suppose $\alpha' < \alpha$. Take a ray at P making an angle α' with PQ. Since $\alpha' < \alpha$, by definition of limiting parallel, this ray must meet l at a point R. Lay off R' on l' so that $QR \cong Q'R'$. Then by (SAS) the triangles PQR and $P'Q'R'$ are congruent. Hence the angle $Q'P'R'$ is equal to α', so the ray $P'R'$ is equal to m, which is a contradiction.

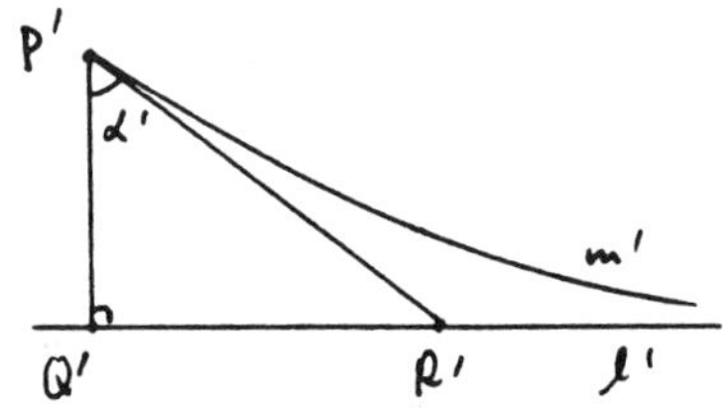

Remark 40.1.1. In particular, the angles made by the two limiting parallel rays to l at P are equal, and this angle depends only on the congruence equivalence class of the segment PQ. We call it the *angle of parallelism* of the segment PQ, and denote it by $\alpha(PQ)$. (Lobachevsky uses the notation $\Pi(PQ)$.)

Since the perpendicular to PQ at P is one parallel to l through P (I.27), and by (L) the two limiting parallel rays do not lie on the same line, the angle of parallelism α is a strictly acute angle: $\alpha <$ RA.

We will see later a converse to this proposition, namely if $\alpha(PQ) = \alpha(P'Q')$, then $PQ \cong P'Q'$ (40.5). We will also see that for every acute angle α there exists a segment PQ with $\alpha(PQ) = \alpha$ (40.14).

Proposition 40.2 *Let m be a limiting parallel ray from a point P to a line l. Then for any other point A on the line containing the ray m, the ray from A along m is also a limiting parallel to l.*

Proof. We divide into two cases. If A lies on the ray m, let n be any ray from A between m and l. Let Q be a point on n. Then PQ lies below m, so PQ meets l, and hence n must also meet l.

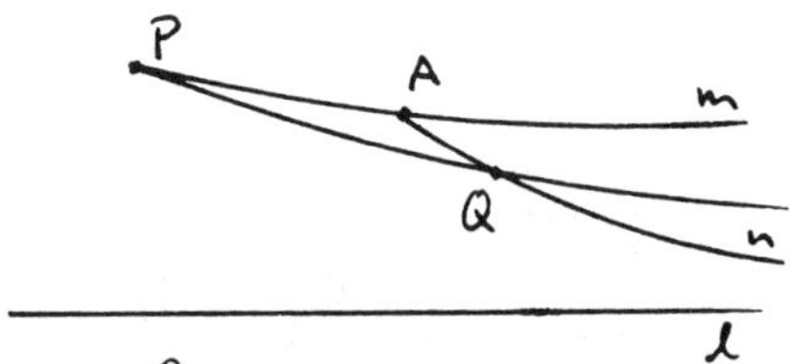

If A lies on the ray m extended behind P, again let n be any ray from A between m and l. Take a point Q on the ray n extended behind A. Then QP gives a ray through P below m, so QP must meet l, and hence n also meets l.

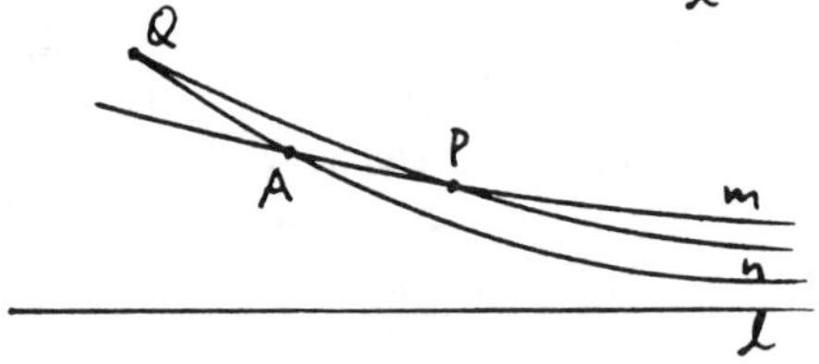

Remark 40.2.1. Because of this result, it makes sense to say a line m is limiting parallel to another line l, without specifying the point P on m. We still have to show that the relation of being limiting parallels is an equivalence relation.

Lemma 40.3 *Let AB and $A'B'$ be two segments, with $AB > A'B'$. Then $\alpha(AB) \leq \alpha(A'B')$.*

Proof. Take $PQ = AB$, let l be the perpendicular at Q, and let m be the limiting parallel from P to l. Then m makes an angle $\alpha = \alpha(AB)$ with PQ. Take R on PQ so that $PR = A'B'$. Any ray from P inside the angle α will meet l, and hence also it will meet the ray n through R perpendicular to PQ. Therefore $\alpha(PR) = \alpha(A'B') \geq \alpha$, as required.

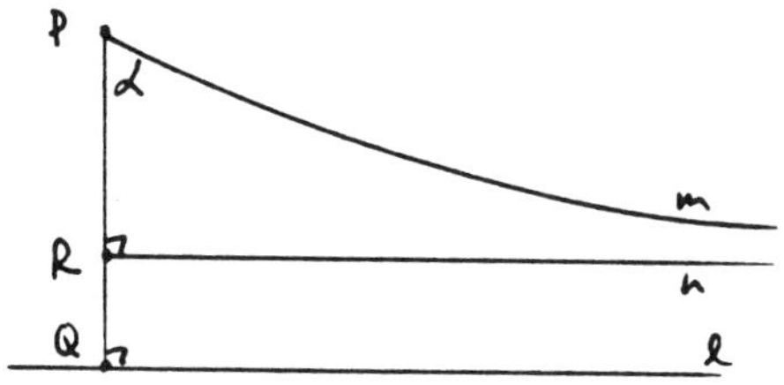

Remark 40.3.1. This result is only provisional, because we will improve it below (40.5.1).

Proposition 40.4 *If m is a limiting parallel ray to l through P, then l is a limiting parallel to m through Q (the foot of the perpendicular from P).*

Proof. Let n be any ray through Q lying above l. Let PR be the perpendicular from P to n. Then $PR < PQ$ since the leg of a right triangle is less than its hypotenuse. Therefore by (40.3), $\alpha(PR) \geq \alpha(PQ)$. But on the other hand, PR makes an angle with m which is strictly less than $\alpha = \alpha(PQ)$. Therefore m must meet n. We conclude that l is limiting parallel to m through Q, as required.

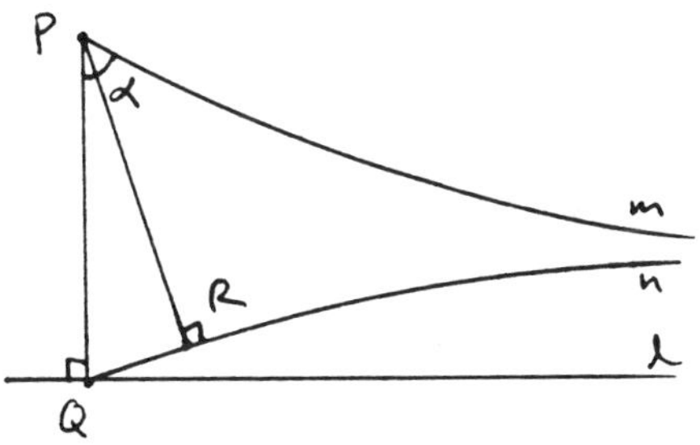

Proposition 40.5 *Let AB and $A'B'$ be two segments. If $\alpha(AB) = \alpha(A'B')$, then $AB \cong A'B'$.*

Proof. If not, suppose $AB > A'B'$. Take $PQ = AB$, let l be the perpendicular at Q, and choose R such that $PR = A'B'$ and let n be the perpendicular at R. The ray m forming the angle $\alpha = \alpha(AB) = \alpha(A'B')$ will then be limiting parallel to both l and n. By (40.4), l is then limiting parallel to m.

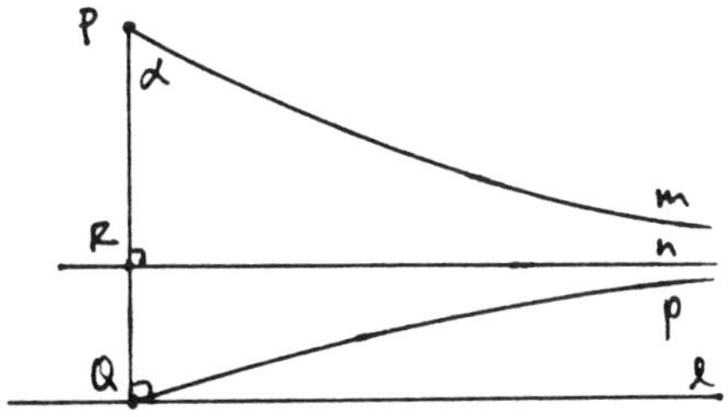

Now the angle of parallelism of RQ is an acute angle (40.1.1), so there is a ray p from Q above l which is limiting parallel to n. Since it is below n, it does not meet m. But this contradicts the fact that l is a limiting parallel to m. We conclude that $AB = A'B'$ as required.

Remark 40.5.1. Combining this with (40.3), we see that if $AB > A'B'$, then $\alpha(AB) < \alpha(A'B')$.

Proposition 40.6 *If l, m, n are three rays, with l limiting parallel to m, and m limiting parallel to n, then also l is limiting parallel to n.*

Proof. If l and n are on the same side of m, then n is also limiting parallel to m by (40.4). Therefore l and n cannot meet because, using (40.2), that would contradict the definition of limiting parallels at this meeting point. So suppose n is in between l and m. Then from a point of l any ray between l and m will meet m, and hence also n. Therefore l is also limiting parallel to n.

If l and n are on opposite sides of m then from a point of l let r be a ray between l and m. Since l is limiting parallel to m, this ray meets m at a point B. Since m is limiting parallel to n, the continuation of this ray from B must meet n. Hence l is limiting parallel to n.

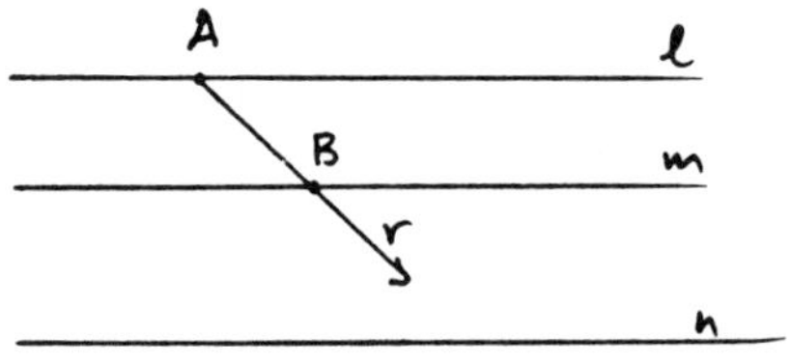

Remark 40.6.1. Combining the results so far, we see that being limiting parallels is an equivalence relation for rays, or for lines, always keeping a direction on each in mind. An equivalence class for this relation will be called a *pencil of limiting parallel rays*, or an *end*. Thus to say two lines have *an end in common* means that they support limiting parallel rays.

Now that we have established the properties of limiting parallels, our next goal is to prove congruence theorems for triangles involving limiting parallels, analogous to those for ordinary triangles in Euclid's *Elements*, Book I.

Proposition 40.7 (ASA) *Suppose given two segments AB and $A'B'$, with rays l, m, l', m' emanating from A, B, A', B'. Assume that the angle at A equals the angle at A', the side AB equals the side $A'B'$, and the angle at B equals the angle at B'. Then l is limiting parallel to m if and only if l' is limiting parallel to m'.*

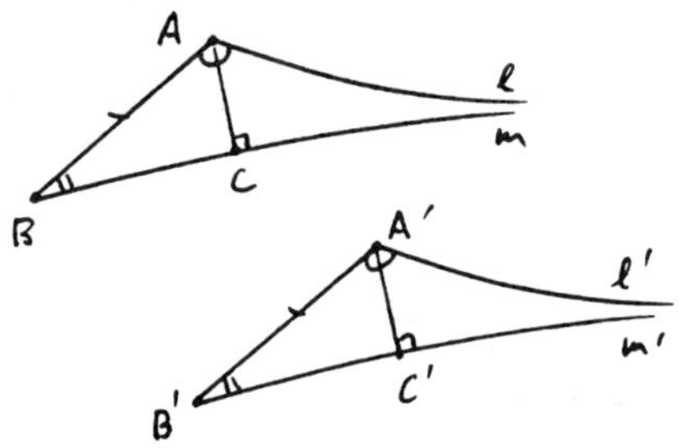

Proof. Let AC be the perpendicular from A to m, and let $A'C'$ be the perpendicular from A' to m'. Then by (AAS) = (I.26), the triangles ABC and $A'B'C'$ are congruent. It follows that $AC \cong A'C'$, and by subtraction the angles $\angle CAl$ and $\angle C'A'l'$ are equal. If l is limiting parallel to m, then this angle is the angle of parallelism for AC, so by (40.1) it follows that l' is the limiting parallel to m'.

Proposition 40.8 (Exterior angle theorem) *If AB is a segment, with limiting parallel rays emanating from A and B (we call this figure a* limit triangle*) then the exterior angle β at B is greater than the interior angle α at A.*

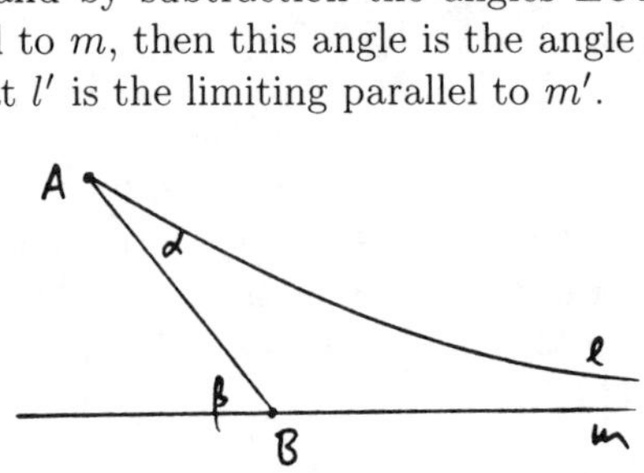

Proof. Because the ray through A making an angle β with AB is one parallel to l (I.27) we know at least that $\alpha \leq \beta$.

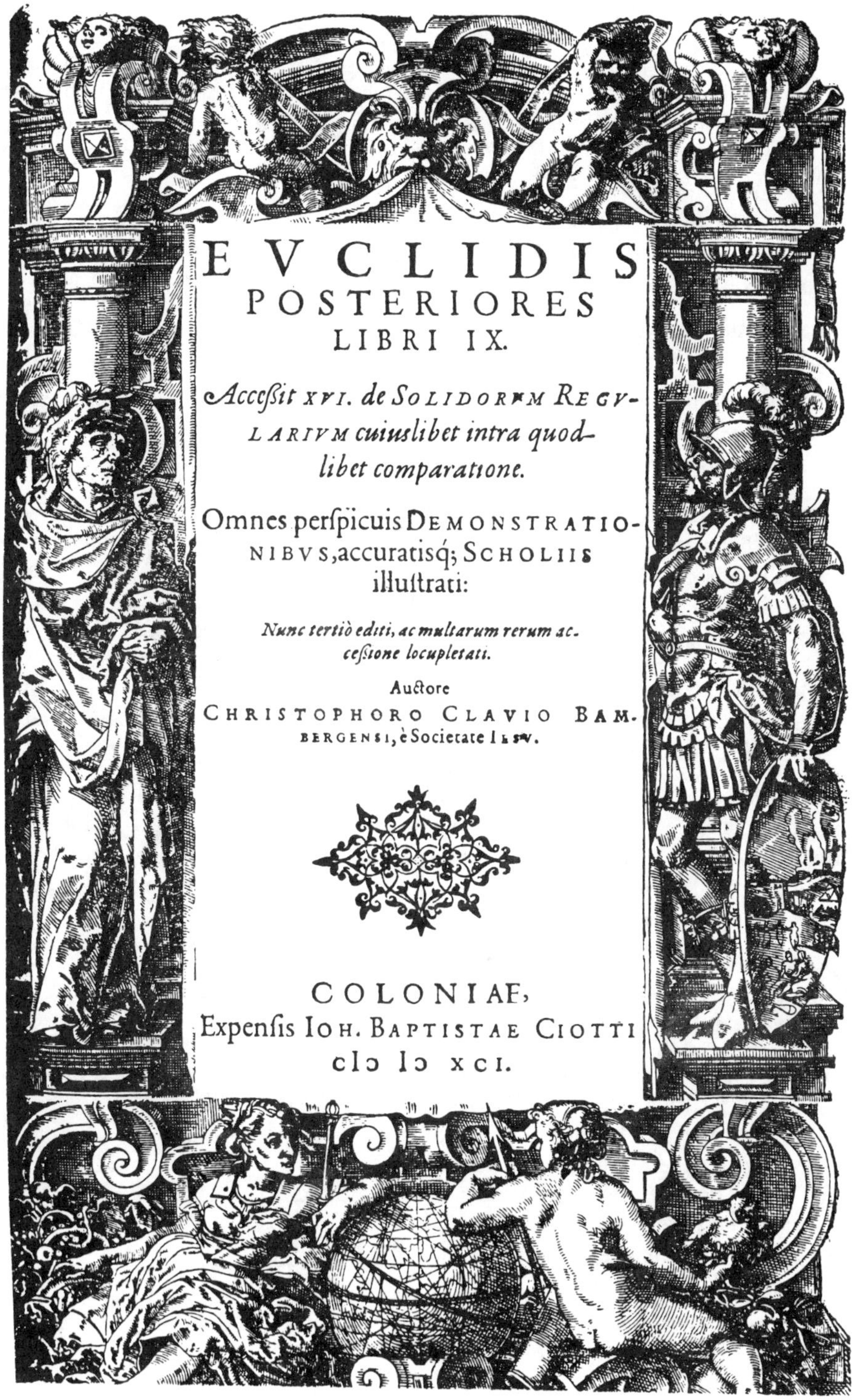

EVCLIDIS
POSTERIORES
LIBRI IX.

Acceßit XVI. de SOLIDORVM REGVLARIVM cuiuslibet intra quodlibet comparatione.

Omnes perſpicuis DEMONSTRATIONIBVS, accuratisq; SCHOLIIS illuſtrati:

Nunc tertiò editi, ac multarum rerum acceſsione locupletati.

Auctore
CHRISTOPHORO CLAVIO BAMBERGENSI, è Societate IESV.

COLONIAE,
Expenſis IOH. BAPTISTAE CIOTTI
CIↃ IↃ XCI.

Plate XII. Title page to Clavius' Euclid (1591), Vol. 2.

So suppose $\alpha = \beta$. Let l' and m' be the opposite rays to l and m. The supplementary angles at A and B will also be equal. Since AB is equal to itself, we can apply (40.7) to AB, l, m and BA, m', l'. We conclude that l' is also limiting parallel to m'.

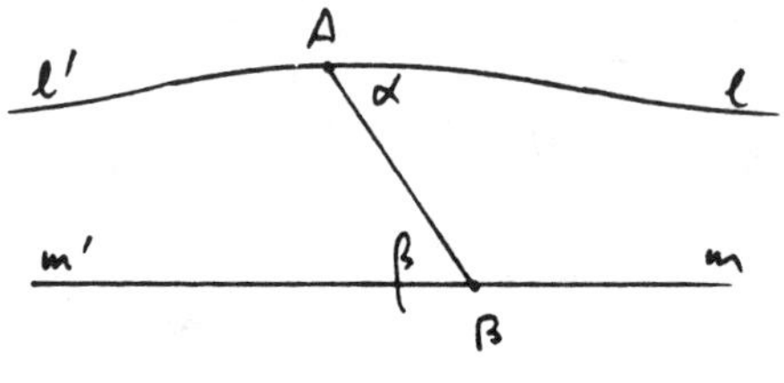

But this contradicts the axiom (L) which says the two limiting parallels from A to m do not lie on the same line. Therefore $\alpha < \beta$, as required.

Corollary 40.9 *In a hyperbolic plane, the sum of the angles of any triangle is less than two right angles.*

Proof. According to (34.6), for any triangle, there is a Saccherei quadrilateral whose top two angles are equal to the angle sum of the triangle. So we have only to prove that the top two equal angles (34.5) of any Saccheri quadrilateral are acute.

Let the Saccheri quadrilateral be $ABCD$, with base $AB = l$. Draw limiting parallels from C and D to l, with end ω by axiom (L). Then by (40.1) the angles of parallelism α are equal.

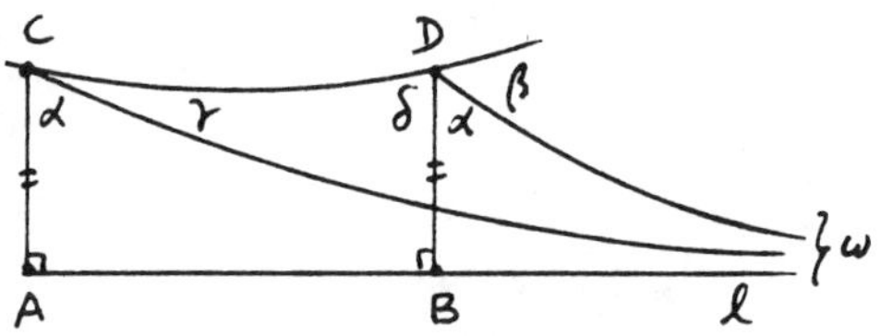

Looking at the limit triangle $CD\omega$, by the exterior angle theorem (40.8), $\beta > \gamma$. On the other hand, by (34.5), the top angles $\alpha+\gamma$ and δ of the Saccheri quadrilateral are equal. We conclude that $\alpha + \beta > \alpha + \gamma = \delta$, and so δ must be acute.

Remark 40.9.1. Note how different this proof is from the proof of the Saccheri–Legendre theorem (34.2) which reaches the same conclusion under different hypotheses. There we made use of Archimedes' axiom and a countable limiting process. Here we do not need (A), but we use instead the powerful axiom (L) on the existence of limiting parallels.

Proposition 40.10 (ASL) *Let $ABlm$ and $A'B'l'm'$ be two limit triangles. Assume that the angle at B equals the angle at B', and the side AB equals the side $A'B'$. Then also the angle at A equals the angle at A'.*

Proof. Drop perpendiculars AC and $A'C'$ from A to m and from A' to m'. Then by (AAS), the triangles ABC and $A'B'C'$ are congruent. Hence the angles $\angle BAC$ and $\angle B'A'C'$ are equal, and the sides AC and $A'C'$ are equal. Since l and m are limiting parallel, the angle $\angle CAl$ is the angle of parallelism for the segment AC. The same holds in the second triangle, so by (40.1) the angles $\angle CAl$ and $\angle C'A'l'$ are equal. Adding equal angles, we find that the angles at A and A' in the limiting triangles are equal.

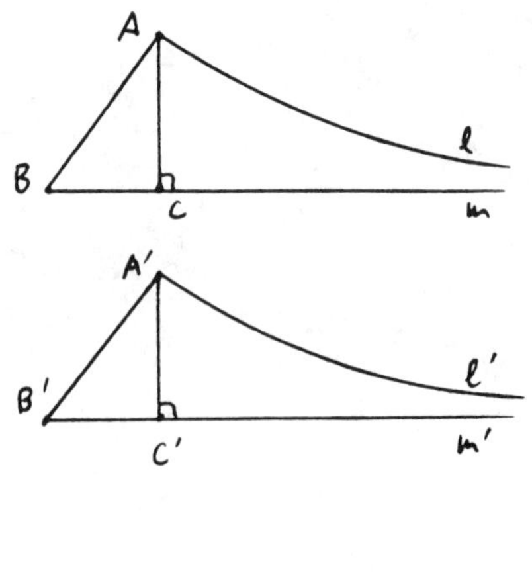

Proposition 40.11 (AAL) *Given two limit triangles $ABlm$ and $A'B'l'm'$, suppose that the angles at A and B are equal respectively to the angles at A' and B'. Then also the sides AB and $A'B'$ are equal.*

Proof. If not, let us suppose that $AB > A'B'$. Choose a point C on AB so that $CB = A'B'$, and draw a ray n at C, on the same side of AB as l and m, making an angle equal to the angle at A', which is also equal to the angle at A. Now comparing C, B, n, m to the limit triangle $A'B'l'm'$, it follows from (40.7) that n is limiting parallel to m.

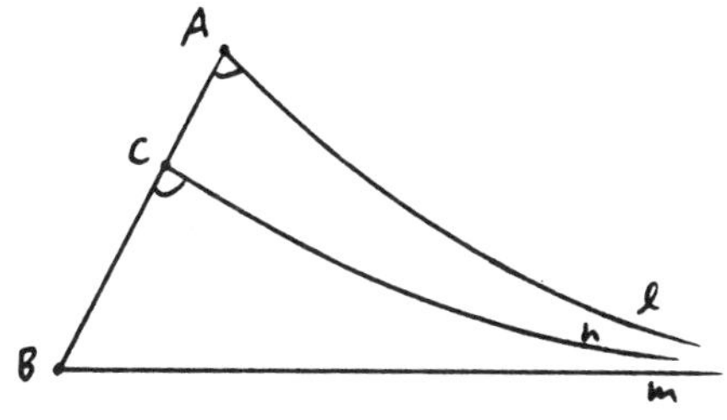

Then by transitivity (40.6) it follows also that l is limiting parallel to n. But this contradicts the exterior angle theorem (40.8) because the angle at C, which is exterior to the limit triangle $ACln$, is equal to the angle at A.

We conclude that $AB = A'B'$, as required.

Remark 40.11.1. For some results about triangles with two or three "limit angles", see Exercises 40.2 and 40.11.

Theorem 40.12 *In a hyperbolic plane, if l and m are two parallel lines which are not limiting parallels, there is a unique line in the plane which is perpendicular to both of them.*

Proof. Let l and m be two parallel lines which are not limiting parallels. Let AB and CD be two perpendiculars from points A, C or l to m. If $AB = CD$, then $DBCA$ is a Saccheri quadrilateral, and hence the line joining the midpoints of AC and BD will be perpendicular to both l and m, by (34.5).

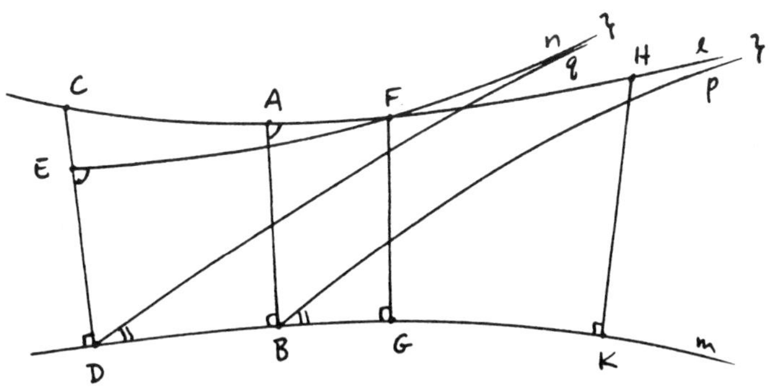

If $AB \neq CD$, we may assume $CD > AB$, and we proceed as follows. Take E on CD so that $AB = ED$. Let n be a ray through E, making the same angle with ED as l makes with AB. I claim that n will meet l in a point F. Indeed, let p

be a limiting parallel from B to l. Since by hypothesis l and m are not limiting parallels, this ray lies above m. Let q be the ray through D making the same angle with m as p does at B. Then q is parallel to p by (I.28), but not a limiting parallel, by the exterior angle theorem (40.8). On the other hand, applying (40.7) to $ABlp$ and $EDnq$, we find that q is limiting parallel to n. Therefore n is not limit parallel to p, and hence n must meet l at some point F. (In the figure we put F on the far side of A from C, but the proof works equally well if F is between A and C.)

Now take H on l so that $AH = EF$, and take K on m so that $BK = DG$. Then comparing the quadrilaterals $EFDG$ and $AHBK$, two applications of (SAS) show that $FG = HK$ and HK is perpendicular to m. Thus $GKFH$ is a Saccheri quadrilateral, and the line joining the midpoints of FH and GK will be perpendicular to both l and m.

It remains to show the uniqueness of the line perpendicular to both l and m. Suppose to the contrary that AB and CD were two common perpendiculars to l and m. Then $ABCD$ would be a rectangle, which is impossible—cf. proof of (40.9).

Proposition 40.13 *Given an angle in the hyperbolic plane, there is a unique line (called the* enveloping line *of the angle) which is limiting parallel to both arms of the angle.*

Proof. Let O be the vertex of the angle, and choose points A, B on the two arms of the angle, at equal distance from O. It will be convenient at this point to introduce a new notation. We denote by α the end of the ray OA, that is, the equivalence class of all rays limiting parallel to OA—cf. (40.6.1). Then we may draw the line $B\alpha$, meaning, let $B\alpha$ be the ray through B limiting parallel to OA. We may also speak of the limit triangle $AB\alpha$, consisting of the segment AB plus the two limiting parallel rays $A\alpha$ and $B\alpha$.

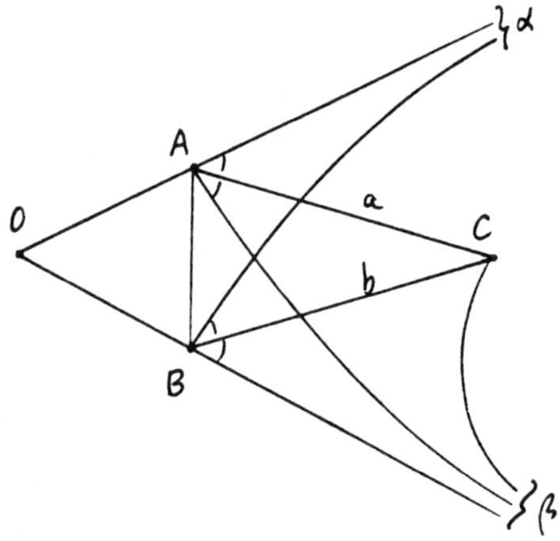

To continue our proof, let α be the end of OA, and let β be the end of OB. Draw $B\alpha$ and $A\beta$. Let a be the ray bisecting the angle $\alpha A\beta$, and let b be the ray bisecting the angle $\alpha B\beta$. We distinguish three cases.

Case 1. The lines a and b meet at a point C. By symmetry (!) $AC = BC$. Draw the line $C\beta$. Then by (ASL) = (40.9) applied to the limit triangle $AC\beta$ and $BC\beta$, the angles at C of these two triangles are equal. But this is clearly not so, so this case cannot occur.

Case 2. The rays a and b are limiting parallel with an end γ. In this case we let C be the intersection of $A\beta$ and $B\gamma$. By (AAL) = (40.11) applied to the limit triangles $AC\gamma$ and $BC\beta$, the sides AC and BC are equal. Therefore by (I.5) the angles BAC and ABC are equal. But this is not so, because the angle BAC is also equal to the angle $AB\alpha$ which is properly contained in the angle ABC. So this case cannot occur either.

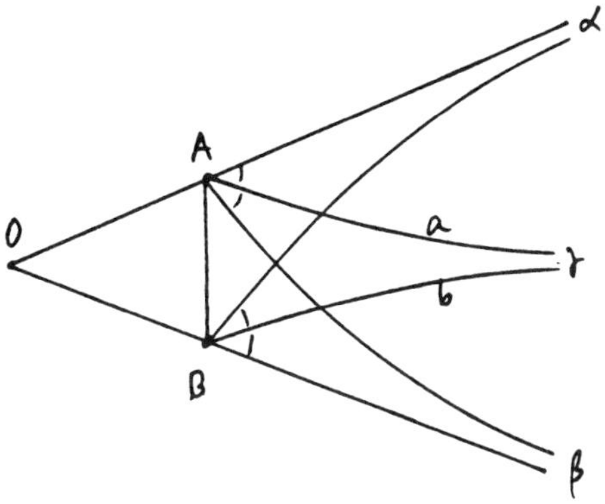

Case 3. The only remaining possibility is that a and b are parallel but not limiting parallels. Then by (40.12) there is a common perpendicular line l, meeting a at C and b at D. I claim l is the required enveloping line, i.e. l has the ends α and β.

By symmetry it is enough to show l has end β. If not, draw the lines $C\beta$ and $D\beta$, which will be distinct from l. We compare the limit triangles $AC\beta$ and $BD\beta$. The angles at A and B are equal, by construction. The sides AC and BD are equal by symmetry (!), so by (ASL) = (40.10) the angles at C and D are equal. It follows that $C\beta$ and $D\beta$ make equal angles with l at C and D, which contradicts the exterior angle theorem (40.8). We conclude that l has ends α and β, as required.

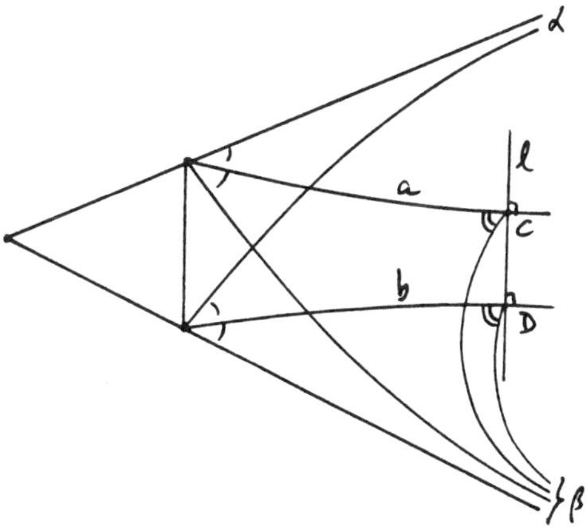

The uniqueness of the enveloping line is clear, because by (L) we cannot have two distinct lines which are limiting parallel at both ends.

Corollary 40.14 *For any acute angle α, there exists a line which is limiting parallel to one arm of the angle and orthogonal to the other arm of the angle. In particular, there is a segment whose angle of parallelism is equal to α.*

Proof. Given the acute angle α, we double it, and consider the enveloping line (40.13) of the angle 2α. This will be orthogonal to the angle bisector of 2α, which is one arm of the original angle α. Thus α becomes the angle of parallelism of the segment cut off on that arm of the angle.

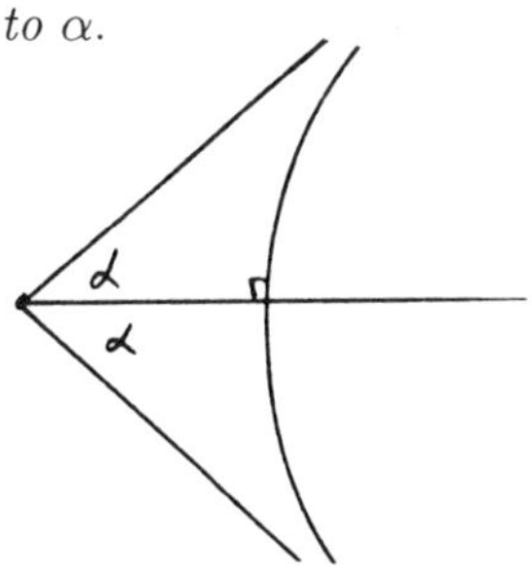

Remark 40.14.1. Combining with (40.1) and (40.5), we see that there is a one-to-one correspondence between the set of congruence equivalence classes of line segments and the set of congruence equivalence classes of acute angles, given by associating a segment AB to its angle of parallelism α. In particular, there is a

uniquely determined standard or absolute segment size corresponding to one half of a right angle.

Be careful, however, because this correspondence does not send sums of segments into sums of angles. There is a more complex relationship which we will see later (Exercise 42.7).

Proposition 40.15 *In a hyperbolic plane, Aristotle's axiom holds, namely, given an angle α and a segment AB, there exists a point C on one arm of the angle so that the perpendicular CD from C to the other arm of the angle is greater than AB.*

Proof. Given the angle α at A, let l be a line limiting parallel to one arm of α and meeting the other arm at right angles at a point F (40.14). Take E on l so that $EF = AB$. Draw a perpendicular to l at E, and let it meet the other arm of the angle α at C (cf. Exercise 40.9). Drop a perpendicular CD from C to AF.

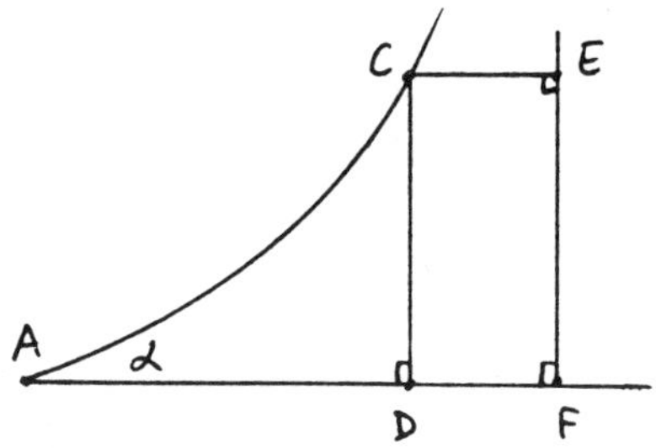

Consider the quadrilateral $DFCE$. Because of (40.9), the angle at C must be acute. Therefore $CD > EF = AB$ (35.3), as required.

Remark 40.15.1. In fact, a stronger result is true, namely, given α and AB as above, one can find C such that $CD = AB$. The proof uses hyperbolic trigonometry (Exercise 42.8).

Now, as an illustration of the techniques of this section, we will give the hyperbolic version of a familiar Euclidean theorem on the angle bisectors of a triangle. The fact that the (internal) angle bisectors of a triangle meet in a point is true in neutral geometry, hence both in Euclidean and hyperbolic geometry, as we have seen before (Exercise 11.6). The following result has to do with the external angle bisectors of a triangle.

Proposition 40.16 *In a hyperbolic plane, let ABC be a triangle, and consider the (internal) angle bisector at A, and the external angle bisectors at B and C.*

a) *If two of these angle bisectors meet in a point, so does the third.*

b) *If two of these angle bisectors have a common perpendicular line l, then the third is also perpendicular to l.*

c) *If two of these angle bisectors are limiting parallels, so is the third, at the same end.*

Proof. a) If two of them meet in a point Y, then Y is equidistant from all three sides of the triangle, hence it lies on the third angle bisector. The proof in this case is the same as the Euclidean case (IV.4).

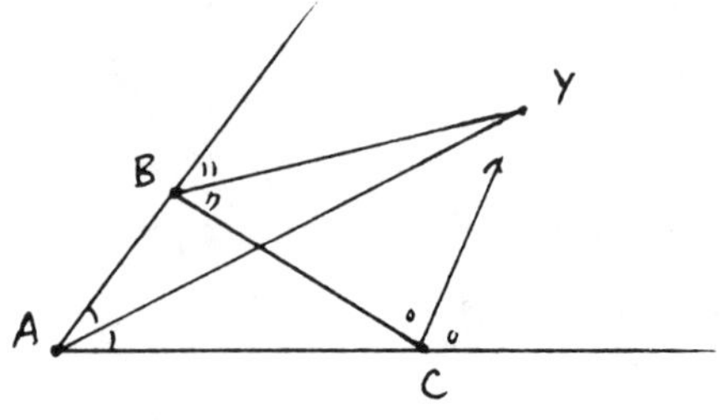

b) Suppose that the angle bisectors at A and B have a common perpendicular line l.

We first claim that l cannot meet any side of the triangle. If it meets one side, then by reflecting in the two angle bisectors, it will meet the other two sides, and it will meet all three at the same angle. Two out of three of these intersections (in the diagram V, W), will have the angles in corresponding positions, so that by (I.28) the lines BC and AC will be parallel. This contradicts their meeting at the point C. Thus l cannot meet any side of the triangle.

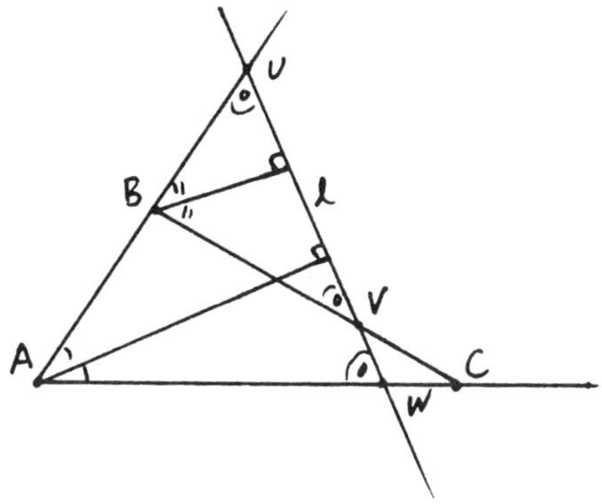

Secondly, we note that l cannot be a limiting parallel to any side of the triangle. If it were, then by reflecting in the angle bisectors, it would be limiting parallel to the other two sides, and so would have three ends, which is absurd.

So l neither meets nor is limiting parallel to any side of the triangle, hence by (40.11) it has a common perpendicular with each side of the triangle. Using the lemma below, the first and second of these common perpendiculars are equal. Similarly the first and third are equal, because the angle bisectors at A and B are orthogonal to l. Therefore the second and third are equal, and using the lemma in the other direction, we see that the angle bisector at C is also perpendicular to l.

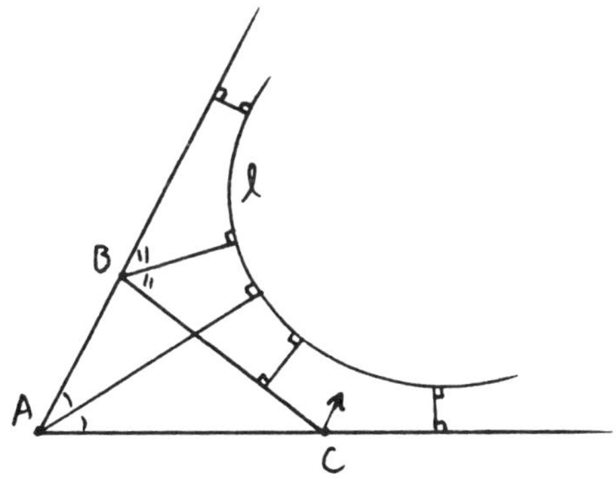

c) This third case of the proposition follows by elimination. Suppose two angle bisectors are limiting parallel. If the third is not, then it either meets one of the others or has a common perpendicular, which puts us in case a) or b), contradicting the first two being limiting parallel.

Lemma 40.17 *Consider a five-sided figure $ABCDE$ with right angles at A, B, C, D. Then $AC = BD$ if and only if the angle bisector at E meets the opposite side at a point F at right angles.*

Proof. First suppose that the angle bisector at E meets AB at a point F, making a right angle there. Then reflection in the line EF sends the line AB into itself, and interchanges the lines CE and DE. So the segments AC and BD are interchanged, because they are the unique common perpendiculars (40.11) between the lines AB and CE and AB and DE. Hence $AC = DB$.

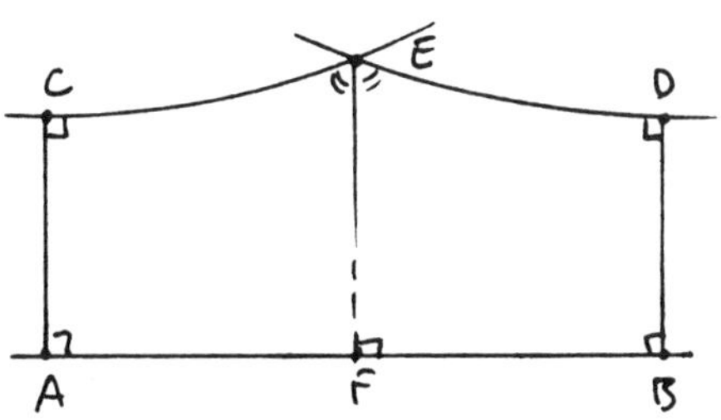

Conversely, suppose $AC = DB$. Draw the line CD. Then $ABCD$ is a Saccheri quadrilateral, and the angles at C and D are equal (34.5). It follows that the base angles of the triangle CDE are equal. Hence it is an isosceles triangle, and the angle bisector at E will meet CD at its midpoint at right angles. Now it follows from (34.5) that this line continued will meet AB at its midpoint F, at right angles.

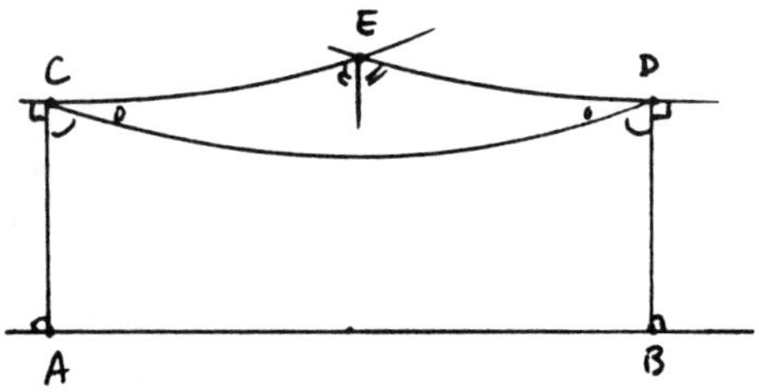

Remark 40.17.1. To make a more unified statement of (40.16) we will define an *ideal point* P^* to be an equivalence class of lines, all of which have a common perpendicular line p. We will say P^* lies on a line l, if $l \perp p$. Then we define a *generalized point* to be either a usual point, or an end (40.6.1) or an ideal point. Using this language, we can say that the three angle bisectors of (40.16) meet in a common generalized point.

Exercises

The following exercises all take place in a hyperbolic plane, that is, a Hilbert plane satisfying (L).

40.1 If l and m are limiting parallel rays, and if n is any ray in between l and m, which meets neither one, then l and n are also limiting parallels.

40.2 (ALL) Suppose given equal angles at two points A and A', and let l and l' be their enveloping lines. Show that the perpendicular AB from A to l is equal to the perpendicular $A'B'$ from A' to l'.

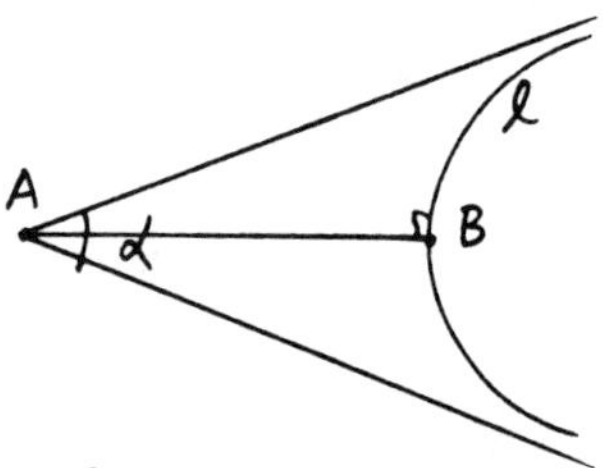

40.3 If l and m are two parallel, but not limiting parallel, lines, show that their common perpendicular AB is the shortest distance between the two lines. Namely, show for any other points $C \in l$ and $D \in m$, that $CD > AB$.

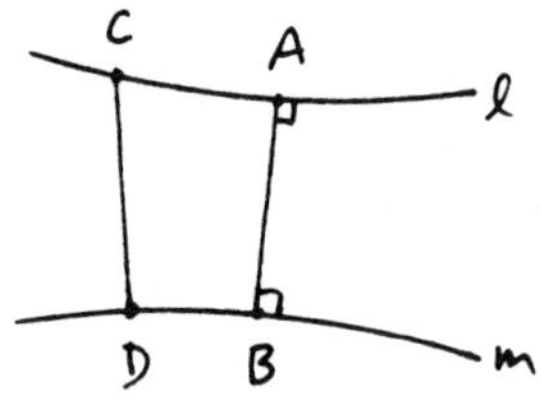

40.4 Show that ends of lines behave somewhat like points, as follows.

a) Given a point P and an end α, there exists a unique line l passing through P and having end α.

b) Given two distinct ends α, β, there exists a unique line l having ends α and β.

40.5 Given two limiting parallel lines l and m, show that there exists a line n, limiting parallel to (the other end of) l, and orthogonal to m.

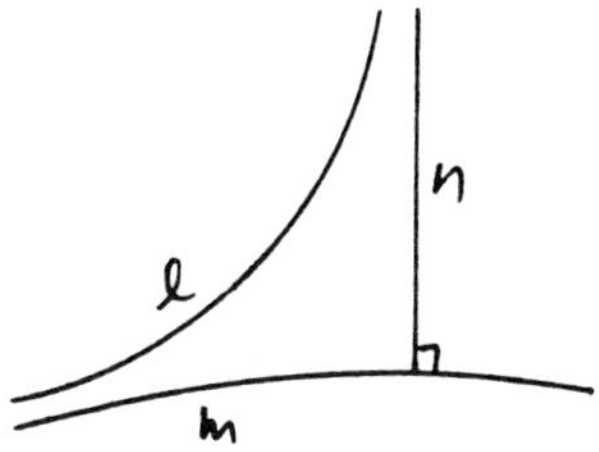

40.6 Given two limiting parallel lines l, m, show that there exists a third line n, limiting parallel to the other ends of both l and m.

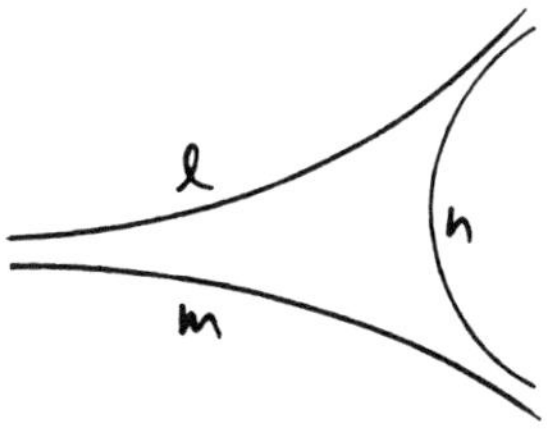

40.7 Given two limiting parallel lines l, m,

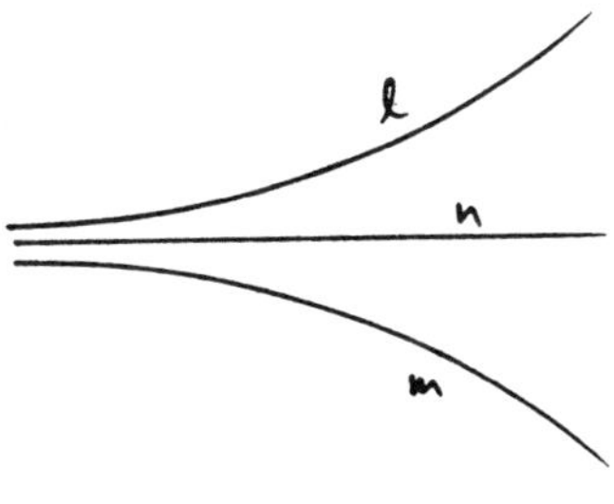

a) Show that there exists a third line n, limiting parallel to l and m, such that the reflection in n interchanges l and m. We call n the *midline* of the pair l, m.

b) Show that every point of n is equidistant from l and m.

40.8 Given two limiting parallel lines l, m, and given a segment AB, no matter how large or how small, there exists a point C on l such that the perpendicular CD to m is equal to AB. *Hint*: Take m' perpendicular to AB through B and let l' be the limiting parallel to m' through A. Apply (Exercise 40.5) both to the pair l, m and the pair l', m', and compare.

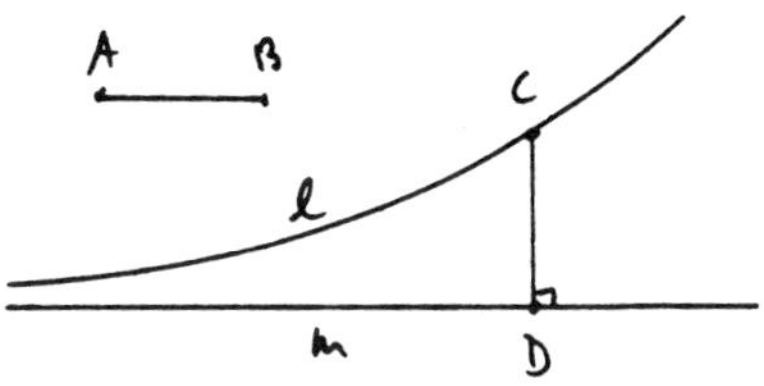

40.9 Show that the analogue of Pasch's axiom (B4) holds for limit triangles: Let $AB\alpha$ be a limit triangle. Let l be a line that does not contain A, B, or α (i.e. l is not limit parallel to $A\alpha$). Suppose that l meets one of the sides AB or $A\alpha$ or $B\alpha$ of the triangle. Then l must meet one of the other sides, but not both.

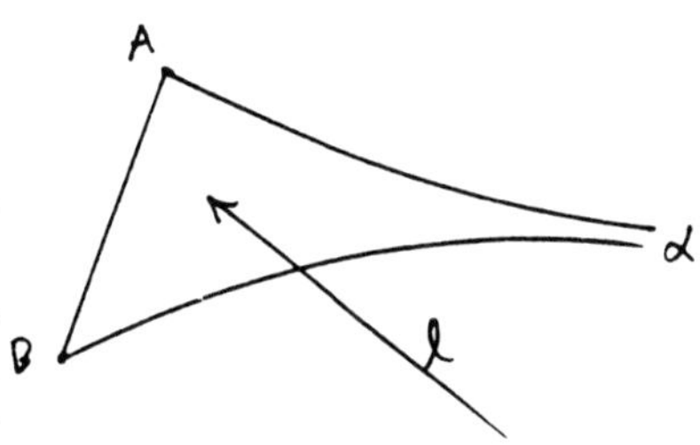

40.10 In a limit triangle $AB\alpha$, show that the two angle bisectors at A and B, and the midline of the pair of limiting parallels $A\alpha$, $B\alpha$ (Exercise 40.7) all meet in a single point. This point will be the center of the *inscribed circle* of the limit triangle.

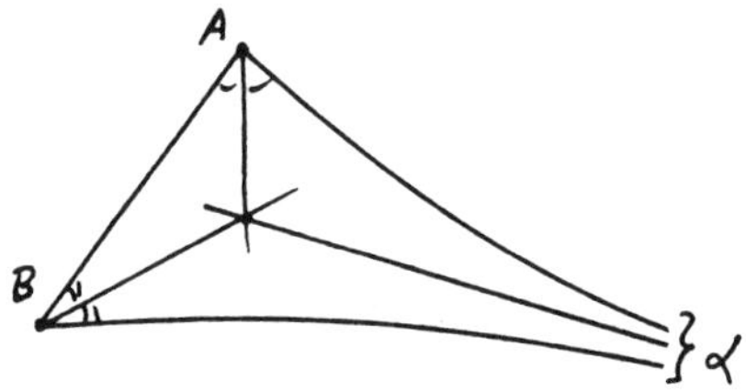

40.11 (LLL). Let l, m, n be three lines, each limiting parallel to the other two.

a) Show that the three midlines (Exercise 40.7) to the three pairs of lines are orthogonal to the opposite sides of the trilimit triangle l, m, n, and all meet in a single point A, which is equidistant from l, m, n.

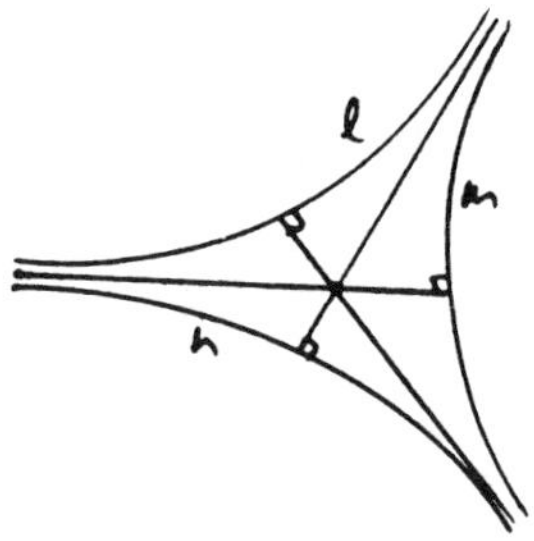

b) If l', m', n' is another such trilimit triangle, with corresponding point A', show that the distance from A to the three sides l, m, n is equal to the distance from A' to the three sides l', m', n' of the second triangle.

40.12 Given two angles α, β, with $\alpha + \beta < 2\text{RA}$, show that there exists a limit triangle with angles α, β.

40.13 A *limit quadrilateral* is a figure consisting of four lines l, m, n, p, with each limit parallel to the next, in cyclic order.

a) If $lmnp$ is a limit quadrilateral, show that opposite sides are parallel but not limit parallel.

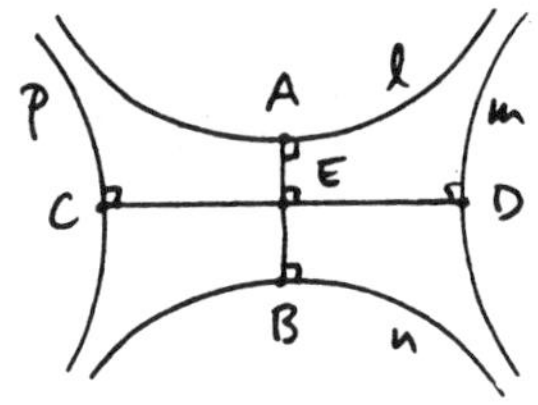

b) Show that the common orthogonals AB of l and n and CD of m and p meet at right angles at a point E.

c) Show that there exists a limit quadrilateral with AB equal to any prescribed segment.

d) Two such limit quadrilaterals can be moved one to the other by a rigid motion of the plane, if and only if the segment AB of the first is equal to one of the segments $A'B'$ or $C'D'$ of the second.

40.14 Show that ideal points (40.17.1) behave somewhat like regular points, as follows.

a) Given a (regular) point P and an ideal point Q^*, there is a unique line containing them both.

b) Given an end α and an ideal point Q^*, and assuming that α is not an end of the defining line q of Q^*, then there is a unique line containing Q^* with end α.

c) Any two distinct lines have a unique generalized point in common.

40.15 You may have noticed, while doing (Exercise 40.14), that two ideal points do not necessarily lie on a line. So we define a *generalized line* to be either

1) a regular line, together with its two ends and ideal points, or

2) a *limit line*, which consists of an end α, together with all ideal points P^* whose defining line p contains α, or

3) an *ideal line*, which consists of all ideal points P^* whose defining line p contains a fixed (regular) point L.

Show that the set of all generalized points of the hyperbolic plane, together with the subsets of generalized lines, forms a *projective plane* (Exercise 6.2). In particular, any two generalized points lie on a unique generalized line, and any two generalized lines meet in a unique generalized point.

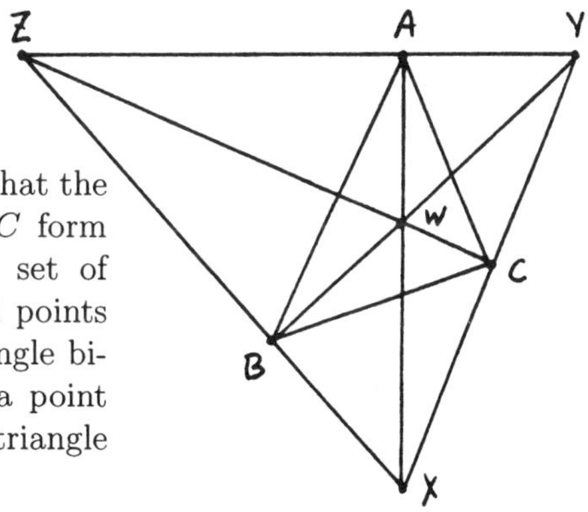

40.16 Let ABC be any triangle. Show that the external angle bisectors at A, B, C form a "generalized triangle", i.e. a set of three lines meeting in generalized points X, Y, Z. Show that the internal angle bisectors of ABC, which meet at a point W, are the altitudes of the new triangle XYZ.

40.17 Reverse the argument of (Exercise 40.16) to prove that in any triangle ABC, the three altitudes will meet in a generalized point.

Hint: Let BD and CE be two altitudes. Reflect the line DE in AB and in AC to get two new lines, which meet at a generalized point F. Show that B is equidistant from the three sides of the (generalized) triangle DEF, and from this conclude that F is a real point (not an end or an ideal point). Now apply (Exercise 40.16) to the triangle DEF. Conclude that BD, AF, CE meet in a generalized point G, and that F lies on BC, and AF is orthogonal to BC, so in fact, AF is the third altitude of the original triangle.

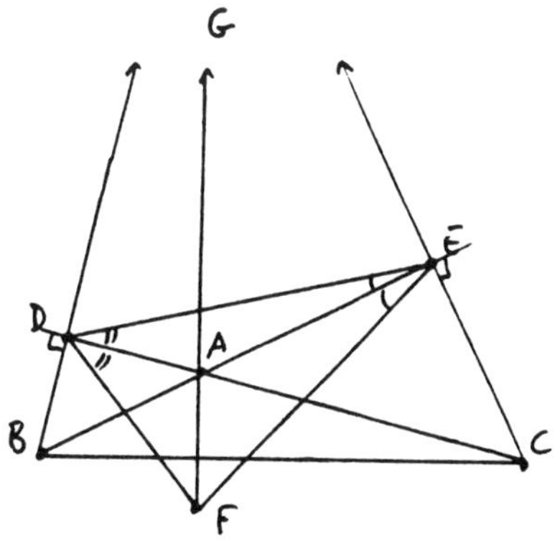

Note that if we assume two altitudes of the triangle meet in a (regular) point, then the entire proof can be carried out in a Hilbert plane with no further hypothesis, i.e. in neutral geometry.

40.18 Extend the theorem on (internal) angle bisectors of a triangle, as follows. Consider a generalized triangle, consisting of three non-concurrent lines (meaning they have no generalized point in common). Let the vertices be generalized points A, B, C.

a) Define the analogue of an angle bisector for two lines meeting at an end or an ideal point.

b) Show that the three (internal) angle bisectors of the generalized triangle ABC always meet at a (regular) point W.

c) Show that W is the center of an *inscribed circle* which is tangent to the three sides of the triangle.

41 Hilbert's Arithmetic of Ends

We come now to one of the most beautiful parts of the theory of non-Euclidean geometry, which is another illustration of the usefulness of abstract algebra. This is Hilbert's tour de force, the creation of an abstract field out of the geometry of a hyperbolic plane. In the same way that the field of segment arithmetic (§19) helped us to understand Euclidean geometry this field will help us in our study of non-Euclidean geometry. Using it we can prove results, such as Bolyai's parallel construction, or the theorem on the three altitudes of a triangle, for which we do not have purely geometric proofs. We also set up a hyperbolic analytic geometry and hyperbolic trigonometry, whereby any geometric problem can (in principle) be translated into an algebraic problem in the field. This is analogous to the analytic geometry of Descartes, but the particulars are all different, having "suffered a sea change into something new and strange". Finally we will be able to show that the hyperbolic plane is uniquely characterized by its associated field, and is in fact isomorphic to the Poincaré model over that field.

We start with a hyperbolic plane, as in the previous section, which is a Hilbert plane satisfying the axiom of limiting parallels (L).

Proposition 41.1 *Let A, B, C be three noncollinear points in a hyperbolic plane and consider the three perpendicular bisectors l, m, n of the sides of the triangle ABC.*

a) *If two of the lines l, m, n meet at a point P, then the third line also passes through P, and in this case A, B, C all lie on a circle with center P.*

b) *If two of the lines l, m, n have a common perpendicular p, then the third is also perpendicular to p, and the three points A, B, C are equidistant from the line p.*

c) *If two of the lines l, m, n are limiting parallel, the third is also, and all three have a common end.*

Proof. a) This is essentially what Euclid proves in (IV.5), the only difference being that we must assume that two of the lines meet.

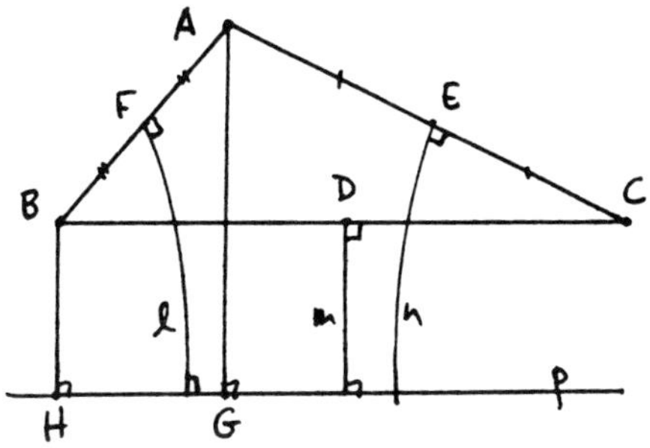

b) Suppose that two of the lines, say l and m, have a common orthogonal line p. Dropping perpendiculars AG and BH to p, one application of (SAS) and one application of (AAS) show that $AG = BH$ (see second diagram). In other words, A and B are equidistant from p.

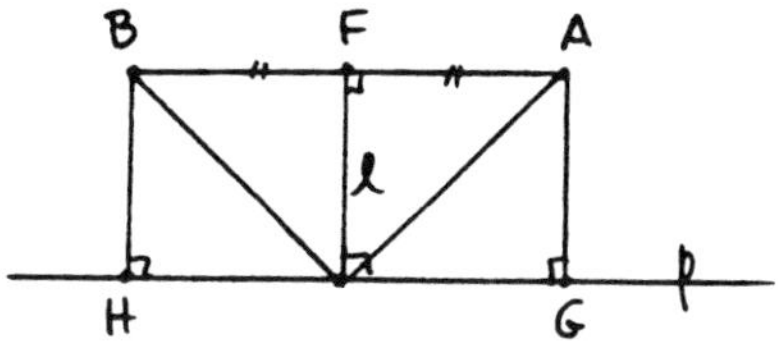

The same reasoning applied to m shows that B and C are equidistant from p. Therefore A and C are also equidistant from p, and then it follows from (35.1) that n is orthogonal to p, as required.

c) Now suppose that l and m are limiting parallels. If the third line n meets l at a point, we apply a) above to find l and m also meet, which contradicts our hypothesis. If n and l have a common perpendicular, then by b) so do l and m, which again contradicts our hypothesis (40.1.1). It follows from (40.11) that the only remaining possibility is that n and l are limiting parallels. (We leave to the reader to figure out why all three lines are limiting parallel at the same end.)

Proposition 41.2 *Given three lines a, b, c in the hyperbolic plane, with a common end ω, there exists a fourth line d with end ω, such that reflection in d is equal to the product of the reflections in a, b, c:*

$$\sigma_c \sigma_b \sigma_a = \sigma_d.$$

Here σ_l for any line l denotes reflection in the line l.

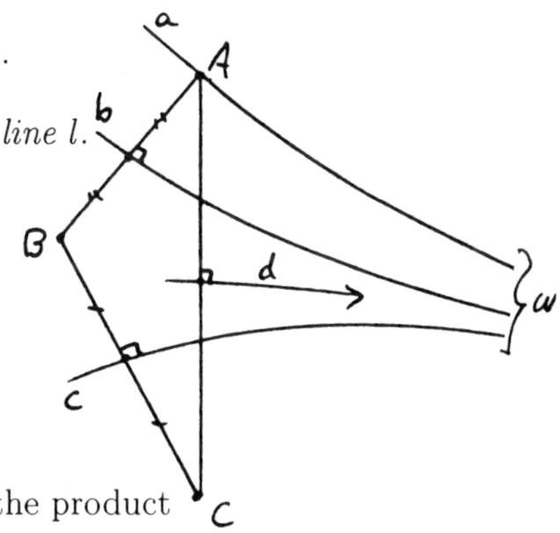

Proof. Take any point A on the line a. Let B be its reflection in the line b. Let C be the reflection of B in C. Draw AC, and let d be the perpendicular bisector of AC. Then d will be the required fourth line.

To show this, it is equivalent to show that the product

$$\varphi = \sigma_d \sigma_c \sigma_b \sigma_a$$

is equal to the identity. Note first that $\varphi(A) = A$ by construction of d. Note also that b and c are two perpendicular bisectors of sides of the triangle ABC and they are limit parallels with end ω. It follows from (41.1) that d also has end ω. Therefore φ preserves the end ω, and because of this φ fixes every point of the line a. Therefore φ is either the identity or reflection in the line a (cf. (17.4) and (Exercise 17.3)).

Suppose that $\varphi = \sigma_a$. Then $\sigma_d\sigma_c\sigma_b$ is equal to the identity. Then for any point $P \in b$ we would have $\sigma_d\sigma_c(P) = P$, so $\sigma_c(P) = \sigma_d(P)$ and this implies $c = d$ which is absurd. We conclude that $\varphi =$ identity, as required.

Now we are ready to define the arithmetic of ends. Fix a hyperbolic plane Π. Recall that an *end* is an equivalence class of limiting parallel rays (40.6.1). Fix one line and label its ends 0 and ∞. We let F be the set of all ends in the plane Π different from ∞, and then we set $F' = F \cup \{\infty\}$, so that F' is the set of all ends of the plane. We will make the set F into an ordered field by defining arithmetic operations $+$, $\cdot$, and an ordering on it.

Note that given any end α and any point P, there exists a unique line m passing through P and with end α. Just take l a line having end α, and use the axiom (L) to find a line m through P, limiting parallel to l. Similarly, given two ends α, β, there exists a unique line l having ends α and β. Just take any point P, consider the angle formed by the rays $P\alpha$ and $P\beta$, and let l be the enveloping line of the angle $\alpha P\beta$ (Exercise 40.4). We will denote the line with ends α, β by (α, β).

Definition. Given two ends α, β, not equal to ∞, we define their sum $\alpha + \beta$ as follows. Take any point C on the line $(0, \infty)$. Let A be its reflection in the line (α, ∞). Let B be its reflection in the line (β, ∞). Then $\alpha + \beta$ is the end of the perpendicular bisector of AB other than ∞.

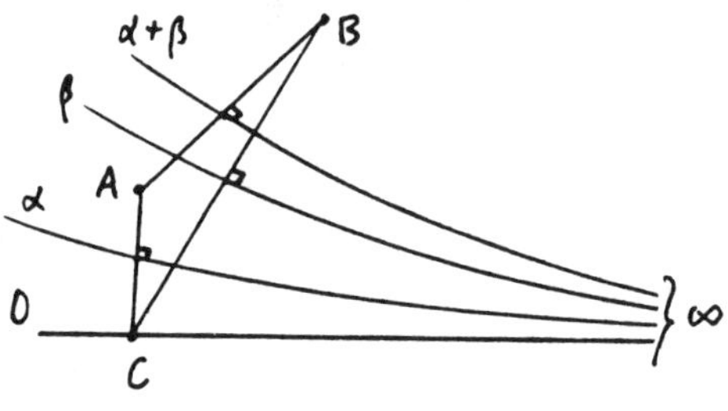

Proposition 41.3 *The addition of ends is well-defined, and makes the set $(F, +)$ into an abelian group with additive identity* 0.

Proof. First note that by (41.1) the perpendicular bisector of AB has one end ∞, so the definition makes sense. Second, note by (41.2) that the sum $\alpha + \beta$ is characterized by the property

$$\sigma_{\alpha+\beta} = \sigma_\beta\sigma_0\sigma_\alpha,$$

where for any end α, σ_α denotes reflection in the line (α, ∞). Thus $\alpha + \beta$ is independent of the choice of C, and so is well-defined.

From the definition, it is clear that $\alpha+\beta = \beta+\alpha$. It is also clear that $0+\alpha = \alpha$ for any α. If we denote by $-\alpha$ the reflection of α in the line $(0, \infty)$, then $\alpha + (-\alpha) = 0$, so we have additive inverses. For the associative law, just note that

$$\sigma_{(\alpha+\beta)+\gamma} = \sigma_\gamma\sigma_0\sigma_{\alpha+\beta} = \sigma_\gamma\sigma_0\sigma_\beta\sigma_0\sigma_\alpha$$

which is independent of the order of the operations.

Definition. In order to define multiplication of ends, we first fix a line perpendicular to the line $(0, \infty)$, and label one end 1. Note that the other end is then -1. Given ends α, β, draw the lines $(\alpha, -\alpha)$ and $(\beta, -\beta)$ which will meet the line $(0, \infty)$ at right angles at points A, B. Let O be the point where the line $(1, -1)$ meets $(0, \infty)$. Find C on the line $(0, \infty)$ such that $OC = OA + OB$, treating these as *signed* distances: if A is on the ray $O\infty$, then OA is positive. If A is on the ray $O0$, then OA is negative.

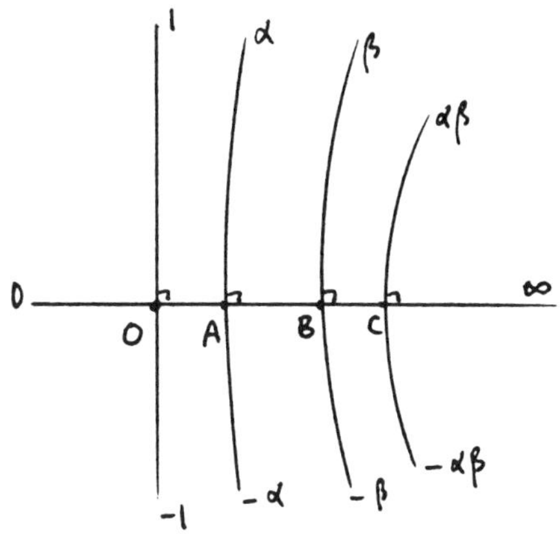

Take the line through C, perpendicular to $(0, \infty)$, and its ends will be $\alpha\beta$ and $-\alpha\beta$. Adjust the sign as follows: ends on the same side of $(0, \infty)$ as 1 will be called *positive*, and ends on the same side of $(0, \infty)$ as -1 will be negative. We choose the sign so that pos $\times$ pos $=$ pos, pos $\times$ neg $=$ neg, and neg $\times$ neg $=$ pos, as usual.

Theorem 41.4 *In a hyperbolic plane* Π, *fix two perpendicular lines and label their ends* $0, \infty, 1, -1$ *as above. Let* F *be the set of all ends of* Π *different from* ∞. *Then* F, *with the two operations* $+$, $\cdot$, *and the notion of positive elements defined above, is an ordered field satisfying* $(**)$: *every positive element has a square root.*

Proof. We have already seen (41.3) that $(F, +)$ is an abelian group with identity 0.

From the definition of multiplication we see immediately that $(F \backslash \{0\}, \cdot)$ is an abelian group with identity 1. Indeed, multiplication of ends corresponds to addition of signed segments on the line $(0, \infty)$, which is an abelian group. Reflection in the line $(1, -1)$ sends α to α^{-1}.

Multiplication by zero was not defined, so we define $0 \cdot \alpha = 0$ for all $\alpha \in F$.

For the distributive law, we proceed as follows. Given and end γ which we may assume to be positive, let the line $(\gamma, -\gamma)$ meet $(0, \infty)$ at C. We define a rigid motion τ of the plane, called *translation along the line* $(0, \infty)$ *by* OC, as follows. For any point P, let PQ be the orthogonal to $(0, \infty)$. Choose Q' on $(0, \infty)$ so that $OC = QQ'$ as signed intervals. Take P' on the line through Q', orthogonal to $(0, \infty)$, on the same side as P, so that $PQ = P'Q'$.

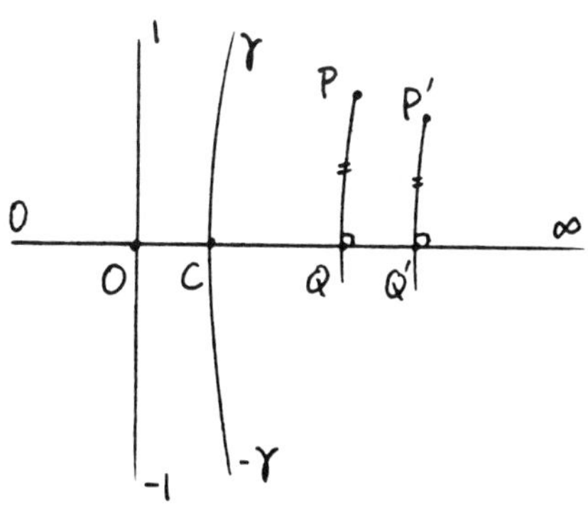

One can verify easily that τ is a rigid motion of the plane (Exercise 41.1). It leaves the ends 0 and ∞ fixed, and sends any other end α to $\gamma\alpha$. Applying this rigid motion to the diagram used to define $\alpha+\beta$, the diagram is sent to another diagram with the same properties, and from this it is clear that $\gamma(\alpha+\beta)=\gamma\alpha+\gamma\beta$.

We have defined the positive ends to be those on the same side of the line $(0,\infty)$ as 1. From our definitions it is clear that sums and products of positive ends are positive. It is also clear by definition that for any $\alpha \in F$, either α is positive, or $\alpha = 0$, or $-\alpha$ is positive. Hence F is an ordered field.

For the last property, let α be a positive end, let the line $(\alpha, -\alpha)$ meet $(0,\infty)$ at A, and let B be the midpoint of OA. Then the line perpendicular to $(0,\infty)$ at B will have an end β with the property $\beta^2 = \alpha$. Thus α has a square root in F. □

To relate this newly constructed field of ends with the geometry of the plane, our first step will be to describe some useful rigid motions and their effect on the ends. Also, to simplify notation, we will now write elements of the field F with Roman letters, even though as ends, they were originally written with Greek letters.

Proposition 41.5 *Let Π be a hyperbolic plane and F its field of ends, as above. We describe various rigid motions and their effect on ends.*

a) *The reflection in $(0,\infty)$ sends $x \in F'$ to $-x$. So we write $x' = -x$.*

b) *The reflection in $(1,-1)$ gives $x' = \frac{1}{x}$.*

c) *Translation along $(0,\infty)$ is represented by $x' = ax$ for any $a \in F$, $a > 0$.*

d) *For any $a \in F$, there is a rigid motion $\sigma_{\frac{1}{2}a}\sigma_0$, the composition of reflection in the line $(0,\infty)$ and reflection in the line $\left(\frac{1}{2}a,\infty\right)$, which we call "rotation around ∞", which gives $x' = x + a$.*

e) *The rotation around the point O, which sends 0 to a for any $a \in F$ gives*

$$x' = \frac{x+a}{-ax+1}$$

on ends. The rotation around O sending 0 to ∞ gives $x' = -\frac{1}{x}$.

Proof. a) We saw in the definition of addition.

b) We saw in the definition of multiplication.

c) We saw in the proof of the distributive law in (41.1).

d) From the definition of addition, we see that σ_a sends 0 to $2a$, and σ_b sends 0 to $2b$. Then σ_{a+b} sends $2a$ to $2b$. Changing variables, let $c = a + b$ and $d = 2a$. Then σ_c sends d to $2c - d$. Using these relations, we see that $\sigma_{\frac{1}{2}a}\sigma_0$ sends any x to $x + a$, as required.

e) We give an indirect proof that rotation around O has the given effect on ends. If $a = 0$ we have the identity. If $a = \infty$, we have the composition of reflection in $(0,\infty)$ and reflection in $(1,-1)$, which is a rotation of 2RA. So we may assume $a \neq 0, \infty$.

Note that

$$\frac{x+a}{-ax+1} = \frac{1}{-\left(\frac{a}{a+a^{-1}}\right)x + \frac{1}{a+a^{-1}}} - a^{-1},$$

which is a composition of operations of the types given in a), b), c), d) above. So there is a rigid motion having the effect

$$x' = \frac{x+a}{-ax+1}$$

on ends. Next we will show that any rigid motion having that effect on ends must be the rotation around O taking 0 to a. Indeed, substituting 0 and ∞ in this expression, we find that 0 goes to a and ∞ goes to $-a^{-1}$, so the line $(0, \infty)$ goes to $(a, -a^{-1})$, which passes through O.

Next, we compute, and find that the line $(1, -1)$ goes to the line $\left(\frac{1+a}{-a+1}, \frac{-1+a}{a+1}\right)$, which also passes through O. Indeed, a line (c, d) passes through O if and only if $d = -c^{-1}$, as we see by applying reflections in $(0, \infty)$ and $(1, -1)$. Hence the point O, which is the intersection of the two lines $(0, \infty)$ and $(1, -1)$, goes to the intersection of the images of those lines, which is O.

Finally, if we try to solve the equation $x = \frac{x+a}{-ax+1}$, we find $x^2 = -1$, which has no solution because F is an ordered field. Thus no end is fixed under this transformation. Now a rigid motion which fixes a point O and has no fixed ends must be a rotation around O (Exercise 17.4), so we are done.

Proposition 41.6 *Let Π be a hyperbolic plane with field of ends F, as above.*

a) *A line is represented by an unordered pair (u_1, u_2) of distinct elements of $F' = F \cup \{\infty\}$.*

b) *A point is given by the equation of all lines containing it, which is of the form*

$$u_1 u_2 - b(u_1 + u_2) + a^2 = 0$$

with $a, b \in F$, $a > 0$, and $|b| < a$.

Proof. a) Indeed, a line is uniquely determined by its two ends, which are distinct elements of F'.

b) First consider the point O. A line (u_1, u_2) contains O if and only if $u_1 u_2 = -1$. Indeed, such a line is stable under the rotation round O by 2RA, which sends $x' = -\frac{1}{x}$ by (41.5).

Now we can move O to any other point by first making a translation along $(0, \infty)$ and then a rotation around ∞. A translation along $(0, \infty)$ gives $x' = cx$, so we get a new equation $u_1 u_2 = -c^2$. Then a rotation around ∞ gives $x' = x - b$ so we get a new equation

$$(u_1 - b)(u_2 - b) + c^2 = 0$$

or

$$u_1 u_2 - b(u_1 + u_2) + b^2 + c^2 = 0.$$

Here b is any element of F and we can set $b^2 + c^2 = a^2$ with a positive, $a > |b|$.

Remark 41.6.1. We can think of the point with equation

$$u_1 u_2 - b(u_1 + u_2) + a^2 = 0$$

as having coordinates (a, b). It is the intersection of the lines $(a, -a)$ and (b, ∞). However, we have found that calculations seem to work out better if we continue to think of a line as given by coordinates (u_1, u_2), and a point as given by an equation. This is the opposite of the analytic geometry we are used to, where a point has coordinates and a line has an equation.

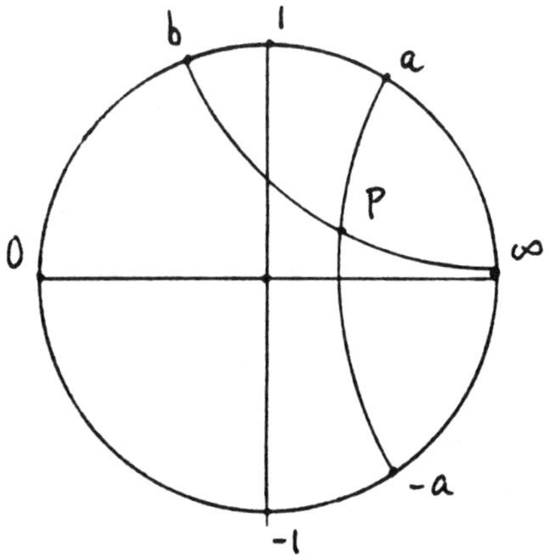

Proposition 41.7 *For any segment AB in the hyperbolic plane, lay off a congruent segment OC on the ray $O\infty$, and let the line perpendicular to $(0, \infty)$ at C be $(a, -a)$ with $a > 0$. We define $\mu(AB) = a$. Then μ is a* multiplicative distance function *on the plane (§39) with values in the field F, namely*

a) $\mu(AB) > 1$
b) $AB \cong A'B'$ *if and only if* $\mu(AB) = \mu(A'B')$
c) $AB < A'B'$ *if and only if* $\mu(AB) < \mu(A'B')$
d) $\mu(AB + CD) = \mu(AB) \cdot \mu(CD)$.

Proof. These properties are all immediate from the definition of multiplication in F.

Proposition 41.8 *For any angle θ in the hyperbolic plane, lay out an equal angle centered at O, and reaching from 0 to a on the positive side of $(0, \infty)$. Then we define* $\tan\theta/2 = a$. *This* tangent function *has the following properties.*

a) $\tan\theta/2 \in F$, *and* $\tan\theta/2 > 0$.
b) $\tan\theta/2 = \tan\psi/2$ *if and only if* $\theta = \psi$.
c) *If* $\theta < \psi$ *then* $\tan\theta/2 < \tan\psi/2$.
d) *If* $\theta = \mathrm{RA}$, $\tan\theta/2 = 1$.
e) *If θ and ψ are two angles for which $\theta + \psi$ is defined, then*

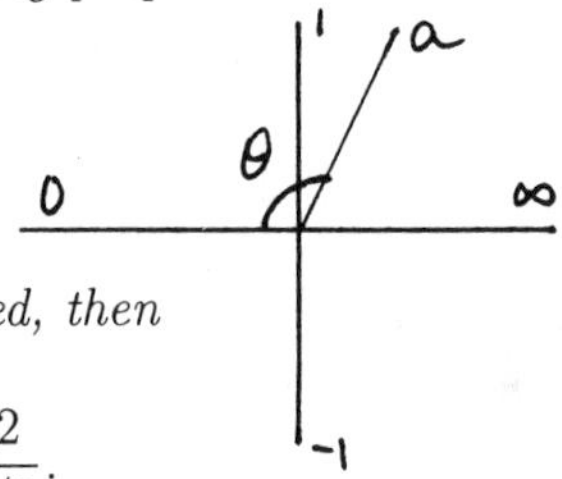

$$\tan(\theta + \psi)/2 = \frac{\tan\theta/2 + \tan\psi/2}{1 - \tan\theta/2 \tan\psi/2}.$$

Proof. a), b), c), d) are immediate by construction. For e), let $\tan\theta/2 = a$, $\tan\psi/2 = b$ and $\tan(\theta + \psi)/2) = c$. Then the rotation around O from 0 to c is the composition of the rotations from 0 to a and 0 to b. Using the formula of (41.5), we have

$$\frac{x + c}{-cx + 1} = \frac{\left(\frac{x+a}{-ax+1}\right) + b}{-b\left(\frac{x+a}{-ax+1}\right) + 1}.$$

A brief calculation gives

$$c = \frac{a+b}{1-ab}$$

which is the desired result.

Remark 41.8.1. We cannot define the tangent of an angle as in Euclidean geometry by using right triangles, because in non-Euclidean geometry, right triangles of different sizes are not similar to each other. Therefore we use this ad hoc definition. However, the terminology $\tan\theta/2$ for this function is justified by the properties it enjoys. In the presence of Archimedes' axiom, if one defined the radian measure of an angle by a limiting process using dyadic rational multiples of a right angle, it is easy to prove that this tangent function is the same as the usual one (Exercise 41.16).

Proposition 41.9 *If α is the angle of parallelism of a segment AB, then using the distance function and the tangent function defined above, we have*

$$\tan\alpha/2 = \mu(AB)^{-1}.$$

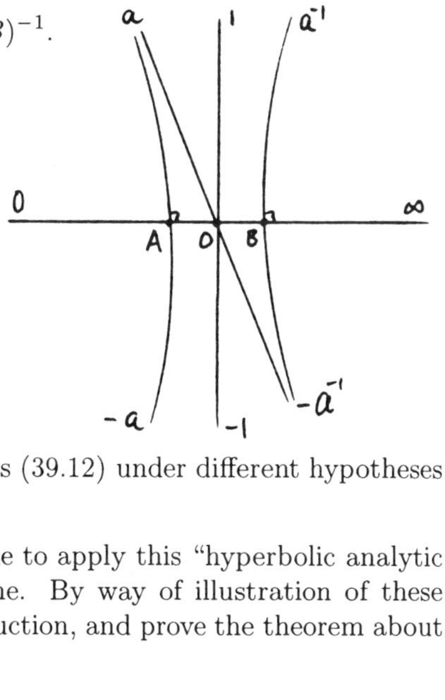

Proof. Lay out the angle α as $0Oa$. Then the line $(a, -a)$ will meet the ray $O0$ at right angles at a point C, and α will be the angle of parallelism of the segment OC equal to AB. To find $\mu(OC)$, reflect in the line $(1, -1)$ to get OD and the line $(a^{-1}, -a^{-1})$. Then $\mu(AB) = a^{-1}$ and $\tan\alpha/2 = a$ which gives the desired result.

Remark 41.9.1. We recover the same result as (39.12) under different hypotheses and different definitions of tan and μ.

Now we have enough basic results to be able to apply this "hyperbolic analytic geometry" to problems in the hyperbolic plane. By way of illustration of these techniques, we will give Bolyai's parallel construction, and prove the theorem about the altitudes of a triangle.

Proposition 41.10 (Bolyai's parallel construction) *Suppose given a line l and a point P not on l in the hyperbolic plane. Let PQ be the perpendicular to l. Let m be a line through P, perpendicular to PQ. Choose any point R on l, and let RS be the perpendicular to m. Then the circle of radius QR around P will meet the segment RS at a point T, and the ray $n = PT$ will be the limiting parallel ray to l through P.*

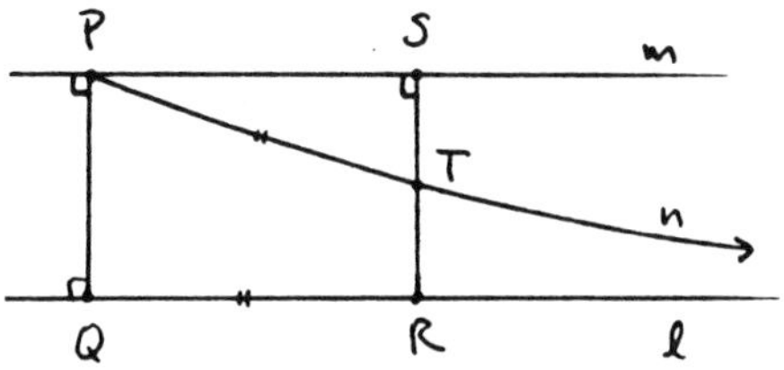

Proof. To prove this, we use the field of ends introduced above. By a rigid motion of the plane, we may assume that P is the center of our coordinate system, so that PQ is the line $(0, \infty)$ and PS is the line $(1, -1)$. We will take n to be the limiting parallel to l, and let it meet RS at T. Then we must prove that $QR = PT$. To show this, we will apply rigid motions to move each of these segments to the line $(0, \infty)$, with one end at P. Then we will verify that the other ends land in the same place. For this we need a lemma.

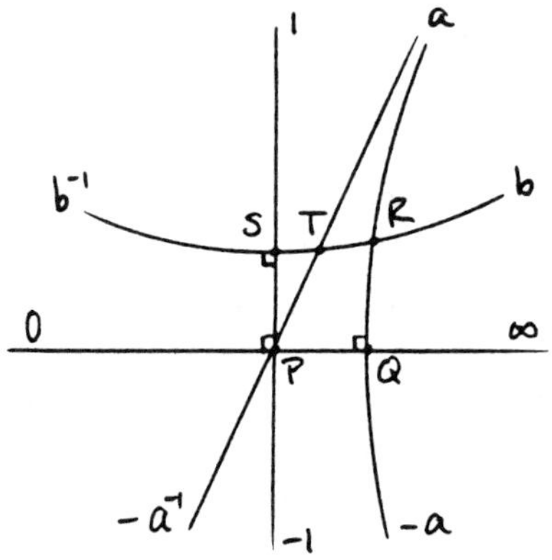

Lemma 41.11 *Two lines* (u_1, u_2) *and* (v_1, v_2) *in the hyperbolic plane meet the line* $(0, \infty)$ *at the same point if and only if* $u_1u_2 = v_1v_2$, *and* u_1u_2 *is negative.*

Proof. Indeed, we have seen in the proof of (41.6) that a point on the line $(0, \infty)$ has an equation of the form $u_1u_2 = -c^2$ for some $c \in F$. So the two lines (u_1, u_2) and (v_1, v_2) pass through the same point of $(0, \infty)$ if and only if $u_1u_2 = -c^2 = v_1v_2$. Since every positive element of F is a square (41.4), such a c will exist if and only if $u_1u_2 = v_1v_2$ and u_1u_2 is negative.

Proof of (41.10)**, continued**. Let us start with the segment PT. We let the ends of the line $l = QR$ be a and $-a$. The line RS we let have ends b and b^{-1}. Then n is the line $(a, -a^{-1})$. We use the rotation around P which sends a to 0. Its effect on ends, by (41.5) is

$$x' = \frac{x - a}{ax + 1}.$$

Hence it sends the line (b, b^{-1}) to the line

$$(u_1, u_2) = \left(\frac{b - a}{ab + 1}, \frac{b^{-1} - a}{ab^{-1} + 1} \right).$$

The image of T will be the point where this line meets the line $(0, \infty)$. In order to apply the lemma, we compute

$$u_1u_2 = \left(\frac{b - a}{ab + 1} \right) \left(\frac{b^{-1} - a}{ab^{-1} + 1} \right) = \frac{(b - a)(1 - ab)}{(1 + ab)(a + b)}.$$

Now consider the segment QR. First we make a translation along $(0, \infty)$ to send Q to P. Its effect on ends will be $x' = a^{-1}x$, so it will send the line (b, b^{-1}) to the line $(a^{-1}b, a^{-1}b^{-1})$ and the line $(a, -a^{-1})$ will become $(1, -1)$. Next we do a rotation around P sending 1 to 0, whose effect on ends is

$$x' = \frac{x - 1}{x + 1}.$$

So this line just mentioned will go to

$$(v_1, v_2) = \left(\frac{a^{-1}b - 1}{a^{-1}b + 1}, \frac{a^{-1}b^{-1} - 1}{a^{-1}b^{-1} + 1} \right).$$

The image of R under the two rigid motions will be the intersection of (v_1, v_2) with $(0, \infty)$. So we compute

$$v_1 v_2 = \left(\frac{a^{-1}b - 1}{a^{-1}b + 1}\right)\left(\frac{a^{-1}b^{-1} - 1}{a^{-1}b^{-1} + 1}\right) = \frac{(b-a)(1-ab)}{(b+a)(1+ab)}.$$

Observing that $u_1 u_2 = v_1 v_2$ and $u_1 u_2 < 0$ since $b > a > 1$ in our situation, and using the lemma, we conclude that $QR \cong PT$, as required.

Remarks 41.11.1. This remarkable result gives a ruler and compass construction for the limiting parallel line. In other words, by making constructions such as dropping a perpendicular from a point to a line, or intersecting a line with a circle—constructions which are possible in any Hilbert plane with axiom (E)—one obtains the limiting parallel ray from a point P to a line l. However, the curious feature of our proof is that we prove this construction works only by first assuming (via axiom (L)) that the object we wish to construct already exists. And in fact, without assuming the existence of the limiting parallel line ahead of time, this result may fail. For example, in the non-Archimedean geometries of (18.4.3) or (Exercise 39.22), the construction gives a line n which is not limiting parallel to l. There remains the possibility that this construction might work in a Hilbert plane satisfying (A) and (E). Indeed, Greenberg (1979) has shown, using the classification of Hilbert planes due to Pejas, that any non-Euclidean Hilbert plane satisfying (A) and (E) is hyperbolic, and so Bolyai's construction works. He adds that if you could find a purely geometric proof of this result, you would probably receive an instant Ph.D.

Proposition 41.12 *If (u_1, u_2) and (v_1, v_2) are any two lines in the plane, and if θ is the angle between the rays u_1 and v_1, and $u_1 < v_1$, then*

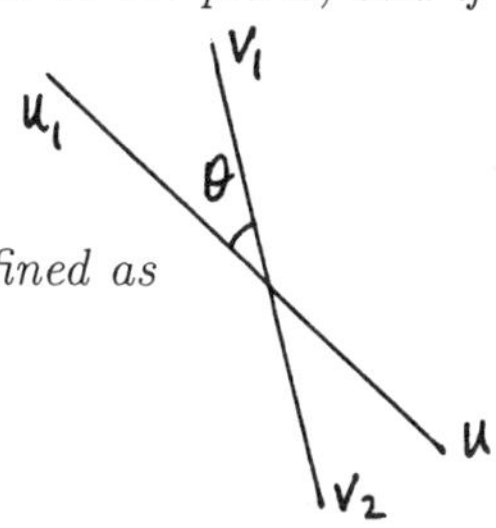

$$\tan \theta/2 = \sqrt{-(v_1, v_2; u_1, u_2)},$$

where the expression under the radical is the cross-ratio *defined as*

$$(v_1, v_2; u_1, u_2) = \frac{v_1 - u_1}{v_1 - u_2} \cdot \frac{v_2 - u_2}{v_2 - u_1}.$$

Proof. First consider the case $(u_1, u_2) = (0, \infty)$ and $(v_1, v_2) = (a, -a^{-1})$. In this case we know $\tan \theta/2 = a$ by definition of the tangent function. We compute the cross-ratio

$$(a, -a^{-1}; 0, \infty) = \frac{a - 0}{a - \infty} \cdot \frac{-a^{-1} - \infty}{-a^{-1} - 0} = -a^2.$$

So the formula holds in this case.

In the general case, we consider a rigid motion which takes (u_1, u_2) to $(0, \infty)$ and (v_1, v_2) to $(a, -a^{-1})$ for suitable $a \in F$. Since a rigid motion preserves angles, we only need to show that a rigid motion preserves cross-ratios of ends. In fact, we can accomplish the rigid motion we need using a composition of rigid motions of types a)–e) of (41.5), and type e) is itself a composition of the four earlier types, so we need only verify that the cross-ratio of four ends is stable under the transformations of types a) $x' = -x$; b) $x' = x^{-1}$; c) $x' = ax$ for $a \in F$, and d) $x' = x + a$ for $a > 0$. These verifications are immediate from the definition of cross-ratio, so we are done.

Proposition 41.13 (Altitudes) *If two of the altitudes of a triangle meet in a point, so does the third. If two of the altitudes are limiting parallels, then all three have a common end. If two of the altitudes have a common perpendicular, then all three have the same common perpendicular.*

Proof. By a rigid motion of the plane, we may assume that our triangle ABC is so placed that AB is the line $(1, -1)$, and c lies on the line $(0, \infty)$, so that OC is one of the three altitudes. Our method is to let (u_1, u_2) be one of the other altitudes, find the intersection of this line with $(0, \infty)$, and then show the third altitude passes through this same point.

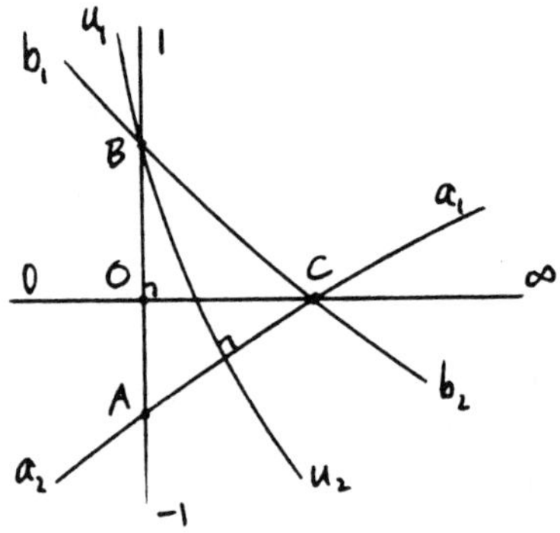

Let AC be (a_1, a_2) and let BC be (b_1, b_2). Since these two lines meet at C on the line $(0, \infty)$, we must have $a_1a_2 = b_1b_2 < 0$ by (41.11).

First we will find the equation of the point B. By (41.6) it is of the form

$$u_1u_2 - d(u_1 + u_2) + e^2 = 0.$$

Since B lies on the lines $(1, -1)$ and (b_1, b_2) we find

$$\begin{aligned} -1 - d \cdot 0 + e^2 &= 0 \\ b_1b_2 - d(b_1 + b_2) + e^2 &= 0. \end{aligned}$$

Solving for d and e^2, we find $e^2 = 1$, and $d = (1 + b_1b_2)/(b_1 + b_2)$, so the equation of B is

$$u_1u_2 - \frac{1 + b_1b_2}{b_1 + b_2}(u_1 + u_2) + 1 = 0. \tag{1}$$

Now the altitude (u_1, u_2) through B satisfies this equation. It is also orthogonal to (a_1, a_2), so by (41.12) the cross-ratio

$$(u_1, u_2; a_1, a_2) = -1.$$

This gives

$$2u_1u_2 - (a_1 + a_2)(u_1 + u_2) + 2a_1a_2 = 0. \tag{2}$$

We solve the equation (2) for $u_1 + u_2$ and substitute in the equation (1). We find

$$u_1u_2 - \frac{2(1 + b_1b_2)(u_1u_2 + a_1a_2)}{(a_1 + a_2)(b_1 + b_2)} + 1 = 0. \tag{3}$$

Since $a_1a_2 = b_1b_2$ as noted above, this equation is stable under interchanging a's and b's. So if (v_1, v_2) is the third altitude of the triangle, then v_1v_2 satisfies the same equation (3), and so $u_1u_2 = v_1v_2$.

Now if the altitude (u_1, u_2) meets the line $(0, \infty)$, then by (41.11), $u_1 u_2 < 0$, and (v_1, v_2) meets $(0, \infty)$ in the same point. If (u_1, u_2) is limiting parallel to $(0, \infty)$, then one of u_1 or u_2 is equal to 0 or ∞, in which case $u_1 u_2 = 0$ or ∞. The same is true of (v_1, v_2), so one of v_1, v_2 is equal to 0 or ∞, respectively, so all three have a common end.

We leave as an exercise (Exercise 41.2) to show that if $u_1 u_2 = v_1 v_2 > 0$, then all three altitudes have a common perpendicular.

Exercises

41.1 Verify that translation along a line, defined in the proof of (41.4), is a rigid motion.

41.2 In the proof of (41.13), show that if two of the altitudes have a common perpendicular, then all three altitudes have a common perpendicular.

41.3 Consider the pencil (set) of all lines (u_1, u_2) having a fixed common orthogonal line. Show that the lines of such a pencil satisfy an equation of the form

$$u_1 u_2 - b(u_1 + u_2) + a = 0$$

with $a < |b|$, or an equation of the form

$$u_1 + u_2 = c$$

for some $c \in F$.

41.4 Find the two ends (as elements of F) of the angle bisector of the angle $1O\infty$.

41.5 Find the multiplicative distance μ between the points $O = (1, 0)$ and $A = (1, a)$, using the point coordinates of (41.6.1).

41.6 In the trilimit triangle of (Exercise 40.11), find the multiplicative distance μ from the center point A to one of the sides.

41.7 In the limit quadrilateral of (Exercise 40.13), express $\mu(CD)$ in terms of $\mu(AB)$.

41.8 Prove that an angle of $\frac{1}{3}$RA exists in the hyperbolic plane.

41.9 Prove that an angle of $\frac{1}{5}$RA exists in the hyperbolic plane.

41.10 In this exercise we describe the group G of all rigid motions of the hyperbolic plane, as follows.

a) Show that every rigid motion of the plane can be expressed as a composition of motions of the four types a), b), c), d) of (41.5). In other words, elements of those four types generate the group G.

b) Show that an element of G is uniquely determined by its effect on the set F' of ends.

c) If $\varphi \in G$ is any rigid motion, then φ acts on F' as a *fractional linear transformation* of the form

$$x' = \frac{ax+b}{cx+d}$$

for suitable $a, b, c, d \in F'$, with $ad - bc \neq 0$.

d) Every fractional linear transformation of F' arises as the action of some element $\varphi \in G$ on the set of ends. Thus G is isomorphic to the group of fractional linear transformations of F'.

Because of this result, we can say that the group of rigid motions of the hyperbolic plane is composed of four components: a copy of the additive group of the field corresponding to (41.5d); a copy of the multiplicative group of positive elements of the field (41.5c); a copy of the circle group of the field (41.5e), cf. (Exercise 41.11) below; and possibly a reflection.

This is in contrast to the Euclidean situation, where the group of rigid motions is made up of two copies of the additive group of the field, corresponding to the translations, and one copy of the circle group for rotations, plus a possible reflection.

41.11 To each rotation ρ around the point O, let us associate that element $a \in F'$ to which ρ sends 0. Show that this correspondence gives an isomorphism of the group of rotations around O with the circle group of the field F, as defined in (Exercise 17.6). In particular, we find the curious result that the group of rotations around a point in the hyperbolic plane is isomorphic to the group of rotations around a point in the Cartesian plane over the field F.

41.12 Let l, m be two lines with a common end α. Generalizing (41.5d), we call the composition $\sigma_l\sigma_m$ of the reflections in l and m, a "rotation around the end α". Fix an end α and a point A. Then the set of all points $\rho(A)$, where ρ ranges over all the rotations around the end α, will be called a *horocycle*. We think of it as analogous to a circle, but one whose center is an end instead of a point.

a) Show that a horocycle could also be defined as the set of all points $\sigma(A)$, where σ ranges over all the reflections in lines with end α.

b) If m is the line through A, orthogonal to the line $l = A\alpha$, then m meets the horocycle only at A. We call m the *tangent line* to the horocycle at A. Show that any other line through A, different from l or m, will meet the horocycle in one other point $B \neq A$.

c) In the case of the Poincaré model, show that the horocycles defined here are the same as those defined in (Exercise 39.14).

41.13 Let (u_1, u_2) and (v_1, v_2) be two parallel, not limiting parallel lines. Show that the multiplicative distance between the two lines is

$$\mu = \frac{a-1}{1-a}$$

where a is the square root of the cross-ratio

$$a = \sqrt{(u_1, u_2; v_1, v_2)}.$$

41.14 If F is a subfield of the field of real numbers $\mathbb{R}$, we can define a function $\lambda(AB) = \ln(\mu(AB))$ where μ is the multiplicative distance function of (41.7) and ln is the natural logarithm.

a) Show that λ is an *additive distance function* (§39) with values in the group $(\mathbb{R}, +)$, namely,

1) for any interval AB, we have $\lambda(AB) > 0$

2) $AB \cong A'B'$ if and only if $\lambda(AB) = \lambda(A'B')$

3) $\lambda(AB + CD) = \lambda(AB) + \lambda(CD)$.

Note however that the values of the function λ may not belong to F.

b) Show that Bolyai's formula (41.9) then becomes

$$\tan \alpha/2 = e^{-\lambda(AB)},$$

where α is the angle of parallelism of the segment AB.

41.15 Suppose again that F is a subfield of $\mathbb{R}$. Show that if λ is any additive distance function with values in $(\mathbb{R}, +)$, in the sense of (Exercise 41.14), then there is a real constant $k > 0$ such that for all AB,

$$\lambda(AB) = k \ln(\mu(AB)).$$

41.16 Assume that F is a subfield of $\mathbb{R}$. Show that any function having the properties a)–e) of (41.8) is the same as the usual tangent function.

42 Hyperbolic Trigonometry

The usual trigonometry which we learn in high school could be described as a collection of relationships between the sides and angles of a triangle, together with rules of operation (the trigonometric identities) which allow us to compute all the parts of a triangle from a few given parts.

To be more precise, if ABC is a right triangle, with angles α, β at A and B, and sides a, b, c opposite the vertices A, B, C, then we have the relations

$$\sin \alpha = \tfrac{a}{c}$$
$$\cos \alpha = \tfrac{b}{c}$$
$$\tan \alpha = \tfrac{a}{b}$$

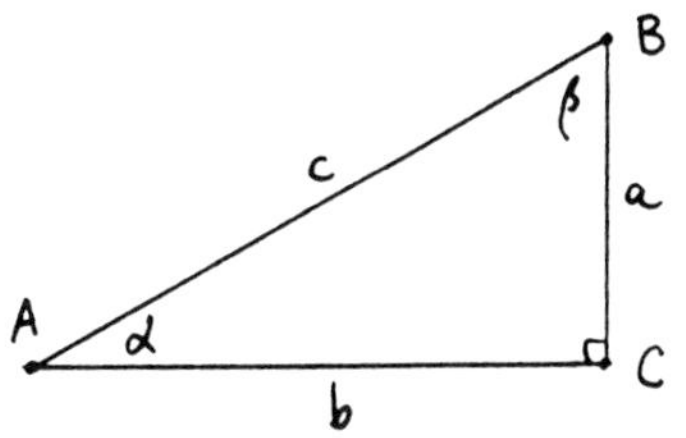

and trigonometric identities such as

$$\sin^2 \alpha + \cos^2 \alpha = 1$$

or

$$\sin(\text{RA} - \alpha) = \cos\alpha,$$

and many others.

Using these relations, if we are given any two sides, or one side and one angle, then we can compute the remaining sides and angles of the triangle. Substituting the formulas for $\sin\alpha$ and $\cos\alpha$ in the relation $\sin^2\alpha + \cos^2\alpha = 1$ we recover the Pythagorean theorem $a^2 + b^2 = c^2$ (I.47).

The hyperbolic trigonometry which we develop in this section accomplishes the same thing in the hyperbolic plane. We develop a series of relationships between the sides and the angles of a right triangle, together with rules of operation on the trigonometric functions, so that if any two of the quantities a, b, c, α, β are given, we can compute the others from those two. This is a slightly stronger result than in the Euclidean case, because in hyperbolic geometry there are no similar triangles, so even giving the two angles α, β uniquely determines the triangle. The situation is also somewhat complicated by the fact that in hyperbolic geometry we have only $\alpha + \beta < \text{RA}$, whereas in Euclidean geometry, β is uniquely determined by α since $\alpha + \beta = \text{RA}$.

To derive the formulas of hyperbolic trigonometry, we put a right triangle in special position as shown, with A at the center of our coordinate system and C on the line $(0, \infty)$. Then the side AB describes the angle α, so that it is the line $(t, -t^{-1})$, where $t = \tan\alpha/2$ (41.8). We let a, b, c represent the multiplicative lengths of the sides opposite A, B, C, so that BC is then the line $(b^{-1}, -b^{-1})$ (41.7).

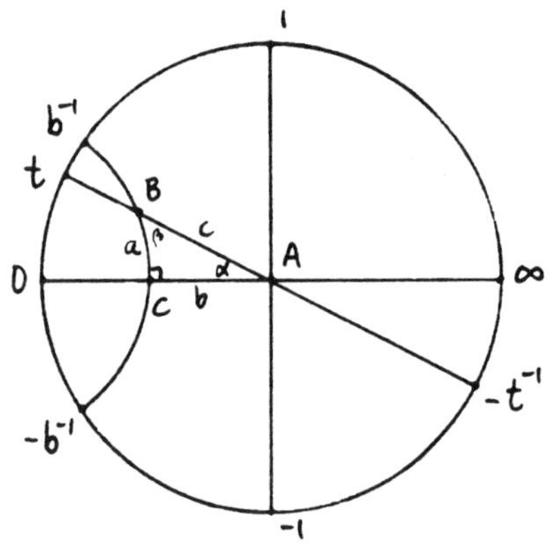

We will use suitable rigid motions to compute the lengths a and c in terms of b and t. For a, we first perform a translation along $(0, \infty)$ to move C to A. This acts by $x' = bx$ on ends, so the line $(t, -t^{-1})$ goes to $(bt, -bt^{-1})$. Now do a rotation by 90° which sends 1 to ∞, and 0 to 1. This acts on ends by $x' = \frac{x+1}{-x+1}$ (41.5), so the line $(bt, -bt^{-1})$ becomes the line

$$(u_1, u_2) = \left(\frac{bt + 1}{-bt + 1}, \frac{-bt^{-1} + 1}{bt^{-1} + 1} \right).$$

This line will meet $(0, \infty)$ at the point which is the image of B under these two maps. To find the multiplicative length a of the segment BC, we need to find the line $(a, -a)$ which meets $(0, \infty)$ in the same point as $u_1 u_2$. Therefore, by (41.11), $u_1 u_2 = -a^2$. Substituting for u_1 and u_2 and simplifying, we find

(1) $$a^2 = \frac{(1 + bt)(b - t)}{(1 - bt)(b + t)}.$$

Next, to find c, we perform a rotation around A which sends t to 0. This acts by $x' = \frac{x-t}{tx+1}$ on ends, so that the line $(b^{-1}, -b^{-1})$ goes to

$$(v_1, v_2) = \left(\frac{b^{-1}-t}{b^{-1}t+1}, \frac{-b^{-1}-t}{-b^{-1}t+1}\right).$$

If c is the multiplicative length of AB, the line $(c^{-1}, -c^{-1})$ will meet $(0, \infty)$ at the same point as (v_1, v_2). Again using (41.11), we find $v_1 v_2 = c^{-2}$, from which we obtain

$$c^2 = \frac{(b+t)(b-t)}{(1+bt)(1-bt)}. \tag{2}$$

To compute the angle β, we use (41.12). If $s = \tan \beta/2$, then $-s^2 = (b^{-1}, -b^{-1}; t, -t^{-1})$. Writing out the cross-ratio and simplifying, we find

$$s^2 = \frac{(1-bt)(b-t)}{(b+t)(1+bt)}. \tag{3}$$

In some sense, the formulae (1), (2), (3) accomplish our purpose, because thereby the quantities a, b, c, t, and s are all related to each other. However, to bring this information into a form which is easier to manage, we will transform these equations.

From (1) an elementary calculation gives

$$\frac{1-a^2}{1+a^2} = \frac{2t}{1-t^2} \cdot \frac{1-b^2}{2b}. \tag{4}$$

From (2) we obtain

$$\frac{1-c^2}{1+c^2} = \frac{1-b^2}{1+b^2} \cdot \frac{1+t^2}{1-t^2}, \tag{5}$$

and from (3),

$$\frac{1-s^2}{1+s^2} = \frac{2t}{1+t^2} \cdot \frac{1-b^2}{2b}. \tag{6}$$

These formulae, which separate the variables in question, suggest the following definition (cf. Exercise 17.6).

Definition. For any angle α, if $t = \tan \alpha/2$, define

$$\sin \alpha = \frac{2t}{1+t^2} \qquad \text{and} \qquad \cos \alpha = \frac{1-t^2}{1+t^2}.$$

Proposition 42.1 *With these definitions, the functions* $\sin \alpha$ *and* $\cos \alpha$ *for angles in the hyperbolic plane enjoy the usual identities of trigonometric function. In particular*

a) $\tan \alpha = \frac{\sin \alpha}{\cos \alpha}$, *and*

b) $\sin^2 \alpha + \cos^2 \alpha = 1$.

Proof. From (41.8e) it follows that $\tan \alpha = \frac{2t}{1-t^2}$, which gives a). The formula b) is a simple computation from the definitions.

Proposition 42.2 *In the hyperbolic plane, let ABC be a right triangle, with angles α, β at A, B, and let $\bar{a}, \bar{b}, \bar{c}$ be the angles of parallelism of the sides opposite A, B, C. Then*

a) $\tan \alpha = \cos \bar{a} \tan \bar{b}$

b) $\cos \bar{b} = \cos \alpha \cos \bar{c}$

c) $\sin \alpha = \cos \beta \sin \bar{b}$.

Proof. By (41.9) we have $\tan \bar{a}/2 = a^{-1}$. Therefore

$$\sin \bar{a} = \frac{2a^{-1}}{1+a^{-2}} = \frac{2a}{1+a^2},$$
$$\cos \bar{a} = \frac{1-a^{-2}}{1+a^2} = -\frac{1-a^2}{1+a^2},$$

and

$$\tan \bar{a} = \frac{2a^{-1}}{1-a^{-2}} = -\frac{2a}{1-a^2}.$$

There are similar expressions for $\bar{b}$ and $\bar{c}$. We use these to interpret the equations (4), (5), (6), which then become a), b), c).

Proposition 42.3 *With the same hypothesis as* (42.2) *we have the further relations*

d) $\tan \bar{c} = \sin \alpha \tan \bar{a}$

e) $\sin \bar{c} = \tan \alpha \tan \beta$

f) $\sin \bar{c} = \sin \bar{a} \sin \bar{b}$.

Proof. We will prove f) and leave d), e) as (Exercise 42.5). To do this we will eliminate α from a) and b). From a)

$$\tan \alpha = \cos \bar{a} \tan \bar{b},$$

and from b)

$$\cos \alpha = \cos \bar{b} \cos^{-1} \bar{c}.$$

Multiplying,

$$\sin \alpha = \cos \bar{a} \sin \bar{b} \cos^{-1} \bar{c}.$$

Substituting these expressions in $\cos^2 \alpha + \sin^2 \alpha = 1$, we find

$$\cos^2 \bar{b} \cos^{-2} \bar{c} + \cos^2 \bar{a} \sin^2 \bar{b} \cos^{-2} \bar{c} = 1.$$

Multiplying by $\cos^2 \bar{c}$, we get

$$\cos^2 \bar{b} + \cos^2 \bar{a} \sin^2 \bar{b} = \cos^2 \bar{c}.$$

Now use $\cos^2 = 1 - \sin^2$ on $\bar{a}$, $\bar{b}$, $\bar{c}$. We get

$$1 - \sin^2 \bar{b} + (1 - \sin^2 \bar{a}) \sin^2 \bar{b} = 1 - \sin^2 \bar{c}.$$

This simplifies to

$$\sin^2 \bar{c} = \sin^2 \bar{a} \sin^2 \bar{b}.$$

Since the sin function on angles is always positive, we conclude

$$\sin \bar{c} = \sin \bar{a} \sin \bar{b}$$

as required.

Remark 42.3.1. The statements of (42.2) and (42.3) plus four more obtained from a), b), c), d) by reversing the roles of A and B, give one equation for each subset of 3 of the 5 quantities a, b, c, α, β. Thus given any two of these, we can calculate the others.

Example 42.3.2. As an application, let us find the relation between the sides and the angles of an equilateral triangle. Let CD be an altitude, let α be the angle at each vertex, and let c be the side. Then ACD is a right triangle with $\beta = \alpha/2$. Using (42.3e) we find

$$\sin \bar{c} = \tan \alpha \tan \alpha/2.$$

Translating back in terms of t and c this gives

$$\frac{2c}{1+c^2} = \frac{2t^2}{1-t^2}$$

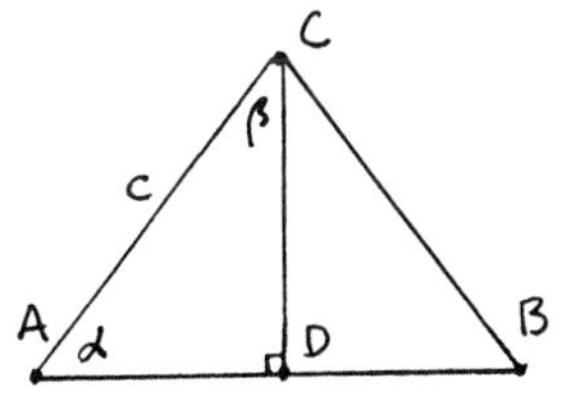

which we obtained by a different method in the Poincaré model (Exercise 39.5).

As another application, we will derive a formula for the area of a circle.

Proposition 42.4 *Let* Π *be a hyperbolic plane whose field of ends* F *is a subfield of the real numbers* $\mathbb{R}$. *To each angle we associate its radian measure, so as to obtain a measure of area function* (36.2) *with values in* $\mathbb{R}$. *Let* Γ *be a circle whose radius has multiplicative length* r. *We define the area of* Γ *to be the limit of the areas of regular inscribed* n*-gons as* $n \to \infty$. *This limit* A *exists (as a real number), and*

$$A = \frac{\pi(r-1)^2}{r}.$$

Proof. First we will express the area of a right triangle in a particular form. Let ABC be a right triangle, with our usual notation, and let δ be its area (as an angle). Then I claim

$$\sin \delta = \frac{\sin \alpha \cos \alpha}{\sin \bar{b}}(1 - \sin \bar{c}).$$

Indeed, since it is a right triangle, $\delta = \pi/2 - \alpha - \beta$, and

$$\sin \delta = \cos(\alpha + \beta) = \cos \alpha \cos \beta - \sin \alpha \sin \beta.$$

ÉLÉMENS DE *GÉOMÉTRIE.*

PREMIERE PARTIE.

Des moyens qu'il étoit le plus naturel d'employer pour parvenir à la meſure des Terreins.

CE qu'il ſemble qu'on a dû meſurer d'abord, ce ſont les longueurs & les diſtances.

I.

POUR meſurer une longueur quelconque, l'expédient que fournit une

A

Plate XIII. Beginning of Clairaut's "Elémens de Géométrie" (1775).

We first use c) and d) of (42.2) and (42.3) to replace $\cos\beta$ and $\sin\beta$ by expressions in $\alpha, \bar{b}$, and $\bar{a}$. Then use f) to replace $\sin\bar{a}$ by an expression in $\bar{b}$ and $\bar{c}$. This gives the formula above.

Now consider a regular n-gon inscribed in the circle, and consider the triangle formed by the radius to one of the vertices, and the orthogonal to the midpoint of one side. Its angle α is π/n, its hypotenuse is r and its side we call b_n. Its area is $A_n/2n$, where A_n is the area of the inscribed polygon. Hence

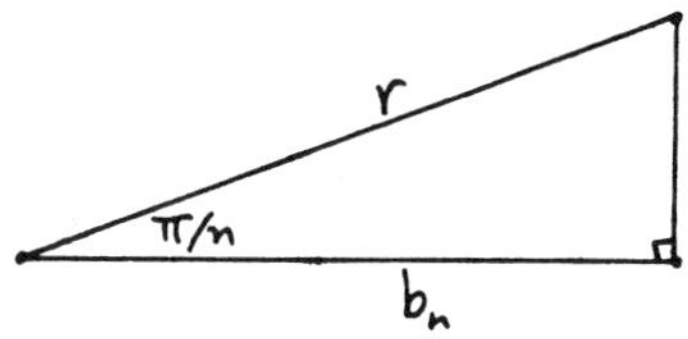

$$\sin\frac{A_n}{2n} = \frac{\sin\pi/n\cos\pi/n}{\sin\bar{b}_n}(1-\sin\bar{r}).$$

Multiply both sides of this equation by $2n$ and take the limit as $n\to\infty$. The limit of A_n will be A, the area of the circle. The limit of b_n is r. We use the convenient results from calculus that

$$\lim_{x\to 0}\frac{\sin x}{x} = 1$$

and

$$\lim_{x\to 0}\cos x = 1.$$

In the limit we find

$$A = 2\pi\left(\frac{1-\sin\bar{r}}{\sin\bar{r}}\right).$$

Substituting

$$\sin\bar{r} = \frac{2r}{1+r^2}$$

we find

$$A = \frac{\pi(r-1)^2}{r}$$

as required. Note that before taking the limit, all our terms were elements of the field F. However in the limit we obtain quantities π and A which are real numbers, but not necessarily in F.

Remark 42.4.1. Using this result, we can study the question of "squaring the circle" in non-Euclidean geometry. Of course there are no squares in a hyperbolic plane, so the appropriate question would be, given a circle in the hyperbolic plane, does there exist a rectilineal figure with area equal to the area of the circle?

In the Euclidean case, if we work over the field K of constructible numbers, for example, the answer is never. Indeed, the formula $A=\pi r^2$ and the fact that π is transcendental imply that if $r\in K$, then $A\notin K$, and conversely, if $A\in K$, then $r\notin K$.

In the non-Euclidean case, we find the surprising answer, sometimes yes and sometimes no. Consider a circle of radius $r\in F$. The question is whether the real number A can be written as a sum of real numbers which are the radian measures of angles in our geometry. For every angle less than a right angle can be made equal to the area of some right triangle (Exercise 42.9).

If for example the quantity $x = (r-1)^2/r \in F$ is a dyadic rational number, then πx will be the sum of radian measures of angles, because we can bisect a right angle any number of times. Thus if $r = 2$, or if r is any power of 2, we obtain circles whose area is that of a rectilineal figure.

On the other hand, if F is the field of constructible numbers, and if $x = \frac{1}{7}$, say, we can find an $r \in F$ with $x = (r-1)^2/r$, for example $r = \frac{1}{14}(15 + \sqrt{29})$, and a circle of that radius has area not corresponding to any rectilineal figure, because an angle of $\pi/7$ does not exist in that plane (cf. (Exercise 41.11) and (29.4)).

Thus the problem of "squaring the circle" is solvable for some circles, and unsolvable for others.

Exercises

42.1 Find the sides and angles of the equilateral triangle formed by the feet of the altitudes of the trilimit triangle of (Exercise 40.11).

42.2 Find the side of an equilateral triangle with angles of 45°.

42.3 Find the sides of an isosceles right triangle with angles $30° - 30° - 90°$.

42.4 Find the side of an equilateral pentagon with all right angles.

42.5 Derive the formulae (42.3d, e) from (42.2).

42.6 Prove the law of sines: in an arbitrary triangle ABC, with angles α, β, γ and opposite sides a, b, c,

$$\sin\alpha \tan\bar{a} = \sin\beta \tan\bar{b} = \sin\gamma \tan\bar{c}.$$

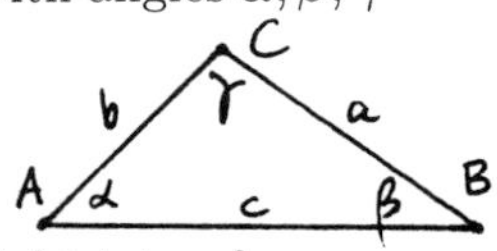

Hint: Use an altitude to divide the triangle into two right triangles.

42.7 Let AB and BC be two consecutive segments on a line, so that $AC = AB + BC$ as segments. Let α, β, γ be the angles of parallelism associated to AB, BC, and AC, respectively.

a) Show that

$$\sin\gamma = \frac{\sin\alpha \sin\beta}{\cos\alpha \cos\beta + 1}.$$

b) Derive analogous formulae for $\cos\gamma$ and $\tan\gamma$.

c) Verify the following formula, due to Lobachevsky, analogous to the law of cosines in Euclidean geometry. Let ABC be any triangle, with angles α, β, γ, and opposite sides a, b, c. Then

$$\cos\gamma \cos\bar{a} \cos\bar{b} + \frac{\sin\bar{a} \sin\bar{b}}{\sin\bar{c}} = 1.$$

Note that if γ is a right angle, this reduces to (42.3f).

42.8 Given an acute angle α at A, and given a segment DE, show that there exists a point B on one arm of the angle, such that the perpendicular BC to the other arm is equal to DE.

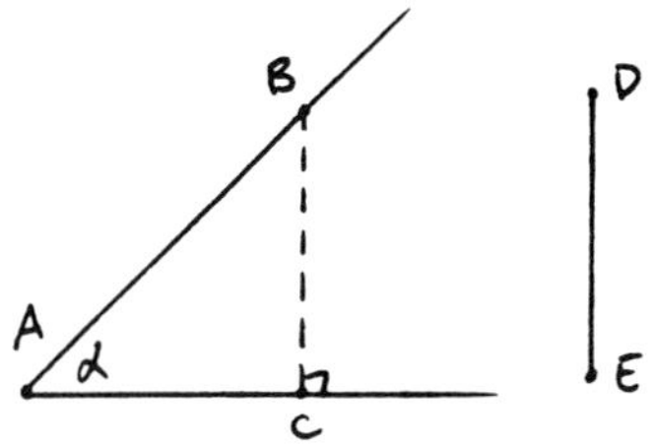

42.9 Given two angles α, β with $\alpha + \beta < \text{RA}$, show that there exists a right triangle with angles α, β.

42.10 Given three angles α, β, γ, whose sum is less than 2RA, prove that there exists a triangle with angles equal to α, β, γ. *Hint*: Glue together two suitable right triangles, and make them fit by solving a quadratic equation.

42.11 Let T be a right triangle with angles α, β and sides a, b, c. Let δ be the area of T (§36). Show that

$$\tan \delta = \frac{\cos \bar{a} \cos \bar{b}}{\sin \bar{a} + \sin \bar{b}}.$$

42.12 Given a triangle ABC, let δ be its area (= defect of ABC), and suppose given an angle δ' with $0 < \delta' < \delta$. Show that there is a point D between B and C such that the area of ABD is equal to δ'. *Hint*: First draw an altitude from A to BC, and thus reduce to the case of a right triangle. Then show that the point D can be found by solving a suitable quadratic equation in F.

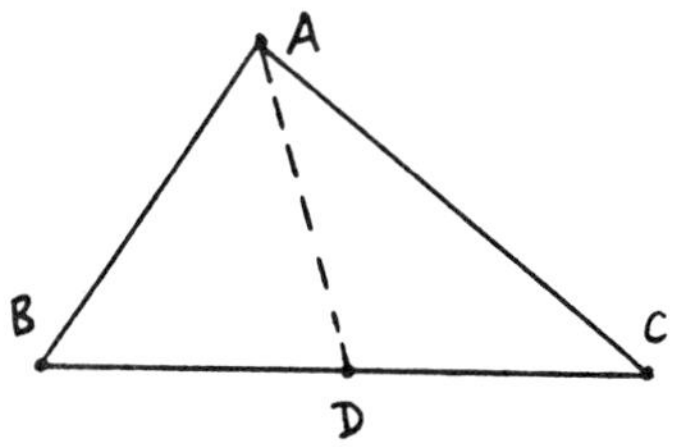

42.13 Given a point P, a line l, and an acute angle α, show that there is a line m through P, making an angle α with the line l.

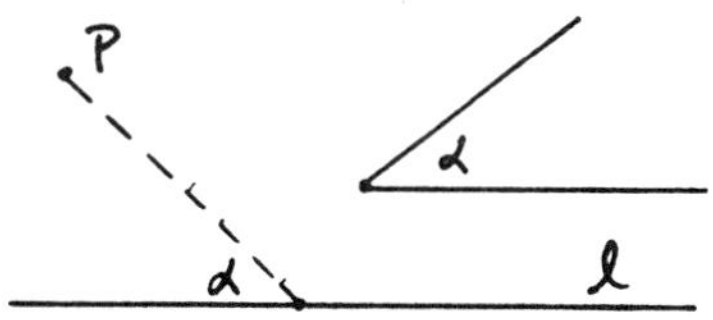

42.14 Given a point A on a line l, and an end λ of l, let Γ be the horocycle defined by A and λ (Exercise 41.12). Let m be a line which meets the ray $A\lambda$ in a point $C \neq A$. Prove that the line m meets the horocycle in two points B, B'. *Hint*: First reduce to the case $m \perp \lambda$. Then find B using trigonometry and solving a quadratic equation.

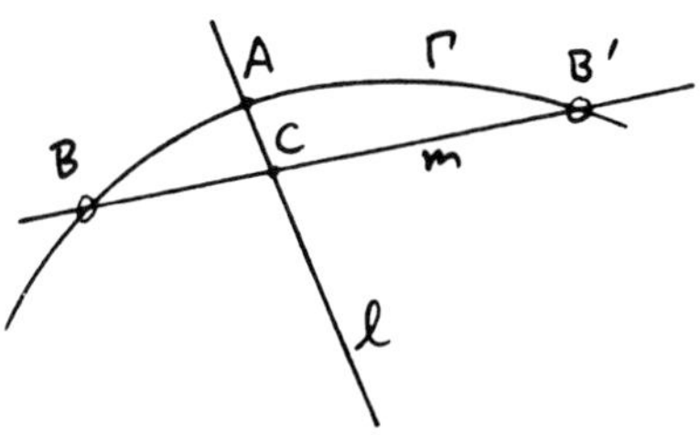

42.15 In this exercise, we show how to translate the formulae of hyperbolic trigonometry into relations among the hyperbolic trigonometric functions, in the case when our field F is a subfield of the real numbers.

For any $x \in \mathbb{R}$, we define the hyperbolic trigonometric functions

$$\begin{aligned}\sinh x &= \tfrac{1}{2}(e^x - e^{-x})\\ \cosh x &= \tfrac{1}{2}(e^x + e^{-x})\\ \tanh x &= \sinh x / \cosh x.\end{aligned}$$

a) Assume $F \subseteq \mathbb{R}$, and let λ be the additive distance function of (Exercise 41.14). With our usual notation for a triangle, let $\tilde{a}$, $\tilde{b}$, $\tilde{c}$ be the additive lengths λ of the sides opposite A, B, C. Show that

$$\begin{aligned}\sin \bar{a} &= 1/\cosh \tilde{a}\\ \cos \bar{a} &= \tanh \tilde{a}\\ \tan \bar{a} &= 1/\sinh \tilde{a}.\end{aligned}$$

b) Use these to translate the formulae of (42.2) and (42.3). In particular, show that

$$\cos \alpha = \frac{\tanh \tilde{b}}{\tanh \tilde{c}} \qquad \text{and} \qquad \cosh \tilde{a} = \frac{\cos \alpha}{\sin \beta}.$$

42.16 Let Γ be a circle of radius r, and let $x = \ln r$ be the additive length of the radius (Exercise 42.15).

a) Show that the area can be written as

$$A = 4\pi \sinh^2(x/2).$$

b) Expand in power series to show

$$A = \pi \left(x^2 + \frac{1}{12}x^4 + \dots \right).$$

Thus the area of a hyperbolic circle is "bigger" than the area of a corresponding Euclidean circle.

42.17 Use a limiting process similar to the one in (42.4) to show that if p is the multiplicative length of the circumference of a circle of radius r, then

$$\ln(p) = \frac{2\pi}{\tan \bar{r}}.$$

Presumably, this implies that the length of a circle over the constructible field is never *rectifiable*, i.e. is not equal to the length of any segment in the geometry.

42.18 Let T be an equilateral triangle with all angles equal to $30°$. Find, and express using square roots in standard form

a) the radius of the inscribed circle

b) the radius of the circumscribed circle

c) the radius of the circle with the same area. .

Check: Decimal answers to two of the above are 1.6093 and 8.1266.

42.19 Let S be a regular quadrilateral (all 4 sides equal and all 4 angles equal) whose area is equal to the area of a circle of radius 3.

a) find the radius of the inscribed circle

b) find the radius of the circumscribed circle

c) find the side of the quadrilateral.

Check: Decimal equivalents to two of these are 7.3276 and 27.8205.

42.20 Find an element $R \in F$ such that a circle of radius r admits a circumscribed hexagon if and only if $r < R$.

43 Characterization of Hyperbolic Planes

Thinking back to our study of Euclidean geometry, recall that we pursued two different logical paths. One was the abstract development of a geometry from axioms. The other was the analytic approach given by the Cartesian plane over an ordered field F. We brought these two paths together by introducing the field of segment arithmetic into the abstract geometry (§19) and then showing that any Hilbert plane with (P) is isomorphic to the Cartesian plane over its field of segment arithmetic (§21). To express this in other words, a Hilbert plane with (P) is *characterized* as the Cartesian plane over a certain ordered field F. It follows that two Hilbert planes with (P) will be isomorphic, as abstract geometries, if and only if their associated fields are isomorphic, as ordered fields.

In this section we will do the same thing for non-Euclidean geometry, at least for those non-Euclidean geometries satisfying (L), which we have called hyperbolic planes. We have developed some properties of the abstract hyperbolic planes in §40. On the other hand, we have introduced the Poincaré model over a field in §39, and have used the ambient Cartesian geometry to investigate some of its properties. Then in §41 and §42 we have introduced the field of ends into an abstract hyperbolic plane, and have used the field to create a sort of hyperbolic analytic geometry and trigonometry.

The last step in this logical progression is to characterize the hyperbolic planes by showing that a hyperbolic plane is determined, up to isomorphism, by its associated field of ends, and that any hyperbolic plane is isomorphic to the Poincaré model over its associated field.

Proposition 43.1 *Given a hyperbolic plane Π, the field of ends is uniquely determined, up to isomorphism. Two hyperbolic planes Π_1 and Π_2 are isomorphic* *as Hilbert planes if and only if the associated fields F_1 and F_2 are isomorphic as ordered fields.*

Proof. Of course the set of ends in the plane Π is uniquely determined. But in order to define the field structure on this set (41.4), we made a choice of two orthogonal lines, and labeled their ends $0, \infty, 1, -1$. If we made a different choice, a different end might become the zero element of the field, so clearly the field structure on the set of ends is not unique. But we will show that the field is unique up to isomorphism.

So suppose l_1, m_1 is a first set of two orthogonal lines with ends labeled 0_1,∞_1,1_1, -1_1, giving rise to a field structure F_1 on the set of ends (minus ∞_1). Suppose l_2, m_2 is a second choice of orthogonal lines with ends labeled $0_2, \infty_2, 1_2, -1_2$, giving a second field structure F_2. We can find a rigid motion φ of the plane which takes l_1 to l_2 and m_1 to m_2, in such a way that the ends $0_1, \infty_1, 1_1, -1_1$ are sent to the corresponding ends $0_2, \infty_2, 1_2, -1_2$. This rigid motion induces a one-to-one correspondence from the set of ends of Π to itself, which therefore gives a one-to-one map of F_1 to F_2 sending the elements $0_1, 1_1, -1_1$ of F_1 to the corresponding elements of F_2. The constructions which we used to define addition and multiplication in F_1 are now carried over to the corresponding constructions for F_2, and the ordering is preserved. Therefore φ induces an isomorphism of F_1 and F_2 as ordered fields, which is what we wanted to prove.

Now suppose that Π_1 and Π_2 are isomorphic hyperbolic planes. If $\varphi : \Pi_1 \to \Pi_2$ is an isomorphism, we can choose orthogonal lines l_1 and m_1 in Π_1 with which to construct the field of ends F_1 of Π_1, and then take $l_2 = \varphi(l_1)$, $m_2 = \varphi(m_1)$ to construct the field of ends F_2 to Π_2. Then it is clear that the induced map $\varphi' : F_1 \to F_2$ on ends will give an isomorphism of fields.

Finally, let Π_1 and Π_2 be hyperbolic planes, and suppose given an isomorphism $\psi : F_1 \to F_2$ of the associated fields of ends. We wish to show that Π_1 and Π_2 are isomorphic Hilbert planes. That is to say, there is a map $\varphi : \Pi_1 \to \Pi_2$ of the points which is one-to-one and onto, and φ preserves lines, betweenness, and congruence line segments and angles.

We construct φ as follows. First extend the map $\psi : F_1 \to F_2$ to the set of all ends by setting $\psi(\infty_1) = \infty_2$. Any line of Π_1 is given by an unordered pair (α, β) of distinct elements of $F_1' = F_1 \cup \{\infty_1\}$. So by sending (α, β) to $(\psi(\alpha), \psi(\beta))$, we obtain a one-to-one map of the set of lines of Π_1 onto the set of lines of Π_2. A point P of Π_1 is determined by the set of all lines passing through it, and this set of lines satisfies an equation of the form

$$u_1u_2 - b(u_1 + u_2) + a^2 = 0$$

with $a, b \in F_1$, $a > 0$, and $|b| < a$ (41.6). Since ψ is an isomorphism of ordered fields, the set of images by φ of the lines containing P will satisfy a similar equation in F_2, and so they will define a unique point which we denote $\varphi(P)$. Then by construction φ is a one-to-one map of the set of points of Π_1 onto the set of points of Π_2, sending lines into lines.

Since betweenness of points can be expressed in terms of the ordering of the field of ends, and ψ is an isomorphism of ordered fields, φ preserves betweenness.

Congruence of line segments can be measured by the multiplicative distance function μ (41.7) and congruence of angles can be measured by the tangent function (41.8). Therefore the map $\varphi : \Pi_1 \to \Pi_2$ also preserves congruence of line segments and angles. Thus φ is an isomorphism of Hilbert planes, as required.

Theorem 43.2 *Let F be an ordered field satisfying* (**), *and let Π be the Poincaré model constructed over the field F (§39). Let F_1 be the field of ends of Π. Then F and F_1 are isomorphic ordered fields.*

Proof. Our strategy is to establish a one-to-one correspondence between the sets F and F_1, and then carry out the constructions of addition and multiplication in F_1 in the geometry of the Poincaré model, to show that the field structures are isomorphic.

We may assume that the Poincaré model is constructed using the unit circle Γ in the Cartesian plane over F (Exercise 39.21). Let us choose the x-axis and y-axis to be the orthogonal lines used in the construction of the field of ends F_1. Recall that the points of the Poincaré model are those points of the Cartesian plane inside Γ. It follows that the ends of the Poincaré model Π are exactly the Cartesian points of the circle Γ. We label the ends of our two axes so that

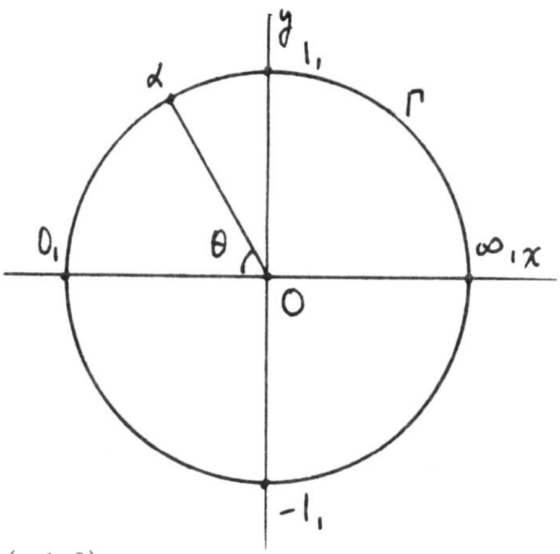

$$\begin{aligned} 0_1 &= (-1,0) \\ \infty_1 &= (1,0) \\ 1_1 &= (0,1) \\ -1_1 &= (0,-1) \end{aligned}$$

in Cartesian coordinates.

The next step is to give a mapping of sets between the elements of the fields F and F_1. The elements of F_1 are all the points of the circle Γ except $\infty = (1,0)$. We define a map $\varphi : F_1 \to F$ as follows. First we set $\varphi(0_1) = 0$. Next, for any end α in the upper semicircle of Γ, we set $\varphi(\alpha) = \tan\theta/2$, where θ is the angle from 0_1 to α subtended at the center of the circle. If $-\alpha$ is the reflection of this point in the x-axis, we set $\varphi(-\alpha) = -\varphi(\alpha)$. Here we understand θ to be the angle in the Cartesian plane, and the tangent function has its usual meaning (§16). As α ranges over the upper semicircle, θ ranges over all possible angles (between 0 and 2RA), so the function $\tan\theta/2$ ranges over all positive elements of the field F. Given any element $a \in F$, $a > 0$, there is an angle θ such that $\tan\theta/2 = a$, so this mapping φ is a one-to-one correspondence between the sets F_1 and F. Clearly φ preserves the ordering on the two sets.

The hard work we must do is to show that φ is compatible with the operations of addition and multiplication in the two fields. We start by computing the Cartesian coordinates of an end $\alpha \in F_1$, which is a point of the circle Γ, in terms of $\varphi(\alpha)$. We will do all our calculations for points α in the upper semicircle, since the results for their negatives will follow immediately.

So let $\alpha \in F_1$, $\alpha > 0$. Let $\varphi(\alpha) = a = \tan\theta/2$. It follows that the Cartesian coordinates (x, y) of α are given by

$$\begin{cases} x = -\cos\theta = \frac{a^2-1}{1+a^2} \\ y = \sin\theta = \frac{2a}{1+a^2}. \end{cases}$$

Here we use the usual trigonometric formulae expressing $\sin\theta$ and $\cos\theta$ in terms of $\tan\theta/2$.

To study addition, we will compute the reflection σ_α in a line (α, ∞). In the Poincaré model, reflection is given by circular inversion in the corresponding Cartesian circle (39.5). So given a point α on Γ, let Δ be the P-line (α, ∞), which is a circle orthogonal to Γ at the points α and ∞. Let A be the center of this circle. Then by a little elementary Euclidean geometry we see that the angle $OA\infty_1$ is equal to $\theta/2$, so the coordinates of A are $(1, a^{-1})$.

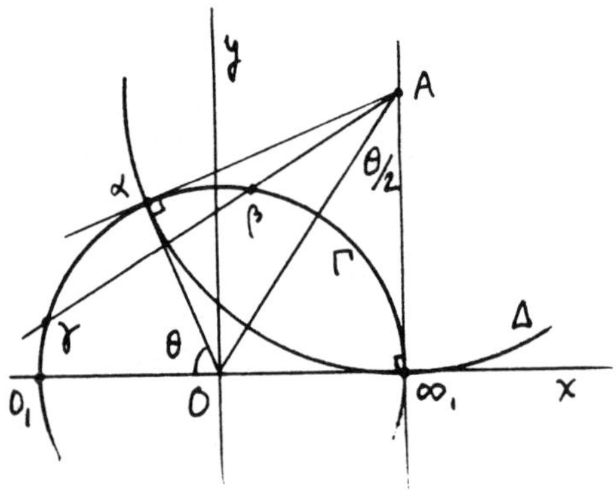

The P-reflection in Δ is circular inversion in Δ in the Cartesian plane. To study its effect on ends, let β be any end, and let γ be its image under reflection in Δ. Since Γ and Δ are orthogonal circles, γ is just the intersection of the lines $A\beta$ with Γ. Let $\varphi(\beta) = b$ and $\varphi(\gamma) = c$. Then β and γ have Cartesian coordinates

$$\beta = \left(\frac{b^2-1}{1+b^2}, \frac{2b}{1+b^2}\right)$$
$$\gamma = \left(\frac{c^2-1}{1+c^2}, \frac{2c}{1+c^2}\right).$$

To express the fact that A, β, γ are collinear, we set the slopes of $A\beta$ and $A\gamma$ equal to each other. This gives

$$\frac{\frac{2b}{1+b^2} - a^{-1}}{\frac{b^2-1}{1+b^2} - 1} = \frac{\frac{2c}{1+c^2} - a^{-1}}{\frac{c^2-1}{1+c^2} - 1}.$$

Simplifying, we get

$$2a(b-c) = b^2 - c^2.$$

Assuming $\beta \neq \gamma$, so $b \neq c$, we can divide out $b - c$, and so

$$c = 2a - b.$$

Thus the reflection in Δ, which is σ_α, has an effect on ends in F_1, which is transformed by φ into the transformation

$$x' = 2\varphi(\alpha) - x$$

for elements of F.

Recall that addition of ends is characterized by

$$\sigma_{\alpha+\beta} = \sigma_\beta \sigma_0 \sigma_\alpha$$

using reflections (41.3). Via the mapping $\varphi : F_1 \to F$, therefore, $\sigma_{\alpha+\beta}$ becomes

$$x' = 2\varphi(\alpha+\beta) - x.$$

On the other hand, $\sigma_\beta\sigma_0\sigma_\alpha$ becomes a transformation which first sends x to

$$2\varphi(\alpha) - x$$

then to

$$2\varphi(0) - (2\varphi(\alpha) - x)$$

then to

$$2\varphi(\beta) - [2\varphi(0) - (2\varphi(\alpha) - x)] = 2\varphi(\alpha) + 2\varphi(\beta) - x.$$

We conclude that

$$\varphi(\alpha+\beta) = \varphi(\alpha) + \varphi(\beta),$$

so that φ is a homomorphism for addition.

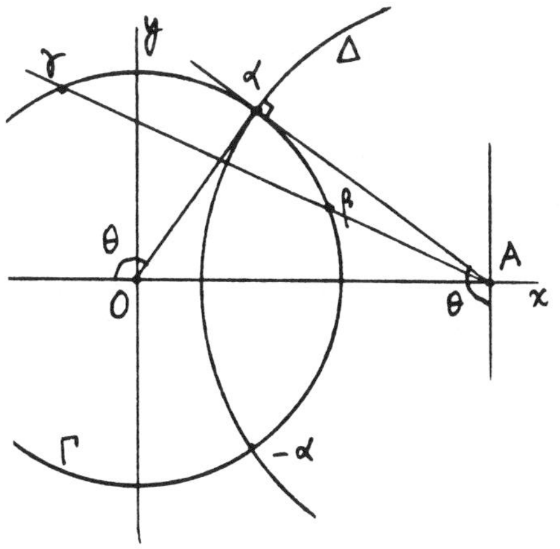

Now let us consider multiplication. To do this (changing notation) we consider the reflection σ_α in a line $(\alpha, -\alpha)$ in the Poincaré model. This is given by circular inversion in a circle Δ, orthogonal to Γ, passing through the points α and $-\alpha$. Let A be the center of this circle. Then the angle $OA\alpha$ is equal to $\theta-$ RA in our diagram. Thus $OA = -1/\cos\theta$. If $\varphi(\alpha) = a = \tan\theta/2$, then

$$A = \left(\frac{1+a^2}{a^2-1}, 0\right).$$

Now let β be another end, and let γ be its image under σ_α which is circular inversion in Δ. Then A, β, γ are collinear. Letting $\varphi(\beta) = b$ and $\varphi(\gamma) = c$ as before, we express the collinearity by equating the slopes of the lines $A\beta$ and $A\gamma$. This time we get

$$\frac{\frac{2b}{1+b^2}}{\frac{b^2-1}{1+b^2} - \frac{1+a^2}{a^2-1}} = \frac{\frac{2c}{1+c^2}}{\frac{c^2-1}{1+c^2} - \frac{1+a^2}{a^2-1}}.$$

Simplifying gives

$$bc(b-c) = a^2(b-c).$$

Assuming $b \neq c$, we get

$$c = a^2 b^{-1}.$$

Thus the reflection σ_α is transformed by φ into the transformation

$$x' = \varphi(\alpha)^2 x^{-1}$$

in F.

Going back to the hyperbolic plane, it is easy to see that the composition $\sigma_\alpha\sigma_1$ is equal to a parallel transformation along the line $(0, \infty)$ which sends 1 to α^2. On the other hand, multiplication is defined by composing parallel translations. So multiplication is characterized by the formula

$$\sigma_{\alpha\beta}\sigma_1 = \sigma_\beta\sigma_1\sigma_\alpha\sigma_1$$

or simply

$$\sigma_{\alpha\beta} = \sigma_\beta\sigma_1\sigma_\alpha.$$

Transporting this by φ, we find that for elements of F, the transformation

$$x' = \varphi(\alpha\beta)^2 x^{-1}$$

is equal to

$$x' = \varphi(\beta)^2\varphi(\alpha)^2 x^{-1}.$$

We conclude $\varphi(\alpha\beta) = \varphi(\alpha)\varphi(\beta)$ as required.

We have now shown that $\varphi : F_1 \to F$ is a one-to-one transformation compatible with addition, multiplication and the ordering. Hence F_1 and F are isomorphic as ordered fields. □

Corollary 43.3 *If Π is a hyperbolic plane with associated field of ends F, then Π is isomorphic to the Poincaré model over the field F.*

Proof. Indeed, the plane Π and the Poincaré model over F both have isomorphic fields of ends by (43.2), so by (43.1) they are isomorphic planes.

Corollary 43.4 *The circle-circle intersection property* (E) *holds in a hyperbolic plane.*

Proof. Indeed, the hyperbolic plane is isomorphic to the Poincaré model over its field of ends, and (E) holds in the Poincaré model (39.9).

Remark 43.4.1. Thus in a Hilbert plane, the axiom (L) of existence of limiting parallels implies (E). We do not know any purely geometric proof of this fact.

Remark 43.4.2. In fact, a stronger result holds. Namely, in any hyperbolic plane, given any two curves, each of which is either a line or a circle or a horocycle or a hypercycle, they will intersect if a certain betweenness condition (which we leave to the reader to make explicit) is satisfied. Indeed, these all correspond to various Euclidean circles in the Poincaré model (Exercise 39.14), and these will meet by (E) in the ambient Cartesian plane.

Exercises

43.1 Show that the result of (Exercise 1.15) holds in an arbitrary hyperbolic plane.

Notes

§**1**. To appreciate this text, you should have a copy of Euclid's *Elements* handy. The most natural choice for an English-speaking reader is Heath's authoritative translation and commentary (Heath 1926) available in an inexpensive Dover reprint.

The "hard problem" in this section is discussed in Coxeter and Greitzer (1967).

§**5**. For the history of the theorems of the three medians and the three altitudes of a triangle, see Tropfke (1923) vol 4, pp. 163, 164. Both are implicit in the work of Archimedes, but were not included in the repertoire of elementary geometry until much later.

The proof given for (5.6) is due to Gauss, Werke, vol 4, p. 396.

We give several different proofs that the altitudes of a triangle meet in a point:

a) Gauss' proof (5.6)

b) using the Euler line (5.7)

c) using cyclic quadrilaterals (Exercise 5.7)

d) by analytic geometry (13.1)

e) in the Poincaré model (Exercise 39.17)

f) using angle bisectors (Exercise 40.17)

g) by non-Euclidean analytic geometry (14.13).

§**6**. The axioms presented in this and subsequent sections 7, 8, 9 are essentially the same as the axioms proposed by Hilbert in his Foundations of Geometry (1899; 2nd ed. of English translation 1971). We have made a few small changes.

First of all, Hilbert postulates a set of points and a set of lines, together with a relation of incidence "a point lies on a line". We however postulate a set of points, and take lines to be subsets of the set of points, so that the incidence of a point and line simply becomes membership in the set.

Secondly, Hilbert formulates his axioms for 3-dimensional space. We have taken only the plane axioms because they exhibit all the essential features of the geometry we need.

Thirdly, Hilbert is a minimalist. For example, in the (SAS) axiom (C6), having assumed $AB \cong DE$, $AC \cong DF$, and $\angle BAC = \angle EDF$, he postulates only that $\angle ABC = \angle DEF$. He then proves as a theorem that $\angle ACB \cong \angle DFE$ and $BC \cong EF$. This degree of minimalism in the axioms seems unnecessary for an elementary text such as ours, so we have simplified in a couple of places by making an axiom slightly stronger than necessary.

Exercise 6.2. Unfortunately we do not have the space to develop the ideas of projective geometry further. But see e.g. Hartshorne (1967) for an introduction.

Exercise 6.5. Kirkman's schoolgirl problem was first published in the *Lady's and Gentlemen's Diary* for (1850). See Ball (1940), Chapter X for an extensive discussion. One solution can be found by taking the fifteen points of the projective 3-space over the field of 2 elements to be the girls, and the 35 lines of 3 points each to be the rows. Then with a little care you can find five lines which fill the space, and an automorphism σ of order 7 which cycles those five lines through the set of all the lines in such a way as to solve the problem.

§**7**. Exercise 7.3. There is a model consisting of 5 points only, due to Landau. See [Liebmann (1923), p. 12].

It is possible to take (7.1) as an axiom and then prove (B4), as is done in Greenberg (1993).

§**14**. I got the idea for Exercises 14.3–14.12 from a paper of Sturmfels.

§**18**. While the possibility of a non-Archimedean geometry was perhaps foreshadowed by the controversy about the angle between a circle and its tangent line, started by Peletier and Clavius in the 16th century, the first serious study of a non-Archimedean geometry is due to Veronese at the end of the 19th century. Our treatment follows Hilbert's approach via fields in his Grundlagen. See Enriques (1907), Chapter VII.

§**19**. For the segment arithmetic, we follow Hilbert's Grundlagen, Chapter III, with simplifications by Enriques described in Supplement II.

§**22**. Awareness of the problem of area grew in the 19th century (see Simon, §15). We follow Hilbert's treatment in his *Foundations of Geometry*, Chapter IV. Hilbert was the first to recognize the importance of Archimedes' axiom in the theory. Terminology is not uniform. What we call "equal content" is sometimes called "equicomplementable" (German: ergänzungsgleich). What we call "equidecomposable" or "equivalent by dissection" is sometimes called "equivalent by finite decomposition" (German: zerlegungsgleich, or teilungsgleich). The axiom (Z) is named after A. de Zolt, who attempted in 1881 to prove this statement geometrically (cf. Simon, §15).

§**24**. For the history of the Bolyai–Gerwien theorem, see Simon, §15. For practical notes on efficient dissections, see Lindgren. For (Exercise 24.8), see Dudeney.

§**25**. There is a vast literature on the problem of squaring the circle and attempts to understand the analytic as well as the geometrical significance of the number π. In fact this single problem has been the catalyst for significant advances in many branches of mathematics over its 4000 year history. A full discussion of the subject would lead way beyond the confines of this book. For an approach to the literature, see for example Beckmann (1971), Hobson (1953), Simon (1906), §6, and Rudio (1892).

§**26**. The correspondence between Gauss and Gerling can be found in Gauss, Werke, vol VIII, p. 241 ff.

§**27**. This treatment of the Dehn invariant is based on Cartier's Bourbaki Seminar talk (1985). See also Boltianskii (1978) for a detailed treatment of the problem.

§**28**. There is an extensive literature on these classical problems and attempts at their solution. See for example Klein (1895), Enriques (1907) or Lebesgue (1950). Descartes (1637) was already aware that the first two required cubic equations, and could not be solved by quadratic equations, but the first rigorous proof of the impossibility is attributed to Wantzel (1837). Our proof using the notion of the degree of a field extension is more recent, because abstract fields and field extensions have been considered only since the turn of the century (cf. Steinitz).

The trisection given in (Exercise 28.6) comes from Archimedes, Book of Lemmas, Proposition 8.

For the squaring of the circle, see notes to §25.

§**29**. Gauss' proof of the constructibility of the regular 17-gon can be found in his Disquisitiones Arithmeticae (1801), or in his Werke, vol I.

The construction given in the text is due to Maywald (1861). See the book of Goldenring, p. 16, who collected more than twenty different constructions.

§**30**. For references on the use of the marked ruler for solving the classical problems of trisecting the angle and doubling the cube, see [Enriques (1907) II, p. 204ff and p. 233ff].

In his history of the conic sections in antiquity, Zeuthen (1886) expresses the opinion that the early geometers accepted the use of the marked ruler (German: Einschiebung) along with the ruler and compass as a legitimate tool of construction, and that it was only from the time of Plato and Euclid that a strict distinction was made of those problems which could be solved with ruler and compass only. Once the theory of conics was well developed, the use of conics to solve the "solid" problems was preferred to marked ruler.

The construction of (30.1) is ascribed by Pappus to "the ancients", but he says the construction of (30.2) is due to Nicomedes. The construction in (Exercise 30.3) is due to Vieta (see Eves, *History*, p. 103).

Another, similar construction of the regular heptagon is given by Gleason (1988), who also gives a construction of the 13-sided polygon.

§**31**. The fact that real roots of cubic and quartic equations could be found using square roots, cube roots, and trisecting angles, was known to Descartes [La Géométrie, Book III]. Cf. also [Enriques (1907), II, p. 266].

The problem of Alhazen (Exercise 31.8) is mentioned by Eves (1969).

§**32**. Consult your favorite algebra book. One I like is Stewart (1989).

§**33**. References to the work of Proclus, Tacquet, Clairaut, Clavius, Simson, and Playfair can be found under their names in the bibliography.

For more details on the history of the theory of parallels and the discovery of non-Euclidean geometry, see the detailed book of Bonola, which includes the texts of Bolyai and Lobachevsky, or the book of Engel and Stäckel, which includes selections from the work of Wallis, Saccheri, Lambert, Gauss, Schweikart, and Taurinus.

Among more recent texts, the books of Wolfe (1945) and Greenberg (1993) have very readable accounts of the development of non-Euclidean geometry.

§**34**. See the works of Saccheri and Legendre in the bibliography, especially Note II to the 12th edition of Legendre's *Eléments*, where it seems that he is attempting to supply in the quantity of his proofs what is lacking in their rigor.

See also Hilbert's *Foundations*, §10 for the material of this section.

§**36**. The non-Euclidean version of the Bolyai–Gerwien theorem (36.6) is apparently due to Finzel (1912). However, the present proof without appealing to continuity arguments appears to be new. See Liebmann, p. 41; Baldus–Löbell, p. 116; Moise, p. 348; and Boltianskii, p. 67.

Also, our definition and use of the unwound circle group to measure areas is new. All the other treatments I have seen use real numbers to measure angles.

§**37**. According to Max Simon, p. 93, the notion of circular inversion first appears in the work of Poncelet (1822), followed by Steiner (1832), Quetelet (1827), Magnus (1832), and Plücker (1844). Since then it has become a useful standard technique.

Constructions with compass alone were studied by Mascheroni (1797) who proved that they suffice to carry out any construction possible with ruler and compass (Exercise 37.21).

§**39**. A note on the consistency of non-Euclidean geometry. The discoverers of these geometries, Gauss, Bolyai, and Lobachevsky seem to have been convinced of the existence of these geometries by the extensive theory they developed and its internal coherence. Lobachevsky also noted that the formulas of hyperbolic trigonometry could be obtained by taking the formulas of spherical trigonometry for a sphere whose radius is imaginary.

Still, a rigorous proof of consistency (or at least consistency relative to Euclidean geometry) is best made by producing a model. Credit for the first such model goes to Beltrami (1868) who found that the geometry on surfaces of constant negative curvature in 3-space behaved like a hyperbolic plane. Beltrami seems also to have been aware of the several representations of hyperbolic geometry in the Euclidean plane, used later by Klein and Poincaré. Klein preferred the "projective" model, where the plane is represented by the points inside a fixed circle, and the lines by chords of Euclidean lines in that circle. This gives the most direct connection with the projective metric of Cayley, and allows one to use the powerful methods of projective geometry.

Poincaré (1882), on the other hand, was led through his investigations of automorphic functions to a conformal model using the points of the Euclidean upper half-plane, in which the lines become half-circles orthogonal to the x-axis. This model can be easily transformed into the model we use, whose points are the interior of a circle, and whose lines are segments of circles orthogonal to the fixed circle, which was developed extensively by Carslaw (1916).

For references, see Bonola, Poincaré's "Science and Hypothesis", and Klein's "Nichteuklidische Geometrie".

We chose the Poincaré model for this book, because it can be developed directly from Euclidean geometry using inversion in circles, and because it seems the most direct and elementary approach. We could not have developed the Klein model fully without a considerable excursion into projective geometry which extends beyond the scope of the present book.

Exercise 39.20. The theorem that the medians of a triangle meet in a point is still true without the hypothesis of a circumscribed circle, but the proof, using projective geometry, is more difficult. See Greenberg (1993), p. 277. It seems that there is no known synthetic (purely geometric) proof of this theorem in the hyperbolic plane—cf. Baldus–Löbell (1953), p. 102, Liebmann (1923), p. 22.

§**40**. The axiom (L) was proposed by Hilbert in his article "A new development of Bolyai–Lobachevskian geometry", reprinted as Appendix III to his Foundations of Geometry (1971). This article also contains the proofs of (40.11), existence of a common perpendicular and (40.13) existence of the enveloping line.

The fact that the notion of limiting parallelism is an equivalence relation was proved by a slightly different method from the one here by Gauss—see Bonola (1955), §34.

The proof suggested in (Exercise 40.17) is due to Liebmann (1923), p. 31.

§**41**. The construction of the field of ends (41.2), (41.3), and (41.4), together with the equation of a line (41.6) is taken from Hilbert's article "A new development..." cited above. After he has verified the field axioms, he says that "the construction of the geometry poses no further difficulties" (p. 147).

In the remainder of the section we have worked out various applications of Hilbert's field. The use of the multiplicative distance function (41.7) to avoid the real numbers is new, as are the proofs of (41.9)–(41.13).

Bolyai's parallel construction first appears in Bolyai (1832), §34. There are several different proofs in the literature:

a) proofs using the relations between right triangles and quadrilaterals with three right angles (called Spitzeck by Liebmann), and using a hyperbolic trigonometry. See Greenberg p. 413, Bonola, App. III, Liebmann (1923) p. 35

b) a proof using a prism in hyperbolic 3-space—Bonola, App. III

c) a purely geometric proof using the Hjelmslev midline theorem, due to Liebmann (1910): see Carslaw p. 73 or Wolfe p. 95

d) proofs in the Klein model, using the projective geometry of the ambient plane: Greenberg, p. 269; Baldus–Löbell, p. 93.

§**42**. The formulae of hyperbolic trigonometry (42.2) and (42.3) are the same as those given by Lobachevsky in his "Theory of Parallels", §35.

Our derivation of these formulae using Hilbert's field of ends, and independent of the real numbers, is, as far as I know, new.

Bolyai, in his "Science of Absolute Space", §43, studied the area of a circle, and recognized that it could be "squared" or not, depending on the arithmetic properties of the number A/π.

§**43**. This characterization of hyperbolic planes is a natural consequence of the ideas set forth in Hilbert's "Neue Begründung..." article. It gives a direct and elementary proof of a result which otherwise could only be derived as a consequence of the much deeper theorem of Péjas characterizing all Hilbert planes (see Hessenberg and Diller). That proof is accomplished via Klein's perspective of regarding the non-Euclidean plane as a subset of a larger projective plane. Delicate work of Hjelmslev and Hessenberg is then needed to establish Pappus' axiom in the projective plane, allowing the introduction of coordinates along the lines suggested in Hilbert's *Grundlagen.*

References

[1] Archibald, R. C., Euclid's book on divisions of figures, with a restoration based on Woepke's text and on the Practica Geometriae of Leonardo Pisano, University Press, Cambridge (1915).

[2] Archimedes, Quae supersunt omnia cum Eutocii Ascalonitae commentariis. Ex recensione Josephi Torelli, Clarendon, Oxford (1792).

[3] Bachmann, F., Aufbau der Geometrie aus dem Spiegelungsbegriff, Springer, New York (1973) (first published 1959).

[4] Baldus, R. and Löbell, F., Nichteuklidische Geometrie, hyperbolische Geometrie der Ebene, 3rd Ed., de Gruyter, Berlin (1953).

[5] Ball, W. W. R., Mathematical Recreations & Essays, revised by H. S. M. Coxeter, 11th edition, Macmillan, London (1940) (first published 1892).

[6] Beckmann, P., A history of π (PI), 2nd ed., Golem, Boulder (1971).

[7] Beltrami, E., Saggio di interpretazione della geometria non euclidea, Giornale di Mat. **6** (1868) 284–312.

[8] Birkhoff, B. D., A set of postulates for plane geometry (based on scale and protractor), Annals of Math. **33** (1932) 329–345.

[9] Blanchet, M. A., Eléments de Géométrie par A. M. Legendre, avec additions et modifications, 7 ème édition, Firmin Didot, Paris (1859).

[10] Blumenthal, L. M., A modern view of geometry, W. H. Freeman, San Francisco (1961).

[11] Boltianskii, V. G., Hilbert's Third Problem, Winston, Washington (1978).

[12] Bolyai, J., Appendix scientiam spatii absolute veram exhibens: a veritate aut falsitate Axiomatis XI Euclidei (a priori haud unquam decidenda) independentem (1832).

[13] Bonola, R., Non-Euclidean Geometry, a critical and historical study of its development. English translation by H. S. Carslaw. With a supplement containing the Dr. George Bruce Halsted translations of the Science of Absolute Space by John Bolyai, and The Theory of Parallels, by Nicholas Lobachevski, Dover, New York (1955) (English translation of Bonola, first published 1912).

[14] Borsuk, K. and Szmielew, W., Foundations of Geometry, Euclidean and Bolyai–Lobachevskian Geometry, Projective Geometry, North–Holland, Amsterdam (1960).

[15] Camerer, J. G., Apollonii de Tactionibus, quae supersunt, ac maxime lemmata Pappi, in hos libros Graece nunc primum edita, e codicibus mscptis, cum Vietae librorum Apollonii restitutione, adjectis observationibus, computationibus, ac problematis Apolloniani historia, Ettinger, Gothae (1795).

[16] Carslaw, H. S., The Elements of Non-Euclidean plane geometry and trigonometry, Longmans, Green, London (1916).

[17] Cartier, P., Décomposition des polyèdres: le point sur le troisième problème de Hilbert, Sém. Bourbaki **646** (1985).

[18] Cederberg, J. N., A course in modern geometries, Springer, New York (1989).

[19] Clairaut, M., Elémens de Géométrie, Cellot & Jombert, Paris (1775) (first published 1741).

[20] Clavius, C., Euclidis Elementorum, libri XV, accessit XVI de solidorum regularium cuiuslibet intra quodlibet comparatione. Omnes perspicuis Demonstrationibus, accuratisque Scholiis illustrati, ac multarum rerum accessione locupletati, 3rd ed., Ciotti, Coloniae (1591) (first published 1574).

[21] Commandino, F., De gli Elementi d'Euclide libri quindici, con gli scholii antichi, tradotti prima in lingua latina, & con commentarij illustrati, et hora d'ordine dell'istesso transportati nella nostra vulgare, & da lui riveduti, In Casa di F. Commandino, Urbino (1575).

[22] Court, N. A., College Geometry, An Introduction to the modern geometry of the triangle and the circle, Barnes and Noble, New York (1925), 2nd ed. (1952).

[23] Coxeter, H. S. M., Non-Euclidean Geometry, University Press, Toronto (1942).

[24] Coxeter, H. S. M., Introduction to Geometry, Wiley, New York (1961).

[25] Coxeter, H. S. M. and Greitzer, S. L., Geometry Revisited, Random House, New York (1967).

[26] Dedekind, Stetigkeit und irrationale Zahlen, Braunschweig (1872).

[27] Descartes, La Géométrie, Hermann, Paris (1886) (first published 1637).

[28] Dodgson, C. L., Euclid and his modern rivals, Macmillan, London (1879).

[29] Dudeney, H. E., The Canterbury Puzzles, and other curious problems, Nelson, London (1929) (first published 1908).

[30] Dudley, U., A budget of trisections, Springer, New York (1987).

[31] Engel, F. and Stäckel, P., Die Theorie der Parallellinien von Euklid bis auf Gauss, eine Urkundensammlung zur Vorgeschichte der nichteuklidischen Geometrie, Teubner, Leipzig (1895).

[32] Enriques, F., Prinzipien der Geometrie, Enz. Math. Wiss. III.I.1 (1907–1910) 1–129.

[33] Enriques, F., Fragen der Elementargeometrie, II Teil: Die Geometrischen Aufgaben, ihre Lösung und Lösbarkeit, Teubner, Leipzig (1907).

[34] Euclid, Elements (see different editions under the names of Clavius, Commandino, Heath, Peletier, Peyrard, Playfair, Simson, Tacquet).

[35] Eves, H., An introduction to the history of mathematics, Rinehart, New York (1953).

[36] Eves, H., College Geometry, Jones and Bartlett, Boston (1995).

[37] Gauss, C. F., Werke, Göttingen (1870–1877).

[38] Gleason, A. M., Angle trisection, the heptagon, and the triskaidecagon, Amer. Math. Monthly (1988) 185–194.

[39] Goldenring, R., Die Elementargeometrischen Konstruktionen des Regelmässigen Siebzehnecks, eine historisch–kritische Darstellung, Teubner, Leipzig (1915).

[40] Greenberg, M., Euclidean and non-Euclidean geometries without continuity, Amer. Math. Monthly **86** (1979) 757–764.

[41] Greenberg, M. J., Euclidean and non-Euclidean Geometries, Development and History, 3rd ed., Freeman, New York (1993).

[42] Hadamard, J., Leçons de géométrie élémentaire, Paris (1901–1906).

[43] Hartshorne, R., Foundations of Projective Geometry, Benjamin, New York (1967).

[44] Heath, T. L., The thirteen books of Euclid's Elements, translated from the text of Heiberg, with introduction and commentary, 2nd ed., 3 vols, University Press, Cambridge (1926) (Dover reprint 1956).

[45] Herstein, I. N., Topics in Algebra, 2nd ed., Xerox, Lexington (1975).

[46] Hessenberg, G. and Diller, J., Grundlagen der Geometrie, 2nd ed., de Gruyter, Berlin (1967).

[47] Hilbert, D., Foundations of Geometry (Grundlagen der Geometrie), 2nd English ed., translated by Leo Unger from the tenth German edition, revised and enlarged by Dr. Paul Bernays, Open Court, La Salle (1971) (first published 1899).

[48] Hjelmslev, J., Neue Begründung der ebenen Geometrie, Math. Ann. **64** (1907) 449–474.

[49] Hobson, E. W., Squaring the circle, Chelsea, New York (1953) (first published 1913).

[50] Klein, F., Vorträge über ausgewählte Fragen der Elementargeometrie, ausgearbeitet von F. Tägert, Teubner, Leipzig (1895).

[51] Klein, F., Vorlesungen über Nicht–Euklidische Geometrie, neu bearbeitet von W. Rosemann, Chelsea, New York, n.d. (first published 1927).

[52] Lebesgue, H., Leçons sur les constructions géométriques, professées an Collège de France en 1940–1941, Gauthier–Villars, Paris (1950).

[53] Legendre, A. M., Eléments de Géométrie, avec des notes, 12 ème éd., Firmin Didot, Paris (1823).

[54] Leistner, J. I. C., Unwiderrusslicher, Wohlgegründter und Ohnendlicher Beweis der wahren Quadratur des Circuls, oder des Durchmessers zu seinem Umcreysz, wie 1225 zu 3844 oder 3844 zu 1225, Heyinger, Wienn (1737).

[55] Liebmann, H., Nichteuklidische Geometrie, 3rd ed., de Gruyter, Berlin (1923) (first published 1905).

[56] Lindgren, H., Geometric Dissections, van Nostrand, Princeton (1964).

[57] Lobachevsky, N. I., Collection complète des oeuvres géométriques de N. J. Lobatcheffsky, Kazan (1866).

[58] Loomis, E. S., The Pythagorean Proposition, its demonstrations analyzed and classified, and bibliography of sources for data of the four kinds of "proof", Nat. Council of Teachers of Math., Washington (1968) (first published 1940).

[59] Lucas, E., Récréations Mathématiques, Gauthier–Villars, Paris (1882).

[60] Lyusternik, L. A., Convex figures and polyhedra, Dover, New York (1963).

[61] Martin, G. E., The foundations of geometry and the non-Euclidean plane, Springer, New York (1975).

[62] Mascheroni, L., Géométrie du compas, traduit de l'italien par A. M. Caretti, Paris (1798).

[63] Millman, R. S. and Parker, G. D., Geometry, Springer, New York (1991).

[64] Moise, E. E., Elementary Geometry from an advanced standpoint, 2nd ed., Addison–Wesley, Reading (1974).

[65] Montucla, J. F., Histoire des Mathématiques, Blanchard, Paris (1968) (first published 1758).

[66] Pasch, M. and Dehn, M., Vorlesungen über neuere Geometrie, 2nd ed. mit einem Anhang: Die Grundlegung der Geometrie in historischer Entwicklung, Springer, Berlin (1976) (first published 1882).

[67] Peletier, J., In Euclidis Elementa Geometrica Demonstrationum Libri sex, Tornaesium, Lugduni (1557).

[68] Peyrard, F., Les Elémens de Géométrie d'Euclide, traduits littéralement, et suivis d'un Traité du Cercle, du Cylindre, du Cône et de la Sphère; de la Mesure des Surfaces et des Solides; avec des Notes, 2nd ed., Louis, Paris (1809) (first published 1804).

[69] Playfair, J., Elements of Geometry, containing the first six books of Euclid, with a supplement on the quadrature of the circle, and the geometry of solids, Lippincott, Philadelphia (1857).

[70] Poincaré, H., Théorie des groups fuchsiens, Acta Math. **1** (1882) 1–62 and *Oeuvres*, Tome II (1952) 108–168.

[71] Poincaré, H., Science and hypothesis, trans. W. J. Greenstreet, London (1905).

[72] Proclus, A Commentary on the first book of Euclid's Elements, translated with introduction and notes by Glenn R. Morrow, University Press, Princeton (1970).

[73] Rudio, F., Archimedes, Huygens, Lambert, Legendre. Vier Abhandlungen über die Kreismessung, Deutsch herausgegeben und mit einer Uebersicht über die Geschichte des Problems von der Quadratur des Zirkels von den ältesten Zeiten bis auf unsere Tage, Teubner, Leipzig (1892).

[74] Saccheri, G., Euclides ab omni naevo vindicatus: sive conatus Geometricus quo stabiliuntur prima ipsa universae Geometriae Principia, edited and translated by George Bruce Halsted, Open Court, Chicago (1920) (first published 1733).

[75] Simon, M., Ueber die Entwicklung der Elementar–Geometrie im XIX Jahrhundert, Teubner, Leipzig (1906).

[76] Simson, R., The Elements of Euclid, viz. the first six books, together with the eleventh and twelfth. The errors, by which Theon, or others, have long ago vitiated these books, are corrected, and some of Euclid's demonstrations are restored. Also, the book of Euclid's Data, in like manner corrected, Robertson, Glasgow (1803) (first published 1756).

[77] Smith, D. E., A Source Book in Mathematics, Dover, New York (1959) (first published 1929).

[78] Sommerville, M. A., Bibliography of Non-Euclidean Geometry, Harrison, London (1911).

[79] Steck, M., Bibliographia Euclideana, Gerstenberg, Hildesheim (1981).

[80] Steinitz, E., Algebraische Theorie der Körper, Chelsea, New York (1950) (first published 1910).

[81] Stewart, I., Galois theory, 2nd ed., Chapman & Hall, London (1989).

[82] Stillwell, J., Mathematics and its History, Springer, New York (1989).

[83] Tacquet, A., Elementa Geometriae planae ac solidae, quibus accedunt selecta ex Archimede theoremata, Manfrè, Patavii (1738).

[84] Thomas–Stanford, C., Early editions of Euclid's Elements, Wofsy, San Francisco (1977).

[85] Tropfke, J., Geschichte der Elementarmathematik, vol. IV, Ebene Geometrie, de Gruyter, Berlin (1923).

[86] von Staudt, K. G. C., Geometrie der Lage, Bauer u. Raspe, Nürnberg (1847).

[87] Wantzel, P. L., Recherches sur les moyens de reconnaitre si un problème de géométrie peut se resoudre avec la règle et le compas, J. Math. **2** (1837) 366–372.

[88] Wolfe, H. E., Introduction to Non-Euclidean Geometry, Holt, Rinehart, Winston, New York (1945).

[89] Zeuthen, H. G., Die Lehre von den Kegelschnitten im Altertum, A. F. Höst, Kopenhagen (1886).

Index of Axioms & Acronyms

P	Playfair's axiom §3, §6, §33
SAS	side-angle-side = (I.4), §17
SSS	side-side-side = (I.8), (10.1)
ASA	angle-side-angle = (I.26), (40.7)
RASS	right-angle-side-side (Exercise 1.7), (3.2), (Exercise 10.5)
2ASA	(5.3)
2SAS	(Exercise 5.1)
2SSS	(Exercise 5.2)
I1,I2,I3	axioms of incidence §6
P1,P2,P3,P4	axioms of a projective plane (Exercise 6.2)
K1–K5	Kirkman geometry (Exercise 6.4)
B1–B4	betweenness axioms §7
C1–C3	congruence for line segments §8
C4–C6	congruence for angles §9
E	circle-circle intersection property §11, (16.2)
LCI	line-circle intersection property (11.6), (16.2)
A	Archimedes' axioim §12
D	Dedekind's axiom §12
A′	Archimedes' axiom for a field (15.4)
D′	Dedekind's axiom for a field (15.4)
(*)	$\sqrt{1+a^2} \in F$ (16.1)
(**)	$\sqrt{a} \in F$ if $a > 0$ (16.2)
ERM	existence of rigid motions §17, (39.5)
Sim AAA	= (VI.4), (20.1)
Sim SSS	(20.3)
Sim SAS	(20.4)
Z	de Zolt's axiom §22, (36.2.1)
	Aristotle's axiom §33
	Clavius' axiom §33
	Clairaut's axiom §33
	Simson's axiom §33
	Wallis' axiom (Exercise 33.5)
	Bolyai's axiom (Exercise 33.11)
ASL	angle-side-limit (Exercise 35.5), (40.9)
AAL	angle-angle-limit (Exercise 35.7), (40.10)
PPP,PLL,PLC,etc.	problems of Apollonius §38
L	existence of limiting parallel rays §40

Index of Euclid's Propositions

I.1 1,2,12,17,19,20,26,82,85,92,199
I.2 18,82
I.3 19,71,82
I.4 1,2,12,27,29,30,31,77,82,123
I.5 6,12,32,82,89
I.6 12,83,84
I.7 12,13,32,81,83,90
I.8 29,32,34,44,47,83
I.9 19,20,84,85
I.10 20,85
I.11 85
I.12 85,92,93
I.13 35,78,85
I.14 85,86
I.15 35,48,85,244
I.16 33,35,49,85,88,235,244
I.17 86,244
I.18 86
I.19 86,88,92,253
I.20 86
I.21 86
I.22 28,86,91,92,93
I.23 35,36,86
I.24 86
I.25 86
I.26 47,86,300
I.27 35,36,86,94,247,300
I.28 35,172
I.29 34,35,46,94,237
I.30 94,98
I.31 35,86,94
I.32 7,43,94,141,170,235,247
I.33 94,172
I.34 47,48,94,170
I.35 36,37,94,155,156,160,167,172
I.36 160,167
I.37 27,38,155,160,161,167,172,174,259
I.38 161,167
I.39 37,41,155,161,167
I.40 161,167
I.41 161,167
I.42 161,167
I.43 40,161,167
I.44 161,167,182
I.45 161,167
I.46 199
I.47 2,4,7,11,34,38,42,92,95,146,161, 167,169,172,173,174,326
I.48 155,161,167

II.1–4 167
II.5 42,167
II.6 42,167
II.7–10 167
II.11 41,42,95,102,148,161,167
II.12 167
II.13 167
II.14 39,161,167

III.1–9 93
III.10–12 92
III.13 93
III.14 92
III.15 93
III.16 92
III.17 93
III.18 92,266
III.19 92
III.20 95
III.21 1,7,49,96,146
III.22 53,95
III.23–30 95
III.31 93,95,172,265
III.32 43,44,53,95,147
III.33 95
III.34 95

III.35 95,146,161,167
III.36 39,42,95,146,161,167,209,266
III.37 42,43,95,161,167,266

IV.2 43
IV.4 14,46,93,144
IV.5 16,46,48,49,96,141,313
IV.10 41,42,95,148,161,167
IV.11 1,4,38,41,43,95,199
IV.15 11,93,199
IV.16 199

V.Def4 137
V.Def5 136

VI.1 39
VI.2 7,137,145
VI.4 142
VI.5 145
VI.6 145
VI.8 146

X.1 95
X.117 136,223

XI.28 179,181
XI.29–31 182
XI.39 182,183

XII.5 182

Index

accuracy 20
acute 117
addition
  of angles 78
  of congruence 70
  of ends 314
  of segments 137
additive distance function 325
additive group 324
affine plane 61
algebra of areas 39,42
algebraic numbers 122,222
Alhazen 220,343
Almamon 134
alternate interior angles 35
altitudes 4,46,48,98,294,311,312,322,341
analytic geometry 97
angle 67,76,116
  between circles 265
angle bisectors 14,19,21,46,147,306,311
angle of parallelism 233,255,290,298,319
angle sum of triangle 291,302
Apollonius 4,51,134,199,233,275,281
Archimedean solids 4
Archimedes 1,46,53,63,177,220,341,343
Archimedes' axiom 4,95,97,113,137,170, 177,233,242,289
area 2,135,154
  function 154,158,160,163,164,256
  non-Euclidean 256
  of circle 177,329,334
  of triangle 164
  theory of 36
Aristotle 134
Aristotle's axiom 234,239,253,306
arithmetic of ends 312,314
arithmetic of line segments 3
Artin, Emil 122
automorphism 59
axiomatic method 1,11,25
axioms 26 see also Index of Axioms and Acronyms
  of betweenness 62
  of incidence 56
Baldus 344
Ball, W. W. Rouse 33
Beckmann 342
Beltrami 242,344
betweenness 32,62,111,284
Billingsley 1
Birkhoff 28,135
Bischof 240
Boltianskii 343,344
Bolyai, J. 233,241,253,256,296
Bolyai, W. 154,169,171,184,262
Bolyai's axiom 239
Bolyai's parallel construction 312,319, 345
Bolyai-Gerwien theorem 259,342
Bonola 344
Brianchon 50
Camerer 275
Campanus 41
Cardan 214
Carslaw 345
Cartesian plane 3,28,57,61,97,106,142,264, 283,337
  rational 75
  real 57,68,73,75,80
Cartier 343
categorical axiom system 60
Cayley 344
Cederberg 135
centroid 47
Ceva 245
Ceva's theorem 147
characteristic 0 226
characteristic 2

circle 75,87
  area of 329,334
  in Poincaré model 294
  tangent to line 23,39
  tangent to three circles 280
  tangent to two lines 39
circle group 129,324
  unwound 257,263,344
circle-circle intersection 90,119,149,288, 340
circular inversion 4,233,263,264,278,291, 344
circumcenter 46,53
circumscribed circle 22,134,275,276,355
Clairaut 114,236,330
Clairaut's axiom 239
Clavius 235,245,342
Clavius' axiom 236,239
collapsible compass 18
Commandino 216,235
common notions 26
common orthogonal line 292,303
compass, collapsible 18
compass alone construction 272
complete axiom system 60
complex numbers 112,200
conchoid 213
conformal 267,295
congruence 26,285
  of angles 76
  of line segments 70,116
consistency 60,283,344
constructible field 121,194
constructible numbers 149,194,331
construction 12
  of 17-gon 205
  of circles 273
  of heptagon 211,212
  of limiting parallel 321
  problems 193
  with compass alone 272
  with Hilbert's tools 86
  with one-inch ruler 23
  with ruler alone 272
  with ruler and compass 1
  with rusty compass 23
content 37,94,154,155,169,256
continuity 233
continuous functions 28
convex set 69,93
cosine 327
counting steps 19
Coxeter 341
cross-ratio 268,269,285,287,321
crossbar theorem 67,261
cube roots 193
cubic equations 193,198,214,221
cyclic quadrilateral 49,341
cyclotomic field 190
cyclotomic polynomial 191,200,229
de Zolt's axiom 160,258
decagon 201
Dedekind 3,28,136
Dedekind's axiom 96,113,262
defect 158,233,244,333
definitions 25
degree measure 135
degree of field extension 194,222
Dehn 4,154,184,189
Dehn invariant 186,187,343
Desargues' theorem 109,148
Descartes 97,100,101,312,343
difference of angles 81
dilation 267,269
Diller 346
direction 240
directrix 197
discriminant 221
dissection 154,169,182
distance
  between parallel lines 235
  Euclidean 116
  from a line 254
  function 126,289,325
  squared function 116
diverging lines 251
dodecahedron 2
Donne 153,176
double of a triangle 47
doubling of cube 3
Dudeney 4,175,342
Dudley 20

duplication of the cube 194,207
dyadic rational number 262,319
Egyptians 8
Einstein 38
Eisenstein's criterion 224,231
enclosing line 292
end 300,308,312,314
Engel 344
Enriques 137,342,343
enunciation 12
enveloping line 304
equal content 37,155,256
equidecomposable 155,159,169,256,259
equidistant curve 259,294
equidistant lines 235
equilateral triangle 1,17,103,148,291, 329,335
Erlanger Programm 31,123
Euclid 1
Euclid's fifth postulate 233
Euclid's *Elements* 1
Euclidean distance 73,116
Euclidean metric 73
Euclidean plane 93,141
Euclidean triangle 245,246
Eudoxus 1,4,182
Euler φ function 191
Euler line 4,46,48,341
Eves 8,343
exhaustion 4,154,177,182,184
existence of rigid motion 124,286
exterior angle 33,85 see also (I.16) in Index of Euclid's Propositions
exterior of polygon 163
extreme and mean ratio 148
fallacy 33,175
Fermat prime 206
field 104,105
  circle group of 129
  constructible 121,194
  extensions 193,222
  finite 130
  Hilbert's 120
  of characteristic 2
  of constructible numbers 331
  of ends 233,315
  ordered 3,97,111,115,137,141
  totally real 122
fifth roots 219
figure 155
finite element 132
Finzel 344
focus 197
fractional linear transformation 269,324
Franceschini 238
fringe mathematicians 179
fundamental theorem of algebra 223
fundamental theorem of Galois theory 227
Galen 134
Galois group 196,219,221,226,231
Galois theory 3,190,193,201
Gauss 3,4,184,193,199,201,233,241,256, 262,341,343
Gauss' axiom 262
Gauss' lemma 229
generalized line 311
generalized point 308
geometrical algebra 42
Gerling 184,343
Gerwien 154,169,171,184,259
Gibbon 134
Gleason 343
golden ratio 95
Goldenring 343
great circle 251
Greenberg 321,342,344,345
Gregory 157
Greitzer 341
group 124
  of rigid motions 323
Grundlagen der Geometrie 95
Hadamard 56
halves of equals are equal 162,182
harmonic points 270
Heath 341
Henrion 13
heptagon 207,209
Herstein 197
Hessenberg 346
hexagon 10
Hilbert 2,4,11,56,95,108,137,233,297,341
Hilbert plane 81,94

Hilbert's axioms 105
Hilbert's field 120
Hilbert's third problem 154,184,185,189
Hippocrates 134
Hjelmslev 345
Hobson 342
homology groups 186
homotopy groups 186
horocycle 294,324,333,340
hyperbola 235
hyperbolic
 analytic geometry 312,317
 geometry 5,296
 plane 234,297,335
 trigonometric functions 234,334
 trigonometry 262,312,325,328
hypercycle 259,294,340
hypothesis
 of the acute angle 246,248
 of the obtuse angle 246
 of the right angle 246
icosahedron 2
ideal line 311
ideal point 308,310
incenter 46
incidence 56
incidence plane 57
independence of axioms 59
inequality 111
inequality of angles 79
inequality of segements 72
infinite element 132
infinitesimal element 132
infinity 117,121
inscribed circle 22,275,276,310,312,335
inscribed square 148
inside see interior
interior
 of angle 116
 of circle 88
 of polygon 163
 of triangle 67,69,155
intermediate value theorem 28
intersection
 of circles 90
 of circles and lines 27,119
introduction of coordinates 149,150,312
inversion see circular inversion
irrational number 136
irreducible polynomial 225
isomorphic models 58
isomorphism of geometries 149
isosceles triangle 6,43,84
Jordan curve theorem 163
Kirkman's schoolgirl problem 62,342
Klein 31,123,242,270,343,344
Klein model 295
Lambert 241,344
Landau 342
law of cosines 332
law of sines 332
Lebesgue 343
Legendre 181,233,241,242,249
Legendre's axiom 248
Leistner 179,180
length 2,135,142
Liebmann 342,344
limit line 311
limit quadrilateral 310
limit triangle 255,310
limiting parallel 233,253,255,289,292,297
 axiom of 297
Lincoln 11,192
Lindemann 178
Lindgren 342
line 26,56,98
line segment 64
line separation 66,69
line tangent to 2 circles 24
line-circle intersection 90,119,149
linear ordering 69
Lobachevsky 233,241,253,296,298,332
Lovecraft 55
Magnus 344
marked ruler 3,193,199,207,208
Marvell 232
Mascheroni 273,344
Maywald 343
measure of area: see area function
medians 46,47,52,263,296,341,345
Menelaus' theorem 147,209
midline 246,258,309

midpoint of segment 21,75
Millay, Edna St. Vincent 45
minimal polynomial 196,207,225
Miquel point 54
model 57,104,149,283,344
Moise 344
Monge 270
multiplication
    of ends 315
    of segments 137
multiplicative distance function 289,318
multiplicative group 324
neutral geometry 5,35,82,241
Nicomedes 213,343
nine-point circle 4,46,50,53
non-Archimedean geometry 96,131,133, 239,296,321,342
non-Archimedean plane 156,168
non-Euclidean
    angles 55
    area 256
    geometry 5,233,242,283
    triangle 245
non-overlapping 155
nonagon 214
norm 130
normal extension 226
obtuse 117
octahedron 191
one-inch ruler 23
opposite side 65
ordered field see field, ordered
ordered group 257
orthic triangle 51
orthocenter 48,53
P-betweenness 284
P-line 283
P-point 283
Pappus 51,270,275,343
Pappus' theorem 54,108
parabola 197,198,207,219
parallel 22,58,98
parallel axiom 284
parallel postulate 4,34
parallelepiped 181,182
parallelogram 36,169
Pasch 56
Pasch's axiom 64,309
Pejas 321
Peletier 30,40,235,274,342
pencil
    of limiting parallels 300
    of lines 61
pentagon 1,4,15,41,102,161,201,332
perpendicular bisectors 15,46,312
perpendicular line 21
pi (π) 177,178,193
plane separation 64
Plato 1,8,134,234
Playfair 58,237
Playfair's axiom 35,58,131,235,237
Plücker 344
Poincaré 242,344
Poincaré model 233,269,283
point 56
polygon 163,177,262,331
    7-sided 207
    11-sided 219
    13-sided 214
    17-sided 3,193,199
    19-sided 214
    constructibility 199,206
Poncelet 50,270,344
positive 111
postulates 26
primitive root of unity 229
prism 181,183
Proclus 11,12,35,38,80,234
projective geometry 233,269,342
projective plane 61,311
projectivity 269
proof 9
proportion 3,136,142
Ptolemy 134,234
pyramid 181,183
Pythagoras 1
Pythagorean theorem 2,4,7,11,38,42,146, 172,173 see also (I.47) in Index of Euclid's Propositions
Péjas 346
quadratic equation 39,100,120,196
quadratrix 178

quadratura circuli 176
quadrilateral, cyclic see cyclic quadrilateral
quartic equations 193,214,218
Quetelet 344
quintic polynomial 215,224
quintisection of angle 219
radian measure 319
RASS 14,34
ratio 136,142
rational Cartesian plane 75
rational functions 131
rational numbers 112
ray 67,116
real Cartesian plane 57,68,73,97,105,198
real numbers 2,96,97,115,136,334,233
rectangle 14,239,246
rectifiable 334
rectilineal figure 155
reductio ad absurdum 12
reduction mod p 228
reflection 31,126,286,313,316
regular pentagon 41
regular polygons 193
regular solids 4
rhombus 14
Riemann integral 177
Riemann sphere 269
right angle 80,117
right triangle axiom 238
rigid motion 29,97,122,158,285,315,316, 323
root of a polynomial 223
roots of unity 200,229
rotation 31,126,128,316,324
  around ∞ 316
  around an end 324
Rudio 342
ruler alone construction 272
ruler and compass 1,16,100
ruler postulate 2
rusty compass 23
Saccheri 233,241,242,245,344
Saccheri quadrilateral 245,246,252,258, 302
Saccheri-Legendre theorem 243,257,261, 302
Sandburg 192
SAS 31,122,145 see also (I.4) in Index of Euclid's Propositions
SAS axiom 77
Savile 241
Scheubel 143
Schweikart 344
segment arithmetic 135,193
segment-connected set
semi-regular polyhedra 4
sides of a line 65
similar triangles 3,46,136,142,238
Simon 342,344
simple closed polygon 163
Simson 29,91,236
Simson line 54
Simson's axiom 239
sine 327
skew field 108
slope 106
SMSG 28,135
spherical geometry 251
splitting field 225
square 45
square roots 103,119
squaring the circle 176,196,331,342
squaring the parabola 198
SSS (side-side-side) 32,83,145 see also (I.8) in Index of Euclid's Propositions
standard form 103
standard model 74
Steiner 344
Steinitz 343
stereographic projection 270
Stewart 196,343
straight line 26
straight-edge 18
stretch-rotation 125,269
Sturmfels 342
Stäckel 344
subfield generated 222
sum of angles 77
superposition 2,29,30,77,97,123,263
supplementary angles 78

Tacquet 235
tangent circle 88
tangent function 117,121,318
tangent line 88
tangent of angle 319
tangent to circle 15
Taurinus 344
taxicab geometry 76
tensor product 186
tetrahedron 103,185,191
Thabit b. Qurra 172
Thales 7
Theaetetus 1
Theon 237
Thibaut 240
totally positive element 122
totally real field 122
transcendental number 196
transformation 264
translation 31,125,129,176
    along a line 315,316,323
transporter of angles 86
triangle 64 (see also altitudes, angle bisectors, medians, perpendicular bisectors)
    angle sum of 291,302
    Euclidean 245
    non-Euclidean 245
triangle inequality 121
triangulation 292
trilimit triangle 323,332
trisection of angle 3,136,193,195,199, 207,208,294
Tropfke 341
truth 9,26
undefined notion 25
unit segment 138
unwound circle group 257,263,344
Veronese 342
vertex of ray 67
vertical angles 78
vertical line 98
Vieta 275,343
volume 179,181,184
Wallis 238,344
Wantzel 343
whole greater than part 3,160
Wolfe 344
Z (de Zolt's axiom) 160,161,162
Zeuthen 343